The ferment of knowledge

The ferment of knowledge

Studies in the Historiography of Eighteenth-Century Science

EDITED BY

G. S. ROUSSEAU

Professor of English and Comparative Literature
University of California at Los Angeles

AND

ROY PORTER

Lecturer in the Social History of Medicine
Wellcome Institute for the History of Medicine
London

CAMBRIDGE UNIVERSITY PRESS

Cambridge

London New York New Rochelle

Melbourne Sydney

CAMBRIDGE UNIVERSITY PRESS
Cambridge, New York, Melbourne, Madrid, Cape Town, Singapore, São Paulo, Delhi

Cambridge University Press
The Edinburgh Building, Cambridge CB2 8RU, UK

Published in the United States of America by Cambridge University Press, New York

www.cambridge.org
Information on this title: www.cambridge.org/9780521225991

First published 1980
This digitally printed version 2008

A catalogue record for this publication is available from the British Library

ISBN 978-0-521-22599-1 hardback
ISBN 978-0-521-08718-6 paperback

for our fathers

CONTENTS

Contributors viii *Acknowledgements* xii *Abbreviations* xiii

Introduction 1

PART I. PHILOSOPHY AND IDEAS

1. Knowledge *Rom Harré* 11
2. Natural philosophy *Simon Schaffer* 55
3. Social uses of science *Steven Shapin* 93

PART II. LIFE AND ITS ENVIRONMENT

4. Psychology *G. S. Rousseau* 143
5. Health, disease and medical care *W. F. Bynum* 211
6. The living world *Jacques Roger* 255
7. The terraqueous globe *Roy Porter* 285

PART III. THE PHYSICAL WORLD

8. Mathematics and rational mechanics *H. J. M. Bos* 327
9. Experimental natural philosophy *J. L. Heilbron* 357
10. Chemistry and the chemical revolution *Maurice Crosland* 389
11. Mathematical cosmography *Eric G. Forbes* 417
12. Science, technology and industry *D. S. L. Cardwell* 449

Index 485

CONTRIBUTORS

HENK BOS is a member of the staff at the Utrecht Mathematical Institute where he teaches courses on the history and social function of mathematics. His research interests include fundamental questions in seventeenth- and eighteenth-century infinitesimal analysis and analytical geometry. His doctorate was on the concept of the differential in the Leibnizian calculus. He is currently engaged in a study of the conceptual aspects of curves (especially transcendental curves) in seventeenth-century geometry and analysis. He has contributed to the Open University course in the history of mathematics and he has also been engaged in studies of the life and work of Christiaan Huygens.

W. F. BYNUM is head of the Unit of the History of Medicine, Department of Anatomy and Embryology, University College London, and Assistant Director (Research) of the Wellcome Institute for the History of Medicine. His medical degree is from Yale University and his Ph.D. from the University of Cambridge. He has published papers on various aspects of the medical, life, and human sciences since the seventeenth century and is presently working on a monograph on the relations between basic science and clinical medicine in nineteenth-century society. His papers have appeared in *History of Science, Bulletin of the History of Medicine, Journal of the History of Medicine, Isis,* and *Medical History.* He is editing, with Dr R. S. Porter, *A Historical Dictionary of Science,* to be published in 1981.

DONALD CARDWELL is Professor at the Department of History of Science and Technology at the University of Manchester Institute of Science and Technology. He is also Deputy Chairman of the Governors of the North Western Museum of Science and Industry and Joint Honorary Secretary of the Manchester Literary and Philosophical Society. He has written five books and numerous articles concerned with the history of technology, especially the steam engine in the age of the industrial revolution; the development of heat physics; and scientific organization in modern Britain.

MAURICE CROSLAND is Professor of the History of Science at the University of Kent at Canterbury and Director of the Unit for the History, Philosophy and Social Relations of Science established there in 1974. His basic qualification in the history of science was a Ph.D. thesis on the history of chemistry, later developed into a book: *Historical Studies in the Language of Chemistry*. That part of his later research which relates to the history of chemistry has been largely concerned with chemistry in the eighteenth and nineteenth centuries, and includes a monograph on chemistry and the Enlightenment, and papers on the work and influence of Lavoisier. Professor Crosland's most recent book is a study of a nineteenth-century French chemist: *Gay-Lussac, Scientist and Bourgeois*. He is also interested in scientific institutions and has edited a collection of essays on the emergence of science in its social and institutional context in different countries of western Europe. He is now beginning a study of the French Academy of Sciences in the nineteenth century.

ERIC G. FORBES is Professor of History of Science at Edinburgh University. He was born in St Andrews, Scotland, and received his education at the local school and university, graduating from the latter with first class honours in astronomy. He began investigating the problem of the solar red-shifts in St Andrews under the supervision of the late Professor Erwin Finlay-Freundlich and, between 1957 and 1960, collected his own solar spectra at the Arcetri and Göttingen observatories, duly obtaining a Ph.D. from St Andrews University in 1961 for a thesis based on these researches. A few months later he was appointed Lecturer in Physics at St Mary's College, Twickenham, and subsequently promoted to Senior Lecturer in Mathematics. In 1965 he returned to Scotland to take up the newly created post of Lecturer in History of Science at Edinburgh University. Since that time he has taught and written extensively on various aspects of that discipline, with a particular bias towards seventeenth- and eighteenth-century European astronomy. He has published several editions of the hitherto-neglected papers of the Göttingen scientist Tobias Mayer (1723–1762), as well as books on the early history of the Greenwich Observatory and the Gresham Lectures of John Flamsteed, England's first Astronomer Royal. He was Secretary of the XV International Congress of the History of Science in Edinburgh (1977), and is currently Secretary-General of the International Union of the History and Philosophy of Science.

ROM HARRÉ studied engineering and mathematics, and taught applied mathematics at the University of the Punjab. Later he read philosophy at Oxford under J. L. Austin. His early work was concerned with the attempt to introduce historical realities into the philosophy of science, and included such books as *The Method of Science, Theories and Things, The Principles of Scientific Thinking* and (with E. H. Madden) *Causal Powers*. More recently he has been developing a critical methodology for the human sciences, in conscious opposition to the positivism of the recent past. His studies in this area have included (with P. F. Secord) *The Explanation of Social Behavior* and, most recently, *Social Being*. In 1960 he succeeded the late F. Waismann as University Lecturer in the Philosophy of Science at Oxford, and he has been a Fellow of Linacre College since 1965. He is also Adjunct Professor of the Social and Behavioral Sciences at the State University of New York at Binghamton.

J. L. HEILBRON is Professor of History and Director of the Office for History of Science and Technology at the University of California at Berkeley. He is interested in the development of the physical sciences since the late Middle Ages. He opposes neither internal nor external history. His current work concerns the changing fortunes of physics and physicists in the universities, higher schools and learned academies of the western world from 1600 to 1900.

ROY PORTER read history as an undergraduate at Christ's College, Cambridge. He completed his Ph.D. in 1975, which was published in revised form in 1977 as *The Making of Geology: Earth Science in Britain 1660–1815*. Since then he has written articles on the history of geology, and on eighteenth-century science in its social locations. He has edited (with L. J. Jordanova) *Images of the Earth* and (with M. Teich) *The Enlightenment in National Context*. He has completed a social history of eighteenth-century England and is presently editing, with Dr W. F. Bynum, *A Historical Dictionary of Science*. He is currently researching into the history of medicine from the point of view of the patient. From 1972 he was Director of Studies in History at Churchill College, Cambridge, and from 1974 University Lecturer in European History. In 1979 he became Lecturer in the Social History of Medicine at the Wellcome Institute for the History of Medicine in London. Since 1972 he has been editor of *History of Science*.

JACQUES ROGER studied at the Sorbonne and taught French literature and history of ideas at the universities of Poitiers and Tours, where he was Dean of the Faculty of Letters. He is currently Professor of the History of Science at the Sorbonne (Université de Paris, I), A. D. White Professor-at-Large at Cornell University, and Director of the Centre International de Synthèse (Paris). He has published *Les sciences de la vie dans la pensée française du 18e siècle* (1963), an edition of Buffon's *Epoques de la nature* (1962), and many articles in the history of ideas and the history of science.

G. S. ROUSSEAU is Professor of English and Comparative Literature in the University of California at Los Angeles, and is interested in the reciprocal relations of literature and medicine. Recently he has been especially concerned with exploring the medical world's legacy from the humanities. Trained by Marjorie Hope Nicolson, Rousseau holds advanced degrees from

Amherst College and Princeton University, and has taught at the universities of Cambridge, Harvard and UCLA. Among books he has written or edited are *This Long Disease, My Life: Alexander Pope and the Sciences* (1968, in collaboration with Marjorie Nicolson); *English Poetic Satire: Wyatt to Byron* (1969); *The Augustan Milieu* (1970); *Tobias Smollett* (1971); *Organic Form: The Life of an Idea* (1972); and *Goldsmith: The Critical Heritage* (1974). Within his effort to explore the life of science and medicine in the eighteenth century, Rousseau has retrieved the life and works of John Hill, the often-dismissed English naturalist. One volume, *The Renaissance Man in the Eighteenth Century,* has appeared (1977); another, *The Letters and Private Papers of Sir John Hill,* is about to; and a third, a critical study and biography of Hill, is in preparation. Rousseau's new book on medicine and culture between the Renaissance and modern period, *Fire in the Soul: An Anatomy of Four Ideas set in their Social Context,* will be published in 1981.

SIMON SCHAFFER is a Research Fellow of St John's College, Cambridge. He has studied history of science at Trinity College, Cambridge, Harvard University, and at the Ecole des Hautes Etudes en Sciences Sociales, Paris. He wrote his thesis at Cambridge University on the political and religious construction of Newtonian cosmology, and has published papers in the history of astronomy on Edmond Halley, Thomas Wright, Immanuel Kant, William Herschel, and others. His current research interests centre on the cosmology, politics and natural history of Herschel and Laplace, the natural philosophy of visionary sects in Britain in the eighteenth century - particularly Behmenists and Swedenborgians - and he is working on a lengthy study of the astronomical profession in England between 1700 and 1850.

STEVEN SHAPIN is Lecturer in the Social History of Science at the Science Studies Unit, Edinburgh University. He co-edited *Natural Order: Historical Studies of Scientific Culture* (1979) and has written numerous papers on the social organization and social uses of science in Britain during the eighteenth and nineteenth centuries. His particular interest is the exploration of sociological approaches to scientific knowledge. Many of the themes in the chapter in this volume are developed in his book on *The Social Use of Nature,* which will be published soon.

ACKNOWLEDGEMENTS

The editors are pleased to acknowledge the many obligations to friends and scholars that they have incurred in the production of this book. They are too numerous to mention all by name here; but we must mention Dr Michael Hoskin, of Churchill College, Cambridge, for offering much wise informal advice; Dr Richard Ziemacki of the Cambridge University Press, who gave continual editorial, professional and personal support throughout the conception and gestation of the book; and Jane Farrell, also of the Cambridge University Press, for her patient copy-editing. George Rousseau is grateful to the Chancellor of the University of California at Los Angeles for granting him a sabbatical leave during which time he could edit these essays in the comfortable academic environment of Clare Hall within the University of Cambridge; and Roy Porter is grateful to Churchill College, Cambridge, for providing him with congenial surroundings in which he, as well, could engage in the many editorial chores required in a volume of this scope. Both editors have profited from the many colleagues who have cheerfully submitted to having their brains picked time and again about the book, both as concept and reality. Above all, the editors are grateful to the individual contributors for their faith and goodwill in the project, and for their cooperation in writing (and rewriting) their papers. Special thanks also are due to Professor Crosland for having produced his essay on 'Chemistry and the chemical revolution' so promptly at short notice.

October, 1979

G. S. Rousseau
Clare Hall, Cambridge
Roy Porter
Churchill College, Cambridge

ABBREVIATIONS

AS:	Annals of Science
BHM:	Bulletin of the History of Medicine
BJHS:	British Journal for the History of Science
BMJ:	British Medical Journal
ECS:	Eighteenth-Century Studies
HS:	History of Science
JHBS:	Journal for the History of the Behavioural Sciences
JHI:	Journal of the History of Ideas
JHM:	Journal of the History of Medicine
JSBNH:	Journal of the Society for the Bibliography of Natural History
MH:	Medical History
NR:	Notes and Records of the Royal Society of London
PT:	Philosophical Transactions of the Royal Society of London
RHS:	Revue d'histoires des sciences
SSS:	Social Studies in Science

Unless otherwise stated, place of publication of all books is London.

INTRODUCTION

In my experience, almost all erroneous views of what went on in the 19th century are related to particular ideas as to what went on in the 18th century; and for the history of science and the history of ideas in the 18th century you can trust almost no one. The amount of 'hard' history of science for that period is so lacking that one simply leaps . . . from Newton in optics to Young in optics. Who else observed what, how well, and how thoroughly, and with what results? It is hard to tell. It is no reproach to my friends who are trying to do something with the 18th century to tell them that their labors have not yet reached the point at which a 19th century historian can confidently go ahead from the stable platform they have erected. And the same is emphatically true of the history of ideas related to the history of science.[1]

With these trenchant words, Susan Cannon has recently placed a bomb under the whole state of scholarship of eighteenth-century science. What lies behind these devastating criticisms?

The last generation has wrought a revolution in the history of science. As it was written thirty years ago, history of science, with honourable exceptions, essentially celebrated the biography of humble genius and the triumphal progress of discoveries along the royal road of truth. The history of science was the spaniel of science itself. This approach had real merits, not least in spotlighting the tremendous power of science as an engine of investigation. But now the discipline – as many contributors to this volume insist – has changed utterly. Contextual scholarship in the history of ideas, methodological 'externalism', new approaches within Marxism and French structuralism, the techniques of historians of art, religion, philosophy and ideology, the seminal writings of anthropologists and psychologists,

[1] Susan F. Cannon, *Science in Culture* (Folkestone, 1978), 133–4.

the anti-science temper of the counter-culture of the late 1960s, and the question-marks hanging over science in an age of demographic, ecological and technological crisis - all these have compelled profound rethinking. Certainties have given way to questions. The history of science is no longer a scientist's hymn to science: it has become part of history itself.

To rephrase this development sociologically: the last generation - as Roger points out in this volume - has seen the writing of the history of science tend to move out of the hands of scientists, and become the province of a separate, professional body of academic practitioners, whose primary allegiance is to history. The history, philosophy and sociology of science are now secure and prestigious academic disciplines. Nowadays it is possible for a work such as Thomas Kuhn's *Structure of Scientific Revolutions* (1962) not merely to exert a seminal influence in shaping research within its own field but also to offer a paradigm widely appropriated by other disciplines. Largely through Kuhn, the history of science has moved stage-centre in debates about the methods and truth-status of the humanities.

The development of science can no longer be served up as the sure tread towards truth. But exactly how it *should* be viewed is a question on which no consensus is in sight. What *is* clear is that - to pluck some choice examples - the Scientific Revolution will never look the same again since Frances Yates rediscovered Hermeticism; Newton has been historicized by McGuire and Rattansi, and Betty Jo Teeter Dobbs, and psychoanalysed by Frank Manuel; and Darwinianism has taken on new dimensions through the ideological approach of Young, quantum physics through that of Forman. Science today is being thoroughly recontextualized. The focus has switched away from the march of science to the sinuous mind of the scientist working within a society and milieu.

This revolution is, of course, very familiar. Its relevance here is that this profound change in the orientation of the history of science - one riddled with methodological anxieties - has as yet done little for the eighteenth century. Acton's advice to historians is classic: 'Study problems in preference to periods'. Yet the period of the eighteenth century *is* a problem. Historians of science have tended to regard it as a tiresome trough to be negotiated between the peaks of the seventeenth and those of the nineteenth century; or as a mystery, a twilight zone in which all is on the verge of yielding. This judgement may seem paradoxical. For even by the most conventional 'internal' standards of evaluation, the eighteenth century was one of noted achievement: it saw the establishment of fields such as electricity and heat, the 'chemical revolution', the new science of gases, the isolation of oxygen, the nebular hypothesis in cosmology, the foundations of rational mechanics, the birth-pangs of biology, geology and psychology. It was indeed an age when knowledge was in ferment.

Yet the century has suffered neglect. It is thus characteristic that the most recent English-language compilation devoted to eighteenth-century science, *Natural Philosophy through the Eighteenth Century and Allied Topics*, edited by Allan Ferguson, appeared as long ago as 1948, and for lack of competitors was thought worth reprinting in 1972. By many historians the century has been deplored for allegedly producing science that was boring, unoriginal, lacking in rigour, and over-speculative – at best, merely fall-out from Galileo, Harvey or Newton. 'The first half of the eighteenth century was a singularly bleak period in the history of scientific thought'; 'science somewhat languished'; the period was 'comparatively undistinguished in its science', which contained 'an element of dullness', due in part to its 'too ambitious schemes' and its 'obstructive crust of elaboration and formality'.[2] This irritation, prominent in many historians' responses to eighteenth-century science, perhaps betrays frustration at the century's lack of towering heroes, of conflicts between titanic geniuses, or of martyrs – a century not of peaks but of trackless bog.

Indifference or antipathy to eighteenth-century science is nothing new. The 'reading' of science's ineptitude in the age of the Enlightenment had taken firm root from early in the nineteenth century. Think, for instance, of Charles Lyell's characterization of the eighteenth century in geology as a period of 'retardation' whose study was 'singularly barren of instruction to him who searches for truths in physical science'.[3] The myth is part, of course, of the Romantic and counter-French-Revolutionary reaction against all facets of the eighteenth-century world: its religion, its art and poetry, its history-writing. So potent does the myth remain that many fields of Enlightenment ideas and culture – perhaps especially gauging the pitch of its religion – suffer from scholarly neglect at least as seriously as its science.

The upshot is that the eighteenth century has been a period in which the currents of science in general, and the pattern of particular disciplines, have lacked strong interpretation. As Bos emphasizes in his essay here, the picture of the eighteenth century lacks structure

[2] These quotations are taken successively from S. F. Mason, *Main Currents in Scientific Thought* (1956), 223; J. D. Bernal, *Science in History* (1954), 358; A. E. E. Mackenzie, *The Major Achievements of Science* (Cambridge, 1960), 122; A. R. Hall, *The Scientific Revolution, 1500–1800* (1954), 339; H. T. Pledge, *Science since 1500* (1939), 101.

[3] C. Lyell, *Principles of Geology* (3 vols., 1830–3), I, 30.

and focus, light and shade. Tellingly, the recent revolution in the history of science has not hitherto presented us with convincing reinterpretations of the age.

Even the best synthetic views have been fatally flawed. Thus Foucault's quasi-structuralist notion of an eighteenth-century 'Classical *episteme*' underpinning scientific discourses was applied illuminatingly to natural history, but then seems to have been abandoned by Foucault himself and now commands little support, even from the author. The hope of demonstrating the coherence of eighteenth-century science by relating all its elements back to some kind of fundamental 'natural philosophy' has been attacked – not least by Schaffer in this volume – for committing the unhistorical sin of 'lumping'. Attempts to show that eighteenth-century scientific achievements flowed out of the programme and consciousness of the *philosophes* have turned out to be premature. Not Buffon and Lamarck, but perhaps Linnaeus and Cuvier now appear to be the real forerunners of Darwin. Similarly, accounts of the growth of eighteenth-century science in terms of 'professionalization theory' break down. As Roger and Porter stress in their chapters, in some senses the eighteenth century rather witnesses the 'amateurization' of science; natural history is the creation *par excellence* of amateurs, and there is nothing incongruous about the conjunction of amateur with high-quality science. Thus all attempts till now to 'totalize' eighteenth-century science have been wrecked upon the reefs of diversity and complexity; and we do not yet have a modern Marxist reading of the century. Marxists interested in laying bare the constitutive role of science as a legitimating ideology for capitalism have concentrated almost wholly on nineteenth- and twentieth-century industrial societies.

None of this is offered to undervalue the substantial recent revival of research on eighteenth-century science. Nor does it deny that new perspectives and techniques over the last generation have enriched our views in certain directions. Thanks to these, we now take it as axiomatic – and correctly – that eighteenth-century science can be properly grasped only if its 'external' relations to other intellectual and cultural systems, such as theology and epistemology, are tackled head-on. We now no longer ignore the fact that the eighteenth century 'geography of knowledge', the relations between the sciences, was then markedly different from our own. It seems elementary to us (now!) that eighteenth-century scientific ideas cannot adequately be translated one-to-one into twentieth-century terminology. Indeed, one of the aims of this book is precisely to distil and evaluate this substantial body of empirical research that has been conducted in the last generation.

Yet it seemed to us, especially in the planning stages of this volume, that no one had hitherto attempted – and certainly not within the covers of one book – to gather together these detailed and discrete publications, in order to survey, synthesize, appraise and criticize new trends and major controversies in eighteenth-century research. As the 1980s approached we wanted to pose directly such questions as:

What are the main scholarly readings of eighteenth-century science?

Which aspects of these are secure, which under fire, which undergoing fusion and fission?

What are the chief priorities for further empirical research and theoretical reconceptualization?

What can historians of eighteenth-century science most learn from scholars working in other fields?

With these questions in mind this book was conceived and written. The following essays aim to provide up-to-date accounts of research and interpretation across a wide range of the natural sciences, with each author offering his own vision of the 'ferment' of his field. The editors have not imposed any restrictions whatsoever on their contributors, not even a definition of the concept of 'field', but they have suggested that two considerations be borne in each author's mind: (1) that the subject or field be set in a large cultural perspective and an approach be developed for generalists as well as for specialists; (2) that the *historiography,* not the history, of the field be the subject of the discourse. There are numerous 'tunnel' histories of the various sciences; there is unfortunately no historiography of many of these fields, certainly not any that covers all the sciences during the eighteenth century.

No apology is needed for the fact that these are essays in exploration, and that the views they present are personal and provisional. Historians of science have only recently awakened to the fact that history is not a 'science' in the popular sense of that word – objective, definite, positive, neutral – rather it is a dialogue of the present with the past. And in becoming thus aware they have also discovered that the science they study is not 'science' in that sense either, but a dialogue between individuals and societies, culture and nature. It would be absurd to squander this precious new sense of the relativity of historical interpretation by claiming that this book represents anything beyond a stab at 'history for the 1980s'. This is not a textbook of eighteenth-century science; it is an exploration of its history seen through its historiography. As such it seeks to illuminate the issues of understanding science in the *siècle des lumières.*

Some apology is perhaps required for the range of sciences covered. Certain specialties in natural science have been almost totally omitted (cosmology, instrumentation, crystallography, etc.). Also, there are no essays explicitly focused on some of the human sciences (e.g. anthropology, political economy), or on the sociology of scientists and their institutions. Furthermore, science throughout means Western science. We are well aware of these and other gaps; but to have included all we should have wished would have meant the writing of several books, not one.

The choice of scientific areas as the subjects of these essays may also seem inconsistent: in some, attempts have been made to capture fields of knowledge as construed by contemporaries (e.g. "Mathematical cosmography', or 'The terraqueous globe', or 'Psychology'), whereas others enshrine more modern classifications. There are no solutions to this problem, and it is as well not to pretend that there are. There is no natural grid of such a cultural artefact as the system of the sciences, and no single scheme of the division of scientific labour held sway throughout the eighteenth century. A book surveying the elective affinities of the sciences in the eighteenth century, especially in relation to the encyclopaedic ideal, would be very welcome.[4]

In choosing topics, however, we have attempted to give special weight to the most fruitful line of research over the last twenty years or so: the attempt to locate eighteenth-century natural science in its fullest philosophical, religious and linguistic context. Hence the opening chapters by Harré, reassessing major interfaces between Enlightenment epistemology and natural science; by Schaffer, undermining the modern historiography which has conjured up the imposing edifice of 'natural philosophy' without adequately relating it to the actual practice of scientific investigation; and by Shapin, decoding the social messages which are transmitted through the language of eighteenth-century natural philosophy. The other essays in the volume have been chosen in part to give greater prominence to the life sciences – to what Carl Pantin has called the 'extensive sciences' – than they are usually accorded.

The reader of these essays may also notice a common thread running through many of them, one in no way planned for: namely, the way in which so many of the authors seem to stray from their so-called assigned 'science'. Thus, a scholar such as Jacques Roger, while surveying the biology of the period, expends a good deal of time and energy delving into its medicine and physiology; and G. S. Rousseau, while ostensibly charting aspects of Enlightenment psychology finds

[4] One important contribution towards filling this gap is David Knight, *Ordering the World: a History of Classifying Man* (1980).

himself discussing its philosophy and social metaphors. The like is true for many of the other contributors. While preparing these essays for publication the editors, rather independently, noticed how typical these amblings are of eighteenth-century science itself. This seeming deflection into other scientific areas than the author's elected one signals the development of the idea of organic form in the sciences of the eighteenth century, an idea whose life in that period has yet to be taken adequately into account in discussions of nineteenth-century science.[5]

Over the last generation, there has been much theorizing and polemicizing over how the history of science should be written. We offer this volume, to specialists and others, as an exploration of the continuing dialogue between research and debate within a particular field.

[5] For a recent example of a volume of collected essays in which it is not duly considered, see G. S. Rousseau (ed.), *Organic Form: The Life of an Idea* (1972).

PART I. PHILOSOPHY AND IDEAS

1

Knowledge

ROM HARRÉ

Contents

I.	Historiography	11
II.	Case 1: British empiricism	15
III.	Case 2: Intellectual sources of the Enlightenment in France	35
IV.	Case 3: Nescience and necessity: Kant's philosophy of physics	47

I. Historiography

Social conditions

It is fashionable these days to link intellectual movements with social conditions, political issues and even economic developments. But while one acknowledges the importance of these matters it is extraordinarily difficult to establish what the links could be. Economic determinism is altogether too naive a point of view. More sophisticated theories based on conceptions such as that of 'ideology' remain vague as to the nature of the relations that obtain between the various interacting factors from which an author's thoughts and their acceptance or rejection by his contemporaries spring.[1] It is far from clear that the invention of doctrines on the one hand, and their reception on the other, are determined by the same influences. Yet it would be folly to ignore the fact that Locke was deeply involved in politics, or that Diderot and d'Alembert played an important part in the promulgation of middle-class values in pre-Revolutionary France. Nor should it be forgotten that Hume's history outraged both Catholics and Protestants; nor that Priestley's house was burnt by a Tory mob. But how all these matters flow together into the controversies about the limits of knowledge I am quite in despair to understand.

I began work on this chapter fully convinced that a strong case for an 'externalist' treatment was possible. But as the work proceeded I felt more and more inclined to submit to the judgement of Peter Gay,

[1] J. W. N. Watkins, 'Historical explanation in the social sciences', in J. O'Neill (ed.) *Modes of Individualism and Collectivism* (1973), 166–78.

that one's vague sense of a relation between social conditions in England and France and the intellectual movements that accompanied them cannot be realized in detailed accounts of specific paths of influence from the one to the other. Specific pathways were personal and intellectual: the reading of books, the exchange of letters and conversation. The influences of one writer upon another are demonstrable. Kant read the works of Hume, which were sent to him directly from Scotland by a friend. Michell entertained Priestley in his country home. Diderot and d'Alembert shared an office, so to speak. In a sense the Enlightenment was nothing but an intellectual movement, whose development must be understood through an internalist account.

Yet the broad differences between the 'movement' in England, Scotland and France can be related to structural features of society. English discussions of the nature and limits of knowledge sheered away from the Baconian tradition which would have related scientific knowledge to the enhancement of useful crafts. It was dominated by the bearing on theology of philosophical arguments concerning the limits of science. This can hardly be detached from the career structure of Oxford and Cambridge dons. They left their fellowships early in life to take up matrimony and country livings. The calm, good-natured realism of the Scottish philosophers could be seen as a product of the club-like atmosphere they delighted to cultivate, in which both believers and sceptics in the Enlightenment mould liked to distinguish themselves from the sour Calvinists of the 'primitive' party.[2] And as I shall be suggesting in later sections, the more overtly radical atmosphere in which the *philosophes* engendered the French Enlightenment accounts to some extent for the pressures that led to the disintegration of the fragile *rapprochement* between a sensationalist groundwork for the theories of knowledge and traditional French Cartesian rationalism.

In much philosophical writing the works of the past figure as a resource rather than a topic. They are used as much to deploy the philosophical ideas of the historian as to display those of his subject. Writings on eighteenth-century epistemology are no exception. However, there have been some changes of importance in the historian's approach to certain crucial authors. Some historians of philosophy have turned to look more closely at the contemporary scene to understand the preoccupations of a period. I am fairly sure that the right way to approach the history of eighteenth-century philosophical controversies is by way of a study of the intellectual influences of one

[2] H. F. Henderson, *The Religious Controversies of Scotland* (Edinburgh, 1905).

writer upon another. The circles of debate were small. Each author seems to have read the works of most others, and in many cases discussed the issues face to face. But recent approaches to this essentially intellectual history have been different depending on the author who is subject to attention. I shall discuss some important authors separately before I draw any general conclusions about the historiography of philosophical studies of the Enlightenment.

In this chapter I want to challenge three historiographic orthodoxies.

1. That British thought on the problem of knowledge can be fully understood as the progressive refinement of the principle that 'nothing is in the mind which is not first in the senses', motivated by a purely internal demand for progressive philosophical analysis.
2. That the Enlightenment in France was no more than the adoption of the sensationalist foundations for knowledge 'discovered' by Locke.
3. That the colossal synthesis of realism and scepticism devised by Kant can be properly understood without reference to the details of his theory of physics.

To establish the first I turn to the controversies of the time concerning the relation of physics and theology; to establish the second I examine the development of the epistemic theories of one key figure, d'Alembert; and to establish the third I offer a comparison between the *Critique of Pure Reason* and the *Metaphysical Foundations of Natural Science,* to unearth the origins of the Romantic movement in physics.

Sketch of the historical development

It hardly seems necessary to point out that the stature of the authors I shall be discussing may have been very different in the eyes of their contemporaries from their appearance in ours. It is difficult for us to think of Locke without Berkeley and Hume. Looked at with respect to the thesis for which I shall be arguing, the writings of Locke must be taken to have been of paramount importance. Though Berkeley and Hume were widely read, neither was taken nearly so seriously, nor were their doctrines as influential.

Locke's views about the limited possibilities for human knowledge were widely known. They provided the basic dynamics of discussion in all three centres of philosophical ferment – Edinburgh, the Osford–Cambridge–London circle, and Paris. As we shall see, Locke convinced himself (and many of his contemporaries) that we can know very little about the world of nature, the world explored by natural scientists. But what we can know we can come by very easily provided we are strict with ourselves, for it is just ordinary experience unclouded by intellectual illusions that determines all we can ever truly know about nature. The use of reason can add nothing to this basic stock. This was Locke's 'nescience'. But happily the abilities the creator has implanted in us are enough for our needs. This was Locke's 'providentialism'. In short, we do not know very much but it is enough. The problem for the pious amongst the inheritors of Locke's theory of science was to protect their intuitions about God against the contamination of nescientist scepticism, a scepticism most were only too happy to hold about the pretensions of natural science. That these men were dons and parsons, comfortably located in the English oligarchy, is almost enough, I think, to account for their efforts in this direction – while the looseness of doctrine, both political and religious, of that very oligarchy, is enough to account for the extraordinary variety of their defences. The manufacturing middle class of Manchester and Birmingham arrived later on this scene. Freed still further by their non-conformism they arrived in the person of that formidable Socinian, Joseph Priestley. By fusing spirit and matter in one active force he abolished both nescience and providentialism at one blow. The arguments developed within an intellectual tradition, but the occasions to use them arose out of socio-economic conditions of the time.

The influence of the Cartesian tradition was no less potent in France and the German states than was Locke's in England, in the early part of the eighteenth century. The transcendentalism of Leibniz can be felt even in the encyclopaedists of pre-Revolutionary France, since d'Alembert believed at least in the possibility of a mathematically founded rational mechanics from which *a priori* knowledge of the laws of nature would follow. The epistemology of this tradition could be summed up as follows: 'We can know a great deal, no less than the basic structure of the world-system. But this knowledge is difficult to arrive at. It must be reached by intellectual effort, since it is logically prior to experience, which is its product, and in which it is only confusedly represented.'

In each tradition there was a foundation for knowledge, an unquestionable starting point. For the empiricists the foundation was in elementary experiences, simple deliverances of sense, which were particular and distinct, at least to that observer whose mind was unclouded by intellectual illusion. The problem of how a mind so furnished could ever form general ideas was never satisfactorily solved by the philosophers of this tradition. In the end in the nineteenth century even laws of nature were to be resolved into catalogues of like particulars. The Cartesian tradition favoured the clarity and distinctness of necessary truths as the foundation of knowledge, from which the general lineaments of the world system might be reached by strict operation of the mind guided by the principles of logic. Since the principles that governed all things were never manifested in experience, which was confused and partial, the source of the particularity of the deliverances of the senses was never satisfactorily accounted for in the tradition.

As we shall see, the profoundest epistemological theorizing resulted from Kant's systematic attempt to set the bounds to both sense and reason, and to account for the universality of the most general laws of nature in the structure of the mind which imposed them upon its raw sensations in the very genesis of experience.

II. Case 1: British empiricism

The thesis

I shall be setting out to make a case for the view that in Britain in the eighteenth century discussions on the nature and limitations of knowledge should be seen as balanced between the poles of natural science and theology - knowledge of the world and knowledge of God. The improvement of the former had always to be reconciled with the possibility of the latter. Every discussion of the epistemology of natural science from this period must be understood against the presumption that its author was almost certainly scrutinizing his own reasoning for its theological consequences in the knowledge that so too was every one of his readers. I shall try to show that it was widely feared that it was all too easy to pass from scepticism with respect to the deliverances of natural philosophy to scepticism with regard to religion, even if the former had originally been developed in an attempt to demonstrate the superiority of religion as a source of understanding of man and nature. In this essay I have been forced to pay attention not only to the epistemology of natural science but to theories about knowledge of God, since I hope to demonstrate that in the works of most eighteenth-century writers they are intimately connected.

The idea that our understanding of the development of the theory of knowledge in Britain should be referred to the religious controversies of the day is a principle I want to emphasize very strongly in the substantial parts of this essay. The history of eighteenth-century epistemology has usually been treated within a closed intellectual tradition. Even the monumental *Metaphysics and the Philosophy of Science*, Buchdahl's superlative treatment of the standard authors, describes a purely intellectual development.[3] But, as G. Bowles has recently argued, the popular reception of Newtonianism must be understood in a context determined by the effect of Lockean epistemology and the power of Newton's vision of a world of active forces, on perennial British religious controversies.[4] More or less the same point has been made by Heimann and McGuire in their elucidation of the 'native tradition' of dynamism in British thought.[5] Only by paying attention to this context can one come to understand the importance assumed by attacks on and defences of the possibility of knowledge of natural powers in the debate of the period, culminating in the revival of accusations of Socinianism against Priestley, accusations he gloried in.

A good deal has become clear since Schofield's pioneer study of 1970.[6] His contrast between positivists and dynamists is perhaps somewhat over-drawn, but it does provide the essential analytic framework within which to discuss the relations between the authors and their public.

Of course all this was to be transformed as the intuitionism of the Romantic period began to infect science and literature, and drastically shifted the foundations of epistemology. The importance of that movement in the history of chemistry and physics has only recently been appreciated, but its pursuit lies outside the scope of this volume. An excellent brief summary of the Romantic conception of knowledge, in England, can be found in Knight's short article of 1972 and recent book,[7] while a detailed investigation of the sources of Romanticism in physics has been made by B. S. Gower.[8]

The changing historiographic tradition

The case of Locke. In one sense little has changed in the assumptions of writers on the works of Locke. Both Aaron in 1937[9] and Yolton in

[3] G. Buchdahl, *Metaphysics and the Philosophy of Science* (Oxford, 1969).

[4] G. Bowles, 'Physical, human and divine attraction in the life and thought of George Cheyne', *AS*, XXXI (1974) 473–88.

[5] P. M. Heimann and J. E. McGuire, 'Newtonian forces and Lockean powers: concepts of matter in eighteenth century thought', *Historical Studies in the Physical Sciences*, III (1970), 233–306.

[6] R. E. Schofield, *Mechanism and Materialism: British Natural Philosophy in the Age of Reason* (Princeton, N.J., 1970).

[7] D. Knight, 'Chemistry, physiology and materialism in the Romantic Period', *Durham University Journal*, LXIV, (1972) 139–45; see also his *The Transcendental Part of Chemistry* (Folkestone, 1978).

[8] B. S. Gower, 'Speculation in physics: the history and practice of naturphilosophie', *Studies in the History and Philosophy of Science*, III (1973), 301–56.

[9] R. I. Aaron, *John Locke* (Oxford, 1937).

1970[10] emphasized the influence of the prevailing scientific interest on Locke's thought. Both commentators distinguished between the early and continuing influence of Locke's friendship with Boyle, and the effect of his late and occasional relations with Newton.

But Yolton makes a good deal more of the subsequent religious controversies aroused by the cautious empiricism of the scientifically inclined. 'The fact that Locke was a member [of the Royal Society] and was clearly associated with the new way of doing science undoubtedly contributed to the quick reaction to the *Essay* by the defenders of traditional religion.' Aaron's interest in the religious issues of the day is confined to a discussion of the accusations of Socinianism provoked by Locke's *Reasonableness of Christianity*.

Every age reads back its own preoccupations into the writers of the past. Historiographically considered, the framework of ideas with which O'Connor approached the interpretation of Locke's epistemology in 1967[11] is very different from the recent work of Mackie.[12] In setting out the overall layout of Locke's work, O'Connor emphasizes the sensationalist epistemology to almost the total exclusion of Locke's metaphysics and theory of language. Mackie, writing very recently, pays most attention to Locke's theory of substance, his view on essences and natural kinds, and his interest in the logical theory of general words. But this is an interpretation and a selection of problems imbued with Mackie's own interests in the philosophy of science and of language. Reviewers have been dissatisfied with Mackie's account of Locke, and indeed he has been taken to task by M. R. Ayers for massive misunderstandings of Locke's position occasioned, argues Ayers, by inattention to the philosophical commonplaces of the time.[13]

The case of Hume. Though I shall be making a case for playing down the influence of Hume in his own time, his work assumed such great importance later that he attracts a good deal of attention from historians of philosophy.[14] In recent times, to be a 'Humean' was to subscribe to the positivistic side of Hume's curious blend of psychology and philosophy. Lately Hume scholarship has tended to emphasize

[10] J. Yolton, *Locke and the Compass of the Human Understanding* (Cambridge, 1970), 4–15.

[11] D. J. O'Connor, *John Locke* (New York, 1967).

[12] J. L. Mackie, *Problems from Locke* (Oxford, 1976).

[13] M. R. Ayers, 'Review of Mackie's *Problems from Locke*', *Philosophical Books*, VIII (1977), 71–3.

[14] D. D. Raphael, 'Hume's critique of ethical rationalism', in W. B. Todd (ed.), *Hume and the Enlightenment* (Edinburgh, 1974), 14–15.

other aspects of his thought. Capaldi, writing in 1975, insists on a purely psychologistic exegesis of the famous sceptical arguments.[15] He asks us to think of Hume as carrying through his avowed intention of building a theory of 'the understanding' on strict Newtonian principles. But, as Capaldi emphasizes, these are Newton's Rules for the conduct of enquiry, his principles of method. Hume would have no truck with Newton's metaphysics or his associated physical theory which depended heavily on assumptions of natural force and agency. Capaldi insists that Hume never doubted that we had experience of ordinary material things.

In another recent work (of 1974) W. A. Wallace points to the essential tension between Hume's theory of practical scientific reasoning, a theory very like that encapsulated in Newton's Rules, and his metaphysical scepticism.[16] In this context Wallace draws attention to Reid's defence of the use of the idea of a causal power as an empirical concept. This is an important indication of the way Hume's contemporaries regarded his philosophy. It is the clue which draws our attention to the long-running theological debates which dogged attempts to enlarge the notion of knowledge in Britain.

Unlike earlier commentators, both Capaldi and Wallace agree in taking the central historical question to be Hume's relation to the Newtonianism of the day, however differently they interpret it. Wallace emphasizes the chemical tradition (our knowledge of which has been greatly enhanced by Thackray's excellent study)[17] while Capaldi emphasizes the methodological tradition which was to culminate in Mill's inductivism. In what sharp contrast is Price's study of 1940![18] He insists on the phenomenalist element in Hume's thought to the extent that he paraphrases Hume as saying 'what we commonly call our consciousness of material objects and events . . . consists largely of *imagination*'. The marked anachronism of Price's treatment extends to the use of such phrases as 'the truth or falsity of material object sentences' to explain Hume's ideas. One can hardly forbear the comment that the history of theories of knowledge has become a great deal more historical in recent years.

God and gravity

I hope to demonstrate that the eighteenth-century struggle to define the limits of knowledge cannot be fully appreciated without some

[15] N. Capaldi, *David Hume* (Boston, 1975), ch. 3.

[16] W. A. Wallace, *Causality and Scientific Explanation* (Ann Arbor, 1974), II, ch. 1, sect. 5.

[17] A. Thackray, *Atoms and Powers* (Cambridge, Mass., 1970).

[18] H. H. Price, *Hume's Theory of the External World* (Oxford, 1944).

attention being paid to the metaphysical principles that tempted men to try to think beyond the bounds of sense to the inner natures of things. Two metaphysical issues intertwined - the standing of the theory of essences and the question of the source of activity, of the origin of the powers and tendencies of material beings. The burden of my argument in this section will be to establish that in England the problem of the reality of natural powers, of the source of natural activity, was the dominant motif of eighteenth-century philosophical thought on the extent and limits of knowledge and an essential precursor to the development of the concepts needed for the enunciation of field theories.

Material beings manifest two ranges of properties: their qualities (colour, shape, and so on) and their powers (acidity, inertia, gravity, and so on). I hope to show how much the issues raised by this distinction preoccupied the philosophers and theologians of the period. Though the distinction between the two ranges of properties was gradually eroded, there were those who wished to abolish the powers of natural bodies, reserving genuine activity for God. Others, for a variety of reasons, wanted to eliminate sensory qualities from a material world composed wholly of powers. Epistemological issues soon came to dominate the discussion, largely, I shall argue, because of the promise of and doubts about the scientific method of acquiring knowledge, represented for example, in the central paradox of Locke's thought. There could be no question, it was argued, of admitting the natural powers of material particulars to the status of real attributes if there were no way they could be certainly known. And how could powers be known if they were apparent only in sensory manifestations as experienced qualities? It is my belief that the question of the knowability of powers became a central topic of concern. This view is shared by several recent authors, for instance McGuire, Thackray and Bowles. There were several parties. Greene and Priestley thought it obvious that the activities of natural and human agents could be objects of human experience. Locke and Kant supposed the reality of powers to be easily demonstrated as necessary conditions for the experience we do actually have. Hume supposed our experience of power and activity to be but an illusion, the reflection into experience of the psychological state of expectancy produced by the mere repetition of like sensory impressions.

The issue of essences and the problem of powers merged in the debates about whether the real essences of material particulars were nothing but self-existent powers. Both Kant and Priestley held such a view. The promulgation of such ideas laid the foundation for the field theories of the nineteenth century, through the development of a metaphysics for the understanding of the phenomena of electricity, magnetism and light.

The theory of ideas

Locke believed that the ideas that make up our experience of bodies can be categorized into ideas of primary and secondary properties depending upon their resemblance to the real qualities of material things. The list of primary qualities was variously compiled by Locke and his contemporaries. Boyle's 'bulk, figure, texture and motion' is perhaps the neatest summary. But in my view historians have paid insufficient attention to the fact that while Boyle's list is of exclusively occurrent properties, Newton's is of dispositions – solid*ity*, motiv*ity*, mobil*ity*, etc. Philosophers have long puzzled over Locke's confidence in this basic distinction. How can one account for what seems on the face of it a rather simple philosophical error? Some historians of philosophy have held that Locke did indeed hold an untenable representationalist theory of knowledge. A very useful critical discussion (and bibliography) of this exegesis can be found in Woolhouse's study.[19]

The really thorough scholarship of Woolhouse, Ayers[20] and Yolton[21] encourages another interpretation. According to Yolton, 'primary qualities [for Locke] can be and most often are sensible . . . they frequently appear in the coexistences which careful observation records. Ordinary objects have a nature both at the observational as well as at the insensible level'. Resemblance of ideas and qualities is not, then, a philosophical mistake. It is the very basis of the science that grounds explanations of phenomena in the science of the mechanical properties of natural bodies. It is pure contingency that forbids our experience and hence our knowledge of coexistence of the insensible parts. The importance of how the distinction is understood can hardly be exaggerated. If there is a range of ideas that resemble the properties of things that produce the ideas, we have a sure path to knowledge of the material world, so one's attitude to Locke's epistemology will depend on how one takes the doctrine of resemblance. Locke himself puts the question in *Essay:* 4, IV, 3:[22] there must be 'a con-

[19] R. S. Woolhouse, *Locke's Philosophy of Science and Knowledge* (Oxford, 1971), 35ff.

[20] M. R. Ayers, 'The ideas of power and substance in Locke's philosophy', in I. C. Tipton (ed.), *Locke on Human Understanding* (Oxford, 1977), ch. 6.

[21] Yolton, *Locke* (see note 10).

[22] J. Locke, *An Essay Concerning Human Understanding* (1690), 4, IV, 3. All quotations from Locke are taken from the Yolton edition (London and New York, 1961, etc.)

formity between our *ideas* and the reality of things' but 'how shall the mind, when it perceives nothing but its own *ideas,* know that they agree with the things themselves?' Obviously, this conformity cannot be known by experience when the causes of a range of ideas are the *insensible* parts of things. The only other resort is reason. Locke argues that since simple ideas cannot be made by the mind, they must be the products of the influence of things. On what is this very strong statement based? Here we reach the providentialist theme we shall find throughout the eighteenth-century discussions of the limits of knowledge. Simple ideas, says Locke in *Essay:* 4, IV, 4, are 'the natural and regular productions of things without us . . . and so carry with them all the conformity which is intended or which our state requires, for they represent to us things under those appearances which they are fitted to produce in us'.

The limits of human knowledge

To justify my historiographic thesis, that the philosophical debates can be understood only against the wider background of the effect of science and its presumed epistemology on the theological controversies of the day, I must set out Locke's theory of the limits of knowledge in more detail. Locke's version of the theory of ideas may seem to have an optimistic air. But in Locke's own commentary upon the possibility of knowledge within this framework it becomes apparent that the optimism is a mere reflection of the unattainable ideal. The limits to human knowledge of nature emerge very rapidly and are severe. They can best be appreciated against Locke's catalogue of what could fall within 'the compass of the human understanding' were our means to knowledge more powerful. There would be (*Essay:* 4, XXI, 1) '*first,* the nature of things as they are in themselves, their relations and their manner of operation; or, *secondly,* that which man himself ought to do as a rational and voluntary agent; . . . or, *thirdly,* the ways and means whereby the knowledge of both the one and the other of these are attained and communicated'.

Unfortunately, studies of the third kind sharply separate the first and the second by showing up the severe limitations our nature imposes on the possibilities of coming to know the real properties of things. All of this derives from the fact that according to Locke we can work only from ideas.

'Knowledge', according to Locke (*Essay:* 4, I, 2), is '*the perception of the connexion and agreement or disagreement and repugnancy of any of our ideas*'. Moral principles, the second kind of knowledge, can, Locke believes, be known with certainty, since they are demonstrable from the internal relations of ideas, just as geometrical principles are.

Could we have a similar kind of knowledge of the natures, powers and properties of material bodies? Only if we could know their natures. 'Had we such ideas of substances as to know what real constitutions produce those sensible qualities we find in them and how those qualities flowed from thence, we could, by the specific ideas of their real essences in our minds, more certainly find out their properties and discover what qualities they had or had not, than we can now by our senses.'

None of this programme can be realized because we are confined to our ideas and our ideas are separate and independent. The trouble is double-edged. Lacking powerful senses we are unable to exploit the resemblance between the ideas of primary qualities and the material properties that cause them in us, to come to know the inner constitution of bodies. And even if we had such powerful senses we would be unable to experience the productive connection between things and experience of things.

This was the epistemology which it is my contention we must see informing the interaction between theology and science if we are to understand the way doctrines of the limits of knowledge developed in the eighteenth century.

If we cannot have knowledge, nevertheless men do assent to and dissent from empirical propositions. According to Locke, our assent and dissent is limited to our judgement of the probability of an opinion. But probability is not the ground of assent, rather assent is a ground of probability. The highest degree of probability, according to Locke, is reached 'when the general assent of all men in all ages, as far as it can be known, concurs with a man's constant and never failing experience, in like cases, to confirm the truth of any particular matter of fact attested by fair witnesses' (Essay: 4, XVI, 6).

For many matters this joint criterion of historical assent and inductive support yields probabilities 'so near to certainty that they govern our thoughts and actions as fully as the most evident demonstrations... we make little or no difference between them and certain knowledge... our belief thus grounded, rises to *assurance*'.

And this assurance is all we need. In understanding how Locke saw his sceptical doubts about knowledge I am convinced the providentialism of the time must be taken into account. I have yet to establish the historical plausibility of this point, but I shall demonstrate the ubiquity of the providentialist assumption even in so unlikely a thinker as Hume.

Knowledge of powers

Just as the problem of real essences is set by Locke in immemorial terms – 'Reason calls for hypotheses about real essences but Experience cannot supply them' – so he set the corresponding problem for powers in a similar immemorial form. He saw that the powers of substance to change the sensible qualities of other bodies are a topic of central importance for our enquiries into nature – but he doubts 'whether *our Knowledge reaches* much further than our experience'. Even knowledge of nominal essences would not help since we could not infer 'that they are in any subject by the connexion with any of those ideas which to us mark its essence'. And since we cannot discover the 'texture and motion of parts' which are the ways of operating of the powers of bodies, 'it is in but a very few cases we can be able to perceive their dependence or repugnance to any of those *ideas* that make our complex one of that sort of thing' (*Essay:* 4, III, 16).

But it is worth noticing that nowhere does Locke cast any doubt on the propriety of speaking of powers as qualities of material beings even though our experience can only be of their effects. That had to await the turn of the Humean screw. Indeed Locke sets out the experiential conditions for knowledge of powers in *Essay:* 2, XXI, 1. Such knowledge is derived from the experience of change and the concomitances of change, from the experiences of like changes associated with like agents, and the consideration of the possibility of simple ideas being changes. Within the general category of powers we distinguish active powers 'able to make a change' and passive powers 'able to receive a change'.

The scene is now set for the great debate to follow, since I believe it was Locke's views above all which seeded the epistemological controversies of the eighteenth century.

Locke's 'nescience' derived from the incompatibility of his realist conception of what a completed science would be like and his sensationalist epistemology. I shall explore several different reactions to the Lockean position, to illustrate the diversity of early-eighteenth-century views about knowledge,[23] and in the hope of proving, by demonstration, the historiographic thesis with which I began.

Reactions can be classified according to whether they took the issue of powers or the problem of essences as the central topic of concern. The theological position is very clear. The source of the active powers of people (to think) and of things (to gravitate, cohere, etc.) had to be found in the Divine Being, at peril of the darkest irreligion. This theme runs through Ditton's *Discourse* of 1712[24] to Beattie's *Essay* of

[23] I owe notice of the wide variety of these reactions to G. Bowles, cf. his 'The place of Newtonian explanation in popular thought, 1687–1727' (D. Phil. thesis, University of Oxford, 1977).

[24] H. Ditton, *A Discourse Concerning the Resurrection of Jesus Christ* (1712), appendix.

1770.[25] To those who wrote early in the century the most alarming form of secular dynamism was that which explained the active powers of people in terms of the active powers of matter. As Ditton says, 'Let him [the Lockean sceptic] begin with making *Matter and Motion think;* and he shall end with making *the Gospel an Imposture' (Discourse,* 474). As we shall see, towards the end of the century the argument broadened and the evident powers of material beings were being cited as the best possible evidence for the existence of God, since powerless matter could not be responsible for the powers of material things.[26] It is my contention that the arguments concerning the limits of scientific knowledge were intimately intertwined with the theological debates concerning the source or origin of powers.

Epistemologically glossed, the most common form of the argument runs:

1. We know that people and things have active powers.
2. The source of these powers cannot be matter, which is essentially passive.
3. The source must be external to the material system and is none other than God.

The founding of nescience on the empirical impossibility of knowing real essences was challenged quite early in the century, though how representative are the authors I will consider, I am uncertain. A very detailed study of the theological literature of the period would be required to confirm the choice. G. Bowles's work is a valuable beginning.

Lee's *Antiscepticism* of 1702[27] takes its start from the issue of whether matter can think. It is clear that if we can 'let Matter have what Figure, Bulk, Motion or Position of its Parts that can be imagined, it can no more perceive or be conscious of its own Actions or Motions than a Stone can rise from the Ground of its own accord' (*Antiscepticism,* 247/8).

But suppose Locke were right and matter had an unknown essence in which the power to think might be materially grounded? This worry could be removed at a stroke if 'our Senses be right', i.e. give us knowledge as good as intuitive and demonstrative knowledge, and if essences were perceptible in clusters of sensible qualities. But

[25] J. Beattie, *An Essay on the Nature and Immutability of Truth in Opposition to Sophistry and Scepticism* (Edinburgh, 1770).

[26] Berkeley's philosophy was only of secondary importance as an influence on the thought of the *philosophes.* He figures in the Jesuit attack on Malebranche. Most British philosophers (cf. Beattie) treat him as a sceptic. H. M. Bracken, *The Early Reception of Berkeley's Immaterialism, 1710–1730* (The Hague, 1965).

[27] H. Lee, *Antisceptism* (1702).

everyone means by 'Genus, Species and Essence', so declares Lee, no more than 'Names of more or fewer Qualities or Properties, by which things agree or differ from each other' (*Antiscepticism,* 251). So our sensory knowledge is science in just Locke's sense.

On the other hand, by distinguishing between the sensible and the intelligible as sources of knowledge, Sergeant[28] argues for knowledge of real essence in Locke's original sense – the inner constitution of bodies responsible for the regularities of their appearances – and that real essences can be 'in the mind' just as much as ideas of sensible qualities – only they come to be in the mind via the intellect rather than the senses. This is the first hint in an English writer of the period of the characteristic rationalism of the Leibnizian approach to theoretical entities. Sergeant distinguishes notions of things gathered by reason from ideas taken for similitudes. The impossibility of checking ideas for similitude to things is what leads to scepticism. Pre-empting Berkeley, Sergeant notices that while we cannot have general ideas we can have general notions

> ... Bodies, which ... are one way or other contained or included in a spirit cannot be included in it Quantitatively; after the manner a Vessel contains *Liquoier* [*sic*], or as a Bigger Box contains a *Lesser.* Wherefore, it must be said, that those Lesser essences of Body are contained in the *Superior* Essence of an *Angel* Indivisible; or after the manner of a *Spirit,* that is *Knowingly,* or as *Objects* of their Knowing Power.[29]

They are there as *essences,* that is as *intensional* objects.

Neither way out gained widespread assent. Ditton's positivism was hardly likely to appeal to those who, whatever their theology, accepted some form of Newtonian corpuscularianism. Sergeant was a Catholic, a fact unlikely to favour the acceptance of so radical a solution to the problem of nescience. The persistent worry came from seeing sensationalism both as a support for theology (relieving us of the dangers consequent on being forced to postulate active and perhaps even thinking matter) and as a support for scepticism (God as an active but hidden being is in no better case than matter). Gravity became the focus of the discussion. It seemed to be an ineliminable real power attesting to a universal natural activity. Could it be domesticated into theology and claimed as a support for religion?

The powers of matter

So central did the problem of the explanation of gravity become in the controversies that we might well speak of the theologizing of the

[28] J. Sergeant, *Solid Philosophy Asserted, Against the Fancies of the Ideists* (1697).

[29] J. Sergeant, *Transnatural Philosophy* (1790), bk. 2, ch. 2.

concept of a gravitational power. The connection between the implications of powers for theology and nescience is made very clear by Samuel Clarke.[30] As he says (*Demonstration,* 83):

> For we see and feel, and observe daily in ourselves and others, such Powers and Operations and Perception, as undeniably evince themselves either to be Properties of Immaterial Substances: or else it will follow that Matter is something of whose innermost Substance and Essential Powers we have altogether as little Idea as we have of Immaterial Beings; and then how are Immaterial Substances more impossible than Material?

And then follows later in the work the oft-repeated argument that gravity depends on an immaterial substance.

> Even the very first and most universal Principle of Gravitation itself in all inanimate matter, since it is ever Proportional, not at all to the *Surfaces* of Bodies or of their Particles in any possible Superstition, but entirely to the *Solid Content* of Bodies, 'tis evident it cannot be caused by Matter acting upon the *Surfaces* of Matter, which is all It can do; but must be caused by something which continually penetrates its *Solid Substance.*

But this argument can be made to cut both ways. On the other hand Whiston[31] concludes:

> that this Gravity is an intirely Immechanical Power, and beyond the Abilities of all material Agents whatsoever. (*Astronomical Principles,* 45)

elaborating the argument somewhat to include the point that its effect is proportional to solid content and not to surface. Whiston also notes that gravity is independent of state of motion and acts at a distance. From this it follows that

> it's not, strictly speaking, any Power belonging to Body or Matter at all . . . but is a Power of a superior Agent, ever moving all Bodies after such a manner, as if every Body did Attract, and were Attracted by every other Body in the Universe.

Since this power has been demonstrated to be immechanical and beyond the abilities of all material agents, 'tis certain that the Author of this Power is an Immaterial or Spiritual Being, present in, and penetrating the whole Universe' (*Astronomical Principles,* 89). So gravity serves as a defence of the existence of God. Indeed, it is a premise from which the existence of God seems certainly to follow according to this line of argument.

[30] S. Clarke, *A Demonstration of the Being and Attributes of God* (1716).

[31] W. Whiston, *Astronomical Principles of Religion, Natural and Revealed* (1717).

However, by 1762, while accepting the premises, Jones[32] can draw the opposite conclusion:

> From these and many other experiments open to common observation it must appear to every unprejudiced philosopher that nature is furnished with a mechanical cause whose activity is not confined to the surface of bodies, but extends to their constituent parts, that is to their quantity of solid matter. (*Essay*, 26–7)

This leads Jones to argue in Hutchinsonian style: 'Shall we allow that God governs the world by a subordinate agency and mechanism in some cases, where that agency appears to us; and deny it in others, merely because we have lost sight of it, or because it would count against us?' (*Essay*, 30–1).

In defence of the idea of a 'native tradition'

Robert Greene's massive anti-Lockean work appeared in Cambridge in 1727.[33] So far as I can discover it caused about as much stir as Hume's *Treatise*. Yet to a student of the epistemological developments of the eighteenth century it is a work of importance, representing what Heimann and McGuire have called 'the native tradition of dynamism', existing, though hardly flourishing, alongside the dominant but paradoxical sensationalism and atomism of the Newtonians and their 'official' philosopher John Locke, the epistemological consequences of which I have argued must be seen in a primarily theological context. To defend the idea of the 'native tradition' I shall expound Greene's views in some detail.

Greene opens his attack with a pointed statement of the deep-lying incoherence in the Newtonian system. 'So that if we compare the original hypothesis of this philosophy with the Conclusions of it, the one seems to be little less than a confutation of the other. Matter is supposed to be entirely passive. The sum of all the mathematical reasoning upon that hypothesis is, that it has in every part of it a Force of Gravitation, that is, that it is entirely Active, and if such a conclusion is not a contradiction to such a Hypothesis, I cannot tell what is' (*Principles*, 39).

His solution was to abolish the passive materialism of the atomic hypothesis. Like Boscovich and Kant (and later Faraday) he proposes a system of forces, the balance between attraction and repulsion producing the material world as we experience it. A solid body, for

[32] W. Jones, *An Essay on the First Principles of Natural Philosophy* (Oxford, 1762).

[33] R. Greene, *The Principles of the Philosophy of the Expansive and Contractive Forces* (Cambridge, 1727). Greene's name is sometimes spelt Green.

instance, is a closed surface at which the nett force is repulsive. 'From a various Mixture of a greater or less proportion of these forces, all the qualities of bodies arise'.

Scientific knowledge on this view must be of forces and their mixtures. It must be formulated for a world 'on the other side', so to speak, of sensations, which are no more than effects. With such a programme the epistemology must be hypothetico-deductive. 'If I can produce', says Greene (*Principles*, 286), 'those actions and forces, which will not only solve all the phenomena of matter which we are acquainted with by our sensations from it, but even those which may possibly arrive to other animals of a distinct nature and species from us, I hope I shall have given a full account, not only of the essence of matter, but of its Real and Essential properties.' And this, in almost infinite detail, he proceeds to do. The thoroughgoing dynamism of Greene's *Principles* poses an interesting (and as far as I know unsolved) historical question. How far did the continental dynamists in the tradition of Leibniz, such as Boscovich and Kant, know anything of the British 'native tradition' exemplified, say, by Greene? For the abolition of Lockean primary qualities in favour of a balance of forces is so similar to Kant's treatment fifty years later that one wonders whether Kant perhaps had some acquaintance with Greene's philosophy. Standard works on Kant's intellectual sources make no reference to Greene as among the authors with whose works he was acquainted.

According to Greene, the sources and the forces required for the dissolution of primary qualities are either those forces themselves or the direct activity of God. The former will not do as an ultimate explanation since, according to Greene, whatever exists of itself will have a necessary principle of its existence. This sets the limit to our knowledge. 'The furthest the human mind can reach in philosophy' is to catalogue the variety of expansive and contractive forces. And this is to catalogue the variety of 'the substratum or essence of matter', since that can only be 'Action or Force'. Does this contradict Locke's limits of knowledge? It is not entirely clear that it does. Locke's epistemological difficulties derived from the impossibility of discovering in sensory experience any simple ideas of those qualities of the inner constitution of bodies on which their powers depended. But that supposes that the powers of bodies depend on some kind of texture of atomic bodies. By denying the necessity for that grounding and basing matter directly on powers or forces Greene abolishes Locke's limits by abolishing the farther region. We can neither reach nor fail to reach that which does not exist. Towards the end of the century dynamism became a more and more popular solution to Locke's problem. I illustrate both the later dynamist conception of nature and the social structure of the English intellectual orders in the person of Michell.

J. Michell (1724–1793) perfectly exemplifies the academic career of the period. As a Cambridge don he was famous for the 'Mathematical Bridge' over the Cam. Later, translated to a country living he brooded on philosophy, conversed with his neighbours and friends but wrote nothing. Most of what we know of Michell's theory comes from Priestley, upon whom he exerted a considerable and lasting influence.[34]

Priestley's account of Michell's path to the conception of matter as active power runs as follows:

> This scheme of the *immateriality of matter,* as it may be called, or rather, the *mutual penetration of matter,* first occurred to Mr Michell on reading Baxter *On the Immateriality of the Soul.* He found that the author's idea of matter was that it consisted as it were of bricks, cemented together by an immaterial mortar.[35]

Carrying this argument through, it seems that the need for an impenetrable and basic matter simply disappears. The form of the argument is close to that adopted by Greene. By comparison of hypotheses the force theory triumphs over the theory of material substance, and since observable effects are more readily and uniformly accounted for by postulating fields of force, reason and experience conspire to support the field point of view.

But the underlying causal powers and forces cannot be perceived. The final apotheosis of the Lockean theory of sensory ideas ran straight into the developing field theory, for Hume's theory of knowledge seemed to wipe out force as mere reflections of psychological phenomena.

The field theory, whose epistemology troubled the philosophers and theologians of our period, is not usually credited to Greene but to the independent discoveries of John Michell and Roger Joseph Boscovich. Joseph Priestley, for example, couples their names when discussing the activity of matter.

Though the field theories of Michell and Greene were perhaps exceptional in being very fully developed, many people of the period held to some sort of force theory of matter, often inconsistently coupled with a vaguely substantialist basis. Preference for forces over substances accounts for one curious incident of the time: Stephen

[34] J. T. Rutt (ed.), *Life and Correspondence of J. Priestley* (1831).

[35] This quotation is taken from A. Geikie, *Memoir of John Michell* (Cambridge, 1918). So far as I can ascertain this is the only biography of this interesting man.

Hales's rejection of John Mayow's 'discovery' of 'oxygen'.[36] Hales, like many chemists and physiologists, had performed experiments on the breathability of air, and knew of the reduction in volume occasioned by a plant or animal using air. Mayow had explained the phenomenon, late in the seventeenth century, as the result of the fixing of a material substance – nitro-aerius vapour, a substance which made up one fifth of the bulk of the air – by the breathing creatures. Hales rejected this view, arguing instead that by breathing the air in and out the creature reduced its elasticity, the force which held the air expanded out to a certain volume. A reduction in the elasticity would lead to a shrinkage in bulk of the original air. Hales had arrived at fields too soon.

To complete my case for the historical importance of British dynamical doctrine, and to show how deeply providentialist assumptions were ingrained in the epistemology, I turn now to an exegesis of Hume's philosophical writings to demonstrate the importance of the eighteenth-century interest in the sources of activity, both natural and human. Only against that background, I argue, can the direction of Hume's investigation of the limits of human knowledge be understood.

The Humean enigma

Hall has demonstrated how little philosophical interest was aroused, until relatively recently, by Hume's sceptical arguments. The bibliographical evidence seems conclusive on this issue.[37] And yet Hume was a considerable figure in his own time, directly influential on many important authors. In a recent essay Jessop has raised in an interesting form the historical problem of the interpretation of Hume's philosophy. Following Kemp Smith, Jessop argues that Hume saw himself primarily as a moral philosopher, and perhaps even as something of a moralizer.[38] His scepticism is directed not so much against particular doctrines as such, but against the uncritical reliance on reason in human affairs.

If indeed the central issues of concern in intellectual circles in eighteenth-century Britain was the extent of our knowledge of powers and activity, then it is hardly surprising that the analysis of causal relations should figure largely in Hume's *'philosophia prima'*.[39] Hume

[36] S. Hales, *Vegetable Staticks* (1727), experiment CVI.

[37] R. Hall, *Fifty years of Hume Scholarship* (Edinburgh, 1978), 1–14.

[38] T. E. Jessop, 'The misunderstood Hume', in Todd, *Hume and the Enlightenment* (see note 14), 1–13.

[39] D. G. C. McNabb, in his *Introduction to 'A Treatise of Human Nature', by David Hume*, bk. 1 (1962), says 'the strength and consequences of this philosophy

does not seem to have doubted that material beings, animate and inanimate, had powers, but he claims to be unable to see how human beings could have any knowledge of powers in themselves. To summarize the well-known argument: the idea of causal connection involves four root ideas – contiguity of the idea of the cause and the idea of the effect; succession of these ideas; the production of the one by the other; and their necessary connection. To discover the status or real meaning of these ideas one looks for the corresponding impressions. Contiguity and succession of ideas do indeed seem to be associated with contiguity and succession of impressions. But Hume can find no impression corresponding to the idea of production. The analysis of the idea of necessary connection is his *tour de force* since he claims to show that there is a corresponding impression, but at bottom it is the psychological effect produced by repeated contiguities and succession of like pairs of impressions. In consequence it is not an idea corresponding to a real relation between things or events. Since, according to Hume, 'the terms of *efficacy, agency, power, force, energy, necessity, connection* and *productive quality* are all nearly synonymous' the same treatment will do for them all.[40]

By linking meaning to experience, the content of ideas to their original impressions, Hume abolished all ideas that purport to refer beyond experience. Science, as the totality of causal principles, can be no more than the record of repeated concomitances of impressions. It seems to me quite clear that Hume uses this line of reasoning to introduce the central theme of his moral philosophy, a kind of secular providentialism, that feelings, passions, instincts, if you will, are an adequate guide to human action.

But I am convinced that the theological motivation of discussions of the extent of human knowledge was as potent at the end of the century as it was at the beginning. I want to try to demonstrate the need to structure one's historical analysis on this assumption through a brief glance at two rather different reactions to Hume's sceptical philosophy: those of Priestley and of Beattie.

Priestley's discussion goes to the heart of Hume's argument.[41] It turns upon the principle that relates an idea to one and only one impression as its original. Priestley simply points out that no reason is given for this arbitrary restriction. As soon as one enquires as to how

were totally unrecognised by almost all Hume's contemporaries and immediate successors'. This seems overly strong. Frequent references to his thought can be found from about 1760 onwards and he is usually billed as the arch-sceptic, revealing the true consequences of Locke's way of ideas.

[40] D. Hume, *A Treatise of Human Nature* (1739), bk. 1, sect. XIV.

[41] J. Priestley, *Disquisitions Relating to Matter and Spirit* (1777).

the idea of a power is acquired one sees that it is correlated with a *set* of impressions, which taken together yield the idea. Following Locke, whom he quotes and discusses extensively, Priestley boldly offers a field theory of matter, quite within the tradition of Greene, Michell, Boscovich and Kant.

Priestley's argument follows the standard pattern of eighteenth-century analysis of matter. He shows how primary qualities, just like secondary qualities, can be accounted for in terms of a theory of paired forces. '[*R*]*esistance,* on which alone our opinion concerning the solidity or impenetrability of matter is founded, is never occasioned by *solid matter,* but by something of a very different nature, viz. a *power of repulsion* always acting at a real but assignable distance from what we call the body itself' (*Disquisitions,* 4). 'No such figured thing can exist unless the parts of which it consists have a mutual attraction, so as either to keep contiguous to, or to preserve a certain distance from each other' (*Disquisitions,* 5). But Priestley was a Socinian. Mind is material and man is active. He needs no immaterial soul to endow him with his capacities, since matter, it is now securely established, is an active being. Theological consequences turn the discussion back to Baxter's troubled spiritual universalism.

Baxter had supposed that the need to analyse matter as power requires that we 'make that Deity himself to *do* and to *be* every thing' (*Disquisitions,* 9). Priestley's theological conclusion is Unitarian in every sense. 'If I be asked how, upon this hypothesis, *matter* differs from *spirit,* if there be nothing in matter that is properly solid or impenetrable, I answer, that it no way concerns me, or true philosophy, to maintain that there is any such difference between them as has hitherto been supposed.'

If we take Beattie as typical of reactions to Hume's sceptical views within the Scottish Enlightenment, it is clear that the knowability of agency is indeed the prime topic of concern. Beattie locates the origin of scepticism in the arguments (but not, of course, the intentions) of Descartes.[42] 'His successors [Locke, Berkeley and Hume], the further they advance in his systems, become more and more sceptical, and at length their reader is told, to his infinite pleasure and emolument, that the understanding, acting alone, does entirely subvert itself and leaves not the lowest degree of evidence in any proposition' (*Essay,* 141).

The crux of the matter is perception; and the point at issue is knowledge of power or energy and of the self. Beattie points out that the doctrine that impressions and ideas have an essential identity,

[42] Beattie, *Essay* (see note 25).

differing only in strength (the basis of Hume's sceptical argument) depends upon a confusion between three distinct uses of perception words: for the thing perceived, for the power or faculty of perceiving, and for the impulse or impression conveyed to the mind. The idea of a thing lacks many of the qualities of the thing – but the senses present, *in their own style,* the way things really are. So, for instance, perspective is not an illusion, but how things look if they are really at different distances.[43]

The powers of things to produce effects are perceived in the activity we experience. Contrary to Hume's thesis, that constant conjunction is the only empirical element in causality, Beattie argues (*Essay,* 199–201) that though it may be a necessary it is certainly not a sufficient condition for identifying causation. We need to experience also the acting of one thing on another, in short the manifestation of its power.

The providentialist theme is present in Beattie's views too (*Essay,* 169–70): 'That matter or body has a real, separate, independent existence . . . has been the belief of men who were not mad, ever since the creation. This is believed, not because it is or can be proved by argument, but because the constitution of our nature is such that we must believe it.' And of course it is implied that it is to our advantage to have such a nature.[44]

I have pointed out earlier in the chapter how importantly a providentialist justification of the actual condition of man loomed in the background to philosophical scepticism. We may not be able to know very much, but what we do know is enough for our needs. The development of a kind of secular providentialism seems to me to have been an essential element in the preparation of the conditions for the appearance of evolutionary theories based upon adaptation. In a way, adaptation through natural selection is a providentialist notion, and the idea that men, like other beings, are adapted to their life conditions is a providentialist thesis. I see a clear adumbration of that idea in the works of several of the members of Hume's Edinburgh circle, in

[43] It is clear that Beattie owes a great deal to Reid. Cf. T. Reid, *Essays on the Intellectual Powers of Man,* (Edinburgh, 1785): 'ingenious men . . . [misled by a theory] confound the operation of the mind with their objects, and with one another, even where the common language of the unlearned clearly distinguishes them' (*Essays,* 166); and 'I have a clear and distinct notion of each of the primary qualities. I have a clear and distinct notion of sensation. I can compare the one with the other; and when I do I am not able to discern a resembling feature . . . Figures, divisibility, solidarity, are neither arts nor feelings' (*Essays,* 215).

[44] Adam Smith, *An Inquiry into the Nature and Causes of the Wealth of Nations* (2 vols., 1776; New York, 1948).

particular those of Adam Smith. Indeed, a kind of optimistic providentialist tone runs through all the 'Moderate' party, and is in sharp contrast to the Calvinism of the 'Populars'.

Hume's providentialism emerges in his psychological theory of belief. Clearly, we would be best constituted for practical life were we to come to believe in the continuance of just those phenomena which have been regularly conjoined in the past. Reason, so Hume argues, cannot make knowledge out of this – but our psychological constitution can make belief. If 'a belief differs from a fiction only in the manner of its being conceived' (*Treatise*, I, 145), that is in its 'force and vivacity', then we must look for the source of that 'force and vivacity'. It is found in the repetition of the impressions upon which the idea is formed, by which it is 'invigorated' (*Treatise*, I, 164). Further assent is procured by this vigour – and providentially, since for practical life we had better believe in just those happenings which have been parts of frequently repeated sequences in the past. Though Hume does not say this, so far as I am aware, it is clearly implied that those who do not form beliefs in this way will run into trouble. Further, in elaborating this theory of action in Book II, Part III, Section III of the *Treatise*, Hume is at pains to demonstrate the ancillary rather than fundamental role of reason in the control of practical action.

The theme of automatic adjustment of the impulses natural to man to the characteristic forms taken by consciously planned activities of men in society, is central to Adam Smith's philosophical economics. Despite references to 'the divine Being whose benevolence and wisdom have, from all eternity, contrived and conducted the immense machine of the universe so as at all times to produce the greatest possible quantity of happiness', in this and similar passages,[45] Adam Smith's optimism in the face of the prevailing nescience about the inner workings of society and nature could be given a quite secular cast. On the one hand there are the instincts implanted by nature and on the other the consciously formulated plans of man. It is through the interaction of these secondary systems (God's order being primary but quite hidden) that this happy world is produced. 'The rules which she [nature] follows are fit for her, those which he [man] follows, for him; but both are calculated to promote the same great end – the order of the world and the perfection and happiness of human nature' (*Wealth*, 197).

In a recent essay of great interest, A. S. Skinner has shown how sophisticated was Adam Smith's conception of scientific method.[46]

[45] Adam Smith, *Wealth* (see note 43), 194.

[46] A. S. Skinner, 'Adam Smith: philosophy and science', *Scottish Journal of Political Economy*, XIX (1972), 307–19.

Like Hume, he advocated the use of the 'Newton's Rules'. But Skinner has shown that Adam Smith was clear that these principles of method had been consciously formulated long before Newton set them out. This is historiographically important. It points to the possibility that there was current in the circles Hume frequented, a distinction between the Newtonian method and the Newtonian metaphysics. And that would help to account for the way Hume adopted the one and used it to develop sceptical arguments against the other.

The solution to the Humean enigma is, I believe, of startling simplicity. It is that Hume must be treated as a man of his time. Thus we would expect him to be providentialist with regard to the conditions and capacities of mankind, and deeply interested in the status of hypotheses about the powers and agency of material beings. Though Hume's arguments and conclusions are distinctive, I would conclude that his interests were in all essentials just the same as those of Whiston and Clarke.

III. Case 2: Intellectual sources of the Enlightenment in France

The traditional view

The traditional view of the origins of Enlightenment thought expressed, say, by Gay[47] or Berlin,[48] sees the *philosophes* engaged in intellectual warfare with the priestly legions of the *ancien régime* and drawing new ammunition from their importation of the philosophy of Locke and the science of Newton. 'By mid-century the philosophers were trying to identify their procedures with the methods of the natural sciences', says Gay (*Enlightenment,* I, 140–1). Voltaire's attitude to his character Dr Pangloss is, I suppose, thought to be typical of Enlightenment attitudes to the philosophers of the immediate continental past. In its emphasis on experience, the philosophy of the Enlightenment 'rejected the assertion that Reason is the sole, or even the dominant spring of action'. According to Gay, 'His [Locke's] impact on the Enlightenment was so pervasive that to analyse it fully would be to write another book' (*Enlightenment,* I, 321). Locke's philosophy was taken to support a sensationalist epistemology and Newton's science, both a practically oriented methodology and a model for a philosophy reduced to a simple psychology of associated sensations. The great popularizers of the age, says Berlin, 'headed by Voltaire, Diderot, etc., . . . were agreed upon the crucial importance of this sensationalist approach . . . The dominant trend is in favour of

[47] P. Gay, *The Enlightenment: An Interpretation* (2 vols., 1966–9).
[48] I. Berlin, *The Age of Enlightenment,* (New York, 1965), introduction.

analysing everything into ultimate irreducible atomic constituents, whether physical or psychological' (*Age*, 20). All this is in sharp contrast to the rationalist idea that there are *a priori* principles to which nature and experience must conform. According to Gay, 'Bacon, Newton and Locke had such splendid reputations on the Continent that they quite overshadowed the revolutionary ideas of a Descartes or a Fontenelle' (*Enlightenment*, I, 11–12).

I shall try to demonstrate that this judgement is quite mistaken. The borrowings Diderot made from Locke are ambiguous, those of Condillac very selective, while d'Alembert's diversion to empiricism was short-lived. Lately, Vartanian,[49] Kiernan[50] and others have challenged this view of the period. Vartanian claims to identify within the movement a continuing Cartesianism and Kiernan demonstrates convincingly that the semi-mechanist relations which are thought to obtain between sensations in Condillac's system were quite foreign to Diderot's general philosophy. I hope to demonstrate how poorly sensationalism was regarded by comparing the views of Condillac with the final apostasy of d'Alembert. I should add, however, that I have no quarrel with Gay's general view. He defines the Enlightenment *historically* as a conflict with traditional religion, using the philosophy of Locke as one among its weapons. The other was history, and history of a particular propagandist kind, as for example that pursued by Gibbon and Hume.

Sensationalism and the naïvety of Diderot

As Gay's citation shows, the traditional picture owes much to the writings of Condillac.[51] His is a thoroughly sensationalist philosophy, a philosophy that is a Newtonian science of ideas exactly according to Hume's prescription. It seems Condillac could hardly have been acquainted with the writings of Hume. Condillac explicitly acknowledges the Lockean provenance of his theory. But, it is a position of great originality since he proposes to show that not only are our ideas generated in sensations but so are our mental powers. '[T]he investigation of truth', argues Condillac, 'consists in ascending to the origin of our ideas, in unravelling their formation, and in compounding or decompounding them in different ways, in order to compare them in

[49] A. Vartanian, *Diderot and Descartes: A Study of Scientific Naturalism in the Enlightenment* (Princeton, N.J., 1953).

[50] C. Kiernan, *The Enlightenment and Science in Eighteenth Century France*, 2nd edn (Banbury, 1973).

[51] E. B. de Condillac, *An Essay on the Origin of Human Knowledge* (Paris, 1746), trans. T. Nugent (1956) repr. in facsimile ed. (Gainesville, Florida, 1971); and *Treatise on the Sensations* (Paris, 1754), trans. G. Carr, (1930).

every light that is capable of showing their relations' (*Essay*, II, 2,3). 'A very slight attention must convince us, that when we perceive light, colours or solidity, these and the like sensations are much more than sufficient to give us all the ideas which we generally have of bodies' (*Essay*, I, 1,2). That is not all. The original of all our ideas, and mental faculties, is to be found in sensations. 'The connexion of ideas, formed by attention, produces the imagination, contemplation and memory.' The only contributions prior to experience are instinctual mechanisms which withdraw the organism from painful stimuli and maintain its contact with pleasant ones. The intellectual powers of reflection are not innate. They develop as the organism copes with sensations. By noticing the difference between truths which change and those which do not, a primitive being could even acquire the distinction between truths of fact and truths of reason. So, argues Condillac, a primitive being, developing solely under the influence of sensations, 'has, by the help of the senses alone, cognitions of every kind'.

Condillac lays out his analysis in the form of a fable. He asks us to imagine a statue that is progressively provided with sensory and motor capabilities. At each stage he shows how much of our mental life, and what level of knowledge, such a being could attain. In the end, he hopes to convince us the statue has acquired all the skills and capacities we know in man. 'The statue is . . . nothing but the sum of all it has acquired. May not this be the same with man?' It is not difficult to draw the revolutionary conclusion that every man could acquire the knowledge or intellectual power of any other man.

Though it is quite certain that Condillac was not influenced by Hume, there are similarities in their views on the dynamics of mental operations. '[P]leasure and pain become the unique principle which determines all the operations of the soul, and will raise it by degrees to all the cognitions of which it is capable' (*Treatise*, I, 21). Reason has not quite yet become the slave of the passions, but clearly, the affective dominates the cognitive. 'An intellectual idea is the memory of a sensation' (*Treatise*, II, V, 13) and 'Pleasure leads the memory' (*Treatise*, I, 21). The point of the whole analytical programme is to found knowledge in practical necessities, and of the actual conditions of living. Nature has implanted the instinctual mechanisms of ideas, of their comparison in judgement. Finally, 'the habits, which it has contracted following its judgements, are what I mean by *practical knowledge*' (*Treatise*, IV, intro.). Foresight is no more than the habit of remedying or relieving needs in good time to avoid the remembered 'torments endured before the remedy was found'.

The thoroughgoing convergence between the theories of Condillac and Hume deserves more consideration than it has received from commentators on either. Once, for whatever reason, the Lockean metaphysics of powers is dropped, it seems inevitable that causation should be reduced to psychological habit and judgement to the domination of reason by the passions.

If the traditional thesis is correct then sensationalism ought to be found at its most fully developed in the main authors of the *Encyclopédie,* Diderot and d'Alembert, since of all documents of the time the *Encyclopédie* would surely represent the distillation of Enlightenment opinion. Not only is it difficult to find the strong sensationalist views of Condillac in the contributions of Diderot but, as I shall demonstrate, d'Alembert in his mature thought returned to a rationalism of a very Cartesian mould.

Though Diderot's epistemology is based upon uncritical empiricism ('Observation', he announces, 'collects the facts'), it does not depend at all on an atomistic sensationalism as a foundation.[52] But the break with Descartes and Leibniz is sharp. The role of reason is reduced to forming combinations of facts to be tested by experiment, and further reduced to the formulation of hypotheses from materials given in observation. 'So long as things are only in our understanding, they are our opinions; these are notions which are true or false, granted or denied. They take on consistency only by being related to externally existing things. This connection is made either by an uninterrupted chain of reasoning which is connected at one end with observation and at the other with experiment; or by a series of experiments dispersed at intervals along the chain of reasoning, like weights along a thread suspended by its two ends. Without these weights the thread would be in the sport of the slightest motion of the air.'[53]

Diderot's scepticism is not the strongly anchored nescience of the British empiricist philosophers descended from Locke. So far as I can see, Diderot owed almost nothing to the sensationalism at the basis of British nescience, even though he read Locke and his close association with Condillac could hardly have failed to make him aware of its ramifications. His doubts about God are those of a simple positivist. 'The supposition of any being whatever placed outside the material universe is impossible. We must never make such suppositions because one can never infer anything from them' (*Diderot,* 132).[54]

[52] J. Kemp, *Diderot: Interpreter of Nature (New York, 1963).*

[53] 'On the interpretation of nature VII', trans. J. Stewart and J. Kemp, cited in Kemp, *Diderot,* 43 (see note 52).

[54] Kemp, *Diderot* (see note 52).

But his doubts about mundane knowledge are merely Cartesian. In *Pensées philosophiques:* xxx[55] he asks '"What is a sceptic?" He is a philosopher who doubts all that he believes, and who believes that a legitimate use of his reason and senses is to demonstrate the truth of that [doctrine].' He goes on to say, in a thoroughly Cartesian vein, 'Scepticism is then the first step to truth. It has to be general, for it is a touchstone.'

With this epistemological foundation Diderot has no trouble with active, natural powers. Each kind of being has its own natural powers. 'I conclude that matter is heterogeneous, that an infinity of divers elements exist in nature, that each of these elements, by its diversity, has its own particular force, innate, immutable, eternal, indestructible' (*Diderot,* 131). These natural powers are invariant potentialities. 'Force which acts on a molecule exhausts itself; force which is a part of the molecule does not exhaust itself' (*Diderot,* 129). To a simple-minded realist as robust as Diderot, to know of these powers is simply to observe them in action. 'I who consider bodies as they are in nature and not as they are in my head, I see them as existing, differing, having properties and actions, and moving in the universe as they do in the laboratory, where a spark cannot be placed beside three particles made of saltpetre, carbon and sulphur, without an explosion necessarily following' (*Diderot,* 129).

Diderot notices the very same relation between theology and the issue of whether matter has active powers as did his English counterparts, but he does not express the issue directly in terms of knowledge, though he does discuss the issue of scepticism and its relation to ontology. 'The subtleties of ontologies exist despite the sceptics. It is to the knowledge of nature that truth is reserved by true deists. Whether movement were to be essential or accidental to matter, I am now convinced that effects lead to growth' (*Pensées philosophiques:* xix).

The *Encyclopédie* was meant to realize two leading principles – the unity of science and the scientific improvability of crafts. Both principles had political connotations for Frenchmen of the period. A unified science would be a comprehensive intellectual system competing with orthodox religion for the allegiance of educated people, and this is the aspect emphasized throughout by Gay. It would necessarily be materialist at bottom, since the unification was to be achieved by the demonstration of a hierarchy of sciences based upon geometrical mechanics. As for the other principle, nothing is so politically subver-

[55] D. Diderot, *Pensées pholosophiques: Oeuvres complètes* (Paris, 1875–7). Unless otherwise cited, passages from the works of Diderot are my translations.

sive as the thought that crafts are the foundation of civilized life – a thought implicit in the Baconian principle, whose assumption is evident throughout the *Encyclopédie,* that the improvement of life follows the development of crafts.

I have emphasized the strongly providentialist tone of much British writing on the limits of knowledge. It was essentially conservative whether from the pen of Locke or Hume. The *Encyclopédie* was concerned, too, with what we need to know for an ideal practical life, but the authors thought we actually fell far short of it. The deficit could be made up by deliberate efforts to improve both the methods and the status of crafts. It is clear that the *Encyclopédie* attitude to science and craft was radical. Given both the social conditions and the administrative follies of the time, it is easy to see why the police were instructed to hinder and even, at times, to suppress it.

Though d'Alembert continued to contribute articles to the *Encyclopédie,* his once close relationship with Diderot became more and more distant. This must be put down in part to a very sharp intellectual disagreement as to the epistemology of science. Diderot faithfully adhered to his Newtonian (even Baconian) respect for experiment as the arbiter of knowledge, but d'Alembert revealed a very deep commitment to the logicism and rationalism of the Cartesian tradition.

The retreat to abstraction: the apostasy of d'Alembert

It seems clear that, intellectually at least, the Enlightenment was very far from a simple absorption by advanced French intellectual circles of the British empiricism of the philosophy of Locke and the scientific method of Newton. French protestations on these matters hardly stand up to an examination of their actual doctrines, though d'Alembert speaks of taking back from the English what the French gave them in the philosophy of Descartes. Unfortunately, Gay seems to accept French evaluations of their intellectual mentors at face value. I have already quoted his remark about the ubiquity and dominance of the reputations of Bacon, Locke and Newton. Of course, as a propagandist declaration the claim to be the descendants of Locke and Newton had some force. But the intellectual biography of d'Alembert illustrates how limited in the end that influence proved to be.[56] The development of his thought shows how hard it was to hold fast to the empiricism he so greatly admired. His early adulation of Bacon, Locke and Newton,

[56] Cf. T. L. Hankins, *Jean D'Alembert: Science and the Enlightenment* (Oxford, 1970) chs. 3 and 4.

amply shown in his *Discours préliminaire* to the *Encyclopédie*[57] is contradicted by the strict positivism of the *Discours préliminaire* to his *Traité de dynamique*.

Even d'Alembert's introduction to the *Encyclopédie* is not fully in keeping with his remark 'We can say [Locke] created metaphysics almost as Newton had created physics, [since] he reduced metaphysics to what it must, as a matter of fact be, the experimental science of the mind' (*Encyclopédie*, 25). D'Alembert's sensationalist theses are different from Locke's. They agree in holding that the origins of knowledge are wholly sensory. D'Alembert, like Condillac, holds to a reduction principle. 'All our direct knowledge can be reduced to what we receive through the senses, hence it follows that we owe all our ideas to our sensations' (*Encyclopédie*, 2). The status of the reduction principle is not at all clear in d'Alembert's thinking. It appears to state a result of 'the experimental science of the mind'. But it could equally plausibly be treated as a criterion for distinguishing direct from reflective knowledge. The mind reasons 'about the objects of direct ideas or it reproduces and imitates them'.

Reflective knowledge is produced by 'memory, reason ... and imagination' [i.e. the talent of creating while reproducing objects] which are 'the three means by which our mind acts upon the objects of those thoughts' (*Encyclopédie*, 12).

In Lockean style d'Alembert distinguishes ideas of sensation and of reflection – a distinction which yields three realms of knowledge ('History, which relates to memory; Philosophy, which is the fruit of reason; and Fine Arts, which are created by imagination'), according to which the *Encyclopédie* was organized.

At this stage of his development d'Alembert shares the comfortable optimism of Diderot and Condillac. 'The first thing that our sensations teach us is our existence', 'The second ... is the knowledge of the existence of external objects'. There is no hint of a sceptical doubt about the verisimilitude of the deliverances of the senses. Locke's nescience in the face of the limitations of sense was so prominent a feature of his *Essay* that one wonders with Grimsley whether the *philosophes* read much beyond book II. Gay, too, has cast some doubt on how far the members of the Voltaire circle actually studied authors whose authority they claim.

The final clearly identifiable borrowing from Locke is a sensuous atomism already implicit in the principle of reduction. It appears most

[57] S. J. Gendzier, *Denis Diderot's 'The Encyclopedia'* (New York and London, 1967), for translations of passages cited from D. Diderot, *Encyclopédie* (Paris, 1746–66).

plainly in d'Alembert's discussion of definition, which consists 'only in disentangling in each notion the simple ideas contained in it: it is necessary - in order to know how to define - to learn how to distinguish the compound ideas from those that are not'.[58]

Even in his earliest philosophical works d'Alembert held to a strong reductionism, for he believed deeply in the essential unity of all science, a doctrine which he intended the *Encyclopédie* to demonstrate. He seems to have believed in two kinds of oneness in science - the possibility of a unified system of laws and the reality of a single kind of experiential origin.[59]

Given a sensationalist epistemology which reduces all sources of knowledge to one, unification of the system of laws is not a metaphysical issue but a technical one. 'One has happily applied Algebra to Geometry, Geometry to Mechanics, and each one of these three sciences to all the others, of which they are the basis and foundation' (*Traité*: iii).

But there is yet another sense in which d'Alembert believed in the possibility - indeed the necessity - of a unity of knowledge. He believed in the common origin of all the sciences in the need to deal with practical affairs. As time goes on an interest in knowledge, independently and gradually detached from practical affairs, begins to develop. This leads to the emergence of a kind of knowledge 'more refined but still relative to our needs and whose chief aim is a deeper study of less sensuous properties: the alteration and decomposition of bodies and the uses to which they can be put' (*Encyclopédie,* 24). This seems to foreshadow his later doctrine of abstraction. The results of the historical development from practices to theory are ordered by the discovery of the actual dependencies in nature.

But d'Alembert's later philosophy is very different. One source of a change of mind seems to have been his recurrent difficulties with the very problem that had led Locke to nescience and Hume to scepticism - the relation of ideas to things. Grimsley points out that d'Alembert quotes Berkeley *à propos* the problem, but in his earlier thought on the matter he held to a fairly strong realism - though never of the robust naivety of that of Diderot. In his *Encyclopédie* article, 'Elements of the sciences', he propounds the principle that 'true or assumed principles are grasped by a sort of instinct to which we must abandon ourselves without reserve'. One such principle is *eorundem effectuum eaedem*

[58] J. d'Alembert, *Traité de dynamique* (Paris, 1796). Passages cited from the work are my translation.

[59] D'Alembert, *Oeuvres,* II, 33, quoted by R. Grimsley, *Jean D'Alembert* (Oxford, 1963), 230.

causae (*Oeuvres:* II, 35). Throwing oneself on the mercy of this principle 'without reserve', he concludes optimistically, 'the sensations we experience have as much force as if these objects really existed; so bodies do exist' (*Oeuvres:* II, 135).

But almost in the same breath his confidence deserts him. 'The essence of matter', he admits, is that 'the idea we form of it' will always be wrapped in obscurity. In the end he argues for the abandonment of 'forces' and 'causes' and for a science only of effects. This leads to the hard-line positivism of his *Traité.*

In the *Discours préliminaire* to his *Traité de dynamique,* d'Alembert sets out the principles of a strong form of positivism – a return to very strict Cartesian logicism, together with an abandonment of all the Locke–Newton power concepts such as mechanical force concepts that Vartanian makes plain the French had never much believed anyway. Theory is transformed from an attempted description of the real but inexperienceable mechanisms of natural processes into a mathematical analysis of appearance. An epistemology of 'illumination' displaces one of 'knowledge'; but it is the illumination vouchsafed to a mathematician. In this work d'Alembert is uncompromisingly anti-sensationalist. 'The most abstract notions', he says, 'those that ordinary people regard as the most inaccessible, are often those that carry with them the greatest illumination.'

The notion of abstraction is intended quite literally in these passages. A scientist is to 'take out' of the complex combinations of sensory elements in experience only those which have to do with shape; that is, with geometrically analysable forms. The process of abstraction leads from obscurity to clarity. 'Our ideas are obscure', says d'Alembert, 'in so far as we treat an object as having all its sensible qualities'.

There is a strong hint in all this that d'Alembert is advocating not just a method for science but a metaphysics of the world as well. It is not clear from the context whether we should read the following passage in a pragmatic or a Platonic frame of mind: 'Not only do we have to apply knowledge derived from the most abstract sciences, but we must think of the objects of the world in as abstract a form as possible.' But at least one can readily discern d'Alembert's general convictions at this time. They mark a return to the native Cartesian tradition.[60]

[60] This schism, leading to a general split in the ranks of the *philosophes* has been repeated often enough in French theory of science as it shows itself in scientific practice. The relationship between the chemists Gerhardt and Laurent was as close and yet as divergent eighty years later, the cleavage appearing along the same fault line.

Cartesian criteria of paucity and clarity are invoked to recommend the geometrizing of natural science. D'Alembert points out with satisfaction that proceeding by 'taking out' geometrical form 'results in our having as clear principles as possible or as few as possible'. Rational mechanics appears as the result of a natural progression of formal sciences. 'Mechanics and geometry are alike in being the investigation of movement' (*Traité:* vi), e.g. the movement of a point generates a line, but differ in that in mechanics we must take into account 'the time that a moving body takes to traverse that space' (*Traité:* vii). This is an account of science about as far from the practical realism of Diderot as one could well get.

The advance to naturalism

If it were true that the borrowing of Locke's and Newton's general philosophy dominated the French Enlightenment, then Voltaire's deism would surely have represented the attitude of most *philosophes*. Vartanian has demonstrated that the internal schism that separated Voltaire from many of his fellow *luminières* lay along a line of cleavage that can be given a Cartesian definition. The issue was naturalism.

There was no question of the piety of Newton and Locke and it is clear from Voltaire's panegyric that this aspect of the British empirical philosophy was well understood in France. In particular, Newton held that the present state of the material world was the product of God's continuing intervention: the very hypothesis of which Laplace assured Napoleon he had no need. In discussing the British tradition of epistemological argument I have already drawn attention to the way in which many of Newton's circle took the existence of gravity as proof of the existence of God. According to Vartanian, what kept Cartesianism alive, despite the particular failures of the vortex theory with which it was so much associated, was the naturalistic element that could be extracted from it: 'To understand in the rise of naturalism among the *philosophes* the crucial contribution of Cartesian science . . . its having made the vortex a necessary consequence of a given definition of matter, rather than dependent like Newton's Law of Attraction on a supraphysical factor.' Even Descartes' curious idea that the starting point for a universe of extended substance and motion was unimportant, since such a universe would always evolve to the kind of world which we presently experience, was pressed into service as a foundation for naturalism. It seemed to suggest that there was no role for God in the creation of this particular world. According to Vartanian, 'LaMettrie, Diderot, Buffon [and] d'Holbach were not primarily concerned with' the elaboration of a naturalistic metaphysics. Central to the meaning of their naturalism was the vast plan of carrying to its completion the kind of unfinalistic, mechanistic science of nature 'of which Descartes had sketched the broad outlines' (*Diderot,* 89).

If Vartanian's judgement is sound, this raises a historical problem of considerable interest, of which I know of no recent discussion. It is clear that the Cartesian tradition was strongly anti-dynamist. One of the central objections to the Newtonian metaphysics of nature was its reification of forces and in particular of gravity. In the Newtonian scheme the substantial matter can be eliminated for the purposes of mechanics by an inertial force located at the centre of gravity of a material body. Newton's dynamism was, as one might say, his own invention. However, the theory of forces and later of fields, developed by the line of thinkers from Leibniz through Boscovich to Kant, depended on a powerful and indeed correct criticism of all theories of nature which depended upon supposing that the fundamental interactions were redistributions of velocity by material contact. Leibniz, Maclaurin and Boscovich had all developed systematic arguments to show the contradictions between the assumptions of a naive, material atomism which assumed that the ultimate bodies were rigid, that forces were finite, that time was continuous, and that action was by contact. It is unclear how far Kant's dynamism was motivated by an understanding of the paradoxes consequential on conjoining the above four principles, but it is quite clear that both Leibniz and Boscovich developed their position in a conscious attempt to resolve the difficulty. How was it, then, that French thinkers were so evidently unacquainted with the substantial, and indeed correct, criticisms of a universalistic contact physics which were being formulated and discussed in just the period in which the naturalists were claiming a support in the contact physics of Descartes? Vartanian, Kiernan and others provide us with very detailed knowledge of what their subjects heard and read. I would like to ask Enlightenment scholars what the figures they discuss did not hear and did not read and how the omissions might be accounted for.

Organicism

Recently Kiernan has proposed a very different framework for analysing Enlightenment thought in France, a framework very similar to that I have imposed on British thought of the period.[61] Kiernan identifies a French debate over the reality of natural agency as central to the dynamics of the time. While the English debate was more or less within the religious oligarchy, the French arguments were an inter-

[61] Kiernan, *Enlightenment and Science* (see note 50).

necine debate amongst those of the anti-clerical party, the *philosophes,* most subscribing to a vague Voltairian deism. Kiernan polarizes the debate around the positivistic empiricism of the mechanists at one pole and those who found their scientific rationale in the *'sciences de la vie'* at the other. But whether Kiernan's view casts more or less light, historiographically the debate lies within the traditional framework, seeing the history of epistemology as an intellectual history.

Kiernan and Vartanian both argue that the contrast between the sensationalism (deriving from Locke) and the mathematicism (deriving from Descartes) was the subject of an internal debate in the physicalist camp: a 'party war' says Kiernan. On this view d'Alembert's slow change from an adherence to sensationalism to an espousal of positivistic mathematicism was a passage within a general thought-frame which was itself in conflict with the 'life sciences' debate over the standpoint of the clerical and official party.

Conclusion

In sum there are four historiographical ideas in play. We are now in a position to adjudicate their plausibility and importance. Recent discussion has turned on three: that the doctrines of the *philosophes* were popularized British empiricism (the standard view); that the Enlightenment was essentially a social and political movement, anti-aristocratic and anti-clerical, seizing upon what authorities it could to roast the opposition (Grimsley and Gay); that the *philosophes,* despite their expressed admiration for the Newtonian tradition maintained a deep-laid Cartesianism.

However, it seems clear that the material I have adduced suggests that Vartanian's conception of the continuing Cartesian tradition is too simple. We need a fourth historiographical principle. There seem to have been two Cartesian doctrines: physical naturalism and mathematical abstractionism, the latter revealed in the apostasy of d'Alembert. It was this duality that separated the ideas of d'Alembert in particular from the admittedly different empiricisms of Diderot and Condillac; then a second line of cleavage separates the naturalists into organicists (Diderot) and mechanists (d'Alembert). It follows from this thesis that d'Alembert's abstractionist doctrines are not only positively Cartesian but anti-organicist. By adding an emphasis on the abstractionist features of d'Alembert's physics to Vartanian's thesis that the naturalism of Diderot and many others drew on a central strand in Descartes' scientific cosmology, I hope to have demonstrated with more force than most the untenability of the standard view.

Recently Fischer has argued for a revision of the basis of studies of the German Enlightenment, again contrary to the importation doctrine, proposing attention to a German 'native tradition'.[62] However, these are really scarcely more than divergences within a historiography, substantially that of Peter Gay, whose cultural eclecticism remains the main source of approaches to our understanding of the thought forms of the period.

But the paucity of Marxist treatments, at least in the work that has come my way, is worth remarking. Grimsley's study of Diderot[63] is not so much an account of origins of the Enlightenment as an attempt to show that Diderot anticipated some of the principles of social criticism that followed from dialectical materialism.

IV. Case 3: Nescience and necessity: Kant's philosophy of physics

Characteristics of the standard treatments

Studies of the most notable philosopher of the period, namely Kant, have not shared in the general trend towards real history. Bennett, for instance, explicitly repudiates an historical interest. In this section I place great emphasis on the synthesizing effect of Kant's critical philosophy. By judiciously blending empiricism and rationalism in the doctrine of the synthetic *a priori* he drew together the main strands of philosophical theories of knowledge in his time. In an account of subsequent developments we would have to see a divergence of viewpoints. Both idealism and transcendental realism were fathered by his thought as it interacted with the radical developments in physics brought about by the absorption of certain ideas from the Romantic movement. But that is another story. Contemporary Kantian scholarship barely touches on these matters. The astonishing neglect of Kant's *Metaphysical Foundations of Natural Science* by most writers on Kant's epistemology is an indication of this.

The received assumptions of Kantian scholarship are nicely summed up in the collection, *The First Critique,* edited by Penelhum and MacIntosh.[64] It includes Beck's important essay setting Kant's thought within the intellectual movements of the eighteenth century. But by way of contrast Körner's discussion of Kant's theories of mathematics is an easy-going exercise in a-history, commenting upon non-Euclidean geometry and even quantum mechanics. This way of

[62] K. P. Fischer, 'John Locke and the German Enlightenment an interpretation', JHI XXXVI (1975), 432–46.

[63] R. Grimsley, *From Montesquieu to Laclos, Studies on the French Enlightenment* (Geneva, 1974).

[64] T. Penelhum and J. J. MacIntosh (eds.), *The First Critique* (California, 1969).

dealing with Kant's ideas has become quite settled. There has been little change in the style of Kant studies between Ewing's *Short Commentary* of 1938,[65] through Strawson's *Bounds of Sense* of 1963,[66] to Bennett's latest volume, *Kant's Dialectic* of 1974.[67] All these works, some of great merit, can be summed up historiographically as telling how the text can be made to speak the ideas of the commentators themselves.

It is clear that Kant thought of himself as providing the final apotheosis of Enlightenment thought through a critical synthesis of the empiricist and rationalist traditions. Beck speaks of 'Kant's strategy', to devise a theory of knowledge that answers Hume's criticism of the power of reason at the same time as it answers to Leibniz's view that the predicates of the laws of nature are not merely accidentally concatenated. In summary Beck[68] says that there are, according to Kant, two facts involved in knowing: 'sensibility and understanding. Neither alone can give us knowledge; either alone is blind or empty . . . Knowledge comes from the application of one to the other . . . So what makes an answer to Hume possible – the rules of relating representations to each other introduce a synthetic element *a priori* into our empirical knowledge – also makes an answer to Leibniz possible – there is a perceptual and intuitional element in all *a priori* knowledge that is not merely and emptily logical'.

Kant describes his work in similar terms, though it seems clear that he thought of the Enlightenment as much in moral as epistemological terms. 'Dare to be free', he says, in summing up the movement. The question 'How is agency possible in the world experienced as causally determined?' preoccupies him, and with each of his enunciations of his solution to the problem of knowledge (with the exception of the *Metaphysical Foundations of Natural Science*), goes an explicit application to the issue of human freedom.

Steps to a more historical treatment

Kant's importance for the history of the philosophy of science can hardly be exaggerated. Directly through Coleridge and Whewell and indirectly through the *Naturphilosophie* of such distinguished electricians as Oersted and Ritter, Kantian notions bore heavily on the work of both Faraday and Maxwell. But this aspect of Kant's influence has been ill-served by commentators. Of recent authors, neither Bennett nor Strawson examines the relation between the doctrine of *Critique of*

[65] A. C. Ewing, *A Short Commentary on Kant's Critique of Pure Reason* (1938).
[66] P. F. Strawson, *The Bounds of Sense* (1963).
[67] J. Bennett, *Kant's Dialectic* (Cambridge, 1974).
[68] L. W. Beck, 'Kant's strategy', *J H I*, XXVII (1967), 224–36.

Pure Reason and the *Metaphysical Foundations of Natural Science.* Indeed, as I have remarked, Bennett explicitly repudiates an historical interest. It seems that Kant tends to engender exegesis without historiography; whereas historiography without exegesis is empty, the former combination is blind.

I want to urge here the necessity for studying Kant's works in the light of the historiographic thesis that they were strongly influenced by the critical problems of the *physics* of the day. I condense these problems into two: the British problem; how is knowledge of natural activity (and/or the universality) of the laws of nature possible? To the best of my knowledge only two major studies of these issues exist. There is Buchdahl's excellent *Metaphysics and the Philosophy of Science*[69] (a model in its way, though it suffers from the defect of concentrating on the 'great figures', somehow letting the problems of physics slip through). G. Martin, in his *Kant's Metaphysics and Theory of Science,*[70] astonishingly, pays no attention to Kant's own *Metaphysical Foundations of Natural Science.*

I venture an outline of an exegesis of my own, with considerable forebodings, but within the spirit of Buchdahl's treatment. I locate the main weight of the problem of understanding how Kant understands the epistemology of science in the relation between the *Metaphysical Foundations* and the *Critique of Pure Reason.*

In Kantian terms the problem for an advocate of a dynamicist physics, as Kant most certainly was, is the necessity he is under to theorize with concepts for which there could be no intuitions - concepts which do not reflect categories which are revealed through the schematisms of phenomena. Geometrical figures can be circular and plates can be round. According to Kant, the heterogeneity of the entities is bridged by the schematism, and so we can think geometrical figures. But for force-fields there is nothing corresponding to the round plate. To this end Kant introduces the possibility of concepts for physics to which there correspond no human intuitions. Upon these he grounds his philosophy of physics.

The primary layout

Kant believed that the world was really a system of active beings. It is a system because these beings must be supposed to interact. Their activity is not given in experience; it is required to account for experience. But this system of things in themselves, of selves, of God and so on, is *absolutely* unknowable experientially, nor can it be thought.

[69] See note 3.
[70] G. Martin, *Kant's Metaphysics and Theory of Science* (Manchester, 1955).

Human beings as cognitively able and sentient creatures experience a phenomenal world laid out in space and time and ordered and stabilized through the regularity of causal processes and the permanence of substances. Kant takes it for granted that there is no reason why the world as experienced should represent the world as it really exists. 'Such knowledge [*a priori* knowledge of reason] has to do only with appearances, and must leave the thing in itself as indeed real *per se,* but as not known by us' (Bxx).[71] All we can know about the world of beings in themselves is that they are an active source of the unstructured content of experience – that content which it is part of the task of the critical philosopher to define. The primary layout entails a stronger nescience than that of Locke, who allowed at least a representational relation between ideas of primary qualities given in experience and the real properties of things. 'Criticism', says Kant, 'has . . . established our unavoidable ignorance of things in themselves, and has limited all that we can theoretically *know* to mere appearances' (Bxxix). But the nescience of this layout is not the subjective phenomenalist scepticism of the inheritors of Lockean sensationalism, since Kant takes pains to demonstrate that the traditional distinctions between ideal and material, illusory and real, are sustained in the world as experienced. 'I do not mean that these objects as they appear are a mere illusion . . . Since . . . in the relation of the given object to the subject, such properties depend upon the mode of intuition of the subject, this object as *appearance* is distinguished from itself as object, *in itself*' (B69). Science is the study of a real, natural and public world, but not of the ultimate world system of things in themselves.

The secondary layout

Kant, it seems, held that the world of experience has a secondary layout which is in important respects a model of the primary layout of the world. The experienced or phenomenal world is stratified in ways parallel to the way the primary layout stratifies the 'whole' world. In consequence, the phenomenal world is richer than it appears in immediate experience. These are crucial principles for Kant's realist theory of science, set out in his *Metaphysical Foundations of Natural Science,* I want to argue that they are prepared for in principle in the *Critique of Pure Reason*.

[71] I. Kant, *Critique of Pure Reason* (Riga, 1781). All cited passages are drawn from the translation by N. Kemp Smith, *Immanuel Kant's 'Critique of Pure Reason'* (1952).

The case of the raindrop. Stratification of causal mechanisms is introduced with the example of the rainbow and the raindrop, proposed quite deliberately as a model of the primary layout. 'The rainbow in a sunny shower may be called a mere appearance and the rain the thing in itself. This is correct if the latter concept be taken in a merely physical sense' (B63). In physical science we explain appearances with the help of the active intervention of real things, such as raindrops. The explanatory power of raindrops is related to the ubiquitous role they play in a variety of solutions to problems of explaining appearances. But Kant makes clear that they could not serve as final explainers since 'if we take this empirical object in its general character, whether it represents an object in itself... we then realise that... the drops of rain are mere appearance... and that the transcendental object remains unknown to us' (B63). Even their primary qualities must be produced by some active property of things in themselves quite inexperienceable by us. But the world of experience is richer than given experience.

The case of the magnetic field. In the *Metaphysical Foundations* Kant argues for a field theory of matter based upon a system of attractive and repulsive 'forces' (in modern terms 'potentials') that is closely modelled on Leibniz's general physics and resembles the influential dynamicist approach of Boscovich. But it is clear from the *Critique of Pure Reason* that though this requires a physicist to postulate a finely detailed structure of active beings which is not given in his experience, Kant considers the subject matter of theoretical physics to be still within the world as constituted in experience, that is it is the phenomenal world.

The argument depends upon a generalization of the characterization of the phenomenal world beyond that realm which our senses, however refined, would reveal. The phenomenal world must include the deliverances of the possible senses of the experience of possible beings. Fields are phenomenal (as opposed to noumenal) because they could be experienced by beings different from ourselves. Thus 'from the perception of the attracted iron filings we know of the existence of a magnetic matter pervading all bodies, although the constitution of our organs cuts us off from all immediate perception of this medium. For, in accordance with the laws of sensibility and the context of our perceptions, we should, were our senses more refined, come also in our experience upon the immediate empirical intuition of it. The grossness of our senses does not in any way decide the form of possible experience in general' (B273). Though field concepts are not mediated by schematisms for human beings, there could be beings for whom there were corresponding intuitions.

The analysis of experience

So far Kant's account of knowledge has strongly favoured Leibniz. But careful consideration of the relation between the secondary and primary layouts with respect to the problem of what can be attributed *a priori* to the world, raises the issue of the necessity of the structural principles of space and time, and of the most general laws of nature. 'If intuition must conform to the constitution of the objects, I do not see how we could know anything of the latter *a priori*, but if the object (as object of the senses) must conform to the constitution of our faculty of intuition, I have no difficulty in conceiving of such a possibility' (Bxvii).

The importance of this idea derives from the immediacy and centrality of intuition in empirically interacting with the world, that is, as opposed to thinking (of) it. 'In whatever manner and by whatever means a mode of knowledge may relate to objects, *intuition* is that through which it is in immediate relation to them... but intuition takes place only in so far as the object is given to us' (B34).

Kant's solution of how *a priori* knowledge of the world is possible is well known: 'even our empirical knowledge is made up of what we receive through impressions and of what our faculty of knowledge (sensible impressions serving merely as the occasions) supplies from itself' (B2). That which we identify in experience as the regularities and structures which are described in the laws of nature and the theories of space, time and kinematics are our own contribution. And so to deal with these contributions we must analytically separate the matter of experience from its form. But it is important to follow Kant in keeping clear the distinction between the specifically cognitive aspects of human experience ('understanding') and the phenomenal aspects ('sensibility') and to keep firmly in mind the point that the *a priori* principles of experience are not found in the understanding (except the temporal) but in sensibility. Kant's epistemology is not idealism. 'There are two stems of human knowledge, namely *sensibility* and *understanding*, which perhaps spring from a common, but to us unknown, root. Through the former objects are given to us, through the latter they are not. Now in so far as sensibility may be found to contain *a priori* representations constituting the conditions under which objects are given to us, it will belong to transcendental philosophy' (B30).

Again he insists 'Objects are *given* to us by means of *sensibility*, and it alone yields us *intuitions;* they are *thought* through the understanding and from the understanding arise concepts' (B34).

If the structuring principles of experience derive from a fixed set of categories then the laws of nature should reflect them. I want to demonstrate that the specificity of the laws Kant deduces from the table of categories is radically different in the *Critique* from that in the *Metaphysical Foundations*. In the *Critique* the table of categories is used to justify only very general natural principles as synthetic *a priori* truths. For instance, from modality the schema of possibility is 'the agreement of the synthesis of different representations with the conditions of time in general' (B184); and of actuality is 'the existence of some determinate time' (B184). Similarly, for relation we have the very general 'the schema of substance as the permanence of the real in time' (B183).

But in the *Metaphysical Foundations*[72] the table of categories yields much more specific laws. The four divisions of the categories - Quantity, quality, relation and modality - correspond to (engender?) four aspects of natural science: phoronomy, the metaphysics of motion; matter not considered in its extension; dynamics, the metaphysics of matter considered as filling space; mechanics, the general theory of material things in interaction; and phenomenology, how matter is presented in external sense. This relation, comprising inherence-subsistence, causality and dependence, and community, is made to yield (*Metaphysical Foundations*, ch. 3, propositions 2, 3 and 4) the general law of the conservation of matter *and* both Newton's first and third laws. Since Kant had already demonstrated, in the chapter on phoronomy, that the quantity of uniform motion is relativized to arbitrarily chosen frames, it follows that only change in uniform motion calls for a dynamic explanation. Hence the category of causality and dependence immediately yields the first Newtonian law.

The most astonishing employment of the categories is surely modality. Possibility yields the principle that rectilinear motion is relative (and hence not caused, by the previous phoronomic argument); existence yields the principle that circular motion is real because it can be known to be non-relative by reference to the fact that forces are required to maintain it (Newton's own argument of course); and necessity yields the principle that relative motion is necessary. As Kant remarks in the 'General observation on phenomenology', 'this . . . like everything adequately provable from mere concepts, [is] a law of an absolutely necessary countermotion'.

[72] I. Kant, *Metaphysical Foundations of Natural Science* (Riga, 1786), trans. J. Ellington (Indianapolis and New York, 1970).

I conclude that Kant's general physics of the phenomenal world involves an employment of his general philosophical framework that is perfectly isomorphic with his application of the scheme to the 'whole' world. But in the general physics 'forces' (or rather as we should now call them, potentials) are real, and the primary properties of matter (extension, impenetrability) demonstrably reducible to them. And this is how Kant came to be the father-in-law of the physics of the field. Like Hume, Kant is in a great 'native tradition': the continental development of the philosophy of physics. His work completes the development of dynamism initiated by Leibniz[73] and elaborated by Boscovich.[74]

[73] G. Gale, 'The physical theory of Leibniz', *Studia Leibnitiana*, IX (1970), 114–27. For a more detailed account see Buchdahl, *Metaphysics and the Philosophy of Science* (see note 2), ch. 7.

[74] R. J. Boscovich, *A Theory of Natural Philosophy* (Venice, 1763; repr. Boston, 1966).

2

Natural philosophy

SIMON SCHAFFER

Contents

I. Introduction 55
II. Natural philosophy as Newtonian matter-theory 58
III. Natural philosophy as the negation of science 72
IV. Natural philosophy as the object of an archaeology and anthropology 86

I. Introduction

The dominant historiographical tradition of eighteenth-century physical science presents natural philosophy as a coherent, unified body of theory and practice. The seminal texts of this tradition, Metzger's *Newton, Stahl, Boerhaave* (1930), Brunet's *L'introduction des théories de Newton en France* (1931), Cohen's *Franklin and Newton* (1956), and others, all picture natural philosophy as a set of discourses which are unified by their debt to Newton and their base in matter-theory.[1] In this picture, Newton's legacy is constituted by the collection of his texts which the eighteenth-century natural philosophers absorbed, albeit selectively and in a highly interpretative fashion. The *Principia,* the *Opticks,* the 'Queries' to the successive editions of the *Opticks* issued in Newton's lifetime (1704, 1706, 1717), and the letters to Oldenburg, Boyle and Bentley, not published till the mid-century, have all been taken by historians as starting points for the various aspects of what has nevertheless remained a single embracing tradition.[2]

[1] H. Metzger, *Newton, Stahl, Boerhaave et la doctrine chimique* (Paris, 1930); P. Brunet, *L'introduction des théories de Newton en France au xviiie siècle* (Paris, 1931); I. Bernard Cohen, *Franklin and Newton* (Philadelphia, 1956).

[2] For a tabulation of the publication and dissemination of these texts, see A. Thackray, *Atoms and Powers* (Cambridge, 1970), 12. Discussions of the Newtonian image in the eighteenth century may be found in H. Guerlac, 'Where the statue stood: divergent loyalties to Newton in the 18th century', in E. R. Wasserman (ed.), *Aspects of the Eighteenth Century* (New York, 1965), 317–34; P. Casini, 'Le "newtonianisme" au siècle des lumières', *Revue du dix-huitième siècle,* I (1969), 139–59; Y. Elkana, 'Newtonianism in the eighteenth century', *British Journal of the Philosophy of Science,* XXII (1971), 297–306; G. Buchdahl, *The Image of Newton and Locke in the Age of Reason* (1961).

Major eighteenth-century interests in the aether, in interparticulate forces, in the role of activity in matter, have all been interpreted in terms of various differing selections from these original texts. In this way, the problem of the object of discourse, and of the relation between that object and the natural philosophical enterprise, has been circumvented. Historians have adopted an approach whose method is profoundly unhistorical. Because of the conventional picture of natural philosophy, historians have been able to avoid examining natural philosophy as *practice* – its unity in Newtonian matter-theory has been taken as axiomatic.

We are told that from Newton's death in 1727 (and from 1738 in France, after the efforts of Maupertuis, Bouguer, Voltaire and others) his disciples laid down the basic concerns of some of the principal traditions of natural philosophy: to understand the properties of nature by reducing them to some combination of the activity of a rare aetherial fluid or fluids, and of definite and quantifiable forces between essentially atomic particles.[3] Thereafter, individual natural philosophers adopted some combination of these elements. Other elements were then added, so that, for example, Boerhaave's magisterial texts on chemistry, pirated, translated and disseminated in England by Peter Shaw and others after 1727, stressed the importance of *fire* as the bearer of activity in nature.[4] Roger Boscovich's *Theory of Natural Philosophy* (1758) developed an original notion of matter as a set of

> indivisible points, that are non-extended, endowed with a force of inertia, & also mutual forces represented by a simple continuous curve . . . defined by an algebraical equation.[5]

Contrasting developments, such as these by Boerhaave and by Boscovich, are pictured as so many deviations from a 'Newtonian' norm.

[3] The major expositions of Newtonianism at this period were J. Keill, *Introduction to Natural Philosophy*, trans. of 4th (Latin) ed. (1720); J. T. Desaguliers, *A Course of Experimental Philosophy* (2 vols., 1734–44); H. Pemberton, *A View of Sir Isaac Newton's Philosophy* (1728); W. J. 'sGravesande, *Physices elementa mathematica, experimenta confirmata* (2 vols., Leyden, 1720–1); Voltaire, *Elémens de la philosophie de Newton* (Amsterdam, 1738).

[4] A pirated version of Boerhaave's lectures, *Institutiones et experimenta chemiae* (Paris, 1724) was translated by Shaw as *A New Method of Chemistry* (1727). For Boerhaave as a Newtonian see Metzger, *Newton, Stahl, Boerhaave* (note 1), 50, 66, 198; for his doctrine of fire see R. Love, 'Some sources of Hermann Boerhaave's concept of fire', *Ambix*, XIX (1972), 157–74; for his textbook, see T. L. Davis, 'Vicissitudes of Boerhaave's textbook of chemistry', *Isis*, X (1928), 33–46.

[5] R. Boscovich, *A Theory of Natural Philosophy* (Cambridge, Mass., 1966), 183 (sect. 516 in original). On Boscovich, see L. L. Whyte (ed.), *Roger Joseph Boscovich* (1961).

Each individual natural philosopher can be singled out, as (in Britain), for example, Desaguliers, Rowning, Knight, Robinson, Colden, Watson, Franklin, Jones, Michell, Priestley or Hutton,[6] and his detailed theory explicated as stemming from some combination of Newtonian *thèses* with a superadded dash of creative reinterpretation. While no actual coherent picture of natural philosophy has emerged, an all-embracing methodology has been adopted which is enough to allow the appropriation of any natural philosopher into the tradition without seeming to over-tax the historical categories employed.

More recent historians, since 1960, have attempted certain substantial revisions of this picture without fundamentally challenging this (by now) conventional historical practice. For example, essays by Guerlac and others have pointed out that Malebranchists such as J. P. de Molières and J. J. d'Ortous de Mairan were at least as responsible as some more obvious candidates for the absorption of Newtonian ideas in France from before 1728.[7] Home, Cantor and Quinn have each shown the difficulty of extending the Newtonian matter-theory base to electricity or optics, where explicitly anti-Newtonian beliefs were upheld by perfectly orthodox philosophers. The wave theory of light, for example, was far more commonly adopted than a traditional picture of Newtonian optics would suggest.[8] Schofield's *Mechanism and Materialism* (1970), which claimed that natural philosophers adopted either 'mechanist' or 'materialist' versions of Newtonian matter-theory through the eighteenth century, in itself represented a confirmation of the unifying vision. Schofield's conclusions, however, have since been seriously challenged by historians who wish to connect the understanding of natural philosophy with a much broader philosophical and conceptual context. Yet even Schofield's critics still seem to wish to analyse each philosopher as a specific case of (an admittedly broader) Newtonianism.[9]

[6] This is of course only a partial list. For some idea of the variety of natural philosophers see E. G. R. Taylor, *Mathematical Practitioners of Hanoverian England* (Cambridge, 1966), and M. Rowbottom, 'The teaching of experimental philosophy in England, 1700–1730', *Actes du XIe Congrès internationale d'Histoire des sciences* (Warsaw, 1968), IV, 46–63.

[7] H. Guerlac, 'Some areas for further Newtonian studies', *HS*, XVII (1979), 75–101. On some details of this see T. L. Hankins, 'The influence of Malebranche on the science of mechanics during the 18th century', *JHI*, XXVIII (1967), 193–210.

[8] R. Home, 'Newtonianism and the theory of the magnet', *HS*, XV (1977), 252–66; G. Cantor, 'The historiography of "Georgian" optics', *HS*, XVI (1978), 1–21; A. J. Quinn, 'Evaporation and repulsion: a study of English corpuscular philosophy from Newton to Franklin' (Ph.D. thesis, Princeton, 1970).

[9] R. E. Schofield, *Mechanism and Materialism: British Natural Philosophy in an Age of Reason* (Princeton, 1970). Typical attacks on Schofield are P. Heimann,

The most important task for the critical assessment of these recent revisions is to discover what effect they have had on the conventional method which earlier historians adopted, and on the version of Newtonian matter-theory which this method has been used to construct. Three distinct problems may be isolated here. First, many historians have preserved the alleged coherence of a single 'Newtonianism' by appealing to the philosophical base of the concepts which natural philosophy used. For example, concepts such as 'force' and 'power' are now admitted to have widely varying connotations in Priestley or in Hutton as compared with their alleged source in Newton. Yet a *tradition* has still been preserved by tracing a *philosophical* genealogy from Newton even if their natural philosophical application varied. Secondly, some historians, such as Cantor, Wilde and others, have documented the existence of strong anti-Newtonian themes in natural philosophy – but this has not been enough, by itself, to damage the apparent need for the 'tradition-seeking' method. Thirdly, there is an important need for alternative attitudes to natural philosophy as an historical category, not merely revisions of one or other of the unifying assertions which contemporary historiography has made. This necessarily involves a genuine confrontation with the philosophical debates on the discursive place of history of science which, significantly enough, in the work of Bachelard, Kuhn and Foucault, have all drawn on natural philosophy in the eighteenth century for much of their evidence. Such a confrontation is overdue.

II. Natural philosophy as Newtonian matter-theory

I have indicated, firstly, that an attempt has been made to preserve the fragile unity of the 'Newtonian' tradition by an analysis of the categories of natural philosophy as products of eighteenth-century epistemology. Historians such as Heimann and McGuire have criticized any attempt to limit natural philosophy to an essentially scientific discourse.[10] This limitation has been cheerfully taken for granted in most accounts of natural philosophy in standard histories of the Enlightenment. Heimann and McGuire single out Peter Gay's claim that after the Scientific Revolution science was sundered forever

'Newtonian natural philosophy and the scientific revolution', *HS*, XI (1973), 1–7, and J. McEvoy, 'Joseph Priestley, natural philosopher: some comments on Professor Schofield's views', *Ambix*, XV (1968), 115–33.

[10] P. Heimann and J. E. McGuire, 'Newtonian forces and Lockean powers', *Historical Studies in Physical Science*, III (1971), 233–306; I also consider P. Heimann, '"Nature is a perpetual worker": Newton's aether and 18th century natural philosophy', *Ambix*, XX (1973), 1–25.

from philosophy, theology, or any other belief-system. Other historians, such as Peter Linebaugh, have also attacked Gay's account of the effects of the Scientific Revolution on science.[11] In contrast to Gay, then, recent historians have argued that an understanding of natural philosophy can be obtained through an analysis of the philosophical production of the concepts philosophers used: specifically, the concepts of *activity, force, spirit* and *power.* Alongside natural philosophy, Heimann, McGuire and their colleagues have constructed a parallel discourse inhabited by practitioners like Locke, Berkeley, Andrew Baxter, Hume, Reid, and Dugald Stewart, who provide categories for natural philosophical use and in turn respond to this use. The original work by Metzger, the substantial texts of Cohen, Thackray, and Schofield, together with investigations of classical analytical mechanics, the problems of lunar motion, and the *vis viva* controversy, all have provided plentiful support for this philosophical project.[12] Increasing study of the social and philosophical impact of the Scottish Enlightenment from the work of the Rankenian Club onwards, has drawn historians' attention to the close connections between natural philosophy and such discussion of basic epistemological terms.[13]

However, natural philosophy is now to be interpreted as a system of concepts which mediate between science and philosophy. These concepts have been very easily used as the constituent elements of a Newtonian 'tradition', in which the differences in use by natural

[11] Heimann and McGuire, 'Newtonian forces' (see note 10), 234n. For another attack see P. Linebaugh, 'The Tyburn riot against the surgeons', in D. Hay *et al.* (eds.), *Albion's Fatal Tree* (Harmondsworth, 1977), 69n. See P. Gay, *The Enlightenment: An Interpretation* (2 vols., 1966–9).

[12] A Thackray, *Atoms and Powers: an Essay on Newtonian Matter Theory and the Development of Chemistry* (Cambridge, Mass., 1970); C. Truesdell, *Essays in the History of Mechanics* (New York, 1968); P. Chandler, 'Clairaut's critique of Newtonian attraction', *AS*, XXXII (1975), 369–78; T. L. Hankins, 'Eighteenth century attempts to resolve the *vis viva* controversy', *Isis*, LVI (1965), 281–97; L. Laudan, 'The *vis viva* controversy, a post mortem', *Isis*, LIX (1968), 131–43; C. Iltis, 'The decline of Cartesianism in mechanics', *Isis*, LXIV (1973), 356–73; P. Heimann, "'Geometry and Nature", Leibniz and Johann Bernoulli's theory of motion', *Centaurus*, XXI (1977), 1–26.

[13] J. B. Morrell, 'Professors Robison and Playfair and the "Theophobia Gallica"', *NR*, XXVI (1971), 43–63; *idem*, 'Reflections on the history of Scottish science', *HS*, XII (1974), 81–94; G. T. Bryson, *Man and Society: the Scottish Inquiry of the Eighteenth Century* (Princeton, 1945); L. Laudan, 'Thomas Reid and the Newtonian turn of British methodological thought', in R. E. Butts and J. W. Davis (eds.), *The Methodological Heritage of Newton* (New York, 1970), 103–31, G. Cantor, 'Henry Brougham and the Scottish methodological tradition', *Studies in History and Philosophy of Science*, II (1971), 69–89.

philosophers in the eighteenth century have been categorized as mutations, problems of transmission, or constructive reinterpretation. In this way, the basic coherence of the Newtonian tradition is unambiguously preserved. Heimann and McGuire have argued that in the Newtonian term *force* and the Lockean term *power* historians can find 'common themes which were fundamental to the problems of scientific explanation'. They acknowledge that such themes by no means exhaust the natural philosophical enterprise, and, furthermore, that this tradition consists of a set of *rejections* of Newtonian axioms.[14] This calls into question the notion of 'tradition' which lies at the base of this historical approach. For example, Heimann has recently persuasively argued that British natural philosophers came to reject the Newtonian notion that motion in nature stems immediately from God, and espoused the more 'philosophical' theory that activity was inherent in matter.[15] We can, in fact, locate various formulations of this problem in writers as diverse as Samuel Clarke (1723), John Keill (1734), Colin Maclaurin (1748), and Cadwallader Colden (1751). Whereas Clarke argued strongly that there are in nature active principles 'which are not occult properties of matter, but original and general laws impressed by God', Keill claimed that pure mechanism was a completely adequate basis for natural philosophy. 'If the true causes are hid from us, why may we not call them occult qualities?', he asked.[16] Maclaurin, the leading Scottish disciple of Newton of his age, argued that causal intermediaries, such as 'a rare elastic aetherial medium', must be judged as 'instruments' of God. He claimed that 'it does not appear to be a very important question . . . whether these great changes are produced by . . . instruments or by the same immediate influences which first gave things their form'. Arguing against the Leibnizian conception of pre-established harmony, Maclaurin stated that the world would have been 'incomparably inferior to what it is' if it had been 'governed only by those principles which

[14] Heimann and McGuire, 'Newtonian forces' (see note 10), 305.

[15] P. Heimann, 'Voluntarism and immanence: conceptions of Nature in 18th century thought,' *JHI* xxxix (1978), 271–83; *idem*, 'Science and the English Enlightenment', *HS*, xvi (1978), 143–51.

[16] S. Clarke (ed.) and J. Clarke (trans.), *Rohault's System of Natural Philosophy* (1723), i, 54–5, and ii, 96–7; J. Keill, *An Examination of Dr. Burnet's Theory of the Earth*, 2nd edn. (Oxford, 1734), iii, 3–4. For Clarke, see M. A. Hoskin, '"Mining all within": Clarke's notes to Rohault's *Traité de physique*', *The Thomist*, xxiv, (1951), 353–63, and J. H. Gay, 'Matter and freedom in Clarke', *JHI*, xxiv (1963), 85–105. In his attack upon Thomas Burnet's *physica sacra* Keill stresses the *in*capacity of Newtonian mechanism as a complete explanation of Biblical processes: what cannot be explained mechanically must be a miracle, not an 'occult' force.

arise from the various motions and modifications of inactive matter by mechanical laws'.[17] It is clear that the switch from divine voluntarism to the immanence of activity in matter is by no means a simple, continuous development of a single idea.

The problematic causation of such 'switches' is amply demonstrated by the cases of John Harris, Boyle Lecturer and editor of the *Lexicon technicum,* and George Cheyne, fashionable physician and friend of such religious thinkers as William Law, John Byrom, and the Countess of Huntingdon. Bowles has demonstrated that between the first and second parts of his *Lexicon,* Harris converted to the belief that all activity is due to 'power' in matter.[18] This transformation implied major changes in Harris's attitude to theology and politics, since 'power' was a theological and political term: I believe that this switch of Harris's can only be explained as part of the political polemic waged by Harris against the early Jacobites such as Charles Humphrey, a lecturer in Harris's own parish, and the rivals of Harris's friend John Woodward, the geologist and antiquarian.[19] The conversion is no mere philosophical decision about the use of a natural philosophical term. Similarly, the unfamiliar arguments and the adventurous aether theory developed by Cheyne have been difficult for some historians to handle. Thackray argued that ideas on the aether would have been of 'little interest' to Cheyne, whilst Schofield, who did notice Cheyne's development of an aether theory from at least 1706–11 onwards, stated, however, that this 'was a typical failure to

[17] C. Maclaurin, *An Account of Sir Isaac Newton's Discoveries* (1748), 84, 390. See a discussion of Maclaurin's epistemology in G. Buchdahl, 'Explanation and gravity', in R. Young and M. Teich (eds.), *Changing Perspectives in the History of Science* (1973), 167–203, 197–8. In a letter to Colin Campbell in 1721, Maclaurin commented, significantly, that Newton's *Opticks* rendered the properties of light obviously *non*mechanical. For another view, see H. Odom, 'The estrangement of celestial mechanics and religion', *JHI,* XXIX (1966), 533–48 (pp. 546–7).

[18] G. Bowles, 'John Harris and the powers of matter,' *Ambix,* XXII (1975), 21–38; see also D. McKie, 'John Harris and his *Lexicon Technicum*', *Endeavour,* IV (1945), 53–7.

[19] John Harris was appointed Boyle Lecturer by Evelyn on Woodward's recommendation in 1698. He then stated that 'Passive matter and Active Power... were perfectly distinct'. He defended Woodward against his many critics from 1695, and began to attack the Tories, including the physicians Freind and Mead from 1709, when appointed secretary to the Royal Society. His break with the latitudinarians and his attack on the Jacobite Charles Humphrey are discussed in M. C. Jacob, *The Newtonians and the English Revolution* (Hassocks, 1976), 179ff. On the Woodward controversies, which involved a combination of natural philosophical, medical and political polemic, see J. Levine, *Dr Woodward's Shield* (Berkeley, 1977).

see contradictions in the style of exposition'. Schofield went on to claim that although Cheyne's ideas on matter-theory were original, 'other natural philosophers had moved in a parallel direction for better or at least more scientifically defensible reasons'.[20] He condemned Cheyne's text *Philosophical Principles of Natural Religion* as an 'incoherent' demonstration of its own distance from genuine natural philosophy: Cheyne himself explained that this text, at least in part, had been designed

> To shew, that all our Knowledge of Nature was by Analogy, or the Relations of Things only, and not their real Nature, Substance or internal Principles.[21]

Once again, Bowles has demonstrated that for Cheyne, with this theory of method and a complex political and medical-theological system of nature, concepts such as 'attraction' were by no means simply philosophical terms given meaning in the realm of science.[22] This is evidently the case for much of natural philosophy: the goal of the historian is not to assimilate these different categories into one vague field, but to delineate different discourses and their articulations.

Important examples of the problems raised by the approach which treats philosophical and natural philosophical problems as essentially identical can be found when this method is applied, as it had been, to less ambiguous cases than those of Cheyne and Harris. The work of Joseph Priestley and James Hutton, both of whom stand at the centre of eighteenth-century natural philosophy, has received intense attention in the past fifteen years. In the picture of natural philosophy which stresses the unity of the Newtonian tradition, Priestley occupies a crucial place. This is in distinct contrast to the judgement of earlier historians, who condemned Priestley's errant empiricism and his lack of a coherent programme of research. In order to recuperate Priestley's reputation, therefore, contemporary historians have searched for and located the grounds of a unity of thought in his

[20] Thackray, *Atoms and Powers* (see note 2), 49–51, and Schofield, *Mechanism and Materialism* (see note 9), 57–62. On Cheyne's connections with religious enthusiasm and the Platonist revival, see C. F. Mullett (ed.), *Letters of Dr Cheyne to the Countess of Huntingdon* (San Marino, Cal., 1940); H. Metzger, *Attraction universelle et religion naturelle chez quelques commentateurs de Newton* (Paris, 1938); and P. Casini, *L'universo-macchina* (Bari, 1969), 175–203.

[21] *Dr Cheyne's Account of Himself and his Writings*, 2nd edn. (1743), 10. Compare his interesting poem on Platonism, B.L.MS Burney 390 fol. 8b, and the text of *Philosophical Principles of Religion, Natural and Reveal'd* (1715), ii.

[22] G. Bowles, 'Physical, human and divine attraction in the life and thought of George Cheyne', *AS*, XXXI (1974), 473–88.

natural philosophy, since it is the condition of such a recuperation that some unifying feature be found. Schofield's detailed work has uncovered the important connections which Priestley maintained with industrialists and other natural philosophers through the Lunar Society in Birmingham, while also stressing that Priestley's *scientific* search for dynamic interparticulate forces unified the whole of his research programme.[23] This unification is thought to be too simple by Heimann, McGuire and McEvoy. Heimann and McGuire have demonstrated that a *philosophical* coherence was maintained by Priestley in his work on the concepts of power and experience. In response to the work of Locke, Berkeley and Andrew Baxter in this area, they argue, Priestley developed an epistemology which based itself on the relative status of primary qualities, and denied the distinction between matter and spirit. Matter, for Priestley, was characterized by its attributes of attractive or repulsive power, and, for Heimann and McGuire, this characteristic was a philosophical and epistemological claim which informed all of his natural philosophy.[24] John McEvoy, in turn, has found even this claim for coherence insufficient, and has stated that 'Priestley's main concern was with religion', and that Priestley's *theological* doctrine of rational dissent generated a 'conceptual framework that manifests the synoptic unity of epistemological, metaphysical, methodological, theological, and strictly scientific parameters in his thinking'.[25] The thrust of contemporary historical understanding has therefore been to broaden the coherence in order satisfactorily to pull together all, or as much as possible, of Priestley's work.

[23] R. E. Schofield, 'Joseph Priestley, natural philosopher', *Ambix*, XIV (1967), 1–15, and *idem*, *The Lunar Society of Birmingham* (Oxford, 1963). For a representative set of condemnations of Priestley see F. Gibbs, *Joseph Priestley* (1965), 117, and J. Partington, *History of Chemistry* (3 vols., 1962), III, 246–57. The best bibliographical survey is in J. McEvoy, 'Joseph Priestley: philosopher, scientist and divine' (Ph.D. thesis, Pittsburgh, 1975), 1–16. The best sense of Priestley's own view of himself may be gathered from the *Autobiography* in the edition edited by J. Lindsay (Bath, 1970).

[24] Heimann and McGuire, 'Newtonian forces' (see note 10), 268–81.

[25] McEvoy, *Joseph Priestley* (see note 23), 20, and *idem*, 'Electricity, knowledge and the nature of progress in Priestley's thought', *BJHS*, XII (1979), 1–30 (p. 2). See also *idem* and J. E. McGuire, 'God and nature: Priestley's way of rational dissent', *Historical Studies in Physical Sciences*, VI (1975), 325–404, and 'Joseph Priestley, "aerial philosopher"', *Ambix*, XXV (1978), 1–55, 93–116, 153–75 *et seq*. McEvoy is keen to argue that the 'rational dissent' of Priestley provides a methodological and epistemological unity which then guided Priestley through a practice constructed by his predecessors in the field – Hales, Brownrigg, Black and Cavendish, for example – who provided this 'conceptual framework'. There is certainly no appeal to any extra-intellectual factors.

This series of analyses of Priestley's natural philosophy has itself posited a set of traditions at whose intersection Priestley situated his work. Schofield has emphasized the importance of Rowning and Boscovich for Priestley. Rowning's declaration that 'the particles of all Bodies... attract and repel each other *alternately* at different distances', can be cited, and, noting the close similarity of this idea to that of Boscovich, Priestley, John Michell and William Herschel, this declaration can be used to construct a tradition of spheres of forces situated in particulate matter.[26] Priestley himself, in his *Disquisitions relating to Matter and Spirit* (1777), insisted that

> Father Boscovich and Mr Michell's new theory concerning matter... was calculated, as will be seen, to throw the greatest light on the constituent principles of human nature.[27]

This declaration enabled McEvoy to connect this matter-theory to a more general theological doctrine of the soul. In this way, a coherence can be given to natural philosophy through the characterization of specific scientific, philosophical and theological traditions.

In the case of an equally central natural philosopher, James Hutton, the integration of his matter-theory into some kind of Newtonianism has set even more intense problems. Hutton connected his seminal work in geology with a complex theory of matter, which involved the

[26] J. Rowning, *Compendious System of Natural Philosophy* (2 vols., Cambridge, 1735), II, 6. Compare *ibid.*, I, 12: 'Matter has certain Powers or active principles... probably not essential or necessary to its existence, but impressed upon it by the Author of its Being'. Schofield suggests this as a source for some of these ideas: see Schofield, *Mechanism and Materialism* (see note 9), 245n. Priestley cites Michell and Boscovich's ideas in a letter to Bretland, 1773, published in J. Rutt (ed.), *Life and Correspondence of Joseph Priestley* (2 vols., 1831–2), I, 189–91; *Disquisitions on Matter and Spirit* (1777), 112 and elsewhere; and in *History... Relating to Vision, Light and Colours* (1772), 390–2-ff. Herschel's work on this problem may be found, with diagrams, in J. Dreyer (ed.), *Scientific Papers of William Herschel* (1912), lxviiiff, in particular in a paper read in Bath on 28 January 1780 on Priestley's question 'What becomes of light?'. See also Herschel's list of books in his commonplace book (Linda Hall library, Kansas City, Mo.), pp. 83–5. For John Michell see C. Hardin, 'The scientific work of John Michell', *AS*, XIII (1957), 148–63, and R. McCormmach, 'Henry Cavendish and John Michell: weighing the stars', *BJHS*, IV (1968), 126–55.

[27] Priestley, *Disquisitions* (see note 26), xii. A fascinating sidelight on Priestley's impact is in Edward Gibbon's *Autobiography* (ed. H. Trevor-Roper, 1966), 44–6: Gibbon reports the immense effect of all the public lectures of Bryan Higgins and others in 1777–8, while simultaneously advising the radical Priestley to stick to natural philosophy: 'instead of listening to this friendly advice, the dauntless philosopher of Birmingham continued to fire away his double battery against those who believed too little, and those who believed too much... his trumpet of sedition may at length awaken the magistrates of a free country', declared Gibbon, the stoutest of Tories in this respect.

circulation through the cosmos of two kinds of matter: gravitating matter, which had weight and formed massy objects, and solar matter, which was repellent and manifested its effects through fire, light, electricity and heat.[28] For Hutton, the task of the natural philosopher was the *penetration* of matter: chemistry served as the fundamental science. Already in 1741, Peter Shaw had stressed that it was 'by means of chemistry that Sir Isaac Newton has made a great part of his surprizing discoveries in natural philosophy', and that 'chemistry, in its extent is scarcely less than the whole of natural philosophy'. Playfair, Hutton's biographer and disciple, echoed this point on Hutton's achievement:

> He began from chemistry: and it was from thence that he took his departure in his circumnavigation of the material and the intellectual world. The chemist, indeed, is flattered more than anyone else with the hopes of discovering in what the essence of matter consists; and Nature, while she keeps the astronomer and mathematician at a great distance, admits the chemist to a more intimate acquaintance with her secrets ... He shall be able to bind Proteus in his cave, and finally extort from him the secret of his birth.[29]

[28] James Hutton, *Dissertations on Different Subjects in Natural Philosophy* (Edinburgh, 1792), 388–403, 491–503. This matter theory invites comparison with those of his contemporary natural philosophers: it played a central role in his conception of cyclical process in nature and in the *geological* role of central heat. See P. Gerstner, 'James Hutton's theory of the Earth and his theory of matter', *Isis*, LIX (1968), 26–31; *idem*, 'The reaction to James Hutton's use of heat as a geological agent', *BJHS*, V (1971), 353–62; A. Donovan, 'James Hutton, Joseph Black and the chemical theory of heat', *Ambix*, XXV (1978), 176–90. For the crucial importance of *circulation* and the temporal continuity of system, see, for example, 'Theory of the earth; or an investigation of the laws observable in the composition, dissolution and restoration of land upon the globe', *Transactions of the Royal Society of Edinburgh*, I (1788), 209–304 (p. 215); D. R. Dean, 'James Hutton on religion and geology', *AS*, XXXII (1975), 187–93 (who reproduces an unpublished statement by Hutton on the demonstration of system and the revelation of God); J. Playfair, 'Biographical account of the late Dr James Hutton', *Transactions of the Royal Society of Edinburgh*, V (1805), part 3, 39–99 (p. 55), on a revised statement of this position.

[29] Peter Shaw, *A New Method of Chemistry* (2 vols., 1741), 173n.; Playfair, 'Biographical account' (see note 28), 74. It was in the discussion of the action of phlogiston that Hutton stated that there 'would also appear to be in the system of this globe, a reproductive power, by which the constitution of this world, necessarily decaying is renewed' (*Dissertations* (see note 28), 214). In contrast to Hutton's reference to 'the astronomer', see Playfair's demand that the 'liberty of speculation' of astronomy and mathematics be extended to geology. Important, too, is Davy's chemical critique of Hutton's Plutonism, which in fact covered an objection to the time-scale: see R. Siegfried and R.

Given the importance of chemistry, and the strong influence of Joseph Black's investigations into matter and heat, at the heart of Hutton's project, Heimann and McGuire have found it necessary to integrate his work, too, into the tradition of Newtonian matter-theory.[30] Here too the problems have been intense. Recent studies have demonstrated that Hutton's concept of central heat in the earth was widely *misinterpreted* by his contemporaries, and a paper which has claimed for Hutton priority over William Herschel for the discovery of infra-red radiation has also shown how aberrant much of Hutton's optical and radiative theory was.[31] As Wilde and others have pointed out, the work of Hutton, like that of Patrick Leslie, Adam Walker, Robert Harrington, Bryan Higgins and William Herschel, bears a closer similarity to that of the arch anti-Newtonian John Hutchinson and his disciple William Jones than to alleged 'Newtonians' such as Gowin Knight or Joseph Priestley.[32] Heimann and

H. Dott Jr, 'Humphry Davy as geologist', *BJHS*, IX (1976), 218–27 (p. 222); and L. Pearce Williams, 'Boscovich and the British chemists', in Whyte, *Boscovich* (se note 5), 159, on latent heat, Boscovichean chemistry and the penetration of matter.

[30] Heimann and McGuire, 'Newtonian forces' (see note 10), 281–99, and Heimann, "'Nature is a perpetual worker'" (see note 10), 17–23. Compare Schofield, *Mechanism and Materialism* (see note 9), 273–6.

[31] For the response to Hutton, see Gerstner 'Reaction to Hutton's use of heat' (see note 28); D. R. Dean, 'James Hutton and his public, 1785–1802', *AS*, XXX (1973), 89–105. Under the belief that 'physical science furnishes us with the means of ascending . . . through the whole chain of causes', De Luc, for example, argued against Hutton that 'whatever portion of time may be allowed' causes like those Hutton used were possible explanations. Gerstner has stressed the 'qualitative' gap between Hutton and many of his predecessors and contemporaries: these controversies reveal some of the genuine motives at the root of natural philosophy (J. A. de Luc, *Geological Letters Addressed to Professor Blumenbach* (1793, publ. 1831), 49, 69–70). For Herschel's warm response to Hutton in comparison with his muted treatment of his friend De Luc, see C. Lubbock, *The Herschel Chronicle* (Cambridge, 1933), 243. For infra-red radiation, see M. Watanabe, 'James Hutton's "obscure light"', *Japanese Studies in the History of Science*, XVII (1978), 97–104. Playfair's assessment of Hutton's matter theory and natural philosophy was highly interpretative: compare his Priestley-esque account in 'Biographical account' (see note 28), 76, and his reference to the account of pressure and heat in the sun in query 11 to Newton's *Opticks* in his *Illustrations of the Huttonian Theory of the Earth* (Edinburgh, 1802), 181–2, 187.

[32] C. B. Wilde, 'Hutchinsonianism, natural philosophy and religious controversy in eighteenth century Britain', *HS*, XVIII (1980), 1–24. Wilde argues for a careful distinction between those ideas which are specific to one group and those which may be held by members of that group in common with other writers. In this case, all these philosophers share the idea of the circulation of a form of active matter – which is sometimes identified with fire – through the

McGuire assimilated Hutton into a tradition of Newtonian problems, in fact, by stressing the similarly Newtonian premises of the matter-theories of Knight and of Hutton, and simultaneously reassessing the natural philosophy developed by William Jones.[33] Hutchinson and William Jones did not, as has been claimed, make nature subservient to the spiritual superintendence of the Deity, nor were their ideas due to those of Boerhaave, whose texts were available in Britain only after Hutchinson's ideas were formed. Instead, Hutchinson argued that the operations of nature were due to the triune fluid of fire, air and light, and that the prime evidence for such agency was to be found in a correct reading of Scripture.[34] Hutton and many of his contem-

cosmos and stress the central of the *Sun* in this process. See my 'The phoenix of nature: fire and evolutionary cosmology in Wright and Kant', *Journal for the History of Astronomy*, IX (1978), 180–200, and the following texts: P. Leslie, *Philosophical Inquiry into the Cause of Animal Heat* (1778), 104–20; A. Walker, *System of Familiar Philosophy* (1799), 1–10; *idem*, *Epitome of Astronomy* 13th edn (1798); Robert Harrington, *New System on Fire and Planetary Life* (1796), 30–1; B. Higgins, *Philosophical Essay Concerning Light* (1776), xlvii–xlix; W. Herschel, 'On the nature and constitution of the Sun and the fixed stars', *PT*, LXXXV (1795), 46–72; *idem*, 'Observation tending to investigate the nature of the Sun', *PT*, XCI (1801), 265–318; *idem* 'Observations of a comet with remarks on the construction of its different parts', *PT*, CII (1812), 115–43. Walker claims (p. 534) that Newton had failed to deal with cometary theory, Harrington (p. 14) that Newton's system 'is forced and arbitrary', and Higgins that Newton's 'assumption of one or more aetherial fluids answers no purpose in Natural Philosophy' (p. 58). The responses to Newtonianism are critical: and this demonstrates the excessive *ease* with which any historian can construct a common idea from the otherwise allegedly chaotic natural philosophies, merely by noting specific ideas which some natural philosophers share.

33 Heimann and McGuire, 'Newtonian Forces' (see note 10), 296–7, and Heimann, "'Nature is a perpetual worker'" (see note 10), 14–16. Wilde argues that Heimann mistakes the import of William Jones's attack upon Newton and also his theory of the relation between God and cosmos, at least partly by neglecting the political and cabbalistic elements intrinsic to Hutchinsonianism. Heimann, however, does acknowledge that (*a*) there is some kind of transformation from Newtonian to Huttonian views of nature; (*b*) that the 'assessment of the importance of traditions other than those deriving from the Newtonian corpus is a crucial problem which awaits analysis' (p. 24). See G. Knight, *An Attempt to Demonstrate that All the Phaenomena in Nature may be Explained by Two Simple Active Principles, Attraction and Repulsion* (1754); Knight's and Hutton's theories differ markedly over the issue of primary qualities and the intervention of God, while both differ markedly from those of Newton and of Hutchinson.

34 For Hutchinsonianism, see G. Cantor, 'Revelation and the cyclical cosmos of John Hutchinson', in L. Jordanova and Roy Porter (eds.), *Images of the Earth* (Chalfont St Giles, 1979), 3–22; E. R. Wasserman, 'Nature moralized', *English Literary History*, XX (1953), 39–76; A. J. Kuhn, 'Glory or gravity: Hutchinson versus Newton', *JHI*, XXII (1961), 302–22. Hutchinson's commitment to

poraries could have borne no explicit debt to these ideas: but (and this is important) the notion of the circulation through the cosmos of active, often triune or heterogeneous forms of matter, stands clearly in contrast with many of the most fundamental postulates of conventional natural philosophy.

Heimann's claim that 'the eclecticism of these eighteenth-century theorists is paralleled by Newton's own varying views on the nature of force' is persuasive.[35] Nevertheless, it is necessary to recognize that a stress on the philosophical–epistemological method adopted in their theories by natural philosophers is not, by itself, enough to justify the construction of a 'Newtonian' tradition. This is amply demonstrated by the emphases on the political and anthropological characteristics of rival eighteenth-century cosmologies which several historians have recently deployed. Hutchinson's attitudes, and those of his disciples, can really only be understood as part of an entire *political* attitude to the place of and to the values deployed by natural philosophy.

Distinctive political and religious sects within natural philosophy can be isolated in other areas of scientific practice. Under the influence of William Law's text *Appeal to All in Doubt* (1740), strong Behmenist influences were introduced into natural philosophy, and writers such as Richard Symes, Richard Lovett and John Freke constructed matter theories in which electricity and power in matter were understood in a new way, both epistemologically and ontologically.[36]

mechanism is explicit and anti-Newtonian: 'it would be the strongest Evidence the Devil could produce', he believed, if the alternative were true (*Works of J. Hutchinson* (12 vols., 1748), v, 61). See also W. Jones, *Essay on the First Principles of Natural Philosophy* (1762), 26, the passage quoted by Heimann in this context: Jones himself quotes such sources as Pluche, the colleague of pietists in France such as Fénélon and Andrew Ramsay, in support of his anti-Newtonian theory of matter (Abbé Pluche, *Spectacle de la nature* (1732–50); see J. Ehrard, *L'idée de la nature en France dans la première moitié du 18e siècle* (Paris, 1963), 647ff.).

35 Heimann and McGuire, 'Newtonian forces' (see note 10), 302.

36 On sectarianism and natural philosophy, see the case studies of N. Garfinkle, 'Science and religion in England 1790–1800; the critical response to the work of Erasmus Darwin', *JHI*, xvi (1955), 376–88; M. Neve and R. Porter, 'Alexander Catcott: glory and geology', *BJHS*, x (1977), 47–70; R. E. Schofield, 'John Wesley and science in eighteenth century England', *Isis*, xliv (1953), 331–40; R. B. Rose, 'The Priestley riots of 1791' *Past and Present*, xvii/xviii (1960), 68–88; R. Shackleton, 'The *Encyclopédie* and Freemasonry', in W. Barber (ed.), *Age of Enlightenment* (Edinburgh, 1967); R. Darnton, *Mesmerism and the End of the Enlightenment in France* (Cambridge, Mass., 1968). On Behmenism and its scientific influence, see D. P. Walker, *The Decline of Hell* (1964); N. Thune, *The Behmenists and the Philadelphians* (Uppsala, 1948); C. Walton, *Notes and Materials for an Adequate Biography of... William Law* (1854); A. J. Kuhn, 'Nature

Richard Symes, a Bristol disciple of Jacob Boehme and William Law, stressed (as did Law and his contacts Cheyne and Byrom) that Newton had been influenced by Boehme too. He wrote that 'if Jacob Behmen's and Mr Law's philosophy can be proved by electricity, and vice versa electricity can be explained by their writings, there is scarce any phenomenon in Nature but what may be made intelligible'.[37] In Law's book, Symes and Freke could find ample ground for a developed theory of nature. John Byrom demonstrates in his *Journal* that this theory of nature was shared by a group of committed practitioners. Law wrote that 'a flint is dead . . . because its fire is bound, compacted, shut up, and imprisoned': Symes in describing the action of the electrical machine, talked in terms of 'making visible what is hidden in Nature, just as the striking together a flint and steel is the means of bringing forth to our sight the fire that lay hidden within them'.[38] To attempt to assimilate such natural philosophers to the Newtonian tradition would be no more difficult than it has been for Priestley, Hutton, Knight, or indeed the heretic Robert Greene: but to do so would be to violate the possibility of understanding the profound religious and political implications which are at work in this programme.

spiritualized: aspects of anti-Newtonianism', *English Literary History*, XLI (1974), 400–12; and a brief and contemptuous account of Freke in Schofield, *Mechanism and Materialism* (see note 9), 159–60.

[37] R. Symes, *Fire Analysed, or the Several Parts of Which it is Composed Demonstrated by Experiment* (Bristol, 1771), quoted in Walton, *Materials* (see note 36), 409–10. On the claim (hitherto unsubstantiated) of Newton's debt to Boehme, see S. Hobhouse, *Selected Mystical Writings of William Law* (1948), appendix 4, 397–422. Law told John Byrom that Cheyne was a source for much of his ideas: see H. Talon (ed.), *John Byrom: Selections from his Journals and Papers* (1950), 221ff., and the remarkably similar connections now partly established with Andrew Ramsay and the Camisards and Newton, in D. P. Walker, *The Ancient Theology* (1972), and M. C. Jacob, 'Newton and the French prophets: new evidence', *HS*, XVI, (1978), 134–42. Jacob and others have done much to claim the importance of Masonic networks for the development of natural philosophy, and indeed the Enlightenment as a whole: a similar investigation of the Philadelphians and the Behmenists from this viewpoint is also needed. See J. M. Roberts, *The Mythology of the Secret Societies* (1974), 38ff., on some connections between Masonry, enthusiasm, and philosophy.

[38] William Law, *Appeal to All that Doubt*, in G. Morgan (ed.), *Works of Law* (9 vols. Canterbury, 1892), VI, 89; Symes, *Fire Analysed* (see note 37), 411. On the philosophers who composed this group around Law, see Walton, *Materials* (see note 36), 404–5. Richard Lovett and John Freke figured in this group: see J. Freke, *Treatise on the Nature and Property of Fire* (1752) and R. Lovett, *Electrical Philosopher* (Worcester, 1774), which also cites the Hutchinsonian text of William Jones (see note 34). Compare Talon, *John Byrom* (see note 37), 220 and 281, for example, on the connections with Law.

Hitherto, as I have indicated, the most powerful critique of the prevailing assumption that the historical task is to assimilate all natural philosophy to a single reductive base, is to be found in much wider philosophical debates on the character of the history of science. An immediate (philosophical) connection with the examples which I have cited is found in the status of certain obviously 'Newtonian' axioms in natural philosophy. For example such an axiom is that of inertial homogeneity – the claim that all matter is intrinsically uniformly dense, and resists motion in proportion to its weight. Metzger has emphasized the prolonged struggle to establish and clarify this point: such figures as J. P. de Molières and J. J. d'Ortous de Mairan evidently were confused by this idea, and such eminent authorities as Euler, Lomonosov, Woodward and Higgins all reject the axiom.[39] Robert Greene was a notable critic of this essential claim: in his text of 1727 Greene argued that 'Matter is not Homogeneous in its own Nature . . . and that its Powers and Qualities are not derived from [its Parts] in Conjunction with Motion, but from the peculiar Expansive and Contractive Forces of it'. For Heimann, this idea is best understood 'in terms of [its] relation to Newtonian forces and Lockean powers': historically, Greene's ideas were developed explicitly because of the threat he thought was posed to the Revolution settlement by philosophical arguments drawn from 'Popish countries'. 'I will leave the World to judge', Greene wrote, 'whether Superstition and Foppery in Religion in any Nation, will not make an Impression of

[39] Metzger, *Newton, Stahl, Boerhaave* (see note 1), 20–34 (la conception newtonienne . . . transformée en axiome évident et primordial en loi de nature'); on Molières' and de Mairan's confusion over mass and weight, see E. J. Aiton, *Vortex Theory of Planetary Motions* (1972), 210 sq., and Guerlac, 'Further Newtonian studies' (see note 7). On the concept of inertial homogeneity, see Thackray, *Atoms and Powers* (see note 2), 16; rejections and mis-statements of this allegedly 'natural' axiom include those of John Harris, *Lexicon technicum* (1704), I, art. 'Matter', and Robert Greene, *Principles of Natural Philosophy* (Cambridge, 1712), 117 sq.; M. Lomonosov, 'On the relation of the amount of material and weight' (1758) in H. M. Leicester (ed.) *M. V. Lomonosov on the Corpuscular Theory* (Cambridge, Mass., 1970), 225; Higgins, *Philosophical Essay* (see note 32), 139–41; L. Euler, 'Recherches sur le mouvement des corps célestes en général', in *Opera omnia* (Leipzig), 2nd ser., XXV, 1–44 (pp. 3–4); E. Forbes (ed.), *The Euler–Mayer Correspondence* (1971), 47; L. Euler, *Lettres à une Princesse d'Allemagne* (Paris, 1787), ed. Condorcet, letter 50, 197. H. M. Leicester, 'Boyle, Lomonosov, Lavoisier and the corpuscular theory of matter', *Isis*, LVIII (1967), 240–4, is an interesting study of the connections between the rejection of this axiom and some of Lavoisier's insights. The best account of the emergence of this axiom is J. E. McGuire, 'Body and void and Newton's "De mundi systemate": some new sources', *Archive for History of Exact Sciences*, III (1966), 206–48.

Atheism or Deism upon the Philosophy of it, as it most certainly has done'. Attacks upon Greene's ideas make very clear the political context of Greene's theory, which his contemporaries recognized.[40]

This point may be briefly summarized: because historians have felt that their task is to integrate natural philosophy and to give it an apparent unity, such philosophical moves by natural philosophers as the rejection of the most fundamental Newtonian axioms, the explicit commitment to widely different religious and political systems, or to notions of the epistemological relations between the penetration of matter and the practice of natural philosophy, have all been assimilated into one common tradition. In fact, in this section of my discussion, I have restricted myself to the consideration of a 'Newtonian' reduction base: I believe that the same criticism can be made of 'Leibnizian', 'Cartesian', or indeed 'Wolffian' or 'Kantian' reduction bases in Europe-wide natural philosophy. There have been two consequences of any such assimilation: first, attempts to demonstrate rival, anti-Newtonian groups of natural philosophers remain defined by the contrast with Newtonianism rather than being seen as representative of a distinct philosophy in their own right; secondly, the specificity of natural philosophy as a distinct discourse is never confronted historically. Natural philosophy becomes an inevitable, automatic form of scientific expression. Some historians have attempted to define this specificity: McEvoy has written on Priestley's concept of 'philosophical history' and his studies of electricity and optics which were composed in this mode; other historians have dealt with experimental practice, instrumentation, and the popular lecturing tradition.[41] However, natural philosophy still subsists in a shadowy world between science and other discourses: the problem of definition, as I have stated, remains the exclusive province of the philosophers.

[40] Robert Greene, *Principles of the Philosophy of the Expansive and Contractive Forces* (Cambridge, 1727), 23. Greene coupled Cartesian and Newtonian philosophy together as foreign and Papist: the attack upon him in *A Taste of Philosophical Fanaticism* (1712) couples him with 'a notorious rationalist, one Mr Whiston' (p. 33) and described him attempting to 'prove there's a God after he has evinc'd there's no Occasion for one' (p. 25). See Thackray, *Atoms and Powers* (see note 2), 126–34, who notes the political and theological roots of Greene's thought; Schofield, *Mechanism and Materialism* (see note 9), 117–21; and Heimann and McGuire, 'Newtonian forces' (see note 10), 254–61.

[41] On 'philosophical history', see McEvoy, 'Electricity in . . . Priestley's thought' (see note 25), 6; on the popular lecturers and natural philosophy see Taylor, *Mathematical Practitioners* (see note 6); F. W. Gibbs, 'Itinerant lecturers in natural philosophy', *Ambix*, VIII (1961), 111–17; Rowbottom, 'The teaching of experimental philosophy' (see note 6); A. Hughes, 'Science in English encyclopedias', *AS*, VII (1951), 340–70 *et seq.*

III. Natural philosophy as the negation of science

I have described the work of a group of historians who do not treat natural philosophy as a discourse which can be analysed historically. Instead, natural philosophy is analysed as a system of connected concepts and axioms with no definite associated form of practice. This has enabled these historians to avoid any comparison with other forms of science, and to steer clear of the permanent problem of *progress* which has haunted the history of science, significantly enough, since the mid-eighteenth century.[42] But the circumvention of this problem has been purchased at a price: natural philosophy has acquired the character of permanent philosophical exegesis of nature, itself a 'natural' activity. By contrast, earlier historiography emphasized the *static* character of natural philosophical work. In his survey of Georgian optics, for example, Cantor has noted the common assumption of the 'sterility' of research into the properties of light in the eighteenth century.[43] A. R. Hall, in his survey of eighteenth-century physics, described the 'brittle' and 'dry' character of eighteenth century science, and asserted that one can 'never be interested' in many of the more commonplace activities of natural philosophy. More charitably, he conceded that the natural philosophers did make several advances over the more heroic seventeenth-century scientists, but that these advances were essentially consolidations, that they were derivative in tone and character.[44] C. C. Gillispie, too, argued that 'in science all that declined was the rate of progress'. He attributed this decline to the magnitude of the advance made in the previous century', and to the 'accidental circumstance that the period from 1720 to 1760 produced no really great British scientists'.[45]

In response to declarations like these, Barry Barnes and others have

[42] Classical statements of this problem, from distinctly opposed viewpoints, are David Hume, 'Of the rise and progress of the Arts and Sciences' (publ. 1741), in *Political Essays* (New York, 1953), 111–22; J. d'Alembert, 'Discours préliminaire', *Encyclopédie* (Paris, 1751), I; J. A. N. Condorcet, *Esquisse d'un tableau historique des progrès de l'esprit humain* (Paris, 1793).

[43] Cantor, 'Georgian optics' (see note 8), 3. The targets for Cantor's comment are P. A. Pav, 'Eighteenth century optics: the age of unenlightenment' (Ph.D. thesis, Indiana, 1964); V. Ronchi, *The Nature of Light: an Historical Survey* (1970); H. J. Steffens, *The Development of Newtonian Optics in England* (New York, 1977).

[44] A. R. Hall, *The Scientific Revolution, 1600–1800*, 2nd edn. (Boston, 1966), 341.

[45] C. C. Gillispie, *Genesis and Geology* (New York, 1959), 7 sq. One might compare H. Butterfield's chapter 'The delayed scientific revolution in chemistry', in his *Origins of Modern Science* (1949).

denied that 'we should... seek to define science ourselves'. They have stressed the importance of the categories which were *used* in the eighteenth century, for example, thus reinstating the validity of the category 'natural philosophy' as autonomous from that of science. In the hands of cultural historians this has not necessarily implied the historical reification of such disciplines: this is due to new philosophies of science which these historians have confronted.[46] That is, while the historians whose work I have surveyed have ignored the problem of the progress of knowledge and of the demarcation of types of knowledge, philosophers as sharply opposed as Kuhn, Bachelard, Popper, Pecheux and Foucault have all placed this problem at the centre of their work.[47] Bachelard, the principal influence upon the theories of science developed in France since the early 1960s, upon Althusser as much as upon Canguilhem and Foucault, saw his main role as the demarcation of scientific *value* from the murky, alchemical, pre-scientific beliefs of a pre-rational age. He mobilized concepts which he saw as 'forever' scientific, which a 'sanctioned' history would show to be points of no return.[48]

[46] S. B. Barnes, *Scientific Knowledge and Sociological Theory* (1974), 100–1. See D. Bloor, *Knowledge and Social Imagery* (1976), and A. Thackray, 'Science: has its present past a future?', in R. Steuwer (ed.), *Historical and Philosophical Perspectives of Science* (Minneapolis, 1970), 112–27, and the collection S. Shapin and S. B. Barnes (eds.) *Natural Order* (1979), for some suggestive approaches which at least *acknowledge* the philosophical problems inherent in the history of natural philosophy.

[47] T. S. Kuhn, *Structure of Scientific Revolutions*, 2nd edn (Chicago, 1970); *idem*, 'The function of measurement in modern physical science', *Isis*, LII (1961), 161–93; G. Bachelard, *La psychanalyse du feu* (Paris, 1938); *idem*, *La formation de l'esprit scientifique* (Paris, 1938); M. Pecheux, 'Idéologie et histoire des sciences', in his *Sur l'histoire des sciences* (Paris, 1974), 15–47; M. Foucault, *Histoire de la folie à l'âge classique* (Paris, 1961); *idem*, *Naissance de la clinique* (Paris, 1963); *idem*, *Les mots et les choses* (Paris, 1966). All centre their attention upon natural philosophy and its associated discourses in the eighteenth century.

[48] G. Bachelard, *L'activit é rutionaliste de la physique contemporaine* (Paris, 1951), 25–7: 'Il faut sans cesse former et reformer la dialectique d'histoire perimée et d'histoire sanctionée par la science actuellement active. L'histoire de la théorie du *phlogistique* est perimée puisqu'elle repose sur une erreur fondamentale, sur une contradiction de la chimie ponderale ... Au contraire... d'autres travaux comme ceux de Black sur le *calorique* affleurent dans les experiences positives de la determination des *chaleurs specifiques*. Or la notion de la *chaleur specifique* ... est pour toujours une notion scientifique. Les travaux de Black peuvent donc être décrits comme des éléments de l'*histoire sanctionnée*.' For the influence of this idea of history of science and 'passé actuel' – the concept of permanent division between science and its past – see D. Lecourt, *Marxism and Epistemology* (1975), 129–41; S. Gaukroger, 'Bachelard and the problem of epistemological analysis', *Studies in History and Philosophy of Science*, VII (1976), 189–244; T. Counihan, 'Epistemology of science – Feyerabend and Lecourt',

This mobilization, as Bachelard's Marxist disciples such as Althusser and Lecourt argue, problematized the historical understanding of the difference between '*connaissance vulgaire*' and '*rationalisme scientifique*'.[49] For the classical natural philosophy which is the object of this chapter, Bachelard's intervention results in a series of highly denigratory comments on the psychologistic, closed, commonsensical pretensions of matter-theory. This is directly relevant to the *Newtonian* base of Anglo-American understanding of eighteenth-century matter-theory. Bachelard denied this scientific base: he emphasized the *absence* of exact experiment. For example:

> Ce qui manque, peut-être, le plus à l'esprit prescientifique c'est une doctrine des erreurs expérimentales.[50]

Citing Mme de Châtelet's apparent anticipation of Joule's result that agitation of water produces heat, Bachelard rightly contrasted the exhaustive, but relatively simple, measurements Joule made with the total failure of his eighteenth-century predecessor to make such determinations and her ability to remain satisfied with a qualitative,

Economy and Society, v (1976), 74–110. Althusser's classic statement on his relation with Bachelard, Foucault and Canguilhem is in *For Marx* (1977), 257–8 ('A letter to the translator'). Althusser's primary application of this understanding to the history of science is found in *Philosophie et philosophie spontanée des savants* (Paris, 1967): his use of Bachelard's valorization of the epistemological break in the eighteenth century is explicit in his treatment of the Kantian critique and its effect on the Enlightenment (p. 106). The most informative critique of Althusser's mobilization of this view is in H. Lefebvre, *L'idéologie structuraliste* (Paris, 1971), 191–251.

49 Bachelard, *Le rationalisme appliqué* (Paris, 1949), 101: 'Les sciences physiques et chimiques... peuvent être caractérisées épistémologiquement comme des domaines des pensées qui rompent nettement avec la connaissance vulgaire. [Au contraire] la science de Lavoisier... est en liaison continue avec les aspects immédiats de l'expérience usuelle... Dans la chimie lavoisienne on pèse le chlorure de sodium comme dans la vie commune on pèse le sel de cuisine'. The most familiar use of this idea is by Koyré, 'Galileo and Plato', *JHI*, IV (1943), 400–28, on Galileo's rupture with the 'common-sensical' Aristotelian philosophy. This is how his mentor, Bachelard, pictures the emergence of sciences in the eighteenth century.

50 Bachelard, *Rationalisme* (see note 49), 105–9: in attacking the 'science naturelle de l'électricité au XVIIIe siècle' and its *cosmologie du feu*, bachelard states that 'on trouverait sous les phénomènes pourtant si rares de l'électricité, les qualités profondes, les qualités élémentaires: le feu et la lumière. Ainsi enracinée dans les valeurs élémentaires, la connaisance vulgaire ne peut évoleur. Elle ne peut pas quitter son premier empirisme. Elle a toujours plus de réponses que de questions. Elle a réponse à tout.' On the psychoanalytic character of pre-science, see *Le matérialisme rationnel* (Paris, 1953), 103, where Bachelard cites Jung's studies of alchemy; on the doctrine of experimental error, see *Formation* (see note 47), 216–17.

impressionistic and subjective report of an idea of common experience.[51]

The 'alchemical' aspects of matter-theory effectively determined Bachelard's philosophical critique. Alchemy justified the penetrative, occultist aspects of chemistry which we discovered in the statements of Playfair and Shaw. Science, by contrast, plumbs *genuine* problems in a highly technical fashion, or so Bachelard claimed.[52] Natural philosophy manifested these excavations of matter as *obstacles* which carried the psychological *complexes* from which natural philosophy could never dissociate itself. Amongst these obstacles, Bachelard singled out three specific problems for matter-theory: the obstacle of *primary experience* (the obsession with common-sense experience); that of *realism* (the appreciation of the attributes of a material object as an avaricious appropriation of that object); and, finally, that of *animism* (seeing life and vitality as the concentration of immense power in a small volume of matter).[53] Using these ideas, Bachelard concentrated on precisely those aspects of natural philosophy which Anglo-American historiography ignores now. He argued that the conversational style of natural philosophical texts revealed their ob-

[51] *Formation* (see note 47), 217: 'On s'étonnera moins de cette absence de perspicacité éxpérimentale si l'on considère le mélange des intuitions de laboratoire et des intuitions naturelles'. Contrast this observation by Bachelard with T. S. Kuhn, 'Conservation of energy as an example of simultaneous discovery', in Marshall Clagett (ed.), *Critical Problems in the History of Science* (Madison, 1959), 321–56, on allegedly 'external' factors, and P. Heimann, 'Conversion of forces and conservation of energy', *Centaurus*, XVIII (1974), 147–61, for a representative critique of Kuhn's 'group analysis'. Neither approach is dealing with the problems Bachelard sets for history in its confrontation with natural philosophy.

[52] For Bachelard's obsession with the 'alchemical' as the diseased values of non-science, see *Formation* (see note 47), 51; *Matérialisme* (see note 50), 215; and Lecourt's analysis of this obsession in *Marxism and Epistemology* (see note 48), 68–70. Lecourt, noting Bachelard's debt to Metzger at this point, states that Bachelard argues that (*a*) contemporary (non-scientific) philosophies import (non-scientific) value-systems, just like natural philosophy did; (*b*) the job of the historian is to sort out for ever these external, *alchemical* aspects from 'sanctioned' (scientific) discourse, in philosophy now and in natural philosophy then.

[53] On epistemological obstacles and the psychological complexes see *Psychoanalysis of Fire*, trans. A. C. Ross (Boston, 1964), 59–82; *Formation* (see note 47), 16–25, 138–9, 154–5; 'C'est en termes d'obstacles qu'il faut poser le problème de la connaissance scientifique.' This, for Bachelard, is because *no scientific practice is given:* obstacles must be overcome in order to be in the scientific domain. This is the most important intervention by Bachelard: and the one which, via Canguilhem, has had most effect upon Foucault's 'archaeology' of knowledge (M. Foucault, *L'ordre du discours* (Paris, 1971), 72–4).

session with common-sense experience, and dealt with the problem that matter-theory tended to address itself to issues like the fear of thunder or the sense of loss when gold dissolves, and showed how the common obsession with power immanent in matter and the properties of fire manifested these *psychological* obsessions with realism and animism.[54]

The classic example of the way in which these obstacles worked was Bachelard's account of the image of the sponge: he traced the way in which such natural philosophers as Réaumur, Franklin, Lémery, and J.-P. Marat could use the *image* of the sponge as (for them) a satisfactory *explanation* of several superficially related phenomena, merely by mobilizing the obvious experience of the action of a sponge and applying it to the occult powers manifest in such widely different cases as the dissolution of air in water, the action of electrical matter, the translucence of some substances, and the absorption of heat.[55] This is a classic example of the positive historical impact of Bachelard's philosophy: it provides a coherent method for the analysis of a different, quite distinct grammar of science which needs to be considered as fully demarcated from science itself.

However, Bachelard's philosophical attack upon natural philosophy did more than highlight its distinct grammar: it defined this grammar exclusively through *negativity*, and that has had dire consequences for historical understanding. This is dramatized, for example, in the work of some of Bachelard's disciples: Michel Pecheux, for example, analysed the effect of the seventeenth-century revolution in mechanics upon the eighteenth-century emergence of the science of electricity. He argued that after the Galilean break in physics, a 'point of no return' for mechanics, the impact of this break upon electricity took the form of an obstacle, the '*obstacle imaginaire mécaniste*': the obsession of classical electricians with mechanist *images*. This obstacle, in turn, could only be overcome by a further break within electricity, the '*coupure Coulombienne*'. Once this further break had been made, the 'moral, political, religious and literary connotations' which

[54] *Formation* (see note 47), 138–9, citing Malouin, *Chimie medicinale* (3 vols., Paris, 1755), III, 5–6, and *ibid.*, 154–5, citing La Cépède, *Essai sur l'électricité naturelle et artificielle* (2 vols., Paris, 1781), II, 32; *Psychoanalysis of Fire* (see note 53), ch. 5 ('The chemistry of fire: history of a false problem'), citing Boerhaave, Scheele, and J.-P. Marat, *Découvertes sur le feu, l'électricité et la lumière* (Paris, 1779), 28, etc. 'Sans aucune preuve, par la simple séduction d'une affirmation valorisante, l'auteur attribue une puissance sans limite à des éléments. C'est même un signe de puissance que d'échapper à l'expérience.'

[55] *Formation* (see note 47), 74–6. Réaumur (1731) on the expansion of air; Franklin (1752) on electrical fluid; Lémery on phosphorescent matter; Marat (1779) on cooling bodies in water.

had infected classical electrical theory had been left definitively *'hors jeu.'*.[56] We have already seen, however, how inadequate a picture this is of the development of the connections between matter-theory and electricity from Newton onwards: the development of instrumentation, for which Franklin, Coulomb, Freke, Musschenbroek and many others were responsible, needs more subtle historical approaches than that of a psychological obsession with the image of the machine and of mechanism.[57] Bachelard's value for the historian of matter-theory, therefore, can be confined to his characterization of natural philosophy as a distinctive kind of practice: its popularizing trends, its concern with properties of matter, power and force, and, above all, its estimation of the individual, the dramatic, the *paradoxical*, as the prime object of natural philosophy. In this sense the heroic declarations of Priestley on the philosophical-historical development of natural philosophy can be read as a popular appeal on behalf of a discourse which would provide a dramatization of the bizarre in the inner construction of matter:

> Hitherto philosophy has been chiefly conversant about the more sensible properties of bodies: electricity, together with chymistry and the doctrine of light and colours, seems to be giving us an inlet into their internal structure, on which all their sensible properties depend. By pursuing this new light, therefore, the bounds of natural science may be extended beyond what we can now form an idea of. New worlds may

[56] Pecheux, 'Idéologie et histoire des sciences' (see note 47), 16–32 ('L'effet de la coupure galiléenne en physique'): using the work of Bachelard on electricity in the eighteenth century, Pecheux's lectures (winter, 1967–8: with Althusser, Macherey, Balibar, Fichant and Badiou) argue that the effect of one already constituted *rational science* on another pre-science can be anti-progressive. As Bachelard had argued, the instrumental ensemble of the Leyden jar appears here as *monstrous*, that of Coulomb as a *progressive* rupture. The contrast with Kuhn, who drew for his study of the same phenomenon upon Roller and Roller and Cohen (see note 70), though not entirely in Pecheux's favour, is clear.

[57] On the instrumental-practical component in natural philosophy (matter-theory–electricity), see B. S. Finn, 'The influence of experimental apparatus on 18th century electrical theory', *Actes du XIIe congrès internationale d'histoire des sciences* (Paris 1968, publ. 1971), xa, 51–5; R. McCormmach, 'Henry Cavendish: a study of rational empiricism in 18th century natural philosophy', *Isis,* LX (1969), 293–306; M. Daumas, *Scientific Instruments of the 17th and 18th Century* (1972); on microscopy, see especially S. Bradbury and G. l'E. Turner, *Historical Aspects of Microscopy* (Cambridge, 1967), and, fascinatingly, G. l'E. Turner, 'Microscopical communication', *Journal of Microscopy,* C (1974), 3–20. Unsurprisingly, there is little explicit literature on the connection between 'conceptual' and practical and instrumentational development in natural philosophy from the Anglo-American tradition.

> open to our view, and the glory of the great Sir Isaac Newton himself, and all his contemporaries, be eclipsed by a new set of philosophers and quite a new field of speculation.[58]

To summarize: Bachelard presents natural philosophy as a discourse which is the *negation* of science, structured by the dialectic of the dramatic and the anomalous in contrast with the common-sensical. I believe this is an accurate characterization of the grammar of this discourse – I deny that this can be an adequate base for an historical (as opposed to philosophical) analysis of natural philosophy. Nevertheless, it is strongly suggestive of the grounds for the obsession with natural philosophy which has been displayed by a more recent philosophical theory of science – that of Kuhn.

Kuhn's reconstruction of the history of science involves the isolation of a set of quite distinct and highly differentiated 'paradigms' which are separated from each other both temporally (in that one paradigm succeeds another in the same field) and also socially (in that two rival paradigms can certainly coexist in this field). These paradigms mutate through the emergence of 'anomalies': like Bachelard, therefore, Kuhn focuses on the role of the anomalous, and also pictures the natural philosophical project as a form of negation.[59] The vital differences from Bachelard need no stress: Kuhn denies the coherence of the concept of permanent revolution, while for Bachelard such a label perfectly describes the mechanism of change in science. Bachelard's materialism has enabled him to grasp the social, cultural and instrumentational constitution of scientific discourse. Barnes has noted the puzzling absence of these factors in the Kuhnian account.[60] But it is the central importance of the anomaly which lies at

[58] Joseph Priestley, *History of the Present State of Electricity* (1767), xiii.

[59] Kuhn, *Structure* (see note 47), 52, 13–14. Kuhn states that 'Discovery commences with the awareness of anomaly, i.e. with the recognition that nature has somehow violated the paradigm-induced expectations that govern normal science': the example chosen is the discovery of oxygen in 1776, apparently *both* a discovery within the 'paradigm' of pneumatic chemistry and yet also the crisis which produced the new 'paradigm' of Lavoisian chemistry. We may note, also, that the psychologism of Bachelard (which is distinctly Jungian, and focuses on the relation between conceptual and practical aspects of science) is matched by a Kuhnian psychologism (which is derived from behaviourism, and excludes all reference to any context or *specific* relation with practice: see Kuhn's discussion of the anomalous playing-card experiments in *Structure*, 62–5 and 112–13, and the response in Lecourt, *Marxism and Epistemology* (see note 48), 17–18, which attacks Kuhn's basis of paradigms in '*the nature of the mind*' (p. 64)).

[60] Bachelard, *Matérialisme* (see note 50), 103: 'La pensée scientifique repose sur un passé réformé. Elle est essentiellement en état de révolution continuée'. Kuhn, 'Reflections on my critics', in I. Lakatos and A. Musgrave (eds.) *Criti-*

the heart of the appreciation of natural philosophy and of the way in which it negates itself by the transformation to science. Once again, this is no adequate historical category, but the role of the paradox in natural philosophy has evidently provided historians and philosophers, principally Foucault and Canguilhem, with the basis for a complete reassessment within the history of science.

Most of Kuhn's examples of the emergence of paradigms, as I have indicated, are drawn from the transformation of natural philosophy into science in the eighteenth century. For historians, there are three fundamental consequences of this transformation. First, Kuhn's attention to this field is due, as Bachelard might have noticed, to the importance of the anomaly in eighteenth-century natural philosophy. For Kuhn the accumulation of these anomalies is the mechanism by which science emerges – for Bachelard it is the sign that science has not yet emerged.[61] Secondly, the basic activity of pre-paradigm science (Bachelard would categorize this as non-rational discourse) is taxonomic, fact-gathering, ordering and systematic natural history.[62] Finally, for both Kuhn and Bachelard, the instrumentational, experimental dimension of natural philosophy is definitive. This is where most debate has focused. Followers of Bachelard have paid particular attention to the instrumentation of natural philosophy: such practical

cism and the Growth of Knowledge (1970), 242–3: '[The] phrase 'revolutions in permanence' does not... describe a phenomenon that could exist'. Barnes, *Scientific Knowledge* (see note 46), 107–8: 'There seems to be no particular reason why Kuhn should not have referred to the institutional context as well as the intellectual milieu'.

[61] Compare Kuhn's discussion of the Leyden jar and the anomalies it demonstrated (*Structure*, (see note 47), 61–2) with the condemnatory remarks of Bachelard (*Rationalisme* (see note 49), 148–9).

[62] Kuhn, *Structure* (see note 47), 17, on natural history as the typical pre-paradigm stage; Bachelard, *Formation* (see note 47), 30–1: 'La pensée prescientifique... cherche non pas la variation, mais la variété'. For natural philosophy as taxonomy, see M. Foucault, *The Order of Things* (1974), 144ff.; H. Daudin, *Les méthodes de la classification et l'idée de série en botanique et en zoologie de Linné à Lamarck* (Paris, 1926); M. P. Crosland, *Historical Studies in the Language of Chemistry* (1962), 114–30; K. M. Baker, 'Unpublished essay of Condorcet on technical methods of classification', *AS*, XVIII (1962), 99–123; G. Canguilhem, 'Du singulier et de la singularité en épistémologie biologique', in his *Etudes d'histoire et de philosophie des sciences* (Paris, 1968), 211–25; J. L. Larson, *Reason and Experience: the Representation of Natural Order in the Work of Carl von Linné* (Berkeley, 1971); F. A. Stafleu, *Linnaeus and the Linnaeans* (Utrecht, 1971); D. C. Goodman, 'The application of chemical criteria to biological classification in the 18th century', *MH*, XV (1971), 23–44; W. R. Albury and D. R. Oldroyd, 'From Renaissance mineral studies to geology in the light of Michel Foucault's *The order of things*', *BJHS*, X (1977), 187–215.

ensembles as the Leyden jar or the reflecting telescope have been seen, in Bachelard's terms, as *'un théorème réifié'*.[63]

The contrasts between Kuhnian and Bachelardian theories have been as important. For Kuhn the negation which separates pre- and post-paradigm science takes place *within* science: it is an historical act which does not take the new paradigm into a new realm of values. Ultimate canons of historical definition remain undamaged: thus we can speak of successive 'paradigms' in chemistry, or in astronomy, or (as some of Kuhn's disciples have suggested) in economics or the human sciences. But for Bachelard this negation *creates* science: there is no prehistory of science. In the Kuhnian account there is therefore no point appealing to extrascientific factors. In his revisionist texts, he now acknowledges that 'paradigms have throughout been possessed by any scientific community . . . including the schools of the "pre-paradigm period"'. Paradigm activity commences, in a manner scarcely borne out by an historical analysis of the development of matter-theory, when these anomalies of *experience* are distinguished according to some criterion, which Kuhn pictures as implicit, or from some other science, or 'by personal or historical accident'. (Once again, we notice a complete exclusion of any social, political, economic or extra-scientific sources, for such mediations must, for Kuhn, be accidents.)[64] For natural philosophers, anomalies were the object of experience: through anomaly the correct sense of wonder was to be developed. Priestley writes of 'the discipline of the heart, and . . . benevolent and pious sentiments upon the mind', paraphrasing David Hartley's insistence that these emotions are 'the only keys which will unlock the mysteries of nature, and clues which will lead through her labyrinths. Of this, all branches of . . . natural philosophy afford abundant instances'.[65] In his preface to the *Encyclopédie,* d'Alembert, in a particularly Baconian mood, wrote that

> We might say that the study of Nature lavishly serves our pleasures at least, even though it withholds from us the necessities of life. It is, so to speak, a kind of superfluity that

[63] Bachelard, *Les intuitions atomistiques* (Paris, 1933), 140. Bachelard distinguishes between the 'fictional' production of a concept and the 'technical production' of a phenomenon by instrumentation: 'Le technique [peut] fournir une syntaxe susceptible de rélier entre eux les arguments et les intuitions.'

[64] Kuhn, 'Reflections on my critics' (see note 60), 272n.; note the way (*ibid.*, 260) in which Kuhn deals with some criticisms of his statement that 'Priestley, for instance', after Lavoisier's 'discovery of oxygen', 'ceased to be a scientist', and Kuhn's denying ordinance against using the term 'unscientific'.

[65] Priestley, *History and Present State of Electricity* (1767), xix–xxi; David Hartley, *Observations on Man* (2 vols., 1749), II, 245–6.

> compensates, although most imperfectly, for the things we lack.[66]

In his view, this is explicitly connected with the 'amusement' derived from the penetration of matter and the elucidation of its properties. Adam Smith, most notably, in his essays on the history of the sciences, insisted that it was in the human sense of wonder that the natural philosophical compulsion was located.

> Though it is the end of Philosophy to allay that wonder, which either the unusual or seemingly disjointed appearances of Nature, excite, yet she never triumphs so much as when, in order to connect together a few, in themselves perhaps, inconsiderable objects, she has, if I may say so, created another constitution of things, more natural indeed, and such as the imagination may more easily attend to, but more new, more contrary to common opinion and expectation than any of those appearances themselves.[67]

The experimental (what Bachelard would call the 'experiential') imperative in natural philosophy is grounded by a set of philosophical, methodological and cultural constraints based on 'superfluity' and 'wonder'. In his posthumous text, edited by Hutton and Black, Smith explained that this 'wonder' was responsible for the 'first theism that arose among those nations that were not enlightened by divine Revelation'.[68] For natural philosophy as a whole, most obviously in electricity and in matter-theory, instrumentational practice was the most obvious manifestation of this wonder.

Such instruments as the Leyden jar, therefore, provide the impor-

[66] J. d'Alembert, *Preliminary Discourse to the Encyclopedia of Diderot* (1751; New York, 1963), 16ff.

[67] Adam Smith, *Essays on Philosophical Subjects*, ed. J. Black and J. Hutton (1795), essay I, 55. For the stress by Smith on wonder, sympathy, and the essentially epistemic and aesthetic categories of natural science, see V. Foley, *Social Physics of Adam Smith* (W. Lafayette, Ind., 1976), 22–3; H. F. Thomson, 'Adam Smith's philosophy of science', *Quarterly Journal of Economics*, LXXIX (1965), 212–33 (p. 226); A. Skinner, 'Natural history in the age of Adam Smith', *Political Studies*, XV (1967), 32–48. It is in precisely this respect that Smith comes closest to Hutton in his conception of philosophy and nature; see A. McFie, 'The invisible hand of Jupiter', *JHI*, XXXII (1971), 595–9. (The essays were composed from 1750.)

[68] Smith, *Essays* (see note 67), essay II, 107. Compare p. 44: 'A system is an imaginary machine invented to connect together in fancy those different movements and effects which are in reality performed.' For an analysis of Smith's *political oeconomy*, which examines these problems from within a frame which owes much to Bachelard, see K. Tribe, *Land, Labour and Economic Discourse* (1978), 101–9: Tribe stresses, as historians of science rarely do, the discursive practice into which Smith's work is inserted.

tant connection between the disparate elements of practical discourses which constitute natural philosophy. For Kuhn, the Leyden jar was *strange:* so it provided a progressive anomaly for Franklin to deal with. For Bachelard, the jar was *monstrous:* so it provided a piece of pseudo-science which historians might now psychoanalyse.[69] In reporting the experiment to his colleagues in Paris, Musschenbroek wrote of 'a new but terrible experiment which I advise you not to try for yourself'. Like his predecessor Kleist, Musschenbroek was most impressed by the *violence* of the electrical force released from within matter.[70] Priestley reported that it was possible to earn a substantial living merely from demonstrating the effects of this machine to the public.[71] For Bachelard, the transformation of such an object to a technical machine was anything but easy: it involved the purging of the bizarre and the common-sensical. For Kuhn, the construction of the Jar, quite irrespective of its interpretation, placed the electricians well on the way to a paradigm shift. So the construction of such experimental apparatus is a sign of disease for Bachelard and of health for Kuhn.

Such experimental equipment, Bachelard claimed, provided merely a *metaphor:* a genuine practical ensemble would provide a *syntax* in which the experimental results could be construed. While recognizing the complex social context in which such experimental technique emerged - and this has been a valuable recognition - Bachelard then condemned these contextual links as pre-rational. Social construction of experimental technique cannot, for an historian, be categorized as a *failing*. This applies particularly, for example, to the historical account of the way in which the seventeenth-century stress upon the agency of fire as purgative and purifying supported the categorization of the products of an admittedly destructive analysis of substances by fire as the *true* constituents of that body. The abandonment of this privileged status for fire in chemical analysis was not merely the realization of the fact that fire was very destructive. Instead, the institution by the academies of extensive botanical taxonomic programmes, in which this technique was inapplicable, demonstrated the multifarious techniques of analysis, and demonstrated that in

[69] Bachelard, *Rationalisme* (see note 49), 148: 'Ainsi se forme une monstruosité que la culture scientifique devra psychanalyser'; Kuhn, *Structure* (see note 47), 61–2: 'the experiments which led to its emergence were also the ones that . . . provided the first full paradigm for electricity'.

[70] Musschenbroek, January 1746, quoted in D. Roller and D. H. D. Roller, *The Development of the Concept of Electric Charge* (Cambridge, Mass., 1954). See also Cohen, *Franklin and Newton* (see note 1), 385–6.

[71] Priestley, *History of Electricity* (see note 65), 243.

time the use of solvent techniques would complement more destructive methods.[72] This transformation, therefore, a fundamental one for the development of the practice of matter-theory, was determined by social, institutional and cultural changes as much as by an objective decision. This is a matter of historical analysis, not moral condemnation.

The essence of this natural philosophical attitude, I believe, was the dramatization of the anomalous and the critical. Experimental drama was played out in a theatre constrained by factors as social as they were 'scientific'. This is demonstrated most clearly in the role of matter-theory in experimental practice. The status of fire as a privileged analyst was called into debate: fire itself, and its manifestations as electricity, light, and other forms of active matter, were not merely the conceptual tools of a philosophical discourse, they provided the base for a complete cosmology of nature and of practice. On the one hand, in matter-theory in the eighteenth century, fire carried problematic contradictions between the preservation of nature and the transformation of nature, since fire had this dramatic effect upon matter in experience.[73] On the other hand, fire and its manifestations provided a vocabulary of analogy and metaphor for natural philosophers to use: a clear example of this can be found in a letter from William Watson, Jr, to William Herschel in 1785, suggesting that the way in which 'powdered Rezin sprinkled on an Electrophorus is formed into Stars like those of Snow in the electric field' was applicable to the regular distribution of the stars, a problem which Herschel was sure was connected with central forces in matter.[74] Similarly,

[72] F. L. Holmes, 'Analysis by fire and solvent extractions: the metamorphosis of a tradition', *Isis,* LXII (1971), 129–48; A. Debus, 'Fire analysis and the elements in the 16th and 17th centuries', *AS,* XXIII (1967), 127–47.

[73] D. M. Knight, 'The vital flame', *Ambix,* XXIII (1976), 5–15; Schaffer, 'The phoenix of nature' (see note 32). Compare the statements of William Hamilton, *Observations on Mount Vesuvius, Mount Etna and other volcanoes* (1772), 161: 'May not subterraneous fire be considered as the great plough . . . which Nature makes use of to turn up the bowels of the Earth, and afford us fresh fields to work upon, whilst we are exhausting those we are actually in possession of?', with William Cleghorn's *De igne* (1779), published in a text edited by D. McKie and N. Heathcote, *AS,* XIV (1958), 1–82.

[74] William Watson to William Herschel, 21 January 1785 (Herschel archive, Royal Astronomical Society manuscripts, Churchill College, Cambridge, W.1/W.36 (i)). 'I accidentally met with a fact which perhaps may be found to have some relation to the distribution of the fixt stars, and to form an analogy which may assist in ascertaining it. It is that powdered Rezin sprinkled on an Electrophorus is formed into Stars like those of Snow by the electric fluid.' The point was to demonstrate that because of the repulsive power inherent in matter, Watson claimed that 'Nature affects . . . the figure of a Hexagon'. See

Adam Walker, a follower of Herschel in ideas of active power of fire and the possibility of life existing on the sun, used his public lectures on the properties of fire to elaborate on the thesis that 'Nature never sleeps... no two enemies are more inveterate than heat and cold', and applied these lessons quite explicitly to cosmology, the life of the planets and stars, and, most importantly, to the philosophical basis of radicalism in the state – a problem for the Britain of 1799.[75]

Having argued, therefore, that the anomalous and the marvellous are situated at the heart of the natural philosophical project, and that this has been the cause of the interest in this period of such philosophers as Kuhn and Bachelard, we can understand the way in which this model dominates the history of natural philosophy only by further investigation of the act of classification and the systematics of that philosophy itself. Alongside their attention to the role of the individual in natural philosophy, these commentators notice that since natural philosophy 'searches not for variation but variety', the act of systematizing, compiling facts, classifying, is the obverse of the search for the individual.[76] In matter-theory, as Priestley, Boscovich, Lavoisier and many other natural philosophers pointed out, the fundamental task was the classification of types: in this classification a central-force theory of action would specify a particular reduction base for the taxonomy.[77] Canguilhem has pointed out that the death of systematics and of the 'singularity' marked the emergence of biology as a science. Before this break, natural philosophers such as Blumenbach declared that the 'aberrations of Nature out of its course often shed more light on obscure researches than does its ordinary

W. Herschel, 'On the construction of the heavens', *PT*, LXXV (1785), 213–66 (composed 1 January 1785), and S. J. Schaffer, 'Herschel in Bedlam: natural history and the formation of stellar astronomy' (forthcoming).

75 Walker, *System of Familiar Philosophy* (see note 32), viii, 15. 'In the order of nature, we find opposing or antagonistic principles in a state of perpetual warfare ... Nature never sleeps.' For this trend in matter theory, see A. L. Donovan, *Philosophical Chemistry in the Scottish Enlightenment* (Edinburgh, 1975), and D. M. Knight, 'Steps towards a dynamical chemistry', *Ambix*, XIV (1967), 179–97.

76 Bachelard, *Formation* (see note 47), 30. Compare G. Canguilhem, *La connaissance de la vie* (Paris, 1965), 178.

77 Priestley, 'An examination of Dr Reid's Enquiry into the human mind...' (1775), in McEvoy, 'Joseph Priestley' (see note 23), 135; Boscovich, *Theory of Natural Philosophy* (see note 5), 185; Crosland, *Historical Studies in the Language of Chemistry* (see note 62), ch. 5; D. M. Knight, *The Transcendental Part of Chemistry* (1978), 250–1; B. Sticker, '"Artificial" and "Natural" classifications of celestial bodies in the work of William Herschel,' *Actes du Xe Congrès international d'histoire des sciences* (Ithaca, 1962), II, 729–31.

and regular course'.[78] Buffon, who applied this idea to matter-theory as well as to other branches of natural philosophy, pointed out that 'species and classes are nothing but ideas which we have ourselves formed and established'. This point is explicitly connected with the practice of natural philosophy as a whole:

> Nature being contemporary with matter, space, and time, her history is that of every substance, every place, every age, and although it appears at first glance that her great works never alter or change... One sees, on observing more closely, that her course is not absolutely uniform; one realizes that she undergoes successive alterations, that she lends herself to new combinations, to mutations of matter and form; finally, that the more fixed she appears in her entirety the more variable she is in each of her parts; and if we comprehend her in her full extent, we cannot doubt that she is very different today from what she was in the beginning and what she becomes in the course of time. These are the different changes we call her epochs.[79]

Since these epochs above all *classify* natural forms, this declaration stands as a manifesto of natural philosophy as the discourse of order, of the coming-into-being of system through the presence of the mutation and the existence of the individual.

It has often been pointed out that classification and speciation played an important role in natural philosophy. Roller and Roller, Kuhn's main source for his understanding of eighteenth-century electricity, stress the importance of fact-gathering at this stage of science, and note the achievement of Dufay in classifying types of electrical materials.[80] Priestley's announcement that 'the great business of philosophy is to reduce into classes the various appearances which Nature presents to our view'; Condillac's psychology, with its revelation

[78] Canguilhem, 'Du singulier et de la singularité en epistemologie biologique', in *Etudes* (see note 62), 211–25, citing, amongst others, Blumenbach on variation, and d'Alembert on the magnet.

[79] Buffon, *Epoques de la nature* (Paris, 1780), 3–4. On the connections between Buffon's matter theory and his deism, which determined his view of what cannot be described as 'evolution', see F. Courtès, 'Pour une psychanalyse de l'evolutionnisme', in *Hommage à Gaston Bachelard* (Paris, 1957), 168–70; P. J. Bowler, 'Evolutionism in the enlightenment,' *HS*, XII (1974), 159–83 (p. 173); Metzger, *Newton, Stahl, Boerhaave* (see note 1), 60–2; R. Wohl, 'Buffon and new science,' *Isis*, LI (1960), 186–99.

[80] Roller and Roller, *Electric Charge* (see note 70), 71; C.F. de C. Dufay, 'A letter to Charles Duke of Richmond and Lennox concerning electricity', *PT*, XXXVIII (1734), 258–9.

of the connection between symbolic order and the possibility of acquisition of Nature; Lavoisier's seminal reform of natural history; all these are familiar characteristics of natural philosophy.[81] Yet Gillispie had admitted that 'Taxonomy little tempts the historian of scientific ideas. The problems were fussy and practical, but the question whether classifications are natural or artificial did not ultimately prove interesting'.[82] As a consequence, historians have adopted a heroic and conceptual version of natural philosophy, in which exchange takes place between ideas in a unified Newtonian world, while philosophers have merely defined natural philosophy as a negativity, as *not* science, without confronting it as an historic practice.

IV. Natural philosophy as the object of an archaeology and anthropology

Out of the dilemma of historical and philosophical attitudes to natural philosophy, several new approaches have been adopted which, typically, seem to derive from other attendant disciplines. Though these approaches claim to be exhaustive with respect to the history of science – in the sense that they claim to be applicable to all periods and to all social formations – once again, significantly, eighteenth-century natural philosophy figures as the main exemplar. Notably, some historians have used the ideas of Mary Douglas, Robin Horton, and other cultural anthropologists as clues to unravel the *cosmologies* of natural philosophers, while Michel Foucault has constructed an 'archaeology of knowledge' with which to analyse the structure of natural philosophy as a set of *discourses*.[83] These contrasting approaches derive from two opposed epistemologies. The anthropology of natural philosophy emerged as a response to the problems of unify-

[81] On Condillac, see I. F. Knight, *The Geometric Spirit* (1968).

[82] C. C. Gillispie, *The Edge of Objectivity* (Princeton, 1960), 170.

[83] On anthropology in science, see M. Douglas, *Purity and Danger* (1966); *idem*, *Natural Symbols* (1970); *idem*, *Rules and Meanings* (Harmondsworth, 1973); R. Horton, 'African traditional thought and Western science', *Africa*, XXVII (1967), 50–71, 155–87; M. Eliade, *Myth and reality* (London, 1964); S. B. Barnes and S. Shapin, 'Where is the edge of objectivity?', *BJHS*, VII (1977), 61–6. On the archaeology of knowledge, see M. Foucault, *The Order of Things* (1970); *idem*, *The Archaeology of Knowledge* (1972); *idem*, *L'ordre du discours* (Paris, 1971); *idem*, *Language, Counter-memory, Practice*, trans. D. F. Bouchard (Oxford, 1977); *idem*, 'La situation de Cuvier dans l'histoire de la biologie', *RHS*, XXIII (1970), 63–9; *idem*, 'Politics and the study of discourse', trans. C. Gordon, *Ideology and Consciousness*, III (1978), 7–26.

ing natural philosophy under headings like 'Newtonian matter-theory': in the introduction to their representative collection of papers adopting such a cultural approach, Barry Barnes and Steven Shapin have written of 'a wide range of individuals' whose writings have been 'properly characterized as a *movement* within a culture', and have announced that this anthropological–cultural approach 'closes no evaluative or political options: it merely ejects them from historical practice'. On the other hand, the archaeology of knowledge mobilizes Bachelard's epistemological history of the sciences in emphasizing the rupture between natural philosophy and modern science. The consequences of this contrast for a possible history of natural philosophy are obviously complex.

To treat eighteenth-century natural philosophy from the anthropological viewpoint places ultimate stress on the individual philosophers as subjects. Each philosophical subject is analysed biographically, and his particular cosmology (which is to be seen as embracing social, political, theological and philosophical determinations) actualized within the particular social place he occupied. We can therefore understand this technique as a response to the need to connect the unity of Newtonianism with the complexity of the cultural base. Barnes and his colleagues have stressed the closer cultural connections of natural philosophy than of science: Cantor and others have shown how misleading it has been to attempt to unify matter-theory as a *Newtonian* tradition.[84] Such extreme divergences as those represented by the Hutchinsonians, by William Law and his disciples (who compared the use of reason to 'extinguishing fire by reading it a lecture on the nature of water'), and even parallel philosophical attitudes such as those of Leibniz and Descartes, evident in the work of theorists such as Diderot and Euler, make it impossible to enforce this unity.[85] The response has been to adopt the prosopographical tech-

[84] A. N. Cantor, 'The history of "Georgian" optics', *HS*, XVI (1978), 1–21, and above, note 32. For the use of anthropology of culture in the history of science, see S. B. Barnes, *Scientific Knowledge and Social Theory* (1974), esp. 121–2; D. Bloor, *Knowledge and Social Imagery* (1976), 4–5ff.; and the texts in S. Shapin and B. Barnes (eds.), *Natural Order* (1979). The statement quoted here is at p. 10.

[85] A. J. Kuhn, 'Nature spiritualized' (see note 36), 409, for William Law on reason: some aspects of rival traditions are explored in T. L. Hankins, *Jean d'Alembert: Science and the Enlightenment* (Oxford, 1970); A. Vartanian, *Diderot and Descartes: Scientific Naturalism in the Enlightenment* (1953); W. H. Barber, *Leibniz in France* (1955). All show how the allegedly 'Newtonian' Enlightenment drew on remarkably diverse and self-contradictory sources. O. Wade, *The Clandestine Organisation and Diffusion of Philosophic Ideas in France from 1700–1750* (New York, 1938), is still a fascinating account of the *diffusion* of complex and distinct traditions.

nique most successfully used by Namier and his disciples.[86] The fundamental problem is that whereas for Heimann, McGuire, and other historians the basic categories are the *concept* and the *tradition,* for these anthropological and sociological historians the fundamental categories become the individual *subject* and the *cosmology.* It is precisely these two categories – the individual *subject* and the *cosmology* – which Foucault has criticized most forcefully in his attack upon traditional historiography.

Foucault's response to conventional versions of natural philosophy is based upon a critique of the constitutive subject and the way in which this subject is used to unify discourse around categories like 'author', or 'philosopher', or 'scientist'.[87] He acknowledges that when this category is rejected, 'the cry goes up that one is murdering history . . . what is being bewailed is that ideological use of history by which one tries to restore to man everything that has unceasingly eluded him for over a hundred years'.[88] More is at stake for Foucault, therefore, than the mere comprehension of natural philosophy. By defining the discourse as the object of archaeology, Foucault claims to redraw the intellectual map of natural philosophy, where practical criteria of the epistemological construction of the object of discourse take precedence over the subjective predilections of individual authors (that is, in Foucault's terms, 'author-functions'.) Specifically, the order of discourse established in the early seventeenth century was not, for Foucault, broken at the Scientific Revolution, but instead subsisted as the determining structure of practice until a further epistemological break towards the last decade of the eighteenth century. In those empirical fields 'being formed and defined for the first time'

[86] On prosopography see S. Shapin and A. Thackray, 'Prosopography as a research tool in history of science: the British scientific community 1700–1900', *HS,* XII (1974), 1–28. On 'Namierism' see Sir Lewis Namier, 'History', in his *Avenues of History* (1952), 1–10; *idem, The Structure of Politics at the Accession of George III* (1929); R. Walcott, *English Politics in the Early 18th Century* (Cambridge, Mass., 1956); and a rival version in C. Nicolet, 'Prosopographie et histoire sociale', *Annales: économies, sociétés, civilisations,* III (1970); and N. Annan, 'The intellectual elite', in J. H. Plumb (ed.), *Studies in Social History* (1955), 243–87.

[87] Foucault, 'What is an author?', in *Language, Counter-memory, Practice* (see note 83), 114–38.

[88] Foucault, *Archaeology of Knowledge* (see note 83), 14. Foucault sees that 'the inalienable rights of history and the foundations of any possible historicity' have been conventionally situated in the *subject:* to treat discourse as the object is to problematize this identity, and hence Foucault indulges in an immediate critique of, and break with, precisely the ideology which 'collective biography' espouses, and that ideology is unmistakably at the root of anthropologically based, 'naturalist' approaches to science.

in the eighteenth century (chemistry, electricity, forms of life science) Foucault stresses not 'mechanism' or 'mathematicization' but the 'science of order' as constitutive.[89] This constitution takes place at the level of the linguistic and practical conception of the relation between the network of signs developed by natural philosophy and the object in nature. Conventional divisions in historical understanding are broken up: archaeology cannot be distinguished from epistemological history by contrasting 'disciplines that are not really sciences' from 'sciences that have been formed on the basis of – or in spite of – existing disciplines'.[90] Instead, Foucault definitively rejects all the categories which Anglo-American historians (for example) have deployed in their attempt to pull together the contradictory discourses of natural philosophy: 'under cover of the empty and obscurely incantatory phrases "Cartesian influence" or "Newtonian model", our his-

[89] Foucault, *Order of Things* (see note 83), 42–3 (on the division between Renaissance and Classical orders of discourse); *ibid.*, 56–8 (on the emergence of new discourses in the 18th century); *ibid.*, 72–3 (on the structure of classical discourse as a science of order). 'This relation to . . . the general science of order does not signify that knowledge is absorbed into mathematics, or that the latter becomes the foundation for all possible knowledge; on the contrary, in correlation with the quest for a *mathesis* we perceive the appearance of a certain number of empirical fields now being formed and defined for the first time. In none of these fields, or almost none, is it possible to find any trace of mechanism or mathematicisation: and yet they all rely upon a possible science of order.' For Foucault, the break with the constitutive subject also involves a break with Bachelard's *psychologism;* even if the practice which can investigate what lies back beyond the epistemological break of 1795–1815 can only be an archaeology, it is nevertheless *possible* to investigate natural philosophy.

[90] Foucault, *Archaeology of Knowledge* (see note 83), 178. On the system of signs and natural philosophy, see *Order of Things,* 144ff. On language and nature – the root of natural philosophy – Foucault writes (pp. 160–2): 'Between language and the theory of nature there exists a relation that is of a critical type: to know nature, in fact, is to build upon the basis of language a true language.' This true language appeared through natural history: the problem had changed from asking how it was possible to know that a sign *did* in fact designate what it signified, to (in the Classical era) asking *how* it could be linked to what it signified. The answer: through natural philosophy. This understanding yields at least two results: Foucault argues that the instrumentational aspect of natural philosophy should be viewed as *restrictive* on the area of experience ('Natural history did not become possible because man looked harder and harder'), and he also argues that the evolutionism of the eighteenth century must be seen for what it was: a conception of the summoning into existence of a pre-established order (pp. 125–7) and *not* a development of nature through history. Both of these claims follow from his insistence on the imposition of *order* by natural philosophy on nature. An effective application of some of these insights may be found in Albury and Oldroyd, 'From Renaissance mineral studies to geology' (see note 62).

torians of ideas are in the habit of . . . defining Classical rationalism as the tendency to make nature mechanical and calculable'.[91] Even the 'more perceptive' historians who see the play of contrary forces inside rationalism are condemned. Foucault situates the entire discourse at a completely different place with respect to nature: this is the place which, following some of Bachelard's insights, I have characterized in terms of the individual and the system.

The consequences of this for history of science and its confrontation with natural philosophy and matter-theory are profound. Foucault emphasizes the rules and the discipline of the production of knowledge: 'la discipline est un principe de contrôle de la production du discours'. In a more recent text, *Surveillir et punir*, Foucault has investigated the articulation between the investigatory patterns of Inquisition and Prison, the forms of distribution of power which these institutions presuppose, and produce, and, finally, the associated patterns of acquisition of knowledge, in both the natural and the human sciences, which emerge through these systems of power. Foucault showed, therefore, that the social–political formation of an activity like natural philosophy is located at the place where it connects with knowledge as power ('savoir-pouvoir') and the rules of distribution of this power. Where Gillispie has insisted that 'the permeation of culture by science must be a problem in accommodation rather than a study in validity', Foucault insists that validity itself is a matter of the policing of discourse, of cultural formation. For a statement to be 'in the domain of the true' it must conform in multiple ways to a system which distributes the right to state anything and what is to be stated.[92]

We can therefore contrast the conventional historiographical picture of a free terrain in which a natural philosopher selects his position, his cosmology, and his form of expression of that cosmology according to *nature*, or (as would be the anthropological version) according to biography. For Foucault, and other archaeologists, the

[91] *Order of Things* (see note 83), 56–7. Compare his attack upon historiography in 'Politics and the study of discourse', (see note 83), 8–9: 'The discourse cannot restore the totality of its history within a strict framework'. That is, Foucault's argument that there is no necessary correspondence between the actors' categories and those of the historian seemed to make an 'archaeology' necessary. 'Diderot's *La rêve de D'Alembert* does not belong to the domain of scientificity of Natural History, but it does belong to its archaeological territory . . . Archaeological territories may extend to literary or philosophical texts as well as scientific' (*Archaeology of Knowledge* (see note 83), 183).

[92] Foucault, *Ordre du discours* (see note 83), 10 (on policing of discourse) and 32–3ff. (on the constitution of truth); *idem*, *Discipline and Punish* (1977), 'Panopticism'; Gillispie, *Edge of Objectivity* (see note 82), 156.

terrain is constructed such that the enunciation of 'true' statements about nature, whether in matter-theory or in natural history, is produced and organized by specifically *im*personal structures. These are obviously opposed philosophies just as much as they are alternative methods of history. As natural philosophers suggested in the eighteenth century, when they spoke of 'the foundation of morality in the constitution of things', a choice of historical methods in science is a choice of attitudes to the articulation of science and social relations.

3

Social uses of science

STEVEN SHAPIN

Contents

I. Introduction 93
II. The strategy of Newtonianism 96
III. Intellectualist historiography revisited 105
IV. Strategies of Enlightenment matter-theory 111
V. The strategy of physiological thought 124
VI. Conclusions and continuations 130

I. Introduction

Representations of nature in scientific culture have been widely used to comment upon the social order. Such social uses have not, however, greatly interested historians of science. They routinely consider their business to be the study of the generation and evaluation of scientific knowledge. Social uses of scientific knowledge are conceived to occur in another context from the contexts of production and evaluation of science. How science may be socially used can, therefore, be of little material concern either to the scientists who produce the knowledge, or *a fortiori* to the historians who study the scientists. Thus, the significance accorded to studies of the social uses of science rests upon an historiographical demarcation between contexts of social use and contexts more routinely treated by historians of science. And, as is often the case, the demarcation expresses an evaluation – in this instance a negative assessment of the historical significance of studies of social uses of science. These evaluations and the attendant demarcations are faulty. It will therefore be necessary to state as clearly as possible, even at the risk of some oversimplification, what these largely implicit historiographical assumptions are.

These include the notion that individuals in an esoteric sub-culture generate scientific knowledge by contemplating nature and 'rationally' assessing their findings. The context wherein science is produced and judged is argued (or, more commonly, assumed) to be separable from other contexts. Yet, once science is generated and evaluated by individuals in the esoteric sub-culture, it may passively sift into the wider social and cultural context where its manifest truthfulness is a sufficient reason for its acceptance as an accurate account of natural reality. Once thus accepted, scientific accounts of 'how nature is' become candidates for social and ideological 'extrapolation', and, evidently, have been widely used in such contexts. However, the context in which science is produced and judged is both discrete from and formally antecedent to the context of social use. We may call this the 'illumination' model of science and its uses, and its epigrammatist is Alexander Pope. Interestingly, even certain historians not notably sympathetic to some of the classical divides of historical discourse seem still to accept discrete contexts of production and reception. Thus Henry Guerlac has recently argued that study of the reception or 'legacy' of Newton's science has no bearing upon the 'fatuous' internalist–externalist controversy, precisely because it 'projects outwardly from science . . . to observe its reception by society, instead of pointing inwardly to seek social, economic and intellectual influences upon a scientist's work'.[1]

But there is more to the general set of assumptions about the social uses of science than this 'analysis' of demarcations between contexts of use and contexts of production and judgment: social uses of scientific accounts are not only, in this conception, separable from scientific contexts proper; they may also have an actively pernicious character, and are, in general, to be deplored rather than described and explained. (It is this particular assessment which probably accounts for the relative paucity of such studies.) People not privy to the esoteric sub-culture's 'rules' for finding and evaluating knowledge-claims are liable to 'misunderstand' the 'real (esoteric) meaning' of scientific accounts. Or they may wilfully introduce into the esoteric findings 'subjective', 'ideological', or 'irrational' components, thereby corroding scientific objectivity. This could not possibly be described as a passive process. But neither is it a process traditionally regarded as proper for the serious study of historians of science; it belongs, if anywhere in the academic world, to the study of 'error', of man's benightedness, rather than to his capacity for illumination. These views are articulated with unusual clarity in some of the writings of Charles Gillispie (to be discussed below), who eloquently verbalizes what seem to be the implicit operating assumptions of many historians of science. For Gillispie such social uses of science are to be exhibited as cautionary tales, as the history-of-science analogue of Aesop's fables, that they should not happen again, or as illustrative

[1] Henry Guerlac, 'Some areas for further Newtonian studies', *HS*, XVII (1979), 75–101 (p. 75).

materials which serve to display the real and proper demarcations between legitimate science and ideological follies.[2]

These orientations have not contributed to our present understanding of the social uses of science, and, indeed, may have actively discouraged serious description and explanation of such historical phenomena. One can only speculate that the orientations are sustained not by a concern to describe naturalistically science as an historically situated cultural enterprise, but by an interest in celebrating it and defending it from contamination. 'God said "Let Newton be", and all was light' is passable poetry, but lazy historiography.

In fact, it is in the area of Newtonianism and its career in the eighteenth century that such perspectives show their greatest inadequacies and where new notions of science and its uses display greatest promise. For, if the eighteenth century is seen (at least partly) as a period in which seventeenth-century scientific achievements were accredited and implemented, then a central question for historical research concerns the conditions in which Newtonian natural philosophy was received and institutionalized.[3] Was the new philosophy (or antipathetic reactions to it) considered to be of social use? Of what relevance were such social uses to the production, evaluation and institutionalization of new science? Does sensitive consideration of the social uses of Newtonianism shed new light upon the career of natural philosophy in the period from *c.* 1680 to *c.* 1800?

Over the past ten years or so there has appeared a small, but highly

[2] C. C. Gillispie, *The Edge of Objectivity* (Princeton, N.J., 1960), and, especially 'The *Encyclopédie* and the Jacobin philosophy of science: a study in ideas and consequences', in Marshall Clagett (ed.), *Critical Problems in the History of Science* (Madison, Wisconsin, 1959), 255–89.

[3] In the common usage of historians of science the word 'institutionalization' refers to the processes whereby scientific activity is incorporated into formal organizations and socially recognized roles. By contrast the sense 'institutionalization' is intended to convey here is that usually employed by sociologists and social anthropologists, viz. the processes by which knowledge and associated behaviours are established as the basis for *standardized collective* reference and action. There is a great variety of definitions available, but the present usage draws upon what is common to most of them; see, for example, G. Duncan Mitchell (ed.), *A Dictionary of Sociology* (1968), 99–101. In spite of the slight possibility of confusion with the dominant history of science usage, the term is preferred to 'reception' or 'diffusion', which inappropriately suggest *passive* processes. To refer, for example, to Newtonianism as a social institution in the eighteenth century is entirely correct, and its recognition as such stimulates a search for those active historical processes by which it was so established. But, it should be emphasized, to regard eighteenth-century Newtonianism as an institution is *not* to imply that 'everyone believed in it'; think of *marriage,* one of sociologists' favourite examples of a social institution.

significant, body of writings on the production, evaluation and institutionalization of Newtonianism which has been structured around questions like these. This corpus rejects currently dominant notions of the social uses of science. Instead, it explains the career of Newtonian natural philosophy entirely in terms of its uses in specific historical contexts. Here we have what may be called a 'new contextualist' tradition, in which writers have been able to explain the various evaluations of Newtonian science in impressive historical detail.

The purposes of the present account are, firstly, to identify and to make more explicit the operating procedures of contextualist writing; secondly, to relate its findings to some of the basic concerns of the historiography of science, 1680–1800; and, finally, to suggest ways in which changing historical perspectives of the social uses of science in this period may draw upon, and connect with, certain valuable tendencies in the social sciences.

II. The strategy of Newtonianism

It should not by now be surprising to historians of science that some of the bolder speculations about the links between natural knowledge and social developments often come from scholars not mainly concerned with science at all, and, therefore, not well-versed in the accepted categories of our discipline. Fifteen years ago E. P. Thompson, in a typically challenging aside, reckoned that 'the bourgeois and the scientific revolutions in England . . . were clearly a good deal more than just good friends'.[4] We could do worse than to take this remark as our text, for much of the work on the social uses of science to be discussed here refines and expands the nature of this kinship.

Some of the recent studies concerned with the social use and social context of Newtonianism have emphasized the importance of attending to its 'foreign politics', while others have stressed the significance of 'domestic' political considerations. In a brief but provocative paper Arnold Thackray rejected the all-sufficiency of 'intellectual aspects' for explaining the 'development of natural philosophy in the eighteenth century'.[5] The priority disputes between Newton and

[4] E. P. Thompson, 'The peculiarities of the English', in his *The Poverty of Theory and Other Essays* (1978; orig. publ. 1965), 35–91 (p. 60).

[5] Arnold Thackray, '"The business of experimental philosophy": the early Newtonian group at the Royal Society', *Actes du XIIe congrès international d'histoire des sciences* (Paris, 1970–1), III.B, 155–9; cf. *idem*, *Atoms and Powers: An Essay on Newtonian Matter-theory and the Development of Chemistry* (Cambridge, Mass., 1970), 52–3.

Leibniz and, more generally, the evaluation and institutionalization of Newtonian philosophy in early eighteenth-century Britain, cannot be understood without examining the dynastic politics of the period from the 1680s to the 1710s. The controversy over the calculus, the suggestions of Leibniz's plagiarism, the episode of the *Commercium epistolicum*, and the depth of attachment of the Royal Society circle to Newton's natural philosophy, reflect, as Leibniz himself said, not 'a quarrel between Mr Newton and me, but between Germany and England'.[6] Thackray reminds us of the political background of uncertainty concerning the Protestant succession to the crown following the Glorious Revolution of 1688, and of the probability (later, near-certainty) of the succession of the Elector of Hanover in the early years of the eighteenth century. As Newton was the 'autocrat' of English science, so Leibniz was the 'court philosopher' of Hanover. When it became likely that Leibniz would be 'translated', along with the Hanoverian court, to London, Newton set in motion a sustained collective effort to discredit the worth, religious significance, and originality of the German's science. Differences in 'metaphysical style' and scientific worth are simply not adequate to understand the nature and course of the Newton-Leibniz disputes; the institutionalization of Newton's science as the '*Philosophica Britannica*' must be explained by making reference to a context of foreign politics and a contest for control of 'the business of experimental philosophy'. Additional support for Thackray's suggestions comes from Frank Manuel's detailed accounts of Newton 'as autocrat of science'.[7] He shows how Newton energetically manipulated people and institutions in an attempt to establish his natural philosophy in a position of dominance. Manuel's work lends general support to Thackray's contention that the Newton-Leibniz affair reflected the macro-context of the Protestant succession, while it also underscores the vital significance of secure sources of patronage in a pre-professionalized scientific enterprise.

The intriguing notion that Newton's own *motivation* was not merely proprietary but party-political comes from a challenging paper by George Grinnell, who argues that the *Principia* itself should be seen

[6] Letter from Leibniz to Princess Caroline, May 1715, quoted in Thackray, '"The business of experimental philosophy"' (see note 5), 157. For an excellent account of the controversy which, however, makes no reference to the political context, see Carolyn Iltis, 'The Leibnizian-Newtonian debates: natural philosophy and social psychology', *BJHS*, VI (1973), 343–77.

[7] Frank E. Manuel, *A Portrait of Isaac Newton* (Cambridge, Mass., 1968), 264–91; also *idem*, 'Newton as autocrat of science', *Daedalus* (Summer 1968), 969–1001 (p. 997).

as 'Whig propaganda'.[8] During the 1680s, Grinnell says, Newton's Whiggism was primarily expressed in religious terms, specifically in anti-Catholicism. While Newton was a student at Cambridge, the Restoration not only brought Charles II to the throne, but aligned James II as successor. On the Continent the Catholic Church seemed to grow in power, and the fate of the French Huguenots was a matter of acute concern. Grinnell suggests that Newton's anti-Catholic Whiggism is essential to understanding the development of his universal law of gravitation and the specific form it took.

Unfortunately, Grinnell fails to provide the sort of empirical evidence which would be required to convince most historians. Equally regrettably, his attempt at provocation has met with deafening silence from those historians who are in a position to refute or support his case.[9] Nevertheless, the general form of Grinnell's argument ought to be borne in mind, viz. that one cannot understand scientific judgements without attending to the context wherein scientific accounts were deployed. In this case, the claim is that Newton himself took into reckoning the social uses of scientific knowledge in deciding upon the value of various possible accounts.

This historical background of Restoration party- and religious-politics, the Glorious Revolution, and the dynastic and domestic conflicts over the succession also provide Margaret Jacob with the relevant context for her important reassessment of the institutionalization of Newtonianism.[10] She vigorously rejects Pope's 'illumination'

[8] George Grinnell, 'Newton's *Principia* as Whig propaganda', in P. Fritz and D. Williams (eds.), *City and Society in the 18th century* (Toronto, 1973), 181–92.

[9] I have located no explicit and concerted criticisms of Grinnell's essay by Newton scholars. The two historians best equipped to evaluate his claims only say that 'we do not agree with Grinnell's conclusions': J. R. Jacob and M. C. Jacob, 'The Anglican origins of modern science: the metaphysical foundations of the Whig constitution', *Isis,* LXXI (1980), 251–67 (p. 263, n. 58).

[10] Most of the relevant material is to be found in Margaret C. Jacob, *The Newtonians and the English Revolution 1689–1720* (Hassocks, Sussex, 1976). But, among her articles, see especially: 'Bentley, Newton, and providence: (the Boyle Lectures once more)', *JHI,* XXX (1969), 307–18 (with Henry Guerlac); 'John Toland and the Newtonian ideology', *Journal of the Warburg and Courtauld Institutes,* XXXII (1969), 307–31; 'The Church and the formulation of the Newtonian world-view', *Journal of European Studies,* I (1971), 128–48; 'Scientists and society: the saints preserved', *ibid.*, 87–90 (with J. R. Jacob); 'Political millenarianism and Burnet's *Sacred theory*', *Science Studies,* II (1972), 265–79 (with W. A. Lockwood); 'Seventeenth century science and religion: the state of the argument', *HS,* XIV (1976), 196–207 (with J. R. Jacob); 'Millenarianism and science in the late seventeenth century', *JHI,* XXXVI (1976), 335–41; 'Newton and the French prophets: new evidence', *HS,* XVI (1978), 134–42; and others cited in notes 9, 55 and 58. Margaret Jacob's interpretations have already

model of the diffusion of the new natural philosophy and replaces it with the notion that conceptions of nature are *tools*, instruments which historical actors in contingent settings pick up and deploy in order to further a variety of interests, social as well as technical. Disinterested contemplation and the 'manifest' truthfulness of Newtonian natural philosophy are, she shows, inadequate to explain the acceptability of the science to sectors of English society in the years following the Glorious Revolution. Instead, she looks to Newtonian representations of natural reality as actively manipulated, and actively judged, in the hurly-burly of political life. In order to make the features of a defined historical context do explanatory work for her, Jacob dispenses with vague locutions like 'people believed', or 'the eighteenth century needed'. In their place she provides precise identification of practical social problems faced by specific groups with particular social interests.

Newtonian natural philosophy was built upon, and incorporated, the corpuscular theory of matter expounded by Robert Boyle during the Civil War years and adopted by the Royal Society after the Restoration. It is, unfortunately, outside the ambit of this essay to provide a detailed account of the social uses of mid-century matter-theory. Nevertheless, a brief sketch is essential, for Margaret and especially James Jacob have convincingly argued that the development of natural philosophy in the late seventeenth and early eighteenth century was powerfully shaped by the social uses of natural knowledge during Civil War, Interregnum and Restoration.[11] James Jacob's brilliant contextualist work on Boyle and his circle demonstrates that Boyle, representative of latitudinarian circles generally in the 1650s and 1660s, was overwhelmingly concerned to devise a cosmology

provoked some controversy and will likely continue to stimulate critical reaction for some time. In the present connection the aim is to draw out and render more explicit Dr Jacob's general interpretative thrust. Early critical comment on her book suggests that this is sound, while her specific identification of intellectual positions with historical groupings may require refinement; see, for example, Geoffrey Holmes, 'Science, reason, and religion in the age of Newton', *BJHS*, XI (1978), 164–71; Christopher Hill, 'Scientists in society', *New York Review of Books*, XXV, no. 19 (7 December 1978), 37–9; Roy Porter, '*Review*', in *Social History*, III (1978), 246–9; and P. M. Heimann, 'Science and the English Enlightenment', *HS*, XVI (1978), 143–51, who concludes that the 'vital role' of social factors in the institutionalization of Newtonianism 'can undoubtedly be accepted' (p. 149) – an especially pertinent evaluation from a scholar whose own work on Newtonianism has given little or no attention to such factors (see below and note 37).

[11] M. C. Jacob, *The Newtonians* (see note 10), ch. 1; J. R. Jacob, *Robert Boyle and the English Revolution: A Study in Social and Intellectual Change* (New York, 1977), and additional J. R. Jacob papers cited in notes 12 and 15 below.

which might serve to secure a 'moderate', spiritually governed social order against perceived threats arising from radical sectaries, Hobbists, and other types of philosophical and religious heretics inimical to latitudinarian interests.[12] In these contexts, and, indeed, throughout the period with which we are concerned, the social significance of the natural order was a common denominator of cultural behaviour. Nature was endemically liable to be put to social use, for, in actors' categories, the 'world natural' was a pool of meaning relevant to the 'world politick'.

While Jacob's contextualism focuses most strongly on the ideological uses of corpuscularianism, he does not maintain that its production and evaluation are to be referred solely to its wider social significance. In a particularly concise example, he examines the bases for Boyle's explanation of why water rose in a reed subjected to a partial vacuum.[13] Boyle was vehemently opposed to the view, associated *inter alia* with radical heretics, that the fluid rose because 'it abhorred a vacuum'. To ascribe 'abhorrence' to matter would be to attribute to it properties of 'soul'; and, if one were to endow with such qualities entities not 'acutally' possessed of an immortal and rational soul, then this would imply that man himself did not possess one. If man did not have an immortal soul, distinct from his organic nature, then where was the Church's major sanction on human behaviour? In Jacob's account, the corpuscular theory, in which matter was affirmed to be 'brute and stupid', provided the framework for solving technical problems posed to natural philosophers by the behaviour of liquids in vacuums, *and also* solved the moral and social problems presented to latitudinarians by the heresy of mortalism.[14] Thus, Boyle and his circle dealt with technical-predictive and ideological problems in the same account; his adoption and elaboration of corpuscularianism is made comprehensible by referring it to *both* sets of concerns in a specified historical context.[15]

[12] J. R. Jacob, 'The ideological origins of Robert Boyle's natural philosophy', *Journal of European Studies*, II (1972), 1–21; *idem*, *Robert Boyle* (see note 11), ch. 3.

[13] J. R. Jacob, 'Ideological origins' (see note 12), 17.

[14] *Ibid.*, 18; cf. Christopher Hill, 'William Harvey and the idea of monarchy', in Charles Webster (ed.), *The Intellectual Revolution of the Seventeenth Century* (1974), 160–81.

[15] J. R. Jacob, 'Ideological origins' (see note 12), 12; *idem*, 'Boyle's atomism and the Restoration assault on pagan naturalism', *SSS*, VIII (1978), 211–33 (pp. 213–14); *idem*, 'Robert Boyle and subversive religion in the early Restoration', *Albion*, VI (1974), 175–93; *idem*, 'Restoration, Reformation and the origins of the Royal Society', *HS*, XIII (1975), 155–76; *idem*, 'Boyle's circle in the Protectorate: revelation, politics, and the millennium', *JHI*, XXXVIII (1977), 131–40; *idem*, 'Restoration ideologies and the Royal Society', *HS*, XVIII (1980), 25–38.

Present in the Civil War and Restoration materials are structural themes which appear again in treating Newtonianism and opposed natural philosophies later on. First, it is to be noted that philosophies of nature were routinely seen by the actors as imbued with social meaning. This was not because of 'mere' metaphorical glossing, but because in these (and later) cultural contexts nature and society were deemed to be elements in one interacting network of significances. As Boyle put it, 'each page in the great volume of nature is full of real hieroglyphicks, where . . . things stand for words, and their qualities for letters'.[16] Second, groups with conflicting social interests developed and sustained interestingly different natural philosophies; moreover, these philosophies were often produced explicitly to combat and refute those of rival groups. Third, the distribution of attributes between 'matter' and 'spirit' was an issue of intense concern in all these philosophies; the relations between the two entities seemed to be something upon which all cosmologies 'had to' decide, and the boundaries between 'matter' and 'spirit' were treated as having particularly strong social significance.

J. R. Jacob, and also Christopher Hill, provide especially suggestive materials on the cultural productions of Civil War radical sectaries: Diggers, Levellers, Grindletonians, and the like.[17] Neither historian observes a rigid demarcation between these groups' theological and social views on the one hand and their representations of natural reality on the other. What may be termed 'the sectarian cosmology', social and natural, stressed the principle of 'immanence'. God is virtually present in the faithful; He is also immanent in the material world. Thus matter was not mere inert and passive 'stuff'; it was suffused with spirit, and sensuous experience of it provided a direct, unmediated link between the individual and the Deity. In this there was an important homology between the sectaries' social concerns and their cosmology. But it was not a passive 'expression' of their social situation, for their 'nature' was in fact widely invoked in furtherance of their social aims and against those of other groups.[18] A

[16] Quoted in Jacob, 'Ideological origins' (see note 12), 19.

[17] Especially Hill, *The World Turned Upside Down: Radical Ideas during the English Revolution* (Harmondsworth, 1975), but also *Intellectual Origins of the English Revolution* (Oxford, 1965); *Anti-Christ in Seventeenth-century England* (Oxford, 1971); and 'William Harvey' (see note 14).

[18] For theoretical perspectives on the significance of homologies between the social and the natural, see David Bloor, 'Durkheim and Mauss revisited: classification and the sociology of knowledge', in John Law (ed.), *The Language of Sociology*, Sociological Review Monographs, in press; Steven Shapin, 'Homo phrenologicus: anthropological perspectives on an historical problem', in Barry Barnes and Steven Shapin (eds.), *Natural Order: Historical Studies of Scientific Culture* (London and Beverly Hills, 1979), 41–71.

God immanent within the believer dispensed with priestly hierarchies, such as those of the Anglican Church. A spiritually imbued material world provided a usable vision of a self-moving and self-ordering system, independent of superintendence by spiritual intermediaries. Just as the Diggers' insistence that spirit was immanent in matter constituted an element in their religious and political strategy, so Boyle's latitudinarian response to such views was a strategy crafted to further the interests of groups whose conception of proper social order was very different from that of Civil War sectaries.

While Boyle's 'Christianized Epicureanism' provided the foundations for latitudinarian natural philosophy in the Restoration, his death in 1691 left the legacy of the Boyle Lectures, an enterprise which developed Newtonianism as the strategy of latitudinarian divines from the 1690s to the 1710s. Margaret Jacob's studies of these lectures in their political context are the most sensitive and thorough examinations of the institutionalization of science in a concrete historical context.[19] Once more, as in work on Civil War and Restoration corpuscularianism, the social uses to which Newtonianism could be put, by specific groups, are shown to have had an important bearing on the evaluation and institutionalization of science. The 'context of use' was not, as it were, distal to the 'context of evaluation'. The social group which used the Boyle Lectures as its vehicle consisted largely of Low Church Whigs who faced particular problems of ensuring their authority in the period from 1689 to 1720. How were Whig divines to respond to threats to their moral authority arising from the Toleration Act, an increasingly materialistic civil society, a new generation of radical materialists, and other groups hostile to 'court' prerogatives? And what was to be said in defence of providence, a crucial moral resource of these divines, in a political context which had witnessed, within the lifetime of many, the execution of a king who ruled 'by divine right', a bloody and chaotic Interregnum, the decampment of a Catholic king, the importation of an Orange king, and continuing uncertainties over the Protestant succession?

What aspects of the Newtonian philosophy particularly attracted

[19] M. C. Jacob, *The Newtonians* (see note 10), chs. 4–5; *idem*, 'The Church' (see note 10); *idem* and J. R. Jacob, 'Anglican origins' (see note 9). Jacob's is by no means the only available account of the institutionalization of Newtonianism, but is the only one which attempts to relate the establishment of the new cosmology to a wide range of contextual factors, including political considerations. Compare, for example, John Redwood, *Reason, Ridicule and Religion: The Age of Enlightenment in England 1660–1750* (1976), esp. chs. 1, 2–4; Thackray, *Atoms and Powers* (see note 5), chs. 3–4 (cf. *idem*, '"The business of experimental philosophy"' (see note 5); Robert E. Schofield, *Mechanism and Materialism: British Natural Philosophy in an Age of Reason* (Princeton, N.J., 1970).

Whig Anglicans? Continuing Interregnum and Restoration themes, Margaret Jacob shows that the ontological categories of the new science, and especially its theories of matter and active principles, were central to the social uses of Newtonianism from 1692 to 1714, when the most important of the Boyle Lectures were delivered.[20] Thus Samuel Clarke maintained that matter must be conceived of as 'brute', 'stupid', 'dead' and 'lifeless'; it could not move itself; it possessed no qualities of sentience or self-animation; without some external actuating and ordering agency it could do nothing and could not produce the patterned phenomena of nature.[21] It would be a 'dangerous' error to suppose otherwise, for the exclusive source of motion is God.

Intellectualist historians of science have already documented in abundant detail how the technical problem-solving of the *Principia* and *Opticks* was founded upon Newton's ontologies of matter, force, space and time.[22] Recent contextualist work proceeds to show how key features of Newtonian metaphysics were also deemed essential to solving the social problems confronting the groups which used the Boyle Lectures as the forum for creating and establishing the Newtonian world-view. How were such notions thought to serve social interests? If matter was self-sufficient and self-animating, then *external* spiritual agencies (and their delegates, 'active principles') were not required to produce regularity, order and motion. The Newtonians were explicit in their view that the implication of a 'hylozoist' (or 'pantheistic materialist') cosmology was atheism. 'Active principles' were deemed to be the manifestations of God's active powers in the world natural. They were apart from matter, and superior to it, in the same way that God was apart from and above nature, although work-

20 This brief summary sets aside Jacob's treatment of the Boyle Lecturers' social uses of the mathematical principles of natural philosophy and gravitation, in order to concentrate on the theme of matter-theory; for these other aspects, see M. C. Jacob, *The Newtonians* (see note 10), 137, 188–95.

21 The theme of activity vs. passivity is not confined to matter-theory during this period; Jacob's book largely ignores a parallel body of writings treating the earth as active or passive, analogously with matter; for these see Roy Porter, *The Making of Geology: Earth Science in Britain 1660–1815* (Cambridge, 1977), 71–6; *idem*, 'Creation and credence: the career of theories of the earth in Britain, 1660–1820', in Barnes and Shapin (eds.), *Natural Order* (see note 18), 97–123 (pp. 102–6). The activity/passivity dichotomy also structured the highly charged debates in the late seventeenth and early eighteenth century over spontaneous generation: John Farley, *The Spontaneous Generation Controversy from Descartes to Oparin* (Baltimore and London, 1977), ch. 2.

22 Among many examples, see J. E. McGuire, 'Existence, actuality and necessity: Newton on space and time', *AS*, xxxv (1978), 463–508, and the sources cited in note 40.

ing in it and through it. Such a God, and such a nature, was a crucial cultural resource for legitimating the moral authority of those groups who claimed to interpret God's ways to man, and to exercise their delegated right to sanction behaviour.

Intellectualist historians have already demonstrated that the role of providence in the Newtonian cosmology is properly to be integrated into the ontology of matter and force.[23] Contextualist writings go on to show how a providentialist ontology was an important strategy in dealing with the social and political predicaments of groups which accepted and institutionalized the new science.[24] Providence, according to the Newtonians, was manifest not only in nature's irregularities and catastrophes, but also in its regular, lawful operations. This was the role of providence in the world natural and the world politick which was required (and used) to legitimate the continuing moral authority of latitudinarian divines through the events of 1649, 1660, 1688 and 1707. Newton's God and Newton's ontology were evaluated according to their utility in constructing the apologetics of the new order.

The Jacobs have also argued, although less centrally and in less detail, that such considerations of social use bore upon Newton's own *motivations* in constructing his natural philosophy. Evidence on this point, while less abundant than that available for the voluble polemicist Robert Boyle, is nonetheless highly suggestive.[25] Newton rejected the Cartesian ontology of matter because it 'manifestly offer[ed] a path to Atheism', and he explicitly associated the 'vulgar', hylozoist matter-theory of groups like the radical sectaries with irreligion:

> Indeed however we cast about we find almost no other reason for atheism than this notion of bodies having, as it were, a complete, absolute and independent reality in themselves.[26]

By the 1690s Newton was railing against 'the vulgar' materialists whose theory of self-sufficient matter spawned a 'dwarf-god'. Later

[23] E.g., David Kubrin, 'Newton and the cyclical cosmos: providence and the mechanical philosophy', *JHI*, xxviii (1967), 325–46; J. E. McGuire, 'Force, active principles, and Newton's invisible realm', *Ambix*, xv (1968), 154–208.

[24] M. C. Jacob, *The Newtonians* (see note 10), esp. 136–7, 184–7.

[25] *Ibid.*, 136–7, 156; but especially J. R. Jacob and M. C. Jacob, 'Anglican origins' (see note 9). Historical access to an individual's motivations is a far more hazardous and difficult enterprise than is usually thought. In this connection it can only be pointed out that the availability of documentary material of 'self-imputed' motivations does not, in itself, clinch the case. For reflections on this historiographic problem in another context: Steven Shapin and Barry Barnes, 'Darwin and social Darwinism: purity and history', in Barnes and Shapin (eds.), *Natural Order* (see note 18), 125–42.

[26] Quoted in J. R. Jacob and M. C. Jacob, 'Anglican origins' (see note 9), 262.

still, in the context of his polemical disputes with Leibniz, whose particular hylozoist matter-theory he loathed, Newton developed the view that such solid matter as there was in the world did not amount to very much anyway.[27] He and his followers were agreed that the irreducible 'nut-shell' of matter that remained was 'contemptible' and 'inconsiderable' stuff.[28]

What mainly concern us here are the revisionist implications of recent contextualist work for the historiography of science in the period. Yet this perspective also has consequences for political and constitutional history. Few historians of English politics from about 1660 to about 1730 have had anything to say concerning natural philosophy.[29] If the Jacobs and others are correct in their assessments, constitutional history must take into account the array of legitimating resources available to political actors in that context. And, as has been shown, among these resources Newtonian natural philosophy was prominent. Indeed, the Jacobs have recently gone so far as to argue that Newtonianism provided 'the metaphysical foundations of the Whig constitution' and have given further documentation of the identification between those groups which espoused and disseminated Newtonian natural philosophy and those whose social interests lay in defending and supporting Whig political order and the authority of Low Church Anglican clerics.[30]

III. Intellectualist historiography revisited

If recent contextualist writings, or any of their major findings, come to be generally accepted, then historians of science will be forced to evaluate the respective merits of this approach and intellectualist perspectives of the natural philosophy of the late seventeenth and early eighteenth century. This comparative evaluation will take place in its own time, with a natural stress upon the potential of each perspective for the handling of empirical materials and the generation of fruitful research programmes. Nevertheless, there seems to be a certain historiographic tunnel-vision which hampers explicit confron-

[27] Arnold Thackray, '"Matter in a nut-shell": Newton's *Opticks* and eighteenth-century chemistry', *Ambix*, xv (1968), 29–53.

[28] *Ibid.*, esp. 38–9; McGuire, 'Force, active principles' (see note 23), 192–3.

[29] Among many examples, see J. P. Kenyon, *Revolution Principles: The Politics of Party 1689–1720* (Cambridge, 1977), which makes no mention of Newtonianism or of the Boyle Lectures; cf. the sources cited in note 58.

[30] J. R. Jacob and M. C. Jacob, 'Anglican origins' (see note 9).

tation between distinct perspectives,[31] and it may expedite that process briefly to spell out those features of intellectualist historiography which most clearly distinguish it from the contextualist perspective discussed here.

Ever since Koyré, or even since Burtt and Dijksterhuis, historians of science have generally accepted that the natural philosophy of this period has to be referred to its wider intellectual context. At present it is no longer contentious to maintain that the production and evaluation of science proceeded in a cultural context which included non-natural philosophy, bodies of natural knowledge usually regarded as 'irrational' or 'magical', metaphysics and, especially, religious thought. What has not been widely accepted is the necessity of referring scientific thought to its social and political context. Nothing is quite as handy for the wholesale dismissal of a disliked mode of practice than the availability of an exemplar which may be pointed to at need as particularly crass and sterile. In this connection Boris Hessen's 'Social and economic roots of Newton's *Principia*'[32] has performed yeoman service as straw-man, although whether rejection by *fiat* counts as refutation is another matter.[33] And even Merton's cautious sociological approach to scientific 'foci of interest' has scarcely been confronted, although perhaps he brought that fate upon himself by his reluctance to extend his contextual techniques to the contents of scientific knowledge.[34] However, it ought to be clear that the newer contextualism cannot fairly be tarred with the same brush used for Hessen, for it shares neither his individualism, his positivism nor his determinism, and its sensitivity to the particularities of historical context is unquestionably superior.

The operating procedures and assumptions of the new contextualism are particularly visible, as those of new bodies of practice usually are. Those of institutionalized and routinely applied historical practices are less easy to apprehend, precisely because they count as 'business-as-usual'. Any attempt to elucidate the underlying his-

[31] For example, the central features of the Jacobs' contextualism have been visible in print since 1971, yet intellectualist studies of corpuscularianism and Newtonianism continue to appear which make no reference, even of a critical nature, to their views; e.g., Ernan McMullin, *Newton on Matter and Activity* (Notre Dame, Indiana, and London, 1978).

[32] Orig. publ. 1931; repr. in P. G. Werskey (ed.), *Science at the Crossroads* (1971), 147–212.

[33] E.g. Joseph Ben-David, *The Scientist's Role in Society: A Comparative Study* (Englewood-Cliffs, N.J., 1971), 12.

[34] Robert K. Merton, *Science, Technology and Society in Seventeenth Century England* (new edn, New York, 1970); A. R. Hall, 'Merton revisited, or, science and society in the seventeenth century', *HS*, II (1963), 1–16.

toriographic theories of post-Koyréan intellectualist practice is bound to result in regrettable over-simplification. Yet some such effort ought to be made, for its procedures are undoubtedly different from those of the new contextualist writers. Some of the writers in the post-Koyréan intellectualist tradition whose work may be most fruitfully juxtaposed with that of the new contextualists include: Gerd Buchdahl,[35] Henry Guerlac,[36] P. M. Heimann,[37] Robert Kargon,[38] David Kubrin,[39] J. E. McGuire,[40] Ernan McMullin,[41] P. M. Rattansi,[42] and Richard Westfall.[43]

[35] E.g. Gerd Buchdahl, *The Image of Newton and Locke in the Age of Reason* (1961); *idem*, 'Explanation and gravity', in M. Teich and R. M. Young (eds.), *Changing Perspectives in the History of Science* (1973), 167–203; *idem*, 'Gravity and intelligibility: Newton to Kant', in R. E. Butts and J. W. Davis (eds.), *The Methodological Heritage of Newton* (Toronto, 1970), 74–102.

[36] E.g. Henry Guerlac, *Newton et Epicure* (Paris, 1963); see also note 53.

[37] E.g. P. M. Heimann, '"Nature is a perpetual worker": Newton's aether and eighteenth-century natural philosophy', *Ambix*, XX (1973), 1–25; *idem* and J. E. McGuire, 'Newtonian forces and Lockean powers: concepts of matter in eighteenth-century thought', *Historical Studies in the Physical Sciences*, III (1971), 233–306; *idem*, 'Voluntarism and immanence: conceptions of nature in eighteenth-century thought', *JHI*, XXXIX (1978), 271–83.

[38] Robert H. Kargon, *Atomism in England from Hariot to Newton* (Oxford, 1966).

[39] David Kubrin, 'Providence and the mechanical philosophy: the creation and dissolution of the world in Newtonian thought' (unpubl. Ph.D. thesis, Cornell University, 1968); *idem*, 'Newton and the cyclical cosmos' (see note 23).

[40] E.g. J. E. McGuire, 'Body and void and Newton's *De mundi systemate:* some new sources', *Archive for History of Exact Sciences*, III (1966), 206–48; *idem*, 'Neoplatonism and active principles: Newton and the *Corpus hermeticum*', in Robert S. Westman and J. E. McGuire, *Hermeticism and the Scientific Revolution* (Los Angeles, 1977), 93–142; *idem*, 'Force, active principles' (see note 23); *idem*, 'Existence, actuality' (see note 22); *idem*, 'Boyle's conception of nature', *JHI*, XXXIII (1972), 523–42.

[41] Ernan McMullin, 'Introduction: the concept of matter in transition', in *idem* (ed.), *The Concept of Matter in Modern Philosophy* (Notre Dame, Indiana, and London, 1978), 1–55; *idem*, *Newton on Matter and Activity* (see note 31).

[42] E.g. P. M. Rattansi, 'Some evaluations of reason in sixteenth- and seventeenth-century natural philosophy', in Teich and Young (eds.), *Changing Perspectives* (see note 35), 148–66; *idem* and J. E. McGuire, 'Newton and the "pipes of Pan"', *NR*, XXI (1966), 108–43; *idem*, 'The social interpretation of science in the seventeenth century', in Peter Mathias (ed.), *Science and Society 1600–1900* (Cambridge, 1972), 1–32. Rattansi's writings illustrate the dangers of lumping authors in this way, as some of his work is highly sensitive to social uses of knowledge in context: 'Paracelsus and the Puritan Revolution', *Ambix*, XI (1963), 24–32; 'The Helmontian–Galenist controversy in Restoration England', *Ambix*, XII (1964), 1–23. *Writings* are at issue here, not writers.

[43] E.g. Richard S. Westfall, 'The foundations of Newton's philosophy of nature', *BJHS*, I (1962), 171–82; *idem*, *Force in Newton's Physics* (1971); *idem*, *Science and Religion in Seventeenth-century England* (New Haven, Conn., 1958).

Much of this work on the science of the seventeenth and eighteenth centuries is founded upon three implicit historiographic 'theories' relating ideas to other ideas and to 'context': (i) intellectual contextualism; (ii) footnote contextualism; and (iii) metaphysical determinism.

(i) It is widely recognized that the most brilliant achievements of writers like McGuire, Rattansi and Kubrin have been to situate scientific thought in the seventeenth and eighteenth centuries firmly within the intellectual context of metaphysics and religion. The ontological categories of corpuscularianism and Newtonian natural philosophy have been shown to take a large part of their meaning from general philosophical and theological concerns. Indeed, one of the classic exercises of post-Koyréan intellectualism has been to show that apparently discordant elements in an individual's scientific thought are rendered 'coherent' by referring them to religious preoccupations.[44] What has not been done by writers in this tradition is to refer religious considerations to the wider social and political context. The result is the historiographic analogue of the Javanese cosmology in which the earth rests upon the elephant which stands upon the back of the tortoise which stands upon . . . ? Thus intellectualist practice is 'firmly' anchored in the rarefied medium of ideas.

And, while these writers have unquestionably been correct in their view that historical actors' cultural contexts did not manifest any rigid boundaries between natural knowledge and religion,[45] they have not gone on to mount a concerted and explicit *defence* of the boundaries *they* implicitly draw between religious and political concerns. The latter boundaries appear as historians' conventions rather than as demonstrations of actual demarcations in the actors' contexts. Whether such boundaries between religion and political concerns in fact represent actors' categories of thought seems doubtful. In the case of Boyle's writings, it now appears quite certain that they do not, and in the case of Newton increasingly unlikely.

(ii) While intellectualist writers have not explicitly *argued* that developments in the wider social and political orders are irrelevant to the understanding of scientific culture, their practice has made this case for them. Thus, taking one example among many, McGuire's study of 'Boyle's conception of nature' makes no mention of the Civil

[44] Among many examples, see: McMullin, *Newton on Matter and Activity* (see note 31), esp. 54–6; Kubrin, 'Newton and the cyclical cosmos' (see note 23).

[45] Cf. Mary Hesse, 'Reasons and evaluation in the history of science', in Teich and Young (eds.), *Changing Perspectives* (see note 35), 127–47; *idem*, 'In defence of objectivity', *Proceedings of the British Academy*, LVIII (1972), 3–20.

War, Interregnum, radical sectaries, or Restoration.[46] These omissions have to be taken, in lieu of argument, as an implicit assertion that such developments are not relevant to understanding why Boyle's philosophy of nature took the form it did, and why it changed over time in the way it did.

However, it would be inaccurate to say that intellectualist writers *never* make mention of developments in society and politics; they *sometimes* do, but it is the manner in which such mention is made that is interesting. The social context, when allusion is made to it, appears in the form of 'footnotes' or 'asides', an apparent stipulation that such a context impinged peripherally or in some unspecified, but insignificant, way. An implicit historiographic theory about contexts appears in virtual form at the bottom of the scholarly page in the gap between the large-type narrative of the text and the small-type allusions of the note. Footnote contextualism is, then, a technique for 'pointing to' the social context of scientific thought without risking the stipulation of the manner in which such contexts impinge upon scientific culture.

Nevertheless, footnote contextualism rests upon an implicit theory about the wider social context. Theory and theory-change, especially in the thought of an individual, are sometimes addressed in self-sufficient terms, so that, for instance, the development of Newton's views on active principles is linked to *his* theological concerns, *his* technical problems, 'influences' upon *him*.[47] Sometimes theoretical change in an individual's thought is referred to rival theories maintained by other intellectuals, mainly in the scientific sub-culture. However, intellectualist writers rarely, if ever, refer theory-change in science to rival conceptions of nature used in contexts of practical social action; nor do these writers refer to the work that such conceptions are doing in society at large. Thus, only *social* contextualist writers concerned with the wider uses of conceptions of nature have demonstrated that the development of Boyle's and Newton's science proceeded partly by processes of opposition to hylozoist and pantheistic materialist accounts deployed by radical sectaries and 'freethink-

[46] See note 40; cf. Kargon, *Atomism in England* (see note 38), ch. 9.

[47] A particularly interesting, but by no means unique, instance of footnote contextualism is McGuire, 'Force, active principles' (see note 23), 156, n.8. In this note the author refers to the 'various social and religious tensions' of the 1670s and 1680s, but carefully warns against any unwarranted inference from this allusion by stipulating that 'the internal circumstances of Newton's thought' were '[o]f primary importance'. Is this insistence that ideas have a self-moving and self-ordering life of their own, independent from superintendence by external *material* agencies, a modern historiographic hylozoism?

ers'.[48] The autonomy of intellectualist practice is underpinned by demarcations between scientific culture and the wider context.

(iii) The relations between metaphysics and religion on the one hand, and an individual's scientific thought on the other, have been treated in a variety of ways by post-Koyréan intellectualists. However, one of the most pervasive posited connections between the two domains has been that of *influence*. Thus, McGuire writes of 'the theological framework which *guided* so much of [Newton's] thought', and argues 'that the entrenched theological commitments of Newton's thought had a deep *influence* on his conception of space, force and matter'.[49] McMullin claims that the passive matter of Newton's science derived from 'the philosophical traditions that most *influenced*' him, and that neo-Platonism and alchemy were 'both very *influential* in shaping Newton's thought'.[50] Kubrin's excellent paper is written to show how Newton was '*led* to' his cyclical cosmos by his belief in a providentialist God, the absence of which from nature might in turn '*lead* men to believe that the world had always been... without Creator and Creation'.[51]

Naturally, too much should not be made of mere semantic preferences. But such usages are the manifestations of a consistent model of the cultural actor in intellectualist writings. The scientific actor is treated as passively shaped by metaphysical and religious notions, rather than actively using them as resources and employing them for a variety of ends, in a variety of contexts. To Lévi-Strauss's *bricoleur*, the intellectualists juxtapose an automaton.[52] It could even be said that the charge of determinism, so often used to criticize those writings which take into account social and political contexts, is more

[48] 'Thus, Kargon, *Atomism in England* (see note 38), 96–7, identifies Boyle's condemnation of 'modern admirers of Epicurus' with Hobbes and Lady Margaret of Newcastle'. Elsewhere (pp. 80–3) he briefly refers to sectarian automism and the Civil War setting, but the insight is not developed in any detail; compare J. R. Jacob, 'Ideological origins' (see note 12).

[49] McGuire, 'Force, active principles' (see note 23), 154, 193; italics added. It should be emphasized that these examples (as well as those in notes 50 and 51) are in no way atypical; they are merely randomly selected instances of usages pervasive in the genre.

[50] McMullin, *Newton on Matter and Activity* (see note 31), 29, 54; italics added.

[51] Kubrin, 'Newton and the cyclical cosmos' (see note 23), 327, 344; italics added.

[52] For models of man underpinning different forms of sociological and historical inquiry: Barry Barnes, *Scientific Knowledge and Sociological Theory* (1974), ch. 4; *idem, Interests and the Growth of Knowledge* (1977), chs. 1–2; Steven Shapin and Barry Barnes, 'Head and hand: rhetorical resources in British pedagogical writing, 1770–1850', *Oxford Review of Education*, II (1976), 231–54; Martin Hollis, *Models of Man: Philosophical Thoughts on Social Action* (Cambridge, 1977).

properly applied to post-Koyréan intellectualism. McGuire's Boyle and McGuire's and McMullin's Newton are different sorts of cultural animals from, for example, the Jacobs' Boyle and Newton. In the Jacobs' accounts the actor uses culture; in intellectualist writings culture uses the actor.

Whichever path is taken by historians of late-seventeenth- and eighteenth-century science, two points ought to be made clear. First, even if the 'theories' of post-Koyréan intellectualism are rejected, there is no reason whatever for contextualists to dismiss its empirical findings. Indeed, demonstrated connections between one set of ideas and another are the necessary starting-points for historians who would put an additional set of contextual questions to the materials. Contextualists need not accept the model of the cultural actor which intellectualists employ, but they must build upon intellectualists' empirical findings. Second, it would seem that some decision about the respective merits of the two historical perspectives must soon be taken, for the current situation appears highly unstable. Neighbouring academic disciplines presently deploy techniques which can only be sustained by indifference or hostility to each other's findings. Political historians take it as proven that religious and philosophical thought has to be referred to a context of political action; historians of science have successfully linked natural knowledge to its religious and philosophical setting. Thus, one community's practice connects *a* to *b*, the other links *b* to *c*. Each community *may* persist with its preferred techniques, but each is bound to offer some *justification* for failing to do the sort of work which associates *a* with *c*. Present arrangements have the appearance of reifying late-twentieth-century academic boundaries into divisions in the historical past.

IV. Strategies of Enlightenment matter-theory

If the new contextualism prompts methodological reassessments of the role of social uses of science generally, it also holds out the prospect of a highly challenging revision of a particular field of intellectual history. This section explores the potential of a number of recent contextualist studies for a new historical understanding of Enlightenment natural knowledge. What we begin to see in work of this kind is a sensitivity to the *variety* of conceptions of nature distributed among different social groups. We see how divergent bodies of natural knowledge were used to further social interests and were produced in processes of social conflict. Against an older view that the 'new science' (and especially the 'Newtonianism') of the early and mid-eighteenth century was the underpinning of 'the Enlightenment', we now have a developing perspective which points out the existence of a number of species of natural knowledge, and a number of opposed 'Enlightenments'.[53]

This approach is particularly apparent in studies of anti-Newtonian natural philosophies and the social groups which produced, sustained and used them. As Margaret Jacob has shown, the Boyle Lectures were constituted partly as a response to radically differing conceptions of matter, motion and God's role in the world actually being proffered by other social groups in the 1690s and early decades of the eighteenth century.[54] 'Freethinking' groups continued and elaborated late-Renaissance and Civil War sectarian conceptions of matter which ascribed to it inherent animating principles.[55] Their most celebrated spokesman, John Toland, followed Giordano Bruno in deploying a hylozoist cosmology in which 'outside, immaterial forces are unnecessary to move matter'.[56] Properties of God and properties of matter were so intermingled that the hierarchical relations between the two, so insisted upon by corpuscularians and Newtonians, were erased. Nature was a self-sufficient system. Hylozoist 'freethinkers' and Newtonians formulated their philosophies in explicit opposition to each other.

Moreover, their opposed philosophies of matter and spirit were not merely expressions of differing metaphysical preferences, nor are they to be regarded as 'caused' by contrasting metaphysical 'influences'. Where the Newtonian cosmology of the Boyle Lectures was developed partly as a defence of the Protestant succession and the court which underpinned the moral and social authority of the latitudinarian Low Church, Toland's hylozoism was the voice of conflicting social tendencies.

Margaret Jacob's case for associating Toland with 'radical republican politics'[57] would, however, seem in need of some refinement. Historians of eighteenth-century British political thought have convincingly argued that Toland is properly treated as an ideologue of the neo-Harringtonian 'Country' party, whose conception of social

[53] One can see the beginnings of this kind of sensitivity in Henry Guerlac, 'Newton's changing reputation in the eighteenth century' and 'Where the statue stood: divergent loyalties to Newton in the eighteenth century', in *idem*, *Essays and Papers in the History of Modern Science* (Baltimore and London, 1977), 69–81, 131–45.

[54] M. C. Jacob, *The Newtonians* (see note 10), ch. 6; *idem*, 'John Toland' (see note 10).

[55] See notes 14 and 17 above; also M. C. Jacob, 'Newtonianism and the origins of the Enlightenment: a reassessment', *ECS*, XI (1977), 1–25 (pp. 5, 18).

[56] M. C. Jacob, *The Newtonians* (see note 10), 236.

[57] *Ibid.*, 208.

order stressed the rights of propertied landowners against those of king and court, and who argued for the 'militia' against the standing army.[58] The historical connections between the ideology and interests of the 'Country' party and those of radical anti-court capitalists remain fully to be traced. Nonetheless, it is evident that Jacob's 'freethinking' hylozoists and Newtonian corpuscularians spoke for social groups with significantly conflicting interests and that matter-theory underwrote social theory for both sides.

Where Jacob's work presages its greatest revisionist potential is her identification of a 'radical Enlightenment' whose conception of nature owed nothing to Newtonianism and much to philosophical tendencies generated in explicit opposition to Newton.[59] Criticizing Cassirer[60] for his inattention to these aspects of Enlightenment thought, she has sketched the contours of a 'pantheistic and materialistic natural philosophy' in early- to mid-eighteenth-century Holland, France and England developed and sustained by radical political groups concerned, she says, to free civil society from interference by court and priestcraft. These thinkers, including many involved in early Freemasonry, were at odds with the Newtonians because they perceived them to be 'propagandizers for a science of God that would enhance the authority of ruling oligarchies and established churches'.[61] Moreover, she stresses that the pantheistic materialism of the radical Enlightenment was a tool in practical political action, in distinction to the 'comfortable heresies adopted by the d'Holbach salon'.[62] More explicitly than previously, linking her findings to

[58] See especially J. G. A. Pocock, 'Machiavelli, Harrington and English political ideologies in the eighteenth century', in *idem, Politics, Language and Time: Essays on Political Thought and History* (1972), 104–47; H. T. Dickinson, *Liberty and Property: Political Ideology in Eighteenth-century Britain* (1977), parts 1–2; Caroline Robbins, *The Eighteenth-century Commonwealthman* (Cambridge, Mass., 1959), 125–33; J. M. Beattie, *The English Court in the Reign of George I* (Cambridge, Mass., 1967), ch. 7. Jacob moves some way towards this view of Toland in her 'Newtonian science and the radical Enlightenment', *Vistas in Astronomy*, XXII (1979), 545–55 (p. 547, and n.17).

[59] M. C. Jacob, 'Newtonianism and the origins of the Enlightenment' (see note 55); *idem*, 'Newtonian science and the radical Enlightenment' (see note 58); *idem, The Radical Enlightenment: Pantheists, Freemasons, and Republicans* (in press).

[60] Ernst Cassirer, *The Philosophy of the Enlightenment* (Princeton, N.J., 1951); cf. Lucien Goldmann, *The Philosophy of the Enlightenment: The Christian Burgess and the Enlightenment* (Cambridge, Mass., 1973).

[61] M. C. Jacob, 'Newtonian science and the radical Enlightenment' (see note 58), 552.

[62] *Ibid.*; Alan Kors, *D'Holbach's Coterie: An Enlightenment in Paris* (Princeton, N.J., 1976). On d'Holbach's matter-theory see the very interesting essay by Ernest Gellner, 'French eighteenth-century materialism', in *idem, The Devil in Modern Philosophy* (1974), 113–48 (pp. 128–31).

Marxist historiography, Jacob portrays the anti-Newtonianism of the radical Enlightenment as a 'practical' cosmology, a body of knowledge generated and evaluated mainly in terms of the social work it was designed to do. Here we have an eighteenth-century anti-Newtonianism of 'the left'.

A valuable counter-point to Margaret Jacob's findings is provided by Christopher Wilde, who deploys similar historiographic techniques to identify an important English anti-Newtonianism of 'the right'.[63] The writings of John Hutchinson (1674–1737) have already been the object of some scholarly attention;[64] what Wilde has done is to identify the social groups for whom Hutchinson spoke, the social work Hutchinsonian natural philosophy performed in context, and the concrete circumstances in which this variety of eighteenth-century anti-Newtonianism was produced.

Hutchinson and his followers were, like the Newtonians, concerned to construct a natural philosophy which would be the handmaiden of correct religion. However, on a number of counts they maintained that the Newtonian philosophy failed to serve that function and, indeed, might serve to comfort and aid Dissenters and deists. In particular, Hutchinson was disturbed by the General Scholium to the second edition of the *Principia,* and with its invitation to see God as immanent in the world and in 'filthy Matter'. Newton's predicament had been to construct a cosmology in which God was *both* transcendent *and* so involved in the material world that He could affect its workings. Especially on the question of the cause of gravity and the relations between God, gravity and matter, Newton chopped, changed and struggled to preserve both the theological and technical functions of his system. In so doing, he laid himself open to Hutchinson's charge that his system was the very thing it was constructed not to be – hylozoist. From Hutchinson's perspective, Newton's God was not transcendent *enough;* He was *too* immanent in the world; the material world was not *sufficiently* mechanical and self-contained. Hutchinsonianism, therefore, introduced into the cosmos

[63] C. B. Wilde, 'Hutchinsonianism, natural philosophy and religious controversy in eighteenth century Britain', *HS,* xviii (1980) 1–24; and his forthcoming University of Cambridge Ph.D. thesis.

[64] A. J. Kuhn, 'Glory or gravity: Hutchinson vs. Newton', *JHI,* xxii (1961), 303–22; Michael Neve and Roy Porter, 'Alexander Catcott: glory and geology', *BJHS,* x (1977), 37–60; G. N. Cantor, 'Revelation and the cyclical cosmos of John Hutchinson', in L. J. Jordanova and Roy S. Porter (eds.), *Images of the Earth: Essays in the History of the Environmental Sciences* (Chalfont St Giles, 1979), 3–22; also Thackray, *Atoms and Powers* (see note 5), 246.

'the Names', modifications of the aether and exceedingly fine matter, to replace Newtonian immaterial forces. It posited a plenum in place of Newton's void; contact action instead of action at a distance; a self-sufficient mechanical universe in the stead of one continually dependent upon God's tinkering; and a *totally* transcendent Deity as against a Deity who made periodic service-calls. Thus, by opposing Newtonianism, the Hutchinsonians' cosmos shared crucial characteristics with those views the Newtonians opposed. Most interestingly, both Hutchinsonians and 'freethinking' hylozoists stressed that nature was self-sufficient.

An intellectualist account of Hutchinsonianism might well have stopped at this point. But Wilde goes on to ask *why* Hutchinson and his followers were hostile to the Newtonians and their natural philosophy. It emerges that Hutchinsonianism was the voice of High Church divines, as opposed to the Whig–latitudinarian alliance of the early eighteenth century. Their power at low ebb during the 1700s and 1710s, the High Church party was continually thwarted by latitudinarian–Whigs in obtaining preferment and in their desire to take more vigorous action against heretics and Dissenters. Perceiving Newtonianism to be the legitimating strategy of those divines who blocked their ambitions, the High Churchmen juxtaposed to Newton's physico-theology one of their own devising. The Hutchinsonian critique of Newtonianism shows the marks of the political processes of opposition which generated it. The anti-deistical Newtonians were said to be *insufficiently* anti-deistical; their God was associated with that of freethinkers and Socinians. The theistic Newtonian natural philosophy was said to involve God too much in the material world. Their epistemology was claimed to lay too little emphasis upon the tools of revelation and scriptural glossing. Thus the Hutchinsonian cosmos and epistemology maximized the competences of High Church clerics, while it impugned the theological purity of their enemies, the Low Church–Whig–Newtonians. Throughout the eighteenth century, and into the nineteenth, Hutchinsonianism was employed by High Churchmen in their conflicts with Low Churchmen and with Dissenters like Priestley.

Intriguingly, Wilde argues that the self-sufficient cosmos of the Hutchinsonians bears striking resemblances to a variety of such schemata which became prevalent from the middle of the eighteenth century. Here we touch upon a widely noticed shift in the general features of natural philosophy from the early to the late eighteenth century. Within the post-Koyréan intellectualist tradition, J. E. McGuire and P. M. Heimann have traced the career of 'active principles' and associated theological and cosmological conceptions from the late seventeenth century into the late eighteenth century.[65] Heimann, in particular, has shown that significant changes occurred during this period in attitudes towards the role of God's will in natural phenomena.[66] Where Newton and the Boyle Lecturers were deeply committed to demonstrating God's providence and His voluntary capacity both in nature's laws and in the abrogation of laws, British natural philosophers of the mid-to-late eighteenth century began increasingly to prefer explanations predicated upon divine causality manifested in active powers *immanent* in the natural order. Newton understood active principles to be both lawful *and* products of providence; he thus made the natural order to be utterly dependent upon external spiritual superintendence. Newton's aether, for example, was formulated as one such active principle subordinate to God's will. But by the 1740s, as Heimann shows, British natural philosophers were tending to conceive of the aether as an active principle 'immanent in the fabric of nature' and were rejecting Newton's own theological rubric:

> The emergence of a theory of nature in which activity was considered as immanent in the structure of nature led to a rejection of Newton's doctrine that all causal activity in nature was imposed by God's power and will.[67]

Theology and matter-theory were once more inextricably intertwined. While discarding Newton's views on the nature of active principles, the eighteenth-century natural philosophers discussed by Heimann also jettisoned Newton's conception of the nature of matter and the theological framework in which it was embedded. Matter was now seen as inherently animated and God's volition was no longer required to explain its activity. Thus, Priestley and Hutton sided more with Leibniz than with Newton in stressing God's perfect foresight and the necessity of His role in nature. In so doing, they also rejected the dualism between matter and spirit insisted upon by the Boyle-Newton corpuscularian tradition. The dualism was eroded by stressing that God's active principles were immanent in matter, and, because this was so, matter was drastically revalued. Where Newton and the Boyle Lecturers made matter subservient to spiritual agencies, Priestley and Hutton (like the Diggers) framed their cosmology

[65] See sources in note 37.

[66] Heimann, 'Voluntarism and immanence' (see note 37).

[67] *Ibid.*, 275; cf. 277. For a pertinent survey: G. N. Cantor, 'The theological significance of ethers', in G. N. Cantor and M. J. S. Hodge (eds.), *A Subtle Form of Matter* (Cambridge, in press).

to show that such dependence was unnecessary to explain observed natural phenomena.

Heimann offers no hints as to why this cosmological revaluation and reordering should have occurred during the eighteenth century. That it was, as he says, a 'shift in theological sensibility' counts as a partial description but not as an explanation. And, it should be noted, while Heimann eschews any mention of the wider social and ideological context, he similarly omits any discussion of a changing *scientific-technical* context to which the shift might alternatively be referred. There is thus nothing in his account which supports the notion that the move from 'voluntarism' to 'immanence' was 'dictated' by changing requirements for prediction and control of natural phenomena, although a role for such factors cannot thereby be ruled out.

In fact, historians of science have not yet seriously begun to notice and to explain why it was that similar shifts in 'sensibility' seem to have occurred in several areas of inquiry during the eighteenth century. Not just natural philosophy and matter-theory,[68] but also geology[69] and physiology[70] seem to have changed their preferences from

[68] Work on Stahl, Lavoisier and Dalton is also worth thinking about in these terms. A few references are: David Oldroyd, 'An examination of G. E. Stahl's *Philosophical principles of universal chemistry*', *Ambix*, xx (1973), 36–52; Thackray, *Atoms and Powers* (see note 5), ch. 6; *idem*, *John Dalton: Critical Assessments of his Life and Science* (Cambridge, Mass., 1972); J. B. Conant, 'The overthrow of the phlogiston theory: the chemical revolution of 1775–1789', in *idem* (ed.), *Harvard Case Histories in Experimental Science* (Cambridge, Mass., 1966), 65–115; Marie Boas [Hall], 'The structure of matter and chemical theory in the seventeenth and eighteenth centuries', in Clagett (ed.), *Critical Problems* (see note 2), 499–514.

[69] See writings of Roy Porter cited in note 21; also Porter, 'The terraqueous globe', in this volume; material on Hutton in Heimann and McGuire, 'Newtonian forces and Lockean powers' (see note 37); and in these, and related, connections, Clarence J. Glacken, *Traces on the Rhodian Shore* (Berkeley, Cal., 1967), parts 3–4.

[70] See discussion below on eighteenth-century physiology. Also: Elizabeth L. Haigh, 'Vitalism, the soul and sensibility: the physiology of Théophile Bordeu', *JHM*, xxxi (1976), 3–41; Sergio Moravia, 'From *Homme machine* to *Homme sensible*: changing eighteenth-century models of man's image', *JHI*, xxxix (1978), 45–60; G. S. Rousseau, 'Nerves, spirits, and fibres: towards defining the origins of sensibility', in A. Giannitrepanni (ed.), *The Blue Guitar* (Rome, 1976), ii, 125–53; R. K. French, 'Sauvages, Whytt and the motion of the heart: aspects of eighteenth-century animism', *Clio Medica*, vii (1972), 35–54; and W. F. Bynum's survey of Enlightenment physiology and medicine in this volume. Also see Farley, *Spontaneous Generation Controversy* (see note 21), chs. 2–3, where late-eighteenth-century acceptance of spontaneous generation is traced to belief in the self-organizing capabilities of matter amongst French materialists and to the vitalism of German *Naturphilosophen*.

'Newtonian' theories which required external animating spiritual agencies to those which placed the principles of animation and pattern within the natural entities. The generality of this shift during the eighteenth century needs first to be acknowledged before an historical explanation can be attempted. But early impressions are that the pervasiveness of the change will make any explanation in terms of the 'immanent logic' of scientific inquiry very difficult to sustain. What is strongly suggested by these findings is that historians ought to look to the concrete contexts in which Newtonian accounts were rejected, and to the social, as well as the technical, bases for that rejection during the eighteenth century. Thus, a broad hint can be found in the work of Margaret Jacob and Christopher Wilde, where groups with different social interests, both of whom rejected aspects of Newtonianism for social reasons, produced cosmologies which shared certain characteristics. 'Dialectical' processes of social conflict in the cultural domain may be needed to account for historical changes in dominant cosmologies.[71]

An historiography of science appropriate to this task remains to be constructed. Certainly, there seems to be no obvious way to build such an historiography out of Heimann's unexplained 'shifts in theological sensibility'. Yet within the intellectualist tradition there does exist material richly suggestive of how bridges to contextualist work might be built and what the road forward might look like. The starting point for this historiographic bridge-building could be McEvoy and McGuire's stimulating account of Joseph Priestley's natural philosophy.[72] From the mid-1770s Priestley set about constructing a conceptual scheme in direct opposition to the matter-theory and theology of Newton and the Boyle Lecturers. Like Margaret Jacob's 'freethinkers', he was concerned to undermine the ontological hierarchy which made passive matter subservient to spiritual superintendence by God's providence. He denied that 'there are two

[71] For an account of eighteenth-century changes from this perspective, see my forthcoming book on *The Social Use of Nature*.

[72] J. G. McEvoy and J. E. McGuire, 'God and nature: Priestley's way of rational dissent', *Historical Studies in the Physical Sciences*, VI (1975), 325–404; also McEvoy's four-part study of Priestley's chemistry and metaphysics: 'Joseph Priestley, "aerial philosopher": metaphysics and methodology in Priestley's chemical thought, from 17[7]2 to 1781', Part I, *Ambix*, XXV (1978), 1–55; Part II, *ibid.*, 93–116; Part III, *ibid.*, 153–75; Part IV, *ibid.*, XXVI (1979), 16–38; also see note 76. There are important differences of interpretation of Priestley's chemical philosophy between these authors and Robert Schofield which bear upon the following discussion, but which cannot be gone into here; see McEvoy and McGuire, *ibid.*, 389–90, n.247; cf. Thackray, *Atoms and Powers* (see note 5), 248–51.

distinct kinds of substance... matter and spirit', and asserted that matter possessed inherent powers, such as those of attraction and repulsion.[73] And, like the social groupings that Boyle was involved in combatting, Priestley claimed that matter has a nature similar to that of 'spirit and immaterial beings'.[74] Here we are again confronted with a revaluation of matter and a hierarchy-collapsing strategy encountered in other contexts.[75]

Obviously, then, it is important to give attention to similarities between Priestley's purposes in elaborating his anti-Newtonian cosmology and those of the groups discussed previously. However, it may be even more important to discuss certain crucial differences in these purposes. For example, as McEvoy and McGuire make absolutely clear, Priestley was not embarked upon any 'atheistical' or 'secularizing' enterprise, such as those attributed to Hobbists and 'Epicureans' in the seventeenth century or to Margaret Jacob's 'freethinkers' in the early eighteenth century. Even less did he share the social interests of Hutchinsonian High Churchmen. Indeed, the thrust of McEvoy and McGuire's study is to show how all elements of Priestley's cosmology are to be referred to his strategy of 'rational dissent', specifically his attempt to resuscitate a purified and 'reasonable' form of Christianity from the 'corruptions' introduced into it by misguided theologians and Newtonian philosophers.

It is obviously not the stated intention of the authors to identify cosmological stances as social strategies, yet they provide sufficient material about Priestley's theological and social views to make such an historical enterprise possible. First of all, it is important to know that Priestley was a Dissenter in an historical context where Dissenters suffered from the power of the established Church of England. McEvoy and McGuire show that Priestley was deeply committed to undermining the authority of the state Church and justifying liberalism and toleration in religious matters. They demonstrate that his materialism was intimately connected with his deterministic account of nature and of man's mental processes. Man is wholly material, no immaterial soul being needed to account for religious feelings

[73] McEvoy and McGuire, 'God and nature' (see note 72), 384; see also McMullin, *Newton on Matter and Activity* (see note 31), 118–19.

[74] McEvoy and McGuire, 'God and nature' (see note 72), 385.

[75] The reciprocal hierarchy-justifying strategy was developed in opposition to Priestley and other materialists and associationists by the Scottish philosophers of common sense; see *ibid.*, esp. 357–82, and also N. T. Phillipson, 'Towards a definition of the Scottish Enlightenment', in Fritz and Williams (eds.), *City and Society* (see note 8), 125–47; J. H. Faurot, 'Reid's answer to Joseph Priestley', *JHI*, xxxix (1978), 285–92.

or mental functioning. But, as we have seen, Priestley's conception of matter was diametrically opposed to that of the dominant Newtonian (and Low Anglican) position; Priestley's matter was inherently endowed with active powers, and, because it was so endowed, a materialistically conceived man could *think*. Thus, separate, superintending spiritual agencies were ejected both from man and from material creation. The effect of this cosmological move was to render redundant any requirement for such governing spiritual entities in natural philosophy, in psychology, and in theology.

McEvoy and McGuire seek to demonstrate that Priestley treated all these cultural activities as if they amounted to the same enterprise. God made a lawful material creation in which His active principles and design are manifest to an individual's senses. Man thus needs no spiritual intermediaries to interpret a reality beyond sensuous apprehension. Man himself is material and mechanical, and is of the same 'stuff' as God's nature, so that natural philosophy is on as solid a foundation as theology.[76] Moreover, Priestley wished vigorously to combat the idea that any individual had privileged access to transcendent truth, for his 'program was to eliminate prejudice, dogma, and superstition, and to oppose the forces of authoritarianism with the spirit of liberalism'.[77] He sought to subvert that kind of authority which found its justification in spiritual domination and in claims to a unique access to truth. Thus, Priestley denounced what he took to be the political implications of the Scottish common-sense philosophers' preservation of a separate, immaterial soul:

> [P]oliticians also possessing themselves of this advantage [Common-Sense principles], may venture once more to thunder out upon us their exploded doctrines of passive obedience and non-resistence [*sic*][78]

Priestley, as McEvoy and McGuire say, 'rejects any form of intellectual and, hence, political or religious authority, for "it depends on ourselves . . . to be as wise, judicious and knowing, as any other person whatsoever"'.[79]

Priestley, in other words, was an individualistic liberal in social, political and religious terms, but to make out his position to be founded on anything other than intense piety would be completely

[76] John G. McEvoy, 'Electricity, knowledge, and the nature of progress in Priestley's thought', *BJHS*, XII (1979), 1–30. Here McEvoy beautifully develops the theme of Priestley's determinism and materialism as the underpinning of his epistemological egalitarianism.

[77] *Ibid.*, 12–13; McEvoy and McGuire, 'God and nature' (see note 72), 374–5.

[78] McEvoy and McGuire, 'God and nature' (see note 72), 377.

[79] *Ibid.*

unjustified. The hierarchy he was most concerned to discredit was that of the Church of England; on McEvoy and McGuire's evidence, his cosmological views, and his position on social and political matters, are to be referred to his strategy of 'rational dissent' from the domination of the Established Church.[80] Thus, explicitly contextualist writers have no monopoly on the valuable insight that cosmological representations and metaphysics may be conceived of as *strategies*, crafted to serve a variety of interests, and fluidly negotiated in concrete historical settings. What Priestley's strategy of 'rational dissent' had in common with the materialism of the 'freethinkers' and the self-sufficient mechanical cosmos of the Hutchinsonians was its basis in social antipathy to the Newtonians and their view of nature. That their respective antipathies were differently founded is clear; what is also clear is that processes of opposition resulted in the elaboration of schemata which had several points of contact.

The argument so far is that processes of opposition and the social uses of science in context provide the groundworks for cultural change. Much of the best empirical work relevant to this perspective deals with the British context. However, the importance of attending to uses in concrete contexts requires us to consider the sparser literature on related developments in France. C. C. Gillispie takes as the text for his study of 'the Jacobin philosophy of science' Diderot's radically anti-Newtonian aphorism: 'La distinction d'un Monde physique et d'un Monde morale lui semblait vide de sens'.[81] The *Encyclopédistes* developed a hylozoist cosmology in which the boundaries demarcating matter from spirit, and the world of matter from the world of meaning, were systematically attacked. What is more, the material world of the *Encyclopédistes* and 'Jacobins' was not one of 'brute and stupid' matter, of matter subservient to spiritual actuating

[80] For more on Priestley's political views, see Clarke Garrett, 'Joseph Priestley, the millennium, and the French Revolution', *JHI*, xxxiv (1973), 51–66.

[81] Gillispie, 'The *Encyclopédie* and the Jacobins' (see note 2), 255; the quotation is from *Jacques le fataliste*. For additional material with strong suggestive parallels to the Jacobs' interpretations, see A. Vartanian, *Diderot and Descartes: A Study of Scientific Naturalism in the Enlightenment* (Princeton, N.J., 1953), and *idem*, 'Trembley's polyp, La Mettrie, and eighteenth-century French materialism', in P. P. Wiener and A. Noland (eds.), *Roots of Scientific Thought* (New York, 1957), 497–516. Margaret Jacob does not refer to Gillispie's essay in her 1977 paper (see note 55) reassessing the role of Newtonianism in the Enlightenment, yet her argument gets support from an unexpected quarter – for Gillispie is concerned to exculpate Newton from involvement in Diderot's folly. Diderot 'never', says Gillispie, 'drew his image of science from Newton'. Gillispie claims that 'he drew it from the classics', particularly from the Stoics. But a more probable immediate 'source' is the 'freethinking' circle discussed by Margaret Jacob (Gillispie, *ibid.*, 269).

principles; matter was to be paramount, sentient, inherently possessed of qualities. A world so constituted was thus profoundly knowable through the use of commonplace techniques of sensory observation; chemistry and natural history were, therefore, the Jacobins' *Urwissenschaften*. What was proposed was no less than 'a citizen's science' and, more fundamentally, a citizen's *nature*.[82] All this Gillispie penetratingly diagnoses as ideology. Radical groups were using conceptions of nature to further particular social interests; their boundaries and their epistemology were strategies to further those interests and were not 'mirrors' of reality or 'constrained by' technical requirements.[83] These social interests consisted in celebrating the experience and competences of the bourgeoisie and artisanate, while systematically criticizing the competences, cosmology and cultural corporations of the *ancien régime*.

Some of the Jacobins discussed in Gillispie's account were participants in Mesmerism, the most spectacular manifestation of a 'citizen's science' in the immediate pre-Revolutionary period. Robert Darnton's admirable account of popular science and radical politics in the 1770s and 1780s provides an extensively documented study of a particular 'Jacobin' science.[84] Mesmer's system relied upon an all-pervading, superfine aether which diffused across the boundary between 'le Monde morale' and 'le Monde physique'. This aether - in the Mesmerists' account, more matter than spirit - provided the basis for a medical system and a popular piece of therapeutic theatre, but it also underpinned a general cosmological orientation.[85] There was to be no rigid distinction between brute matter and animating spirit; a materialist account of mental phenomena was provided, and psychic heal-

[82] Gillispie, 'The *Encyclopédie* and the Jacobins' (see note 2), 270. See also Roger Hahn, *The Anatomy of a Scientific Institution: The Paris Academy of Sciences, 1666–1803* (Berkeley, Cal., 1971), chs. 5–6.

[83] It must, however, be pointed out that Gillispie's account is profoundly asymmetrical; whilst his analysis of Jacobin science as a social strategy is highly sensitive, he refers the cosmology of their opponents in the Academy to its correspondence with 'reality'. Of the Jacobins' morally alive cosmos, Gillispie simply declares 'Nature is not like that' (p. 281). He may, of course, be right, but the consequence of this sort of procedure is to situate one sort of cosmology in an historical context and another in a transcendent domain of manifest objectivity.

[84] Robert Darnton, *Mesmerism and the End of the Enlightenment in France* (New York, 1970).

[85] See also William Coleman, 'Health and hygiene in the *Encyclopédie:* a medical doctrine for the bourgeoisie', *JHM*, XXIX (1974), 399–421; *idem*, 'The people's health: medical themes in 18th-century French popular literature', *BHM*, LI (1977), 55–74. For similar themes in a Scottish context, see C. J. Lawrence, 'William Buchan: medicine laid open', *MH*, XIX (1975), 20–35.

ing was mediated by great vats of concentrated aether. The aether united all men in harmonious brotherhood; it flowed across all boundaries. Individual ill-health, as well as social malaise, was put down to blockages in the free flow of this superfine substance. A convulsive crisis was necessary - in the individual and in the state - in order to purge the blockage. Harmony could only exist when arbitrary obstacles to mobility were removed; harmony was natural. This charming metaphor of revolutionary action as laxative is, in fact, a particularly highly developed version of the hylozoist schema and its social uses which the Jacobs have traced as far back as the late Renaissance.

When the Revolution came, the political uses of hylozoism were not lost upon the elites of other countries proud of their social 'costiveness' and deeply committed to maintaining it. Particularly instructive reactions to 'French materialism' occurred in Britain, and it is to that context we now return.[86] In the especially vicious and repressive reaction to the French events of 1789-93 which Scotland experienced, a number of men of science expressed their view of the connection between materialism and social subversion.[87] In Edinburgh John Robison, professor of natural philosophy in the 1790s, was the most vocal defender of the established social and moral order from the threats posed by 'secret meetings of free masons, illuminati, and reading societies'.[88] Robison detected a causal link between materialist cosmologies and French social upheaval, and, like his common-sense philosopher-colleagues, attacked Priestley's metaphysics, natural philosophy and psychology as the stalking-horse for similar de-

[86] In addition to the example discussed below, there are a number of studies of 'science and politics' during the British reaction to the French Revolution: see especially Norton Garfinkle, 'Science and religion in England 1790–1800: the critical reaction to the work of Erasmus Darwin', *JHI*, XVI (1955), 376–88; Dennis R. Dean, 'James Hutton and his public, 1785–1802', *AS*, XXX (1973), 89–105; *idem*, 'James Hutton on religion and geology: the unpublished preface to his *Theory of the earth* (1788)', *AS*, XXXII (1975), 187–93; Porter, *Making of Geology* (see note 21), chs. 6–8; *idem*, 'Philosophy and politics of a geologist: G. H. Toulmin (1754–1817)', *JHI*, XXXIX (1978), 435–50; Ian Inkster, 'Science and society in the metropolis: a preliminary examination of the social and institutional context of the Askesian Society of London, 1796–1807', *AS*, XXXIV (1977), 1–32 (25–7). With special reference to Priestley, see Thackray, *Atoms and Powers* (see note 5), 250–2; *idem*, 'Natural knowledge in cultural context: the Manchester model', *American Historical Review*, LXXIX (1974), 672–709 (pp. 687–8).

[87] J. B. Morrell, 'Professors Robison and Playfair, and the *Theophobia Gallica:* natural philosophy, religion and politics in Edinburgh, 1789–1815', *NR*, XXVI (1971), 43–63.

[88] *Ibid.* The quoted passage is part of the subtitle of Robison's 1797 *Proofs of a Conspiracy against All the Religions and Governments of Europe.*

velopments in Britain.[89] The Tory professor specifically identified Priestley's matter-theory as subversive of correct moral order, since the great Dissenter attributed to material entities qualities 'proper' to God alone. Thus, Robison recognized Priestley's cosmology as a species of hylozoism and took it to be part of a strategy designed to materialize the universe, to restrict the scope of spiritual powers, and to break down crucial boundaries between mere matter and ruling spirit.[90] Because Robison took Priestley to be limiting the voluntary capacities of the Deity, he saw Priestley's materialism as a socially and morally dangerous strategy. We therefore have another example of historians' analyses of matter-theory as social strategy corresponding to the perceptions of actors in concrete historical contexts.

V. The strategy of physiological thought

We have seen that recent historical accounts of the generation, evaluation and institutionalization of the Newtonian world-view (and opposed cosmologies) are the source of valuable new insights into the historical treatment of social uses of science. However, the full weight of these insights might be better appreciated if they could be shown at work in the examination of a *particular* eighteenth-century scientific discipline. Further support for these general orientations, as well as some particularly interesting additional themes, emerges from several recent studies of developments in eighteenth-century physiology.

Theodore Brown has given detailed attention to the reasons for the adoption of mechanical models by English medical and physiological theorists of the late seventeenth and early eighteenth century, as well as to the causes of the replacement of these models by vitalist conceptions by the middle of the eighteenth century.[91] His arguments are of

[89] Morrell, 'Professors Robison and Playfair' (see note 87), 48.

[90] The general form of this controversy between Scottish academic philosophers and materialist opponents was repeated early in the nineteenth century during the Edinburgh phrenology disputes; see G. N. Cantor, 'The Edinburgh phrenology debate: 1803–1828', *AS*, xxxii (1975), 195–218, and Steven Shapin, 'Phrenological knowledge and the social structure of early nineteenth-century Edinburgh', *ibid.*, 219–43. For an account which develops the anthropological themes involved in the understanding of these, and similar, episodes, see Shapin, 'Homo phrenologicus' (see note 18).

[91] Theodore M. Brown, 'The College of Physicians and the acceptance of iatromechanism in England, 1665–1695', *BHM*, xliv (1970), 12–30; *idem*, 'From mechanism to vitalism in eighteenth-century English physiology', *Journal of the History of Biology*, vii (1974), 179–216. Also relevant, but not discussed here, is his 'Physiology and the mechanical philosophy in mid-seventeenth-century England', *BHM*, li (1977), 25–54. See also sources in note 70 above.

general interest to social studies of scientific knowledge. He asks the question: 'Why did members of the London College of Physicians come to adopt iatromechanical models in the period from the Restoration to the end of the seventeenth century?' Brown is thus posing a question central to the main concerns of the history of scientific change, but it is the role played by socio-political interests in his account which is of special relevance here.

All through the middle decades of the seventeenth century the elite doctors of the College were engaged in proprietary conflict with the apothecaries, whom they were nominally intended to 'supervise' in pharmaceutical matters.[92] Practical problems of insuring their control over the lower-status apothecaries were especially severe in the period just after the Restoration, for Parliament was unsympathetic to the physicians' cause. Full-scale criticism of the College and its practices by 'chemical physicians' (Helmontians and Paracelsians) grew increasingly loud in the 1660s. Moreover, the corpuscularians of the newly founded Royal Society (including Boyle) were unimpressed by the 'barren principles of the Peripatetick school' espoused by the main body of elite physicians in the metropolis. Between the 1660s and the 1690s 'the College was in very deep trouble'.[93]

It was against this background of threat to their corporate interests that the College physicians began to discard their traditional theoretical schema in favour of iatromechanism. As Brown notes, the physicians' enduring interest was not in the defence of their theoretical armamentarium but in the protection of their *practice.* Iatromechanism not only allied the College with the social groupings who were finding the mechanical philosophy ideologically convenient; it could also be made to underwrite existing therapeutic practice. In other words, medical practice could be made to bear mechanical glossing quite as well as it had borne scholastic glossing; in no sense was the physicians' practice *determined* by their theoretical preferences. First Thomas Willis, and later Hodges, Goodall, Charleton and Cole, presented the College with a mechanical theory of fevers, apoplexies, and the like, while "justif[ying] the continued use of long-established therapeutic methods'. Pathological phenomena were quite as well able to sustain the mechanical interpretation as they were able to sustain the Galenic-Aristotelian.[94] Brown shows that the context in which many pro-iatromechanical tracts were generated

[92] Brown, 'College of Physicians' (see note 91), 15; also P. M. Rattansi, 'The Helmontian-Galenist controversy' (see note 42).

[93] Brown, *ibid.*, 17, 20.

[94] *Ibid.*, 22.

was also the context of physicians' attack on surgeons' and apothecaries' ambitions. The adoption of iatromechanism is presented as an episode in which historical actors were unconstrained in theory-choice by the technical demands of their science. Social interests, in this case interests in aligning the College with prestigious groupings while undermining the threat of insurgent elements in medicine, were compatible with technical interests. Theory-choice reflected social interests, but it also responded to actors' requirements in explaining the phenomena. It was not, in Brown's account, a question of 'either/or'.[95]

A similar orientation underpins Brown's analysis of the decline of iatromechanical models in English physiology. Here he asks a question structurally similar to that posed in his study of earlier events: 'Why did English physiologists abandon mechanical models and adopt "vitalist" models from 1730 to 1770?'[96] Again, this intellectual shift has been widely noted in the history of science literature, and has been traditionally accounted for by referring it to other intellectual factors. But the relevance here of Brown's study is once more in the way in which his explanation associates social interests and the social use of scientific knowledge with other factors. The social and political situation of the English medical profession is again the focus of the

[95] Brown's treatment of this episode is the object of some remarkable historiographic pronouncements by Larry Laudan in his *Progress and its Problems: Towards a Theory of Scientific Growth* (1977), 214–17. Laudan takes Brown to task for offering an explanation of scientific change which posits 'external' and 'irrational' reasons for the adoption of a new model. Laudan's preferred historiographical scheme identifies a hierarchy of 'reasons' for theory-change, paramount of which are 'internal', puzzle-solving considerations. Thus, Brown is criticized, *not* because Laudan finds his empirical account faulty, and *not* because Laudan has produced evidence that seventeenth-century physicians behaved in ways other than those which Brown reveals, but because Brown *ought* to have done history the way Laudan *prefers*! Thus: 'It *might* just be that Walter Charleton accepted the mechanical philosophy because . . . that theory was rationally preferable to its alternatives' (217; italics in original). There is much to criticize in Laudan's manifesto, not least his incomprehensible equation of 'rational' courses of action with 'internal' scientific ones. But the most relevant aspect for historians to note is that Laudan, like Lakatos, would make empirical findings about how scientists actually behaved subservient to philosophers' models of how they should behave. One can see no 'rational' reason for historians to abandon the evidence of empirical research in favour of this sort of suspect theorizing. For a sociologist's critique of Laudan's 'historiography': Barry Barnes, 'Vicissitudes of belief', *SSS*, IX (1979), 247–62.

[96] Brown, 'From mechanism to vitalism' (see note 91); for additional background to this episode, see Schofield, *Mechanism and Materialism* (see note 19), 191ff. Cf. sources in note 70 above.

explanation. Iatromechanism having been adopted partly as a strategy designed to rescue the profession's prestige and to secure it against threat, these (and similar) considerations continued to sustain it into the first decades of the eighteenth century. Furthermore, Newton's personal control kept a number of his disciples' noses firmly to the iatromechanical grindstone. Thus, if social interests and personal control sustained iatromechanism, then any shift in these factors might reasonably be expected to bring about a shift in preferred physiological theory.

This is in fact what Brown observes. Firstly, Newton's death in 1727 was shortly followed by the drift of several of his most important iatromechanical followers into other areas of inquiry. Social and intellectual control in English physiology were simultaneously relaxed.[97] But far more central to Brown's explanation is the changing social and political situation of English physicians - the 'client group', so to speak, for physiological theorizing. The social interests which prompted College physicians to plump for iatromechanism in the latter half of the seventeenth century continued, and were augmented by other concerns, in the early eighteenth. London surgeons and apothecaries were presenting even more intractable problems of social control than they had in the past. Assisted by the expansion of the London hospital system, the surgeons grew in numbers and status. The apothecaries were provided by the House of Lords with proprietary legal rights to diagnose and prescribe medicines. And the growth of provincial towns, supplied with doctors by the newly reformed Scottish medical schools, meant that centrifugal social forces were making the exercise of professional hegemony in English medicine extremely difficult. One solution to these practical problems of social control was in fact adopted by the College in the seventeenth century - namely the ideological strategy represented by its embrace of a mechanical medical theory; another 'solution' was now emerging. The College would relinquish its 'responsibility' to exert hegemonic control over the whole of English medicine; it 'turned increasingly inward, looking more and more after its own affairs rather than the general state of medicine in England'.[98] Having abandoned the social aspirations which led them to deploy iatromechanical models, the elite physicians of the College now abandoned iatromechanism. It is not strictly correct to say that they abandoned mechanical theoretical models *for* vitalist ones - although vitalist

[97] Brown, *ibid.*, 189–90; compare Brown's argument here with Thackray, '"The business of experimental philosophy"' (see note 5).

[98] Brown, *ibid.*, 214.

conceptions did tend to take their place. Vitalist notions developed elsewhere, particularly in Scotland and on the continent, were no longer, as it were, having to compete with ideologically sustained mechanical models. *Theory* - in the grand, rhetorically and publicly articulated sense - simply dropped away 'and a nontheoretical empiricism began to take its place'.[99]

Brown's studies thus naturally arouse one's curiosity about the origins of vitalist models and modes of explanation in physiology, especially in the non-English environments where they first developed. Christopher Lawrence's recent examination of the local social context of mid-eighteenth-century vitalist physiology in Edinburgh is one of the best and most explicit discussions of the relations between social interests, social uses of scientific knowledge, and technical considerations bearing upon the evaluation and institutionalization of science.[100] He wants to explain why it was that models of bodily integration and control stressing the vital properties of the nervous system replaced Newtonian (and Boerhaavean) mechanical models in Scotland from the 1720s to the 1750s. In choosing a defined local context for his focus Lawrence exemplifies a growing and valuable tendency in the social history of science to erect arguments of very general significance upon detailed examinations of many factors interacting in one place during a fairly restricted time-span. The 'methodology' is justified by the findings it produces. In this case Lawrence discovers that the reasons for Scottish judgements about physiological models cannot be divorced from the concrete particularities of the social and historical context in which they were made. Appealing to 'universal criteria' does no historical work in explaining the shift to this variant of vitalist physiology. Instead, much is explained by making reference to the social situation and predicament of Scottish lowland intellectual elites during, roughly, the middle third of the eighteenth century.[101]

Briefly, Lawrence finds that related 'theories' of 'sensibility' and 'sympathy' were deployed by Scottish intellectual elites in both 'scientific' and 'social' contexts. In a 'scientific' context, Robert Whytt,

[99] *Ibid.*, 216.

[100] Christopher Lawrence, 'The nervous system and society in the Scottish Enlightenment', in Barnes and Shapin (eds.), *Natural Order* (see note 18), 19-40.

[101] For a discussion of the role of these elites in reference to Scottish science, see especially J. R. R. Christie, 'The rise and fall of Scottish science', in Maurice Crosland (ed.), *The Emergence of Science in Western Europe* (1975), 111-26, and Steven Shapin, 'The audience for science in eighteenth-century Edinburgh', *HS*, XII (1974), 95-121.

William Cullen and others developed a model of the body in which overall integration and control was effected by the 'sensibility' of the nervous system. The related 'doctrine of sympathy' maintained that the nervous system mediated mutual feeling between the various bodily organs, and thus achieved a functional solidarity. In the 'social' context, Lawrence discovers that lowland intellectuals (including men of science, but also philosophers and literati) deployed similar notions in expressing their social situation and in legitimating their aspirations as the guardians of refined feeling and integrators of Scottish political identity during the divisive years of Jacobite rebellions and their aftermath. The model of the body used by professors of medicine in the University of Edinburgh was the same model used to justify the social dominance of lowland 'men of feeling'. The progress of society, and the 'improvement' of the Scottish nation, was conceived as the transition from the 'rude' to the 'refined'. Society advanced insofar as 'sensibility' and sympathetic mutual feeling animated it; the natural leaders of an advanced society were those members capable (by virtue of their nervous constitution) of the most refined feelings. It was thus that society 'held together' and was integrated; and it was thus that the body was integrated and coordinated. The new physiological theory not only did 'scientific' work in explaining the phenomena satisfactorily, but it also did ideological work: it provided a naturalistic basis (in the nervous system) for refined sensibility; it identified intellectual elites as the natural leaders of modern society; and it justified the social and economic improvement from which intellectual elites and their kin amongst the 'improvers' stood most to gain.[102]

The general relevance of Lawrence's study lies in its refusal to create gratuitous 'demarcations' between the various contexts in which physiological theory was produced, used and evaluated. As Lawrence observes, the Scottish neural integration model of the body was 'a brilliantly innovative exploration and interpretation of physiological evidence'. It was also a 'celebration' of a certain social order and a legitimation of the 'natural governing role' of the social groupings which articulated it.[103] It was not, so to speak, 'di-

[102] For a challenging, and possibly complementary, treatment of changing eighteenth-century views on the nervous system and its functions, see Karl M. Figlio, 'Theories of perception and the physiology of mind in the late eighteenth century', *HS*, XIII (1975), 177–212, and *idem*, 'The metaphor of organization: an historiographical perspective on the bio-medical sciences of the early nineteenth century', *HS*, XIV (1976), 17–53.

[103] Lawrence, 'Nervous system and society' (see note 100), 33–4.

minished' as a scientific resource for being an ideological resource as well. The 'scientific context' has to be attended to in explaining the adoption of this model in Scotland. But the 'scientific context' is insufficient to explain the phenomenon. Here it is interesting that Lawrence echoes Brown's observation that nothing was 'gained in therapeutic efficiency in the shift from mechanism to vitalism'; '[c]linical practice remained almost identical under the new theory'. Clearly, the context of social use is essential to understanding the shift to the new model of the body. 'The Edinburgh model of the body', Lawrence concludes, 'was "multi-functional".'[104]

VI. Conclusions and continuations

Most of the studies examined above, and others like them which there has not been space to discuss, offer themselves as revisions to the particular historiography of late-seventeenth- and eighteenth-century science. They open up valuable new perspectives of science and its social uses during the Enlightenment and its antecedents. They are concrete studies dealing with particular scientific episodes, and, for the most part, do not pretend to make general theoretical and historiographic pronouncements. Nevertheless, it may be helpful if certain general attitudes and orientations which are to be found in this newer literature are drawn out, made somewhat more explicit, and connected to historiographical concerns of wider interest.

Three themes of wide relevance to social studies of scientific knowledge arise from this examination of recent literature on Newtonian science and its uses. The first two may be stated briefly, while the third bears more detailed treatment.

Social interests and science

Almost all the literature discussed in this essay identifies an important role for social interests in scientific change or in sustaining scientific accounts. These interests are often manifest in the uses to which scientific accounts were put in different historical contexts. Thus, the authors concerned discern a far more central role for the social uses of science than has hitherto been generally allowed. But it is equally important to note that the authors do not posit social uses as the sole explanation of scientific change or stability. Either in their own work, or by referring to existing historical studies, they also identify a role for the technical interests which we are accustomed to call 'properly scientific'.

104 *Ibid.*, 34–5.

In the past it has been usual to regard the identification of 'social influences' on science as a way of *exposing* science, of portraying the science so influenced as corrupted by the action of these factors. Writers like Brown and Lawrence, in their accounts of changes in eighteenth-century physiology, adopt what ought to be termed a *naturalistic* approach to the range of interests and uses to which scientific representations were put and the array of contextual factors which bore upon scientific change.[105] These writers show no inclination to 'rationally reconstruct' the scientific past, or to relegate such social factors to the purdah of the footnote.[106]

Nor is the identification of social interests sustaining scientific knowledge taken as an indication that the knowledge is thus unfit for esoteric technical uses. As Lawrence says in his summation of the mid-eighteenth-century Edinburgh model of the body, the concept was 'multi-functional'; it was not less good as a technical predictive and explanatory resource for being also good as an ideological legitimating resource. Whether social interests of any particular sort are found to sustain a specific body of science, or indeed whether they are found at all, is a matter for empirical research to discover. But, on the model of these studies, finding them present or absent has no bearing whatever for any particular evaluation of science. Revealing science to be affected by elements outside the esoteric sub-culture is merely what is to be expected from a sub-culture (like the science of the seventeenth and eighteenth centuries) incompletely differentiated from the wider culture and from other esoteric sub-cultures. Finding science to be sustained by social interests is again what might be expected from contexts (like those of the seventeenth and eighteenth centuries) wherein accounts of natural reality were constantly generated with a view to legitimating social arrangements.[107]

[105] A naturalistic sociology of knowledge appropriate to the history of science may be found in Barnes, *Scientific Knowledge* (see note 52); *idem*, *Interests and the Growth of Knowledge* (see note 52); David Bloor, *Knowledge and Social Imagery* (1976); and, in practice, in most of the papers in Barnes and Shapin (eds.), *Natural Order* (see note 18).

[106] For the historiographic consequences of 'rational reconstructionism': Imre Lakatos, 'History of science and its rational reconstructions', *Boston Studies in the Philosophy of Science*, VIII (1971), 91–135.

[107] It is not implied that nature ceased being used as a legitimating resource after the eighteenth century. There are interesting shifts in the intensity and type of such usages through the nineteenth century, but it remains very much an open question why these changes occur and what the consequences may be for scientific culture. For some speculations: Steven Shapin and Barry Barnes, 'Science, nature and control: interpreting mechanics' institutes', *SSS*, VII (1977), 31–74 (pp. 59–64); see also notes 116 and 117 below.

The problems of uses and contexts

The second theme which figures prominently in these studies concerns what might be called a revision and a revaluation of the notion of *use* and the contexts in which scientific accounts are used. As was pointed out in the introduction to this essay, it has been customary to conceive of the 'uses' of science, and particularly of its social uses, as 'mere' use. This valuation is sustained by the notion that scientific accounts are generated and judged in separate contexts from those wherein they are used, and that the context of use is posterior to and (even *morally*) inferior to the other, discrete contexts. This is a view much insisted upon by many philosophers of science and other demarcation-fetishists.

Nothing in the studies discussed here supports such absolute demarcations. And the idea that demarcations of this sort adequately describe the scientific culture of the seventeenth and eighteenth centuries is more interestingly seen as an institutionally sustained convention fostered by present disciplinary divisions and the evaluations of science they express. Empirical work by the Jacobs, Wilde, Brown, Lawrence and others has demonstrated that considerations of social use were intimately associated with the production, judgement and institutionalization of science in the late seventeenth and eighteenth century.

Moreover, their studies of the social uses of science raise more fundamental issues about the historical treatment of use generally. It is no great exaggeration to say that the *technical* and *practical* uses of scientific knowledge have traditionally been as neglected as its social uses. It is as if the practical context in which science has been deployed is an historiographical blind-spot. We can, therefore, take many of the lessons learnt from these studies of *social* use as generally applicable to the historiography of scientific uses. A proper perspective of the uses of science might reveal that the sociology of knowledge and the history of technology have more in common than is usually thought.[108]

Not only do such studies find no rigid demarcations between one context of use and another, but, more fundamentally, they do not reveal science to us except in *some* context of use. The historian dealing with institutionalized scientific thought, such as the Newtonian natural philosophy of the eighteenth century, always encounters it in

[108] For an example of the sort of history of technology which might form profitable links with the sociology of knowledge: D. S. L. Cardwell, *From Watt to Clausius: The Rise of Thermodynamics in the Early Industrial Age* (1971).

a context of use. As the writers discussed in this essay show, there is no empirical support for the view that scientific knowledge is 'first' generated in a disembodied context of pure contemplation, and 'then' put to practical, technical and social uses. If a 'philosopher' for this new and valuable historical orientation is wanted, then the Wittgenstein of the *Philosophical Investigations* will serve more than adequately, for he emphasizes that the meaning of propositions is only to be discerned through their use in concrete contexts.[109] Thus, Wittgenstein's analyses do no violence to widely accepted notions of what must, on reflection, count as sound historical procedure; they encourage sensitivity to what scientific representations are *doing* in context, and they stimulate attention to a wide range of contextual factors as potentially relevant in explicating the meaning of scientific ideas.

Nor is the 'conservative' Wittgenstein the only theoretician who provides resources which may be utilized to these ends. The 'radical' Marx and several writers in his tradition also emphasize that intellectual formulations have to be seen as tools in practical political action, and stress the dialectical relations between ideas and practice.[110] In political history, similar orientations are available from the work of Quentin Skinner[111] and J. G. A. Pocock;[112] and, in the history of art, from E. H. Gombrich,[113] Michael Baxandall,[114] and others of

[109] Ludwig Wittgenstein, *Philosophical Investigations* (Oxford, 1953), esp. part 1. With specific reference to the history and sociology of science, see David Bloor, 'Wittgenstein and Mannheim on the sociology of mathematics', *Studies in History and Philosophy of Science*, II (1973), 173–91; Derek L. Phillips, *Wittgenstein and Scientific Knowledge: A Sociological Perspective* (1977).

[110] Karl Marx, *Economic and Philosophic Manuscripts of 1844* (1973), 170–93; *idem* and Frederick Engels, *The German Ideology* (New York, 1963), 1–78; Engels, *Ludwig Feuerbach and the Outcome of Classical German Philosophy* (New York, 1941); Georg Lukács, *History and Class Consciousness* (1971); Alfred Schmidt, *The Concept of Nature in Marx* (1971); Goldmann, *Philosophy of the Enlightenment* (see note 60); and Karl Mannheim, whose *Ideology and Utopia: An Introduction to the Sociology of Knowledge* (1936) interestingly exempts scientific knowledge from sociological treatment; cf. Bloor, 'Wittgenstein and Mannheim' (see note 109).

[111] E.g. Quentin Skinner, 'Some problems in the analysis of political thought and action', *Political Theory*, II (1974), 277–303; *idem*, 'The limits of historical explanations', *Philosophy*, XLI (1966), 199–215; *idem*, 'Meaning and understanding in the history of ideas', *History and Theory*, VIII (1969), 3–53.

[112] Pocock, *Politics, Language and Time* (see note 58); *idem*, *The Machiavellian Moment: Florentine Political Thought and the Atlantic Republican Tradition* (Princeton, N.J., 1975).

[113] E. H. Gombrich, *Art and Illusion: A Study in the Psychology of Pictorial Representation* (Princeton, N.J., 1960); *idem*, *The Story of Art* (1950).

[114] Michael Baxandall, *Painting and Experience in Fifteenth Century Italy* (Oxford, 1972).

the 'Warburg school'. Writers in the 'new contextualist' tradition acknowledge a variety of theoretical 'debts', or none at all. What matters is their practical empirical achievement – the display of a scientific culture whose uses (including social uses) in context constituted its meaning.

Cosmologies and strategies

Perhaps the most intriguing realization which arises from study of recent writings on seventeenth- and eighteenth-century matter-theory is that many of the authors are implicit anthropologists. The themes they address and the manner in which they construct explanations are closely allied to important tendencies in social anthropology, and the link between the two enterprises is not diminished by the fact that scarcely any of the historical writers 'cite' the work of anthropologists. Instead, what we observe here is an apparent convergence of approaches to representations of nature in two academic fields whose practitioners have either been generally unaware of the other's work or unconvinced that each has much to offer the other.

Social anthropologists have been almost exclusively concerned with the belief-systems of tribal, pre-literate societies and have developed standard techniques for explaining collective representations of nature found in these cultures. Similarly, historians of science, working within the culture that they study, have had their own taken-for-granted methods of accounting for scientific knowledge. Each group of practitioners has tended to assume that the techniques appropriate for understanding modern, scientific beliefs are not applicable to accounting for beliefs about nature found in tribal societies, and vice versa.[115]

Anthropologists like Mary Douglas consider collective representations of nature encountered in tribal societies to be institutions inextricably bound up with the social affairs of the communities which generate and sustain them; they are explained by identifying the 'social work' the beliefs do in these communities. For example, such beliefs may be depicted as legitimating resources and strategies deployed to discourage deviance, to justify existing or desired social arrangements, or to criticize current arrangements.[116] It would not

[115] For a summary of such anthropological views: Shapin, 'Homo phrenologicus' (see note 18); also Yehuda Elkana, 'The distinctiveness and universality of science: reflections on the work of Professor Robin Horton', *Minerva*, xv (1977), 155–73.

[116] Mary Douglas is the modern anthropologist most sensitive to the importance of such usages, although it is an orientation basic to much British social anthropology: Mary Douglas, *Purity and Danger* (1966); *idem, Natural Symbols* (1970); *idem, Implicit Meanings* (1975).

occur to most social anthropologists to account for pre-literate beliefs about nature in the way that historians of science routinely interpret 'scientific' beliefs, viz. as individually generated conceptions which 'correspond to' natural reality and follow 'rational' methodological criteria. And it has not, until recently, occurred to historians of science that institutionalized representations of nature found in 'our' culture might have as one of their important functions the legitimation or criticism of the social order, or that such social uses might be central to the generation and maintenance of natural knowledge in modern contexts. Yet this is precisely the purport of the literature on matter-theory reviewed here.[117]

Consider the accounts of seventeenth- and eighteenth-century matter-theory provided by the Jacobs, Hill, Wilde, McEvoy and McGuire, Heimann, Darnton and Gillispie. The whole of this literature serves to demonstrate that the boundaries dividing 'matter' from 'spirit' (or, indeed, the dissolution of such boundaries), and the attribution of properties to each, were not decided upon by purely 'technical' considerations. Obviously, such beliefs often figured in efforts to account for optical, mechanical and chemical phenomena; but, equally obviously, the diversity of these cosmologies and the fluidity of their change makes a nonsense of any attempt to explain them by how well they correspond to 'reality' or accord with universal criteria of 'scientific method'. Instead, recent writing on matter-theory has been enriched by treating such cosmological representations in the manner usually associated with the studies of anthropologists.

Let us briefly review the various positions identified by writers on matter-theory during our period of interest. Fig. 1(*a*) schematically represents the corpuscularian matter-theory promulgated by Boyle and disseminated by the Boyle Lecturers. Each insisted that 'brute' and 'stupid' matter was dependent upon animating, external spiritual agencies for its ordered activity. Fig. 1(*c*) is a representation of the 'hylozoist' cosmology developed by English sectaries, M. C. Jacob's 'freethinking' circles, and, later in the eighteenth century, by Priestley, Hutton and the French 'Jacobins'. In this schema, spiritual animating principles were immanent in matter and external actuating agencies were dispensed with. We have already seen that in both

[117] Mary Douglas, *Cultural Bias* (1978), is the most concerted attempt to construct a social epistemology appropriate to both Western scientific and tribal thought. See also Shapin, 'Homo phrenologicus' (see note 18); Douglas, 'Environments at risk', in *idem, Implicit Meanings* (see note 116), ch. 15. The relevance of Douglas's ideas to history of science is briefly outlined in Barry Barnes and Steven Shapin, 'Where is the edge of objectivity?', *BJHS*, x (1977), 61–6.

Fig. 1

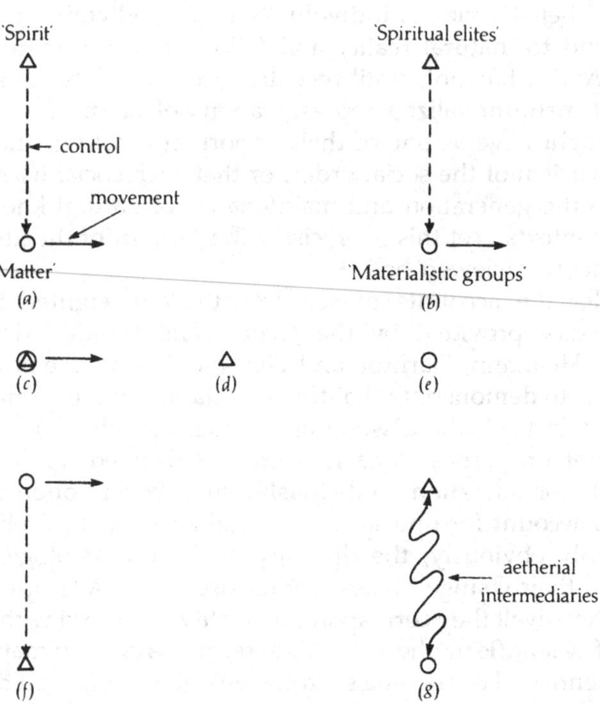

instances cosmological representations were produced and sustained by groups concerned to use them to advance social interests; they were *strategies* crafted to further the groups' conceptions of proper social order.

Fig. 1(*b*) shows how the corpuscularian cosmology legitimated social arrangements. Just as external spiritual agencies were required to superintend passive matter, so civil society was dependent upon control by spiritual elites. Order could not be otherwise guaranteed. It was a strategy which made sense within a cultural context where natural order was routinely used to justify and criticize social order. Similarly, Fig. 1(*c*) is transformed into a depiction of desired social order in which spiritual intermeddling was considered unnecessary and undesirable. Thus, position (*c*) is always available to groups resisting domination by a hierarchy. If (*a*) is a 'hierarchy-justifying' strategy, then (*c*) is a 'hierarchy-collapsing' strategy. The literature discussed here abundantly establishes that these cosmological orientations *were* so construed, both by their proponents and by their opponents in various contexts.

Schemata (*a*) and (*c*) are not the only possible stances concerning matter, spirit and hierarchy. Schema (*d*) in fact appears as the goal of certain of Newton's own writings, specifically where he was attempting to devise an 'almost matterless' universe in opposition to materialistic tendencies.[118] Bishop Berkeley and Boscovich also produced ontologies of pure spirit. Schema (*e*) in which only matter is granted existence is scarcely found during the period, although aspects of the Hobbist, 'Epicurean', and Hutchinsonian cosmologies approach this position, at least when treating the mundane and sub-lunary domains.[119] We can also follow these themes into a later period. The scientific naturalist movement of the late nineteenth century developed a 'reductionist' cosmology similar to (*f*), wherein spiritual (and mental) processes were epiphenomena of 'brute' matter in motion.[120] Frank Turner has brilliantly shown that the naturalist cosmos was the strategy of professionalizing groups concerned to justify the authority of those whose expertise lay in secular nature and to criticize the competences of spiritual elites.[121] Yet the naturalist cosmos itself did not go unchallenged. Traditional spiritual elites, anti-professionalizers, and groups hostile to secular materialism responded by restating schema (*a*), in which superintending spiritual agencies were essential in explaining the natural world.[122] Nor was anti-naturalist 'spiritualism' an exclusive preserve of the late-nineteenth-century British 'right'. A. R. Wallace's strategy for socialist progress also involved the construction of an anti-naturalist teleology in which spirit guided natural processes.[123]

118 Thackray, *Atoms and Powers* (see note 5), 60–7.

119 In a nineteenth-century context Turner intriguingly suggests that pure materialism was a position more commonly *imputed to* individuals by their opponents than actually propounded (Frank M. Turner, 'Lucretius among the Victorians', *Victorian Studies*, XVI (1972–3), 329–48). One wonders whether such an analysis might also hold good for the contexts both Jacobs examine. It might be interesting to see what proportion of their sources for 'Hobbists', 'Epicureans', or even 'hylozoists', derive from the writings of those who saw such positions as serious threats, and what proportion from those who actually espoused these views.

120 Frank M. Turner, *Between Science and Religion* (New Haven, Conn., 1974).

121 Frank M. Turner, 'The Victorian conflict between science and religion: a professional dimension', *Isis*, LXIX (1978), 356–76; see also L. S. Jacyna, 'Naturalism and society, 1850–1914' (Ph.D. thesis, Edinburgh University, forthcoming).

122 L. S. Jacyna, 'Science and social order in the thought of A. J. Balfour', *Isis*, LXXI (1980), 11–34; Brian Wynne, 'Physics and psychics: science, symbolic action, and social control in late Victorian England', in Barnes and Shapin (eds.), *Natural Order* (see note 18), 167–86.

123 Roger Smith, 'Alfred Russel Wallace: philosophy of nature and man', *BJHS*, VI (1972), 177–99; see also note 124 below.

Thus, what we see here is the array of cosmological resources available for commenting upon the genus *hierarchy*. The specific hierarchy justified or collapsed may be *any* hierarchy – the distribution of power in the class structure of society, within a given institution, or between institutions. In any context where the natural order is used to comment upon the social order the boundaries between matter and spirit, and the relations between them, *must* be addressed. But it is not meant that *these* schemata and no others will be encountered. Indeed, the literature on aetherial conceptions is highly suggestive of the ways in which matter–spirit relations can be subtly negotiated by positing hybrid intermediaries, which, in turn, affect the properties of the entities they connect (*g*). The schemata represented here are 'nodal points' in a spectrum of strategies dealing with matter, spirit and hierarchy, all of which are 'possible' and many of which are articulated.

The implicitly anthropological perspective of the writers discussed here represents a non-deterministic sociology of scientific knowledge. By emphasizing that cosmologies are constructed in contexts of use, they replace the 'automaton-actor' of metaphysical-influence studies with an active, calculating actor whose intellectual products are crafted to further the variety of his interests. The array of resources available to him to produce his cosmologies are finite; but no 'logic' constrains him to produce one account or another. Thus, by identifying processes of social conflict, we have seen that groups which did not share significant social experiences or interests nevertheless constructed cosmologies with shared characteristics. And, if we move from straightforward processes of social opposition to examine cases of selective alliance-formation or alliance-dissolution, we can see more finely tuned cultural change resulting.[124]

These perspectives on the social uses of science are termed 'implicitly anthropological' because they are compatible with the work of writers like Mary Douglas, and also because anthropological orientations have begun to make important contributions to the empirical history and sociology of science.[125] However, they might as well have been called implicitly Marxist.[126] As already indicated, Marxist con-

[124] As in John R. Durant's account of Wallace's selective critique of naturalism: 'Scientific naturalism and social reform in the thought of Alfred Russel Wallace', *BJHS*, XII (1979), 31–58.

[125] E.g., D. Bloor, 'Polyhedra and the Abominations of Leviticus', *BJHS*, XI (1978), 245–72; and forthcoming papers by Martin Rudwick and Ken Caneva applying Douglas's 'grid-group' analysis to scientific episodes.

[126] Thus, Margaret Jacob very briefly situates her most recent work within a Marxist historiography: 'Newtonian science and the radical Enlightenment' (see

cerns with the practical uses of intellectual formulations, with dialectical processes, and with the problem of imputing ideas to social groups could well provide the history of science with highly valuable perspectives.[127] Unfortunately, however, explicitly Marxist studies of science have tended to be so highly programmatic and empirically thin that unwarranted doubt is cast on the historiographic value of Marxism itself.[128] However one theoretically situates the work discussed here, it is clear that it is a new perspective on science and its social uses, the final evaluation of which awaits the appearance of many more empirical accounts. The signs are that these will not be long in coming.

All three of the major themes arising out of the literature discussed here have in common a concern to show scientific culture as a human enterprise situated in concrete historical contexts, and actively made and deployed by social groupings to serve a range of interests which cannot be specified in advance of open-minded empirical research. Thus, these studies make an important start towards eroding the 'illumination' model of Enlightenment science. And they provide a constructive programme for future historical work. To the traditional question 'Was ist Aufklärung?', they suggest we add the question 'Wofür ist Aufklärung?'.

Acknowledgements

I wish to acknowledge very considerable debts to a large number of people who read and commented upon earlier versions of this essay. The non-exhaustive list includes Barry Barnes, David Bloor, Geoffrey Cantor, James Jacob, Margaret Jacob, Donald MacKenzie, Jeffrey Sturchio, Christopher Wilde and Peter Wright. Particular mention must be made of Stephen Jacyna, whose work on nineteenth-century scientific naturalism provided insights applied to the earlier period. Students in a post-graduate course at the University of Pennsylvania in summer 1979 also helped enormously by subjecting some of the arguments contained here to ruthless criticism. But my greatest debt is to the authors whose work is discussed.

note 58), 546–7. Her Marxist sensibility seems mainly to consist in attention to the practical deployment of ideas in social contexts.

127 As, for example, discussed in Barnes, *Interests and the Growth of Knowledge* (see note 52).

128 E.g., Alfred Sohn-Rethel, *Intellectual and Manual Labour: A Critique of Epistemology* (1978), esp. part 2; David Dickson, 'Science and political hegemony in the 17th century', *Radical Science Journal*, VIII (1979), 7–37. For an attempt to provide an empirical grounding for Marxist-structuralist imputation in the case of individuals, see D. MacKenzie, 'Karl Pearson and the professional middle class', *AS*, XXXVI (1979), 125–43.

PART II. LIFE AND ITS ENVIRONMENT

4

Psychology

G. S. ROUSSEAU

Contents

I. The 'physick of the soul' 144

II. The 'quadruple alliance': four types of histories 148

III. The problems of coping 161

IV. The impact of philosophy 163

V. Psychology and social history: metapsychology 178

VI. The social history of psychology 186

VII. Mysticism and witchcraft: the dark underbelly of the eighteenth century 192

VIII. Saints and hagiographers: the 'great scholars' 199

IX. The politics of knowledge: future ferments 203

Only connect... (E. M. Forster, *Howard's End*)

In the history of Western medicine, from the school of Salerno to that of Freud, the content and structure of its four principal ingredients – its conception of the nature of man, its technical capacity to do medical research and to give treatment, its sense of religion, and the social structure of medical aid – have changed continually; but the exclusive concern of pathology with the physical side of man's being has not disappeared or changed.
(Pedro L. Entralgo, *Mind and Body: Psychosomatic Pathology,* 1955)

If one poses, for a science such as theoretical physics or organic chemistry, the problem of its relations with the political and economic structure of society, doesn't one pose a problem which is too complicated? Isn't the threshold of possible explanation placed too high? If, on the other hand, one takes a knowledge [*savoir*] such as psychiatry, won't the question be much easier to resolve, since psychiatry has a low epistemological profile, and since psychiatric practice is tied to a whole series of institutions, immediate economic exigencies and urgent political pressures for social regulation? Cannot the interrelation of effects of knowledge and power be more securely grasped in the case of a science as 'doubtful' as psychiatry? It is this same question that

I wanted to pose, in *The Birth of the Clinic* apropos of medicine: it certainly has a much stronger scientific structure than psychiatry, but it is also very deeply involved in the social structures. What 'threw me off' a bit at the time was the fact that the question which I posed did not at all interest those to whom I posed it. They considered it a problem without political importance and without epistemological nobility.
(Michel Foucault in an interview in 1967 with Alessandro Fontano and Pasquale Pasquino, published in Meaghan Morris and Paul Patton eds., *Michel Foucault: Power, Truth, Strategy,* Sydney, Feral Publications, 1979, pp. 29–30)

I. The 'physick of the soul'

My angle of vision in surveying a field of eighteenth-century scholarship may differ from that of other authors in this volume, although I doubt that many of my conclusions will. On the one hand I am a literary, not a scientific, historian, as a consequence of which I tend to see the entire world in terms of 'languages' – printed texts, verbal discourses, symbolic gestures – rather than 'systems of classification'. On the other hand I do not perceive the so-called 'revolution' in the history of science that has occurred in the last generation in precisely the same terms that others do: I view it as neither pronounced to the same degree nor significant to the same extent, although I unequivocally view it as a 'revolution'. Furthermore, and more locally within the so-called 'science' I want to discuss here – psychology – I do not believe we know enough as yet about the history of medicine to differentiate meaningfully between psychology and psychiatry in the eighteenth century: it is easier to contrast the 'philosophy' and 'psychiatry' of the period. But the psychiatry and psychology of the age overlap again and again, sometimes in unpredictable ways, as I attempt to demonstrate below, rendering it almost impossible to chart three distinct areas: psychiatry, psychology, philosophy.

All these obliquities and vantages would seem to leave me in unfortunate and uncomfortable circumstances when placed in the distinguished scholarly company in which I find myself. There are, however, many common grounds I share with the authors in *The Ferment of Knowledge,* among these terrains a sense of 'what actually happened' during the eighteenth century so far as its ideas and practices related to psychology are concerned: to mention just a few, the roots of understanding about the psyche; concepts of soul, spirit and spirits; endless debates about free will versus determinism; further endless debates about nature and man's ability 'to nurture', about the presence of innate ideas and sensations, about natural genius and education, and about the dynamism of active versus passive minds.

The flow of these and other developments then – especially their unfolding and overlapping – is not the subject of this essay, and readers who expect it to be will be disappointed. Many historians named and discussed below have described the life of these ideas and practices. I consider my task here to be an exploration of the ways they have described these developments, not a detailed and comprehensive treatment of the developments themselves.

Why is my essay not titled psychiatry, or, perhaps more comprehensively, psychology *and* psychiatry? First of all, there was no sustained distinction between them at that time, and I see little advantage, except for purely pedagogic reasons, to impose one now. After all, there cannot be a historiography of something that never existed. Moreover, what is today called psychiatry (i.e. study of the administration of a wide repertoire of therapies ranging from chemical injections to psychotherapy and psychoanalysis) enjoyed little, if any, direct correlative then; and an essay that merely tried to compare or contrast the psychiatry (or the psychology) then with ours now would stray very far indeed from the intention of this volume: to survey changing aspects of the scholarship of certain sciences in the eighteenth century.

Instead, it seemed more sensible to admit my own limitations and to take a more catholic view: to call the essay 'psychology', without too great a fear that I would be accused of slighting the period's psychiatry, and then, having made this decision to survey scholarship in several of the areas, not merely psychiatry, which constitute the matrix or nexus of ideas and human practices we today would call psychology *and* psychiatry. Such a procedure seemed plausible and practical for one who had witnessed, even in his relatively short 'history of science' life, a radical alteration in the type of scholarship being written about these areas of human activity; one that in fewer than three decades seems to have moved from mere biographical accounting to detailed analysis, often in depth, of the play of complex ideas, cultural assumptions and practical applications of medical techniques.

Yet there is a caveat, as will soon become apparent. I cannot disguise my bias: the belief that the consequence of all this radical alteration is ultimately linguistic. That is, the sense that one language has been substituted for another, in this case the language of the social sciences for that of biography and the natural sciences. The notion that this great alteration in scholarship has occurred is, of course, unassailable. But, one must inquire, precisely what fields has it opened up and how has it opened them? Those who have expanded the fields of psychological history remain scholars concerned to relate certain developments *in terms of* other developments. They are, in a sense, similar to the early Boyle Lecturers in England who related Newtonian science to audiences who could comprehend it 'in other terms', in this case in religious terms. Such 'translation' or 'relation', whenever or wherever it occurs, necessitates linguistic analysis at a rather deep base-level.

This is why I begin my account with some consideration of the word psychology. There is disagreement about the first use of the term in English. Nevertheless it is clear that it begins to appear at the close of the seventeenth century, perhaps having first been used in 1693 in the English translation of the second edition of Blanchard's *Physical Dictionary*, although it is also found in Nathan Bailey's (the London lexicographer) popular English dictionary: 'Psychologist, one who makes a study of, or is skilled in psychology; a student or teacher of the science of mental phenomena'.[1] The Latin term 'psychologia' was naturally much older – according to the *Oxford English Dictionary* it is a term used in scholarly treatises by the sixteenth century, as in the German-produced *Psychologia anthropologica* (1590–7) – and if one searched ardently enough one could probably find a dozen references to the Greek version. The point here is neither to trace the history of the concept, which has already been done,[2] or to speculate about the conditions under which it originated, but rather to suggest that as an area of human knowledge *psychology* possesses a *historiography*, not only a history, in our period, the eighteenth century, precisely because – and the cause is the important aspect – four or five trends were converging *c.* 1700. These 'convergences' are very much the subject of this essay.

Yet convergences of historical trends – incremental secularism, economic conditions, political developments, scientific societies and

[1] See Nathan Bailey's *An Orthographical Dictionary* (1727), the second volume of his popular *Universal Etymological Dictionary*, 1st edn (1721). The first use of the term 'psychiatry', according to the *New Oxford English Dictionary*, does not occur until the early nineteenth century, but the concept is doubtless known before then. See, for example, the poet Pope's own annotation to *The Dunciad*, III (A), 81–2: 'The *Caliph, Omar I*, having conquer'd Aegypt, caus'd his General to Burn the Ptolomaean library, on the gates of which was this inscription, *Medicina Animae, The Physick of the Soul*.'

[2] See F. H. Lapointe, 'Who originated the term psychology?', *JHBS*, VIII (1972), 328–35, who contends that Wolff, the German philosopher and popularizer of Leibniz, was the first to use the term. For a broad survey of the term, see Lapointe's Origin and evolution of the term "psychology"', *American Psychologist*, XXV (1970), 640–5. Launcelot Law Whyte writes in *The Unconscious before Freud* (New York, 1960), 101–2, that Wolff 'may have been the first influential German writer to give the word *Bewusstsein* its present meaning of awareness, and that his [Wolff's] analysis of unconscious factors is clearer than that of any predecessor'.

schools - are not enough for the student of historiography, though they may be sufficient for the more straightforward historian. The historiographer of psychology - for all periods, not merely for the eighteenth century - must be alive to other considerations as well, especially to written and unwritten assumptions guiding his own age. And then, he must be capable of leaping from the history to the historiography and back again without too great a sense of shock. Not only, for example, was psychology a rapidly developing speculative field throughout the period 1700–1800 (something that cannot be asserted for the previous hundred years - at least I am unaware of any scholar who has tried to make such a case), but also the unprecedented secularization of European culture during the period made tremendous claims on the average person's sense of his own 'ego' and 'selfhood'.

The matter is, of course, even more complicated than this, depending primarily on whether the student holds a dialectical view of the relationship. Not to hold one is to omit its dynamic relationship; to maintain one, on the other hand, can render the relation unwieldy because no 'hard history' of psychology 1700–1800 exists, as I attempted to say at the outset. Even attending to the one (the history of psychology 1700) at the risk of excluding the other (modern historiography construed in a pluralistic age by students racked by radically new versions of intellectual scepticism), and neglecting their dialectical relation (the precise processes used in leaping back and forth), poses severe problems. The reason, simply stated, is that history is not historiography. In the one - the former - the subject of the discourse is the period or era itself; in the other - historiography - discourse replaces history as subject and history recedes to an ancillary position among the various subjects intrinsic to the discourse. In the history versus the historiography of psychology there are other intruding factors as well. Even for the so-called 'hard history' of psychology a question arises regarding 'point of view': politics, economics, science, literature, etc. In the now-popular phrase of the late Sir Herbert Butterfield[3], there is a 'Whig' as well as a 'Tory' history of psychology. There are also others that remain to be written, from the vantage of, for example, the various economic classes (e.g. the poor man's view of psychology in the period of the Georges) or, in contrast, the budding behaviourist putting forward radical new theories. The reader will be able to supply others as well.

[3] See *The Whig Interpretation of History* (1931), and an important application of this theory to literary history during the Enlightenment: H. K. Miller, 'The "Whig interpretation" of literary history', *ECS*, VI (1972), 21–47.

This large problem - history versus historiography - cannot be solved here. Nothing short of a vast research project of scholars or by modern computer methods to compile lexicons of psychological terms would satisfy scholars about the hard facts (what I have been calling 'hard history') of European psychology 1700–1800; and I doubt if even this would satisfy them. Moreover, an equally vast team would be needed to probe ancient and medieval writings to discover the origins of the concept 'psychology' and its cognates. Finally, since many contemporary historians, self-conscious of their methods, now believe there are no 'facts' whatsoever in history and that one reads even the best historian for his style or for his myths and fictions, perhaps we should turn our attentions elsewhere, to areas less hazardous than these.[4] One may lament in this connection that there exists, as yet, no history of the attitude of the masses, but the *Annales* school project of *l'histoire des mentalités* (Mandrou, Febvre and Butke) will some day fill the gap. Given this preliminary caveat, especially the reason why this is the place to scrutinize neither the 'hard history' nor the 'dialectical dynamic', we may cautiously explore the scholarship written about psychology in the eighteenth century.

II. The 'quadruple alliance': the four types of histories

Essentially four types of histories have been written for our period: (1) chronological histories proceeding by year or period; (2) thinker-oriented histories; (3) thematic histories - which necessarily include the works of polemicists who use theme as pretext; and (4) teleological histories that are usually concerned to diminish the novelty of Freud's theories (although some expand it) by demonstrating on how much previous knowledge Freud built his own edifice. Each must be viewed individually since all current accounts of psychology in the eighteenth century, and hence a survey of its historiography, derive from these four types.[5]

[4] For some of the problems involved in this distinction, see Hayden White, *Metahistory* (Baltimore, 1973), and S. Ratner, 'The historian's approach to psychology', *JHI*, II (1941), 95–109.

[5] I intentionally do not cite copious examples of the four types, because the ensuing notes are adequate proof in themselves. The point made in this paragraph is further assisted by these essential tools: George Mora, 'The history of psychiatry: a cultural and bibliographical survey', *International Journal of Psychiatry*, II (1966), 335–56; R. M. Young, 'Scholarship and the history of the behavioural sciences', *HS*, v (1966), 1–51; L. Zusne, *Names in the History of Psychology: A Biographical Sourcebook* (New York, 1975); O. Diethelm, *Medical Dissertations of Psychiatric Interest Printed before 1750* (Basel, 1971); and E.

Chronological histories

These are a province in themselves. Less interesting for their arrangement of materials and logical structure than for the historian's approach to the psychology of the period, they often appear to have been written by emptying index cards on to foolscap sheets. Concern with methodology and scepticism about validity in interpretation are almost universally absent in this genre. Thus a chronological history such as Gregory Zilboorg's *History of Medical Psychology* (New York, 1941), which reports a great deal of information about our period, is arranged diachronically and would seem to command little interest in its approach. After all, Zilboorg was an eminent practising analyst who wrote at a time when the field had been uncharted. More important than an 'approach' is the author's (not merely Zilboorg's) 'reading', as it were, of the psychology of a given epoch, whether ancient, medieval, or Renaissance. Generally speaking, the genre attracts genuine chroniclers far more capable of amassing materials than of theorizing profoundly about them. These histories have generally not been accumulated by powerful analytical minds (Zilboorg was an exception) who have written histories of psychology as a pretext for announcing their own philosophies of history. In point of fact I have not discovered a single chronological historian of psychology who seems to care about problems of periodization (to name one area of concern to theorists), although many historians since 1950 have grown so sceptical of the dangers of generalization based on *a priori* 'period notions' that they have actually neglected their own work and probed that area.[6] Some of those who write in the journal *History and Theory* serve as a good example of the deflection: Louis Mink, Hayden White, John Passmore, W. H. Walsh, Leonard Krieger, Arnaldo Momigliano.

Chronological historians conceived of as a group have had little to say about eighteenth-century psychology, and the explanations for this silence may not be so simple as they appear. Since World War 2 they have written a vast literature about the psychology of the ancient

Clarke, *Modern Methods in the History of Medicine* (1973), discussed in the last section. Of great value also is Richard Hunter and Ida Macalpine, *Three Hundred Years of Psychiatry 1535–1860* (1963).

[6] In American scholarship see E. Heischkel-Artelt, 'The concept of baroque medicine in the development of medical historiography', *Actes du Deuxième Congrès International d'Histoire des sciences* (Ithaca, 1962), IX, part 2, 913–6. In French scholarship see M. Gourevitch, 'La psychiatrie française à l'époque romantique', *Perspectives psychiatriques*, XVI (1978), 5–99.

Greeks[7] and have endowed it, it seems, with every type of label from the 'age of discovery' to the 'age of the birth of the mind';[8] but these same historians have been less excited, for reasons not altogether clear, about psychology in the Enlightenment. The labels they give to our period are an indication of lack of hard analytical thought: Zilboorg and Henry have called it an 'age of reconstruction', while other chronological historians have used similar catch-alls to denote respite and observation rather than to postulate radical activity or radical theory in the period. Such a procedure tallies with that of most contemporary historians of science, who continue to emphasize the seventeenth century as the age of systems and theoretical discovery and who view the eighteenth century merely in terms of collecting and classifying, or, for some, of 'politicking' and 'ossifying'. Considering that most recent historians of psychology/psychiatry have been medical historians of one variety or another, it is not surprising to find this state of affairs. It appears that historians of science tend not to be quite as self-conscious of their methodology as others because they are so preoccupied with composing the 'hard history' of their as-yet unwritten history of science. This generalization is, of course, a matter of great debate. But, the accuracy of the generalization notwithstanding, the relatively conservative attitudes the historians of science embody are surpassed by those of another group: the historians of medicine. These are practitioners of an art – history – for which many are ill qualified, and, at bottom, a group that barely knows what historiography signifies.

Thinker-oriented histories

These histories of psychology are less easily typified; but here again their authors rarely possess an adequate concept of the period that is in touch with diverse types of the best scholarship. Rather than acquire a new view of the era they rely on a traditional concept and inherit as well a sense of the concept's inherent value. As in the case of chronological history, these authors generally invoke metaphors of evolution (*A* flowed from *B*, evolved from *C*, issued from *D*) and rely

[7] A few sources in this vast literature worthy of consultation include: G. Aigrisse, *Psychoanalyse de la Grèce antique* (Paris, 1960); Bennett Simon, M.D., *Mind and Madness in Ancient Greece* (Ithaca, 1977); E. L. Harrison, 'Notes on Homeric psychology', *Phoenix*, XIV (1960), 63–80. A rich resource for the medieval period is E. Ruth Harvey's *The Inward Wits: Psychological Theory in the Middle Ages and Renaissance*, Warburg Institute Survey No. 6 (1975).

[8] See Bruno Snell, *The Discovery of the Mind: The Greek Origins of European Thought*, trans. T. G. Rosenmeyer (Cambridge, Mass., 1953) and two important works by Erich Neumann, both in the Bollingen Series: *The Origins and History of Consciousness* (1970) and *Amor and Psyche* (1971).

on acquired and often dangerous views of cause and effect (*A* resulted in *B*, caused *C*, was the reason for *D*). This genre is a favourite of historians of philosophy as well. It has been popular at least since the early Renaissance, when there arose a tradition of biographical sketches in conduct books and history books (e.g. *A Mirror for Magistrates*).

By the 1850s works such as George Henry Lewes's *The Biographical History of Philosophy from its Origin in Greece down to the Present Day* (1853) – a work containing much we today would call the 'history of psychology' – were being written and continued to be written for another century. More recently the following thinker-oriented histories have appeared: R. I. Watson's *The Great Psychologists: from Aristotle to Freud* (1963) and *Eminent Contributors to Psychology* (1974), organized according to thinker and containing almost no historical discussion; Denis Leigh's *The Development of British Psychiatry*, II, *The Eighteenth and Nineteenth Century* (1961), organized as the separate biographies of three thinkers and written in a type of historical vacuum, the scenario of three lives *sans* a context; and Richard Hunter's and the late Ida Macalpine's *Three Hundred Years of Psychiatry 1535–1860* (1963), a work that reprints dozens of selections and knits them together with brief introductions and prefaces, such 'lace work', lending an impression of historical cohesion – from great thinker to great thinker, or just from selection to selection – and often producing the illusion in the reader of having digested a type of history. But rigorous self-examination reveals that one has substituted 'facts' for 'analyses'. Futhermore, thinker-oriented histories commonly focus on several major figures – usually the empiricists, Locke, Hume, David Hartley, Adam Smith, the Scottish 'common-sense' school, as well as Tuke and Pinel – in their sweeping panoramic views. In traversing large territories there is little chance of presenting a penetrating analysis and none is usually found; instead, the author conceives of his task as the amassing of large amounts of derivative material. He will read the philosophers: Locke, Berkeley, Hume, and some of the Scottish moralists. But he does not burrow in manuscript archives and does not reread primary material. In crude terms, what he produces is a rehash, and a list of works is not needed to prove that the genre is still alive and healthy.

Thematically arranged histories

These are more significant than the two former types and their yield for intellectual history is greater. They often isolate a theme – for example the 'theory of insanity' or the 'rise of madhouses' – and develop it. The old adage about 'wisdom within selection' is certainly evident here: compelled to abandon chronological organization and a history centred on main figures, these authors inevitably produce social history or a type of 'history of ideas'. Thus R. G. Hill, the Victorian surgeon and follower of the non-restraint system for treating lunatics (he followed the lead of William Tuke, as well as the Quaker moralists) wrote a history of insanity entitled *Lunacy, Its Past and Its Present* (1870), a book containing little primary research but one using 'thematic history' to argue that non-restraint is the most successful way to treat the insane.

In practice, thematically arranged histories of psychology are often, as Hill's book illustrates, polemics in disguise. Sometimes the polemical base is not even concealed, as in James M. Baldwin's *History of Psychology: A Sketch and An Interpretation* (1913), which attempts to demonstrate that the eighteenth century was a dull era but does so under the guise of a 'thematic history' of eighteenth-century psychology. Histories of 'emotion' or 'feeling' also fall into this category, although they possess a more interesting aspect: depending upon whether they are written by historians of psychology or by research physiologists, they emphasize, to a lesser or greater degree respectively, the role of the nervous system.[9] Histories of the unconscious in particular fall prey to this polemical tendency, in part because a history of the unconscious can readily form itself into a history of anything, and in part because such vast topics attract historians or scientists with broad cultural interests. As a result, books about the history of the unconscious have fared both worse and better than books on other topics: worse as a consequence of the amorphousness of the subject, better because the author will not confine himself to narrow terrains possibly lacking in significance for a large audience of readers.[10] The books by Whyte and Margetts cited in note 10 are examples: in addition to the willingness of each to survey large terrains, each is brave enough to chart an area where most other historians would fear to tread. Whyte crosses national boundaries, looking at all of Europe, even central Europe and the Slavic countries; Margetts reads primary authors ranging from the Scottish common-

[9] See H. M. Gardiner and R. C. Metcalf *et al.*, *Feeling and Emotion: A History of Theories* (New York, 1937). Some useful remarks on this matter are also found in D. A. Schon, 'Psychiatry and the history of ideas', *International Journal of Psychiatry*, v (1968), 320–7.

[10] For an example of better and worse in one book, see L. L. Whyte, *The Unconscious before Freud* (New York, 1960), and E. L. Margetts, 'The concept of the unconscious in the history of medical psychology', *Psychiatric Quarterly*, xxviii (1953), 115–38.

sense philosophers to the early German romanticists before positing her conclusion:

> There was no clear-cut division between consciousness and unconsciousness [in the eighteenth century]. The principle of continuity was followed; the one shaded into the other - unconscious perception (*petites perceptions,* vague and obscure ideas, subconscious activities) gradually merged with the unconscious apperception.[11]

Another strength of thematic historians is evident in Georges Canguilhem's *La formation du concept de réflex aux xviie siècle* (Paris, 1955), a book often referred to by both general historians and specialist historians. Dedicated to Gaston Bachelard (Canguilhem's teacher and by now the much-discussed structuralist thinker), *La formation* proceeds diachronically from roughly 1550 to 1850, as do the chronological histories. Its strength, though, lies not in its diachronic arrangement but in its narration of the 'life of an idea' (an idea woefully neglected until the 1950s), just as more recently scholars have been surveying the life of another idea, 'organic form'.[12] Canguilhem also resuscitated the work of Thomas Willis amidst a milieu of historians of science who had paid no attention to Willis, and made 'connections' in *La formation* between areas (philosophic, biological, physiological) others would have left underdeveloped. His two earlier books, *Essai sur quelques problèmes concernant le normal et le pathologique* (1943) and *La connaissance de la vie* (1952), which were historico-philosophical studies that also explored large terrains, should have provided a clue to his method. Although Canguilhem was for a time a practising physician, he wrote as a cultural historian and *La formation* still has great value today. It is probably valid to assert that Canguilhem is among the best thematic historians to have written on psychology since 1950.

Teleological histories

These are my last category, and I call them 'teleological' because they 'inexorably lead to' something. They set out to seek sources for Freud's discoveries, not in order to depreciate his work nor to show that he was right or wrong but to demonstrate on what a long tradition his own thought was constructed. Such histories are usually

[11] See Margetts, *art. cit.,* 122.

[12] See G. S. Rousseau (ed.), *Organic Form: The Life of an Idea* (1972), and a second volume in preparation that takes account of recent developments in linguistics.

restricted to a single country, though they occasionally cross boundaries and frontiers. Moreover, teleological histories have a sub-set, i.e. those aiming to show that Pinel was the 'founder of modern psychiatry' - always a dangerous label - and demonstrating his (Pinel's) antecedents: the task originally assigned to Freud's work is thereby transferred to Pinel's, and in this genre Pinel becomes a minor-league Freud. These histories thrive on metaphors such as 'evolution',[13] 'revolution', 'change', 'flow', and other terms designating continuity and discontinuity, although it is perfectly clear that their authors are far more enamoured of the genuine *telos* (Pinel, Freud, Jung) of the history than of any aspect of the historical process. Lancelot Whyte's *The Unconscious before Freud*, already mentioned, is an example of teleological history because it purports to show that the whole history of psychiatry from the ancient Greeks onward results in Freud's theory of the unconscious.[14] This pronouncement is made not to reduce Whyte to simple-mindedness but rather to isolate the assumptions a practising analyst who also writes history of psychology may hold.

Whyte's method and his conclusions reveal how teleological history usually proceeds. He divides the eighteenth century in half - a pre-1730 and post-1730 segment - and postulates that Lord Kames was the first Western thinker (he does not discuss Eastern traditions) to employ the terms 'conscious' and 'unconscious' in a form resembling modern psychological usage. Intellectual historians will be interested in the dates of the segments. 1730 is given because it divides thinkers who stressed the cognitive aspects of the unconscious from those

[13] An example is Richard Lowry's *The Evolution of Psychological Theory 1650 to the Present* (Chicago, 1971).

[14] Other teleological histories include Walther Riese, 'The pre-Freudian origins of psychoanalysis', in J. H. Masserman (ed.), *Science and Psychoanalysis* (New York, 1958), 29–72; Colin Martindale, 'A note on an 18th-century anticipation of Freud's theory of dreams', *JHBS*, VI (1970), 362–4; Richard Lowry, *The Evolution of Psychological Theory 1650 to the Present* (Chicago, 1971); Iago Galdston (ed.) *Historic Derivations of Modern Psychiatry* (New York, 1967); Hadley Cantril, *The Psychology of Social Movements* (New York, 1941); Mark D. Altschule, *Roots of Modern Psychiatry* (New York, 1965). Excellent specimens in French scholarship are René Semelaigne, *Les grands alienistes français* (Paris, 1894); H. Baruk, *La psychiatrie française de Pinel à nos jours* (Paris, 1967), Bruno Cassinelli, *Histoire de la folie* (Paris, 1939). For thirty years at the turn of this century, René Semelaigne wrote 'teleological history' of psychiatry. His classic work is *Les pionniers de la psychiatrie française avant et après Pinel* (2 vols., Paris, 1930–2). For the years 1950–75, a bibliography of more than fifty articles exists.

(post-1730) whose main interest was the unconscious mind as the seat of the passions:

> It will be noted that in the eighteenth century the unconscious mind had already been linked with a primary organizing activity or formative principle, with organs of generation and the *élan* of desire, and with illness. The nineteenth century developed these early speculations in various directions, but added little that was new to the general conception.[15]

Aside from one's attitude to the barrier date or to Whyte's lexicographical research, the conclusion is noteworthy here because it lays great weight on the eighteenth century.

But does it say enough? Presumably even Whyte would concede that the difference between sophisticated and unsophisticated psychological theories rests, perhaps exclusively, on a sense of an unconscious psychic mechanism (the eighteenth century called it 'ruling passion', 'monitor', 'censor', 'demon') guiding the individual's decisions and determining his behaviour; and the eighteenth century is the first to develop it. One can argue, as Boyle did in the Restoration, that physiology is the key to human behaviour and to most 'systems' rationalizing it;[16] but a concept of the unconscious does not ultimately depend on a sophisticated notion of human anatomy or physiology – in this instance a developed science of the brain and nervous system – for either its genesis or amplification. Again, the eighteenth century is the crucial period here. Lord Kames and other Scottish philosophers were important in the rise of psychology because they popularized a concept, however metaphysical at first, of 'the other self', the 'self' that is the shadow of the outer self but nevertheless the 'truer self': in the words of the most lyrical poet of his day, Thomas Gray, the self who is 'the stranger within thee'.[17] No similar concept exists in didactic thought before the eighteenth century, and although in earlier times the philosopher Juan Luis Vives, author of the important psychological study *De anima et vita* (1538),

[15] Whyte, *The Unconscious Before Freud* (see note 2), 104.

[16] The most eloquent of Boyle's many comments on this subject appears in *The Usefulnesse of Experimental Naturall Philosophy* (1663), pt 2, 3, 'Those great transactions which make such a noise in the World, and establish Monarchies or ruin Empires, reach not so many Persons with their Influence as do the Theories of Physiology.'

[17] *Conjectures on Original Composition* (1759; facsimile repr. Leeds, 1966), 52–3. The subject has recently been studied by S. D. Cox in '"The stranger within thee": the self in British literature of the later eighteenth century' (Ph.D. thesis, University of California, 1978).

and certain other authors, especially Cervantes and Shakespeare, have more than an inkling of 'the stranger self', they do not hypothesize about it. But one ought not to turn to Gray and other eighteenth-century sentimentalists – Henry Mackenzie's *The Man of Feeling* – for respectable theories of psychology. These doctrines about the self are reflected in more highly regarded literary forms in Swift, Pope and Johnson. The last author is a reliable, even scientific, commentator on the unconscious, and is quoted by several eighteenth-century physicians in their treatises on insanity.[18] Swift reflects some of these notions, however parodically, in his 'Digression on Madness' in *A Tale of a Tub*, as does Pope in his discussion of the 'Ruling Passion' in *An Essay on Man*, epistle II. After 1740, however, one begins to discover abundant speculation about 'the stranger within', the unconscious, and it may even be (to be historically rather than historiographically discursive) that the poets rather than the scientists generated this and allied concepts, as they did so many others assumed by historians of science to have no literary bearings.

Teleological histories whose *telos* is Pinel are fewer but necessitate commentary in a discussion of the historiography of Enlightenment psychology. Pinel's contribution to psychology and psychiatry is too well known to be discussed here: it is sufficient to note that he was one of the first physicians – Sydenham, Boerhaave, Johann Christian Reil, Cullen, and Benjamin Rush were others – to maintain that lunatics could be cured, and one of the first to erect the social and medical conditions, many taken from his original theoretical writings, necessary for recovery. Pinel is mentioned with good justification in every serious history of psychiatry in France; in social histories, as well as histories of the French Revolution, the poor movement and the humanitarian movement.[19] He has been the subject of numerous

[18] See Kathleen Grange, 'Dr. Johnson's account of the schizophrenic illness in *Rasselas* (1759)', *MH*, VI (1962), 162–8, and R. B. Hovey, 'Doctor Samuel Johnson, psychiatrist', *Modern Language Quarterly*, XV (1954), 321–35.

[19] The typical remark is brief and states that Pinel reformed treatment of the mentally ill. One can readily compile a bibliography of several dozen articles written since 1950 with this as their main theme. Also a humanitarian, Vincenzo Chiarugi is the 'hero' for Italian Enlightenment psychiatry. See, as one example, A. Balli (ed.) *Onoranze a Vincenzo Chiarugi nel secondo centenario* (Empoli, 1961); there are many others. The Americans did not have an equivalent; for a valuable survey of available care, see D. M. Blackman, 'The care of the mentally ill in America, 1604–1812, in the thirteen original colonies', *Nursing Research Conference*, II (1968), 65–113. Recently, Milos Bondy has argued in *MH*, XVI (1972), 293–6, that Johann Ernst Greding (1718–1775), the German physician interested in the classification of mental diseases, has been overlooked. Kathleen Grange has epitomized this trend in her article, 'Pinel or Chiarugi?', *MH*, VII (1963), 371–80.

biographies, and most phases of his life and works have been annotated. It is perhaps not going too far to maintain that he has been to eighteenth-century psychiatry what Newton was to its natural philosophy and Linnaeus to its taxonomy. What has been lacking in the 'Pinelian field of research' is integration of these aspects with broader developments in France: its politics, economics, and intellectual life. There has been no attempt to relate Pinel's accomplishments to 'mob psychology' in the century – to George Rudé's theory of the 'Paris mob', for example. Nor has Pinel been viewed within the context of Andrés Piquér's psychiatric treatises, which antedate Pinel's. Some historians of psychiatry have argued that Spain, not France or Britain, preserved Roman notions (transmitted by the Moors) of humanity to lunatics;[20] and it may be profitable to determine how many of Pinel's concepts are found in Piquér's treatises. In another context – the philosophic – Pinel has found few students, and this may reveal more about the nature of medical history today than about Pinel studies. Studies of the philosophical origins of Pinel's psychiatric thought are needed, that include discussion of eighteenth-century conceptions of sex, the family, and the cycle of life[21], as well as the sociology of French society before and during the Revolution.

There is no point in lamenting these lacunae; an all-important reason why these problems have not been studied involves the realization that there are few scholars anywhere who possess the necessary learning. Only the French structuralists and Marxists today seem ca-

[20] See P. Bassoe, 'Spain as the cradle of psychiatry', *American Journal of Psychiatry*, CI (1945), 731–8. Bassoe notes (p. 731) that Pinel himself had written in the *Traité médico-philosophique*, 2nd edn (Paris, 1809), 238, that 'we must look to a neighbour for an example [of humanitarian reform of lunatics]; not to England or Germany, but to Spain'. This point seems not to have been mentioned again in the secondary literature until E. Ullersperger endorsed the Pinelian attitude toward Spanish reform in *Die Geschichte der Psychologie und der Psychiatrie in Spanien* (Wurzburg, 1871). But the matter turns out to be 'nationalistic' in the ways I described above. For Pinel did not personally view the Spanish asylums until 1808, long after he had written the first version of the *Traité* (Paris, 1801), in which this passage does not appear; and even in the first edition of 1801 he advocates massive reform of existing French hospitals. Outside Spain there is no study of Piquér's medical works. Of particular interest for his theory of insanity is his 'Discurso sobre la enfermedad del Rey Ferdinando', in *Colleciones de documentos inéditos para la historia de España*, XVIII (118 vols., Madrid, 1842–95), 156–226. This work was not published until 1851 and was unknown to Pinel; the original manuscript is still in the personal library of the Duke de Osuña.

[21] Two exceptions are notable: see Lawrence Stone, *The Family, Sex and Marriage in England 1500–1800* (London, 1976), and Randolph Trumbach, *Rise of the Egalitarian Family* (New York, 1978).

pable of making the required connections: leaps from point to point within a framework, or grid, able to contain such a vast edifice; but they have recently turned their attention elsewhere than to these subjects. Even in *Surveiller et punir* (Paris, 1975) and *Histoire de la sexualité*, I: *La volonté de savoir* (Paris, 1976); Foucault is concerned with very different matters; but we must not forget that his very first book, *Maladie mentale et psychologie* (Paris, 1954), though barely read today, is a teleological history in the sense I have been describing.

Teleological histories dealing with Pinel as their fountainhead cannot be abandoned so facilely as this, for Pinel and Tuke are the protagonists in the recent historiography delineated here. As such, they are the 'subjects' of vast amounts of recent scholarship, much of it still undigested. But how are we to conceive of them – whether individually or taken together – as 'subjects'? And how did they, in contrast to other figures, become constitutive subjects in the first place? Perhaps the answer, as usual, lies in the reconstruction of a whole field of study – psychiatry or psychology – through its historiography. This is why the details of Pinel's contribution are so crucial, and why they can never be adumbrated enough often. These so-called Pinelian facts have, of course, been laid out by recent scholars in a literature barely summarized here; and while they have not been delineated at a length sufficient to satisfy all readers, the basic stones have been set. But the stones are solitary and unorganized, as it were, without a surrounding edifice and landscape. The cultural contexts of Pinel's thought – not merely in the sense of causation but of concomitancy with other developments – remain to be explored and laid out by someone able to take account of parallel developments in medicine and who is at the same time informed enough to write sophisticated history. William Bynum, author of the chapter on medicine in this book, has written elsewhere the most penetrating essay on this subject,[22] but does not have space to take the argument far enough. He demonstrates two distinct lines of thought that have developed in England since the Renaissance: first the notion that every mental state depends on functions of the brain (i.e. insanity is a physiological/anatomical aberration in every case) and, second, the notion that the aggregative secularization and gradual acceptance of certain mental

[22] See W. F. Bynum, 'Rationales for therapy in British psychiatry: 1780–1835', *MH*, XVIII (1974), 317–34. Also important as background to Pinel is Kathleen M. Grange, 'Pinel and eighteenth-century psychiatry', *BHM*, XXXV (1961), 442–53; E. A. Woods and E. T. Carlson, 'The psychiatry of Philippe Pinel', *BHM*, XXXV (1961), 14–25; W. Riese, *The Legacy of Pinel: An Inquiry into Thought on Mental Alienation* (New York, 1969); B. Mackler, *Pinel: Unchainer of the Insane* (New York, 1968).

states such as depression and anxiety has shaped our sense of mental illness. Bynum views Pinel within this context, showing how his humanitarian treatment of the insane related to each development. Such an approach renders Pinel important in the evolution of what we would call 'mechanistic medicine' (not to be confused with 'psychosomatic medicine') and also endows him with cultural significance. Moreover, Pinel sedulously argued that what was 'psychologically caused' was most effectively 'psychologically treated', an approach that challenged the physician's authority and categorized Pinel as a member of the tribe of the unorthodox, even the revolutionary.

For these proximate reasons Bynum and others consider Samuel Tuke important: not merely for his proselytization and popularization of Pinel's ideas, especially in translating them from French into English, but also for Tuke's book describing the 'York Retreat' asylum for the insane run by laymen rather than medically trained personnel.[23] 'The success of moral therapy', Bynum argues, 'thus threatened to change the rather newly established place of the medical man in the treatment of insanity'.[24] This approach does not depart altogether from Michel Foucault's in the *Histoire de la folie* (Paris 1961), in that it takes a hard look at the politico-moral dimensions of insanity. But, as Bynum notices, this was also the period (the late eighteenth century) in which 'moral therapy' was practised, and its application was a mixed blessing to 'healthy' and 'sick': the treatment of George III's bouts of madness typifies the mixture.[25] By the Regency, though, medical practitioners such as John Haslam, the prolific psychiatric author, were insistent that the cure of lunatics should be the exclusive right of medically certified men. He argued that 'mental derangement is *always* accompanied by physical derangement of some sort', a valid enough position in light of the two developments surveyed by Bynum; but the more probable reason for Haslam's stand is that medical practice had now become a profession and, consequently, the notion of entrusting lunatics to paramedics constituted a genuine threat.[26]

[23] Samuel Tuke, *Description of the Retreat, an Institution near York, for Insane Persons* (York, 1813).

[24] Bynum, 'Rationales' (see note 22), 324.

[25] The fullest account of the king's 'madness' is found in Ida Macalpine and Richard Hunter, *George III and the Mad-Business* (1969). Sect. 4, 'Georgian psychiatry', 269–347, deals with the subject.

[26] There is a difference of opinion about the moment when medicine became a true 'profession' in Britain. See Susan F. Cannon, *Science in Culture* (New York, 1978), 'Professionalization', esp. 137–41.

Haslam's position hinges, as Bynum shows, on the ties between mental and physical states. Without this concept the exclusive claim that 'lunatic spaces' (the hospital, its landscape, the patient, the mythology connecting them) fell within the sole province of the physician was impossible. Every psychic state had to be correlated with a mechanistic explanation in the body. But whereas this accounts for the development of a segment of the medical profession, it says little about the English scene at the time it received Pinel; and it does not account for the origins of Pinel's thought nor for psychiatry in France from 1740 to 1790, i.e. the half century before he innovated. One can reply that no account is necessary; that Pinel was a humanitarian – as simple as that; that before the 1790s attitudes toward madness in France were morally rather than medically determined, and, more importantly, that these same attitudes coloured medical belief: thus the logical chain becomes a vicious circle of reasoning. One can also postulate that the psychological relativism evident at the end of the century – the notion that mental health is not an absolute state but a relative one depending upon socio-economic norms – is the singlehanded work of Pinel. But I doubt that the matter is so simple, or that attitudes were as monolithically conservative as the 'teleological historians' extolling the work of Pinel would have us believe. Just as Freud had his predecessors in the nineteenth century, so did Pinel in his century, and the major figures should not be given all the credit when some is owing to the minor: in France to the labours of du Petit, Lorry and Saucerotte, in Germany to Stahl, in Italy to Vignoli, and in England to men such as Tyson, Battie, and the Monros (who are an ambiguous case since much of their practice was traditional). To return to France, medical history of the period before the Revolution is still in such a sorry state – as two recent works abundantly testify[27] – that it is no wonder teleological history exists. For example, Singer's widely consulted *Short History of Medicine*, revised by E. A. Underwood and published in 1961, leaps from the sixteenth century to Revolutionary France in the 1790s, proceeding as if nothing whatsoever happened in between. The lament that there is no medical history of Enlightenment France available in English is a theme of this essay.

[27] See George Rosen, *Madness in Society* (Chicago, 1968). Rosen's survey broke new ground but made no pretence of having covered *social* attitudes. Stone's book (see note 21), the more useful, is a goldmine of primary information; but madness is, unfortunately, not among the topics related to the author's constellation of 'marriage and the family'.

III. The problems of coping

The above four types of historical treatment are influenced by, perhaps even determined by, factors demonstrating the difficulty one faces in coping with the abundance of primary material. Chronological histories taken as a class appear to have been written in a tunnel that omits all the light of cultural history. In this genre one proceeds from event to event, from development to development, without any attention paid to the rest of the picture. It is as if the author has painted the bare features, even the dark silhouettes of the characters represented on his canvas, without filling in their shapes, colours, textures. Thinker-oriented histories proceed in almost identical fashion so far as 'coping' is concerned; but their biographical reductionism is so patent (i.e. the mere rehearsal of facts of a life without interpretation: the *ad hominem* approach) and their cultural sophistication so limited that one wonders if Carlyle's theory of history as the mere imprint of the minds of a few geniuses has ever really been laid to rest.[28] It is important (of course) to learn what Kames or Pinel contributed to the history of psychiatry. But when each is considered as if he had been merely the representation of 'some one thing' (Kames his use of a term and Pinel his humanity), and this 'one thing' is isolated as a moment without a background or a future - without an enriched context - then the result is unsatisfactory.

The two categories of 'thematic history' and 'teleological history' fare better because their authors have broader interests and greater intellectual curiosity, yet the way they 'cope' is at best merely different. Thematic histories normally proceed by isolating ideas (e.g. the brain or soul as the seat of madness; social conditions in a given region or country; religious forces in relation to secular ones), but often they reduce competing ideas (antitheses, alternative hypotheses, ideas emerging from radical methodologies) so much that logical falsification as well as historical invalidity obtains. There is another difference too: the 'thematic historians' of psychiatry are interested in change and, at their very best, process; 'static' matters (an event, a moment, a life) claim little of their attention. It is not inappropriate

[28] Actually the Carlylean tendency to view the history of human achievement as a succession of inexplicable geniuses arbitrarily bestowing knowledge upon mankind has never been abandoned. Certain origins of the nineteenth-century concept are found in my 1976 Clark Lecture in the University of California. See G. S. Rousseau, 'Sir John Hill, universal genius *manqué*', in D. J. Greene (ed.) *The Renaissance Man in the Eighteenth Century* (Los Angeles, 1978).

here to consider as an analogy the relation between anatomy (the static study of organs) and physiology (their functioning). 'Chronological' and 'thinker-oriented' historians resemble the anatomists: they have little notion of the life of ideas. 'Thematic' and 'teleological' historians resemble the physiologists: they describe the flow of ideas, the fluidity of medical terminology and the movement of terrains adjacent to their main one in psychology. The difference, as noted, is elusive, and some may pause before they concur that one exists.

Subject areas in the Enlightenment in which thematic histories would be successful (because they lend themselves to this type of treatment and because such studies have not been undertaken) are: (1) study of the madhouses of a locale or region; (2) intensive study of the 'condition' known as hypochondria in men and hysteria in women (especially a study of actual cases, not merely of theory, as in Sydenham's *Epistolary Dissertation to Dr Cole*); (3) delineation of the decline of the larger condition, melancholy, as a peculiarly 'English illness', with analysis of its sociological dimensions and comparison of it in England and other Western European countries.[29] In each case the culture in which the medical condition evolved should not be viewed narrowly.

How authors cope with teleological history is less capable of brief synthesis because, of the four types, it alone addresses itself to philosophic issues in a major sense. Teleological histories inquire into causative questions (in the lowest sense as *cause* and *effect*) by pleading eclecticism. That is, they do not pretend to gaze at every aspect of their province, even fleetingly, but select this or that for the most

[29] Some comment on the historiography of each of these three fields is necessary. (1) In this case the information is difficult to locate and those who attempt to treat it usually say little about the period before 1790 (see, for example, W. Parry-Jones, *The Trade in Lunacy* (1972). A. D. Morris's *The Hoxton Madhouses* (1958, privately printed), is an exception but it is practically unavailable. (2) The two best-known studies are Ilza Veith, *Hysteria: The History of a Disease* (Chicago, 1965), and C. A. Moore, 'The age of melancholy', in *Backgrounds of English Literature* (Minneapolis, 1953), but each is incomplete or defective, Veith because she neglects the eighteenth century, a period that ought to have been her most important, and Moore because he is a literary scholar – and although he valued research and engaged in plenty of it, he was ultimately unaware of the history of medicine. More useful than either of these is John F. Sena's unpublished study, *The English Malady: The Idea of Melancholy from 1700 to 1760* (Princeton University dissertation, 1967) and his *Bibliography of Melancholy 1660–1800* (1970). (3) Melancholy has of course been studied by many, but everyone overlooks its most significant side: the sociology of melancholy in an epoch that made it the most fashionable of diseases (see section IX below).

specific of reasons. Concomitantly, they indulge in constant value judgements: Pinel's reform was more important than A's, Tuke's than B's. More often than not the 'teleological historian' possesses a philosophy of history - perhaps because he is constantly asking questions about change and process, how *E* led to *F* - and coerces himself to view history merely as moving forward in linear time.[30]

The writings of Michel Foucault, the Olympian master of the opaque style within the impersonal mode, defy precise classification, but if one had to choose from our four categories one would probably place him here, amidst the 'teleological historians', although he himself would be the first to deny, and rightly so, that he is an historian of any variety. Nevertheless Foucault synthesizes large terrains, by depicting the changes in language used to describe psychiatric developments: in his *Histoire de la folie* he even glances at the history of these metaphors and at iconographical representations of madness for evidence. Foucault has been faulted, especially by the Americans, for many aspects of his methodology. But objection in itself cannot detract from the worth of the undertaking: to trace the prose language of psychology and psychiatry in France, particularly at the end of the eighteenth century.

IV. The impact of philosophy

Anatomy and physiology would seem to have been the two shaping forces in the history of psychology. All Western theories of the mind, no matter how primitive, ultimately owe their origin and subsequent development to these two domains, although there are moments when political circumstances (e.g. France 1789–98) are as influential. The history of European psychology can be viewed, to echo Whitehead on Plato, as a footnote on the history of this relation between anatomy and physiology, not merely of the one considered in isolation from the other. While religious and secular concepts also shape psychology, they have not done so to a degree that can compare with the role of anatomy and physiology.[31]

True as this precept may be, it overlooks philosophy which, if not as influential in shaping the ideas of Western psychology, certainly

[30] The historian of this variety may also be entrapped unawares by metaphors of flow and culmination. An example is found in the entrapment of Y. Pelicier in *Histoire de la psychiatrie* (Paris, 1976).

[31] I am unaware of any history of medicine, except that of Karl Figlio in *HS*, which studies this particular matrix: the relation of the development of anatomy and physiology to religious and secular patterns. This is what is called for, and what one ought to keep calling for.

has a more problematic role than either anatomy or physiology. The difficulty of philosophy in relation to psychology is one of definition: it is almost impossible to fix the boundaries of each, to say where the one begins and the other ends. So many types of overlap are found in the writings – medical practice is not in question here – of the two groups in the eighteenth century that one wonders if they should be treated in isolation from one another at all, or as distinct from psychiatry. The difference between psychology and psychiatry does not become pronounced until the nineteenth century; therefore, concern about the precise boundaries of these areas can be minimized for the eighteenth century. But the differences between and overlaps of psychology and philosophy in the seventeenth century are crucial. They cannot be dismissed or made light of.

In a sense, the problem of definition and overlap is practical rather than theoretical. Philosophers do not usually treat lunatics, nor did they even in the eighteenth century; but eighteenth-century 'psychiatrists' – Drs Battie, Monro, Cheyne, Whytt – though certified medical men, often wrote as if they were also certified philosophers, defining provinces of knowledge and fixing the boundaries of each. If, by the term 'philosopher' is designated a thinker with a logically rigorous system, then many eighteenth-century physicians qualify for the title, as Lester King has recently shown in *The Philosophy of Medicine: The Early Eighteenth Century*.[32] If the criterion is the resolution of an important logical or epistemological problem that previous philosophers could not solve, then most of these physicians, perhaps all, are eliminated; but most eighteenth-century philosophers would be also eliminated on these grounds. Besides, the last criterion seems to distinguish degrees of importance among philosophers, not whether they are 'philosophic' in the first place.

Thus, the definitional and semantic issue about the meaning of 'philosopher' and 'philosophic' reduces to a practical matter of classification. When Robert Whytt, M.D., often called 'the philosophic doctor', published in 1764 *Observations on the Nature, Cause, and Cure of those Disorders commonly called Nervous, Hypochondriac, or Hysteric,* he thought he was writing an informative treatment of a baffling – perhaps the most puzzling – medical condition.[33] Surely he did

[32] Cambridge, Mass., 1978. King does not address himself specifically to psychology but the kinship is implicit in the medicine he studies, and especially discussed in his chapter on 'Imagination'.

[33] I.e. melancholy and hysterical diseases arising from the structure and sympathy of the nerves. The most authoritative account of Whytt's philosophical medicine – 'philosophical' because the soul plays such an important role and because the tradition of Descartes' dualism looms so large – is R. K. French's

not ask himself, at any stage, whether the book ought to be classified as 'medicine' or 'philosophy'; and even while granting that his classification, if he had made one, would not be the definitive one, it is significant that these categories did not seem of consequence to him. The issue was hysteria, its elusive aspects, and he did not care whether readers or future historians classified his *Observations* as medicine or philosophy so long as he got his medical point across. Nor would he have considered that 'nervous diseases' were more properly classified as 'medicine' than 'philosophy': what counted was the disease, and the possibility that he could develop a system of thought about nervous diseases that would permit new treatment and cure. Of course Whytt knew that as a practising physician and professor of medicine (in the Chair at Edinburgh) his books would be viewed as 'medical books' in essence; yet this fact did not impede him from intruding philosophical content into his books or from commenting on the non-medical aspects of the diseases he studied: the nature of man in general, his human condition, the role of his reason and passion, and so forth.

The crucial issue, then, appears to be not whether the doctor writing about this or that medical condition is also a 'philosopher', but to what degree he is 'philosophic'. Resolution of the point does not reside in the counting of asides and references: the fact that a medical author of the eighteenth century refers to hard philosophy or digresses from his subject into metaphysical disquisitions does not render him a 'philosopher'. His logic and epistemology must be sufficiently sophisticated to permit him to develop a system, as Dr King has shown in *The Philosophy of Medicine*, and the system needs to be airtight even when judged by the most rigorous demands of cause and effect, necessary and sufficient condition, *a priori* and *a posteriori* reason, and the like.

Even these discriminations are insufficient, though, because when considering the overlaps of philosophy and psychology in this

Robert Whytt, The Soul, and Medicine (1969). About the methodology invoked in Whytt's *Observations*, French comments (p. 31): 'Apart from the prefatory chapter on the structure and sympathy of the nerves, Whytt writes primarily as a practising doctor; his interest is in the treatment of the diseases, and his efforts to understand the causes are directed toward that end.' French is right to reduce to two categories all nervous diseases in Whytt's classification: 'Sufferers from nervous diseases... on the whole fall into two categories, those in otherwise good health whose "uncommon delicacy" of the nervous system renders them liable to disturbance from any unusual internal or external circumstance, and those with a constant weakness in one or more of the seats of hypochondria or hysteria' (p. 39).

period, the one (psychology) blends into and seems to emerge from the other (philosophy), philosophy being the much older subject. This admittedly sweeping generalization is only valid so long as the dictates of anatomy and physiology are borne in mind: a philosopher can invent systems by the use of reason and imagination, but that same reason and imagination is determined in the strictest sense by another anatomic system, the brain and its nerves. It is therefore probably accurate to claim, though it is impossible to prove, that psychology in large part 'emerged' (even the verb is problematic) from philosophy; but in doing so anatomy and physiology must not be relegated to a distant background. Such a heavy-handed law would seem to be less valid in times of anatomic stagnation – such as the late Middle Ages – or when physiology is out of fashion or vogue and when, as a consequence, its tenets are given no credence; but when physiology's advocates are active, as in the period of Vesalius or Boyle, or when it is reverently viewed, as it was in the eighteenth century, it must play a rather significant role in shaping even the best philosophy of the day.

To maintain all this is naturally to rehearse the rise of philosophic empiricism, the movement associated with Locke, Berkeley and Hume. But lest one gets bogged down in the quagmire of British empiricism and by so doing avoids the real issue at hand – the boundaries between philosophy and psychology in the eighteenth century – further inquiry must be made into the influence of anatomy and physiology on philosophy. Afterwards, a return to philosophy and psychology can be made. Much of the reason for such procedure is owing to the curious volution of madness: not as an 'idea' tied down to a particular 'field' or 'subject' – say medicine or philosophy – but as a coherent cultural notion transmitted from century to century. Madness, insanity, lunacy, mental derangement – call it by whatever name one wishes – it is the *sine qua non* of psychiatry.

The Renaissance in Europe inherited a theory of madness almost exclusively dependent – to the extent that it was a 'theory' divested of religious and other secondary causes – on Aristotelian concepts of the soul.[34] Long before the sixteenth century, however, there had been speculation about the brain,[35] but little discussion about psychologi-

[34] Herschel Baker, 'The body–soul relationship', in *The Image of Man* (Cambridge, Mass., 1947), 275–92, is an eloquent statement of the Renaissance view. See also R. L. Anderson, *Elizabethan Psychology and Shakespeare's Plays* (1927).

[35] This can best be grasped in papers delivered at the symposium on *The History of the Brain and its Functions*, ed. F. N. L. Poynter (1958). Also see L. G. Ballester, 'Diseases of the soul (*Nosemata tes psyches*) in Galen: the imposition of a

cal health in relation to the brain. By the middle of the sixteenth century Vesalius had made a set of discoveries which were later applied by several physicians (especially Harvey and Willis) to a subject we would call brain theory. Descartes' dualism, especially the dichotomy of soul and brain, significantly influenced this picture. This is not the approriate place to rehearse Descartes' contributions to the development of psychology,[36] but it is necessary to glance briefly at Cartesian historiography as it bears on the eighteenth century.

Crudely speaking, historians of psychology have tended to make too little rather than too much of Descartes' contribution, not because it is essentially metaphysical but because it requires importation from another area, philosophy, and until the 1960s even serious students of the eighteenth century were unable to integrate medicine and philosophy. Some practising contemporary psychologists and psychiatrists are shocked to learn that philosophy now plays or has played any significant role in their 'clinical practice'; this may be a sad situation but is nevertheless true. Most consider themselves the practitioners of a 'science' (although they are cautious in their use of the term) whose roots were recently discovered by Freud. Nothing, however, could be farther from the truth. Philosophy, construed not loosely as unwritten assumptions but as informed printed texts guided by rules of logic and language, remains an intrinsic concern to the best practising psychologists; those who deny it are usually the least knowledgeable clinicians and display their ignorance by the assertion. No thinker in modern European cultural history proves the paradigm better than Descartes. The issue then is not Freud or, more germanely, Pinel, but knowledge of the mind–body problem by historians of psychology. Descartes and physicians contemporary with him acknowledged the proximity of medicine and philosophy – this indeed is one of the themes of this book, as its title and chapters reiterate – but modern historians have been loath to do so, and one reason may be that ours is an age so much less philosophical than the Enlightenment.

Other matters impinge on the philosophic as well as the anatomic bases of eighteenth-century psychology, especially their boundaries and overlaps. During the seventeenth century, anatomy made such rapid advances that almost any scientific hypothesis was soon invali-

Galenic psychotherapy', *Clio Medica*, IX (1974), 35–43; Edwin Clarke, 'Aristotelian concepts of the form and function of the brain', *BHM*, XXXVII (1963), 1–14; A. C. Crombie, 'Early concepts of the senses and the mind', *Scientific American*, CCX (1964), 108–16.

[36] The literature is too vast even to be epitomized, but one will find discussion in any serious work treating of seventeenth-century dualism.

dated. Research on nerves and their interaction with the brain threatened to change everyone's conception of the cause of insanity; now, for the first time in modern culture, it seemed impossible to refute a physiological basis of madness. Thomas Willis, the great English physiologist, was among the first to accept the validity of this concept, viz. of the mechanistic and physiological bases of insanity. He turned away from other pressing work because he believed he could not proceed until he had solved certain riddles about the brain's anatomy.[37] Yet recent historians of psychology have paid little attention to these developments. An example is their analysis of psychosomatic medicine. The first thing to notice is its neglect. No history, however Procrustean, of psychosomatic concepts now exists for psychotherapy,[38] neither for the seventeenth or eighteenth centuries, although no one questions that such concepts were then significant. This is odd as so many practitioners today consider these concepts of crucial importance.[39] Most American postgraduates studying the history of medicine would be hard put to answer questions about the origins of modern psychosomatic medicine, its essential texts and

[37] I have adopted some of T. S. Kuhn's language here from *Structure of Scientific Revolutions* (Chicago, various edns). Willis is often viewed either in the narrow biographical sense – as a scientist without a life or milieu – or as one of several anatomists who contributed to brain theory. I have yet to see an integrated study of his contribution that combines the fields discussed here. It would be useful if Willis were placed in a wider context than he has been in the past. Some clues are afforded by consulting W. J. Dodds, the minor Scottish physiologist, 'On the localization of the functions of the brain: being an historical and critical analysis of the brain', *Journal of Anatomy and Physiology*, XII (1878), 340–63, 454–94, 636–60, valuable for demonstrating how a history of 'brain theory' was envisioned, and Karl Figlio's study of 'Theories of perception and the physiology of the mind in the late eighteenth century', *HS*, XIII (1975), 177–212. The type of approach I am arguing against here constricts the contribution of Willis and unnecessarily limits it; see Horace W. Magoun, 'Early development of ideas relating the mind with the brain,' in G. E. W. Wolstenholme *et al.* (eds.), *Ciba Foundation Symposium on the Neurological Basis of Behaviour*, (1958), 4–22, and A. Meyer and Raymond Hierons, 'A note on Willis' views on the *corpus striatum* and the external capsule', *Journal of Neurological Science*, I (1964), 547–54.

[38] Walter Bromberg's *The Mind of Man: A History of Psychotherapy and Psychoanalysis* (New York, 1959) is no exception: it cursorily dismisses our period and the little it says about it is what every university student already knows.

[39] Most contemporary neurologists are persuaded that 'break-throughs' in brain theory will be discovered in the fields of anatomy and physiology. Such predisposition may change; at the present time there seems to be no doubt that contemporary research has laid its faith there. For a forceful statement of contemporary views, see Stephen Rose, *The Conscious Brain in Modern Research* (New York, 1973).

ideas, or its development from the seventeenth century to the beginning of the nineteenth.[40]

The anatomic bases of Enlightenment psychology are plagued by other problems as well, as it is not enough merely to note gaps. These must be filled, and directions for the future given. One thing is clear: given that the role of the senses is imperfectly understood in Enlightenment thought, most of all their capability of influencing thought processes, it is probably hoping for too much to expect a clear picture of the age's psychosomatic medicine. This may be why the imaginative (i.e. non-scientific) literature of the age is, as I have already indicated, ultimately our most reliable gauge of the degree to which psychosomatic ideas then permeated French and English culture. This may not be a fashionable position to take in a discussion of the historiography of eighteenth-century science, but it is the accurate one. Historians of Enlightenment literature know that all the literary forms – poetry, prose, drama – then invoked are replete with characters whose mental life reflects their physical: in the 'humour' characters in the drama, in Swift's *Mechanical Operation of the Spirit,* in Goldsmith's *Vicar of Wakefield,* everywhere in *Tristram Shandy.* The reciprocity is too blatant to be missed, but it barely exists in previous literature.[41] There are dozens of examples, of which Smollett's irascible hero, Matt Bramble, in *Humphry Clinker* (1771) is a perfect although simple one: when Bramble, whose name reveals his 'ruling passion' (to use psychological jargon of the day), writing to the physician in whose care he has placed himself, complains that he is 'equally distressed in mind and body', we know there is no simpler expression of the paradigm. His nephew, Melford, repeats the assertion over a dozen times in the novel's opening pages lest any reader miss the point. 'Mind and body' were believed to affect each other 'equally'; but the equation in reciprocal interaction was nothing invented by a novelist. It rather derived from anatomic research of the previous century, the seventeenth, that had only just reached the masses – the popular sensibility.

[40] We have no history of psychosomatic medicine, let alone one that emphasizes the Enlightenment. Valuable information is found in L. J. Rather's *Mind and Body in Eighteenth-Century Medicine: A Study based on Jerome Gaub's 'De regimine mentis'* (1965), but not even the author would contend that this specialized study could pass muster as a history of psychosomatic theory in the eighteenth century.

[41] I have surveyed aspects of the reciprocity in 'Nerves, spirits and fibres: towards defining the origins of sensibility – with a postscript 1976,' in A. Giannitrepanni (ed.), *The Blue Guitar* (Rome, 1976), II, 125–53. This essay originally appeared without any postscript in R. F. Brissenden and C. Eade (eds.), *Studies in the Eighteenth Century,* III, *Proceedings of the David Nichol Smith Conference* (Canberra, 1975), 137–57.

Literature demonstrates how science infiltrates culture; this is why literature, or the relations of science and literature, is so vital in any discussion of psychology or psychiatry, not yet a 'science' but striving to be predictive. When the imprisoned Caleb Williams, Godwin's protagonist in his novel by the name (1794), recovers from a violent bout of sickness at the precise moment when his illness is at its worst, Godwin suggests the triumph of mind over body: it is a form of Priestleyanism transposed to the terrain of literary character. For Caleb has just learned that things are looking brighter for him outside jail; and he brings about this amelioration by following a psychosomatic medical tradition almost a century old. Godwin turns the plot at this point so subtly that contemporary readers unaware of popular culture then may miss the detail; but eighteenth-century readers probably did not. They assumed a reciprocity we doubt; and while not all of them were *au courant* with medical theories of psychosomatosis, they believed in mind–body reciprocity in a way we never would. It is undeniable that today no thorough or even partial history of this theoretical relation between mind and body exists, and certainly none that charts its progress down through the eighteenth century.

The reciprocity of science and literature, and the reflection of each in the other, illuminates the semantic issues raised earlier about philosophy and psychology as separate and overlapping domains, but does not resolve their boundary disputes. If an adequate definition of philosophy could be superimposed on the discussion, the problem might be resolved, but no such superimposition is anywhere to be found, and even eighteenth-century philosophers disagree on the nature of the philosophic enterprise, as their disputes over the merits of leading philosophic systems indicate. Furthermore, it is not altogether patent, nor will it soon be, precisely what aspects of philosophy psychological theory absorbs. Eighteenth-century psychology clearly plundered the logical and epistemological areas, but it did so in the name of being 'empirical', of collecting data and not speculating further until its speculations could be verified by repetitive observations. Moreover, the 'rationalism' of both philosophy and psychology is problematic to assess: often it is not what it seems to be. Locke and Berkeley considered themselves 'rationalists' of a type; so too did Battie and Monro, as the prefaces to their treatises on madness indicate. All this dangerous labelling raises the further problematic matter of philosophic texts versus psychological, or psychiatric texts, a distinction almost impossible to substantiate when the contents of both varieties are scrutinized.

These impediments accounted for, one can begin to assess the impact of philosophy on psychology without the charge of lack of definitional rigour. This is exactly the task Lester King recently performed in *The Philosophy of Medicine,*[42] except that King does not discuss psychology. His book is nevertheless germane to this topic because it demonstrates how far afield the scholar must range if his subject depicts the interaction of medicine (whatever its branch) and philosophy. In psychological realms King discusses 'mind and body', the 'imagination', and theories of reciprocal influence in psychosomatic medicine. He prudently observes that all three topics are more or less 'philosophical', especially to the degree that each is based on logical explanations of cause and effect, and he ably argues that no picture of the general medicine of the period is complete without shading in its philosophical dimensions. Presumably King does not believe this to be true of other periods or he would not have confined himself to the 'Enlightenment'.[43] The question for us then is the following: if King's contention about philosophy and medicine obtains for the general medical theory of the age, does it not necessarily obtain for its psychological thought?

The question grows unwieldy unless there is a clear sense of its existence in the Enlightenment. Historians of ideas are right to maintain that in matters regarding the history of concepts over the sweep of time, the inspection of change is a clue. Almost all historians of psychiatry (whether 'chronological', 'thinker-oriented', or of any other type) agree that theories of the emotions were undergoing the most radical and rapid alteration in the Enlightenment: from 'innate ideas' combined at birth according to environmental factors to 'mechanistic chains' of nerves and fibres,[44] the combination of which made for the 'ruling passion' or, the late-nineteenth-century term,

[42] See note 32 above.

[43] Further evidence of this assertion is found in King's papers on the relation of philosophy and medicine. Especially see 'Stahl and Hoffman: a study in eighteenth-century animism', *JHM,* XIX (1964), 118–30; 'Basic concepts of eighteenth-century animism', *American Journal of Psychology,* CCXIV (1967), 797–807; and *The Road to Medical Enlightenment 1650–1695* (New York, 1970).

[44] Those not wishing to use technical language speak of 'the passions' and develop a quasi-scientific theory of the passions and their role in psychological behaviour; others, more physiologically inclined, discuss the ventricles of the brain, often arguing that the front ventricle seats the imagination, the middle ventricle reason, and the rear ventricle memory. The clergy upholding an Augustinian world-view still considers 'psyche-ology' to be the moral *logos* of the soul and its significant choices. Yet an author such as P. Récamier, discusses none of these historical developments in *Le psychanalyste sans divan* (Paris, 1970), purportedly a history of psychiatric institutions. For some significant discussion of these matters, see B. C. Ross, *Psychological Thought within the Context of the Scientific Revolution 1665–1700* (New Hampshire, 1970), who has discovered that '19% (367) of the papers published in the *Philosophical Transactions* from 1665 to 1700 were relevant to psychological thought'.

'the personality'. There was as yet little sense of the relation of domestic environment (from birth onward) to eventual adult personality, but early-eighteenth-century theorists (in England Francis Fuller, Nicholas Robinson, Richard Blackmore, George Cheyne, John Midriff, John Purcell, William Stukeley, John Woodward and others) were beginning to classify human types by their 'nervous' dispositions. Even medically uneducated poets such as Pope had heard a great deal about these notions, as is perfectly evident in the second epistle of *An Essay on Man* (1733), which deals in large part with concepts of the 'ruling passion'.

Now while there is nothing new in the classification of character according to 'character type' or 'humoural type', categorization according to the 'ruling passion' is innovative and points to the type of classification Battie and the Monros make later in the century when they write that, for example, A is principally lecherous or religiously inspired. What signals the 'ruling passion' as an advance is its dependence on the nervous system to an extent that the 'humoural type' never was. The latter, to be sure, was directly tied to the body – to its humours, digestive tract, saline state, the release of recrements – but the nervous system, especially the 'tone of the nerves, fibres and solids' did not loom large. Over the course of the eighteenth century this nervous state was to assume ever greater importance in determining psychological disposition and outlook. The genealogy of psychological advance, then, is this: from 'humoural' theory to 'nervous' analysis to discussions of 'anxiety' and 'insensibility' (Dr Battie's term especially) as the basis of derangement; or, in terms of thinkers, from Burton's legacy to Cheyne to Battie and Cullen. Cheyne, a Newtonian of a curious sort, portends what is to come in eighteenth-century psychological theory when he pronounces at the beginning of *The English Malady* that nearly all infirmities ultimately can be traced to fibres that are too lax or too rigid:

> That there is a certain *Tone, Consistence,* and Firmness, and a determin'd Degree of *Elasticity* and *Tension* of the Nerves or Fibres, how small soever that be... necessary to the perfect Performance of the *Animal Functions,* is, I think, without all Question, from an Excess over or Defect under which, in some eminent Degree, Diseases of one Kind or another certainly arise.[45]

And Battie validates the concept by the weight of his authority and its endorsement in his *Treatise on Madness,* when he contends that 'anx-

[45] George Cheyne, *The English Malady: or, a Treatise of Nervous Diseases of All Kinds* (1733), 66.

iety' and 'insensibility' are the preconditions of mental illness, and that both result from the 'ill conditioned state of the nerve itself'.[46]

Such decoding of the lineage of psychological theory forces one back, again and again, to the old problem of the separation of philosophy and psychology in the eighteenth century. For if mental health and illness depend upon the state of the nerves, as Cheyne and Battie had contended, then the relation of the nerves to one's perception of external objects becomes a matter of paramount importance. Yet these topics are no sooner tapped than one is driven back to empirical philosophy, as it was emerging *c.* 1680–1720. There is no escaping the elenchus in a discussion of the rise of psychology, as there is no refuge from the perpetual question of what it meant to be a 'rationalist' then.

Locke had changed the course of some English and European thinking about perception, as Morris Mandelbaum has shown;[47] and Hartley, the influential philosopher-physician, would probably not have generated his mechanistic theory of the association of ideas ('vibrations') had he been unable to build upon earlier models of memory.[48] These and others were often philosophers as well as physicians; Locke was a trained and certified physician who occasionally practised. They were also 'rationalists' (i.e. speculators who

[46] William Battie, *A Treatise on Madness* (1758), 34. Battie's explanation of madness is almost entirely, if not entirely, anatomic. The 'ill conditioned state of the nerve' derives from a defect in the fibres enveloping the nerves. Poorly constituted fibres simply will not shield the nerves from trauma and excitement. As a consequence, 'sensation of the nervous or medullary fibres, tho' they continue the same, will be in a reverse proportion to the cohesion of those minute particles which constitute the solid and elastic fibres. And in fact we find that Anxiety is almost always the consequence of morbid laxity' (p. 35). It is therefore not surprising that Battie should conclude the following relation between madness and physiology: 'No wonder is it then that the straining or loosening the solid parts of human bodies should frequently render those bodies liable to be violently affected by such objects as are scarce felt or attended to by other men, who enjoy a natural or artificial strength and compactness of fibres' (p. 36).

[47] See 'Locke's realism', in *Philosophy, Science, and Perception* (Baltimore, 1964), and J. Yolton (ed.), *John Locke: Problems and Perspectives* (Cambridge, 1969). Also, see David B. Klein, *A History of Scientific Psychology, its Origins and Philosophical Backgrounds* (New York, 1970). So far as I am aware no scholarly study exists of the reception of Locke's theories in France 1690–1780 and their influence on psychological theory there in pre-Pinelian days.

[48] There is no adequate study of theories of memory in the period, but for Hartley and psychology see R. Hoeldtke, 'The history of associationism and British medical psychology', *MH*, XI (1967), 46–65, and B. Rand, 'The early development of Hartley's doctrine of association', *Psychological Review*, XXX (1923), 306–20.

build systems of thought) with deep empirical proclivities, but they themselves saw less contradiction than we do today in being both rational and empirical; nor did they seem to worry whether they were being styled 'philosophers' or 'physicians'. In the case of some other contemporary rationalists, the act of 'integration' of the disparate realms caused concern. Shaftesbury and Hutcheson (Adam Smith's teacher), primarily rationalistic rather than empirical, continued to care about the 'ruling passion' in relation to human physiology, but did not face the matter squarely in their writing, nor could they accept Cheyne's argument about the supremacy of the nerves.[49] On contrast Adam Smith, Lord Kames, and some members of the Scottish common-sense school extended the 'integration' by combining theories about the private and public self into one system about man in society. Here, of course, are the origins of nineteenth-century sociology; but more pertinent to our discussion is the historical fact that these rationalists -whether primarily empirical, anti-empirical, logical, optimistic, or the purveyors of comon-sense - were generating the very ideas from which the most important practising 'psychiatrists' of the day (the Cheynes, Batties, Pinels, Chiarugis, Tukes, and others) built up their hospitals and eventually unchained the insane.

The conclusion should not be drawn that these psychiatrists were optimists or sentimentalists. They read Hobbes as well as Shaftesbury, and knew all too well that man is not an 'innately good creature', as Shaftesbury would have him in the *Characteristics*, 'engaging in innately good actions'. These doctors of the mind have not usually left behind catalogues of the books they owned or commentaries on the philosophers they read, but their writings make it perfectly clear that they were educated thinkers who had read the pessimistic as well as the optimistic philosophical record. A doctor whose medicine is 'the physick of the soul' will be alive to the anxiety of the 'unconscious'; and the more probing of these physicians went to the works of Mandeville, Swift and Johnson to read about those tensions of the unconscious. In the case of Dr Cheyne, whose long correspondence with the novelist Samuel Richardson survives, the evidence is

[49] I have attempted to isolate some of the problems in the rationalist–empirical trends of the philosophy and medicine of the period in ''Sowing the wind and reaping the whirlwind": aspects of change in eighteenth-century medicine', in P. J. Korshin (ed.), *Studies in Change and Revolution* (1972), 129–59. An example of the treatment received by these empiricists in this century is found in 'Frances Hutcheson and the theory of motives', *American Journal of Psychology*, LXXIV (1961), 625–9, and Thomas Verhave, 'Contributions to the history of psychology', *Psychiatric Reprints*, XX (1967), 111–16.

straightforward: one gleans from his letters exactly what he is reading and sees how literature permeates his medico-psychological ideas.

The ideas articulated in the previous paragraph were, therefore, concepts found in the writings of 'rationalists' of different backgrounds. Who knows if Dr Battie, for instance, did not consider Swift or Johnson as 'influential' a psychologist as Locke or Shaftesbury; even if he did not rank them so highly, he had to be highly knowledgeable about the philosophic legacy from Hobbes onward or his psychiatric writings would not be as they are. For even an early psychiatrist does not conjure ideas merely from other similar thinkers: he learns from such men, but learns as much, if not more, from disparate terrains.

Still another aspect of the 'philosophical' should not go unnoticed. If philosophical impact includes the religious, then the influence of philosophy on psychology must be construed even more widely than we have suggested. In the Restoration and early eighteenth century in England clergymen such as the Cambridge Platonists and latitudinarian divines wrote treatises containing seminal chapters on the memory, senses, imagination, soul; and it was from these books that the 'psychiatrists' of the eighteenth century formulated some of their theoretical ideals and shaped their practice. The degree to which *episteme* was put into *praxis* depended on the doctor's courage and ingenuity; nevertheless there is no question about these writings as a major tributary of psychological knowledge. Thus Locke's critics – especially John Norris and John Sergeant, authors respectively of *A Philosophical Discourse Concerning the Natural Immortality of the Soul* (1708) and *The Method of Science* (1696) – promoted a theory about 'ideas' in relation to 'memory' that was eventually absorbed into medical writings (Robinson, Cheyne, Hartley); and Samuel Clarke, the parsonical and learned Bishop Warburton, and Andrew Baxter, the prolific mid-eighteenth-century Aberdeen tutor,[50] all generated theories about dreams which are repeatedly referred to by early psychiatrists such as Battie and Cullen.

Granted, these and other doctors had little of Freud's sense that the analysis of dreams is a pivotal means by which to cure the patient. Nevertheless, they possessed a sense, however impressionistic, that dreams ought to be discussed when patients were examined. Even

[50] On the extant letter and manuscript remains of these authors one views the true assumptions of the age better than in dictionary definitions of dreams such as E. Chambers's *An Universal Cyclopedia* (1728) or Robert James's *Medicinal Dictionary* (1745).

James Carkesse, a layman-lunatic confined to Bedlam for much of his life, realized that dreams were important and included many in his valuable 'anatomy of a madman's mind'.[51] Anyone who surveys the literature on dreams from *c.* 1680 onwards will note the radical change that occurs in the following hundred years. Such a work as Philip Goodwin's *The Mystery of Dreams, Historically Discoursed* (1658) is altogether different from Andrew Baxter's chapter on dreams in his *Enquiry into the Nature of the Human Soul* (1733; 2nd edn 1750), and these two differ again from mechanistic treatments in which it is held in/that dreams are the result of chemical actions conveyed by the body to the brain (or, as William Godwin affirms in his novel *Caleb Williams* (1794), that bad dreams, which pave the way to anxiety and thereby to insanity, are mechanistically based on 'sensations linked together').[52]

The philosophic assumptions of the age are additionally germane because so much of the 'polite physician's' attention was then devoted to theory, so little to practice. The doctor was still a gentleman, and 'theory' – cast in whatever literary mould – issuing from his pen was the best proof of gentility. He rarely saw patients but often read a great deal. Pope's physician, William Cheselden, was not atypical of the physicians of the time: he assisted Pope in editing Shakespeare and Pope came to rely on him as a serious scholar.[53] Richard Blackmore wrote epic poems – *King Arthur, The Nature of Man, Creation: A Philosophical Poem in Seven Books* – in his carriage on the way to house calls. Other 'gentleman physicians' were interested in discover-

[51] See James Carkesse, *Lucida Intervalla... written at Finsbury and Bethlem...* (1679). Dreams also have an important religious component, although this topic cannot be probed here. In the period in question many of the sermons, religious treatises, even the Book of Common Prayer, offer valuable light on the interpretation of dreams and are in themselves, as a consequence, valuable psychological manifestos. In Catholic France, the power of confession was overwhelmingly important. Throughout Europe, the guilt-laden, patriarchal, orthodox Christian framework – Enlightenment or no Enlightenment – was an extraordinary factor in personal psychological development. Nowhere is this development better documented than in the theological literature of the times.

[52] See Godwin, *Caleb Williams* (New York, Mod. Lib. edn, 1950), 354. The Abbé Richard Jerome's *Théorie des songes* (Paris, 1766) is also worthwhile in this connection.

[53] The interplay of philosophy, psychology and physiology is nowhere clearer than in the case at hand. It was this same surgeon, William Cheselden, whose medical works stimulated the 'philosopher' Edmund Burke, though writing as a 'psychologist,' to conceptualize his particular new theory of vision in the *Philosophical Enquiry into the... Sublime* (1757). See, especially, pp. 144–5 in the 1958 edition, with notes by J. T. Boulton.

ing the nature of emotion, feeling, intuition, of dissecting the most interesting of the passions: fear, lust, anger, love. In part this is why treatises such as Adam Smith's *Theory of Moral Sentiments* (1759) were so popular and were reissued so often (five editions and four translations of Smith appeared by the end of the century) and why these works are found in the libraries of medical men, especially physicians, of the period.[54] As we shall see later, the best application - best in that it could actually touch on real lives - of these theories was made by medical men. Doctors like Battie and Cullen, already referred to, spent much of their energy transforming insanity into a constitutive subject and also persuading their less enlightened medical colleagues that insanity *ought* to be cured; they had to engage in this activity before they could be in a position to effect cures themselves. They were philosophers and propagandists, as it were, before they were physicians, and this progress has not been noted by the historians as clearly as it ought.

It is probably inaccurate to contend that the period from 1680 to 1780 saw any revolution in the *practice* of psychiatry, although the management of madhouses seems to have been improving after *c.* 1770. 'Practical revolution' came much later, in the 1790s, when Pinel unlocked the chains of the insane in France, and when others followed his precedent in other countries. But to view practice alone is to have a narrow view of the history of medicine, or of any practical science. The earlier period, up to *c.* 1780, created a revolution in knowledge about psychology - if not a revolution then certainly a ferment - that was drawn upon at the end of the century when Pinel and others enacted monumental reform. Any deep study of this revolution or ferment will lead the student to the rationalists we have been discussing here, primarily 'philosophers', although by no means philosophers only.[55] Since 1950 there has been considerable interest

[54] The valid question naturally arises of whether these physicians read the books in their libraries. It is difficult to generalize except in individual cases in which a particular library and author is involved and where evidence is discovered about a particular book. Nevertheless, it is reasonable to surmise that some of the philosophical material in the libraries of medical men was being absorbed into their medical thinking, and that some part of it dealt with dreams and the passions. From no other recent source could doctors plunder it. One must also consider the libraries of physicians as status symbols, but that cannot be done here.

[55] Jacob Kantor's unusual theory about Newton should be considered in this context. See his 'Newton's influence on the development of psychology', *Psychological Record*, XX (1970), 83–92, which contends that 'Newton's misleading influence upon psychology is not generally accepted' (p. 83). Kantor's argument is that Newton's theory of perception is 'false', that it misled his con-

in these thinkers, not only from philosophers of science and general historians, but from practising medical men such as Lester King.[56] Yet the medical-rationalists – Robinson, Mandeville, Cheyne, to mention but three among dozens – are always left out. Furthermore, while there is usually a good deal of discussion about abstract ideas such as progress, freedom, happiness, love, and education, the equally abstract but quasi-psychiatric ideas about memory, imagination, the soul, perception, the senses, intuition, and the emotions are rarely discussed in any depth except in obscure history of medicine journals. Those who have written, such as those cited in notes 35, 43, and 48, have scrutinized the rational literature of the age with a fine microscopic eye; but no one has systematically examined this same literature to understand how practising physicians were assimilating this knowledge and putting it to use in medical contexts and physician–patient situations, especially in cases involving the cure of disease. We are so far behind in this area that it would even be helpful at this point to isolate and study the prefaces and conclusions of medical works (of which there is no dearth from 1680 to 1780). Here, again, the minor works should be consulted as well as the major, for they may display the processes of assimilation discussed above far more clearly than the books which everyone read[57] and which are, perhaps, more subtle in this category.

V. Psychology and social history: metapsychology

This relation is so vast that one can best grasp its perplexities by looking at specific texts. Garrison, author of the much-consulted *His-*

temporaries who looked to it as gospel, and that it has unfortunately been accepted 'by such eminent modern physicists as Bohr, Heisenberg, Schrödinger and Bridgman'. All this may be valid, but the scientists mentioned have held little sway over modern post-Freudian psychiatrists who now generate theory used to treat patients. However, the case was different in our period, when theories of perception were plundered by medical men, usually second hand, without questioning any of their philosophical bases. Doctors such as Robert Whytt and William Cullen were exceptions to the rule.

[56] See note 32 above. To cite two non-medical examples, see Peter Gay, *The Enlightenment, An Interpretation* (2 vols., New York, 1966–9) and Laurens Lauden, *Progress and its Problems* (Oxford, 1975). Gay's fourth chapter, 'The science of man', II, 167–215, considers psychological theory solely in terms of rational speculation which, while not inaccurate, is certainly only a part of the total picture.

[57] An example is the preface of William Hillary's *A Rational and Mechanical Essay on the Small Pox* (1735), in which he insists that 'medical systems must be founded on a just mechanical way of reasoning'.

tory of Medicine, devotes a grand total of two and a half pages to the mentally ill in his very long chapter on 'The eighteenth century'.[58] Written in typical Garrisonian fashion it is a passage permeated by concrete detail and assumed to be fact, but the more intriguing aspect of it, and the reason why I have selected it to present the essential problem of 'psychology and social history', is that it comes to an abrupt stop – as if Garrison had feared he was suddenly in dangerous waters. The passage so clearly reveals the kind of writing that has been done since the 1930s in this vein (i.e. charting the social contexts of psychiatry) that it is worth quoting almost in entirety, even at the risk of fatiguing the reader with a very long passage:

> Bad as was the management of hospitals, the treatment of the insane was even worse. They were either chained or caged when housed, or, if harmless, were allowed to run at large, the Tom o'Bedlams of England or the wizards and warlocks of Scotland (Lochiel in Campbell's poem). The earliest insane asylums in the northern countries were Bedlam (1547), the Juliusspital at Wurtzburg (1567), St Luke's in London (1757), the Quaker or County Asylum at York (1792), and the *Narrenthurm,* or 'Lunatics Tower' (1784), one of the showplaces of old Vienna, where, as in ancient Bedlam, the public were allowed to view the insane, like animals in a menagerie, on payment of a small fee. The latter institution was described by Richard Bright in 1815 as a fanciful, four-story edifice having the external appearance of a large round tower, but consisting on the inside of a hollow circle, in the center of which a quadrangular building arose, joined to the circle by each of its corners. The enclosed structure afforded residence for the keepers and surgeons. The circular part contained 300 patients, 'whose condition,' says Bright, 'is far from being as comfortable as in many of the establishments for the insane which I have visited.'[59] It was not closed until 1853. Until well into the nineteenth century, insanity was regarded as not only incurable, but as a disgrace rather than a misfortune. Heinroth (1818) even regarded it as a divine punishment for personal guilt of some kind... In such later asylums as those erected at Munich (1801), Sonnenstein (1811), Siegberg (1815) and Sach-

[58] *History of Medicine* (New York, 1929; rev. edn), 400–2.

[59] See Richard Bright, *Travels from Vienna through Lower Hungary* (Edinburgh, 1818), 87–8. Bright, consultant physician to Guy's Hospital, discovered the class of nephritic diseases that still goes by his name, and travelled extensively in central Europe.

> senburg (1830), the sad lot of the insane was that of Hogarth's engraving and Kaulbach's celebrated drawing.[60] Monkemoller's researches on German psychiatry in the eighteenth century, based on the records of Hanoverian asylums at Celle and elsewhere,[61] confirm what Reil wrote of German asylums in his 'Rhapsodies' of 1803, and go to show that the theoretic part of the science in this period was nebulous philosophic speculation. Insanity was still attributed to yellow and black bile or to heat in the dog days, and symptoms, such as exaggerated self-esteem, jealousy, envy, sloth, self-abuse, etc., were regarded as causes. Kant, in his *Anthropologie* (1798), actually improvised a semeiology of insanity and maintained that, in criminal cases, it was the province, not of the medical, but of the philosophical faculty.

The point about Kant is significant enough, especially the matter of 'medical' and 'philosophical' faculties in criminal consultations. The actual passage from Garrison quoted above has gone unnoticed by almost all historians of psychiatry including Hunter and Macalpine; but the rest of the paragraph reveals how Garrison 'rounded out' his picture of insanity in the Enlightenment with details culled from his German education (though he himself was an American) at the University of Berlin. Conditions and cures are of unusual interest to him:

> The cases treated were all of the dangerous, unmanageable, or suicidal type, and no hope of recovery was held out. There was an extensive exhibition of drugs and unconditional belief in their efficacy. A case that did not react to drugs was regarded as hopeless. Melancholia was treated by opium pills, excited states by camphor, pruritus by diaphoresis, and a mysterious power was ascribed to belladonna: if it failed, everything failed. Other remedies were a mixture of honey and vinegar, a decoction of *Quadenwurzel*, large doses of lukewarm water, or, if this failed, 'that panacea of tartarisatus.'[62] The costly aqua benedicat Rolandi, with three stout ruffians to

[60] Referring to Wilhelm von Kaulbach's 'Das Narrenhaus', in his series of black and white satires, and Hogarth's eighth plate in *A Rake's Progress*. See note 115 below.

[61] Garrison's note reads '[see] Monkemoller: Zur Geschichte der Psychiatrie in Hannover, Halle, 1903', and lists the journals in which portions of Monkemoller's monograph had appeared.

[62] Garrison does not provide a source for the quotation; it may have been taken from one of the popular eighteenth-century pharmacopoeias such as John Quincy, *Pharmacopoeia officinalis* (1726), entry under 'tartar tartarisatus', 349–50.

> administer it, a mustard plaster on the head, venesection at the forehead and both thumbs, clysters, and plasters of Spanish fly, were other resources. Barbarities were kept in the background, but the harsh methods of medieval times were none the less prevalent. A melancholic woman was treated with a volley of oaths and a douche of cold water as she lay in bed. If purgatives and emetics failed with violent patients, they came in for many hard knocks, with a regime of bolts and chains to inspire fear. A sensitive, self-conscious patient was confined in a cold, damp, gloomy, mephitic cell, fed on perpetual hard bread, and otherwise treated as a criminal. The diet - soup, warm beer, a few vegetables and salads - was of the cheapest. There were some attempts at open-door treatment, such as putting the patients to mind geese, sending them to the mineral baths at Doberan, Toplitz, Pyrmont, Vichy, Bath or Tunbridge Wells, or sending them as harvest hands to Holland (*Hollandgeherei*). Marriage was also recommended as a cure.

With these last few remarkable words Garrison concludes his single paragraph on madness in his chapter on the eighteenth century. His 'conditions and cures', enumerated in the last section, tally with the most detailed description of 'conditions and cures' in the period, at least the most detailed in any English work, the one found in Smollett's *Life and Adventures of Sir Launcelot Greaves* (1760–2). This is the story of an ill-fated couple, Sir Launcelot Greaves and Aurelia Darnel, who, though they are 'fated' to marry from the start, are 'doomed' to stray apart and be 'consigned' to a madhouse before they can unite.[63] Garrison's repertory of panaceas is not dissimilar from theirs. Both range from the most common purgatives to the most obscure, such as the 'Roland water' he mentions.

Looking elsewhere, Garrison's last sentence relates that marriage was often considered a cure. He does not say on what authority; the researcher must substantiate or disprove it for himself. There are numerous major and minor female characters in eighteenth-century

[63] Smollett's sources in psychology have not been studied, nor is there an authoritative edition of his work. Suggestions about these sources have been made by R. Hunter and I. Macalpine, 'Smollett's reading in psychiatry', *Modern Language Review*, LI (1956), 409–11, who note that Smollett read Battie's *Treatise on Madness* (1758), and by P. Miles in 'Bibliography and insanity: Smollett and the mad-business', *The Library*, XXXI (1976), 205–22. See also G. S. Rousseau, 'Beef and bouillon: a voice for Tobias Smollett, with comments on his life, works and modern critics', *British Studies Monitor*, VII (Winter 1977), 33–51, who comments on some of these sources.

fiction who are sent off to be married in order to be cured. Arabella in Charlotte Lennox's *The Female Quixote* (1752) may be the parent of the type, but there are dozens of others. Three generations earlier, Sydenham combatted medical authorities who recommended that hysterical women marry; he was controversial in maintaining that the condition would be passed on to the daughters. Sydenham, however sound his intuition in this matter, lost out, and physicians continued to prescribe marriage at least until Richardson's time, and probably afterwards. Gideon Harvey, one of Sydenham's opponents, is an example of the dominant medical attitude. In *Morbus Anglicus . . . To which is added, some Brief Discourses on Melancholy, Madness, and Distraction, occasioned by Love*, published in the year of the 'Great Plague', Dr Harvey demonstrated that Sydenham was wrong. Five months after the South Sea Bubble burst, John Midriff, another physician, argued that economic reversal could be as influential as lost-love in causing insanity. The subtitle of his major medical work reveals the drift of his anti-Sydenham position: 'Containing Remarkable Cases of Persons of both Sexes, and all Ranks, from the aspiring Directors to the Humble Bubbler, who have been miserably afflicted with these Melancholy Disorders since the Fall of South-sea, and other public Stocks'. Young women, it is true, were not carted from the asylums to the marriage bed, but many who suffered the 'vapours' – the Lydia Languishes, Pamelas, cast-off mistresses – and who were portrayed as deranged, required a man in the end if they hoped to get well.[64]

Now all this is nothing Garrison volunteers; even in the last sentence of the long passage quoted above he relates psychological history in descriptive rather than analytical terms. He enjoys no inclination to scrutinize conditions, to probe motives and explore reasons; nor does he have a clue why these deplorable social conditions went unchallenged or take a stand against the progenitors of such intense human misery. If Garrison's mode is compared to Foucault's in *The*

[64] The conclusion is drawn on the basis of a bibliographical search and is confirmed, in part, by Stone, *The Family, Sex and Marriage in England* (see note 21), and by Patricia Meyer Spacks, who has studied female characters in novels written by women; see *The Female Imagination: A Literary and Psychological Investigation of Women's Writing* (1976). Some of the implications of 'female melancholy' for literary tragedy were brilliantly analysed by Walter Benjamin During the years 1924–7 in an essay entitled *Ursprung des deutschen Trauerspiels;* see George Steiner (ed.), *Walter Benjamin: The Origin of German Tragic Drama* (1977), 145–63. In *The Concepts of Illness, Disease and Morbus* (Cambridge, 1979), 88, F. Kräupl Taylor suggests that 'female hysteria' was a nosological concept whose significance so thoroughly disappeared that total reconstruction is necessary to understand the 'morbus' (his term).

Birth of the Clinic,[65] a book dealing with the same subject, the difference between each mode is apparent. Foucault's language substitutes metaphor and metonymy for concrete detail; the narrator hides himself, as it were, behind a veil, and the direction of his paragraphs, written in an opaque style, is so unpredictable that any clear statement of their 'subjects' or 'themes' may suffer from *reductio ad absurdum*. Nevertheless, Foucault states at the start of his book that

> I should like to attempt here the analysis of a type of discourse - that of medical experience - at a period when, before the great discoveries of the nineteenth century, it had changed its materials more than systematic form. The clinic is both a new 'carving up' of things and the principle of their verbalization in a form which we have been accustomed to recognizing as the language of a 'positive science'.[66]

Foucault's rhetorical inventory of themes and concepts is essentially this: the history of medicine construed as a branch of knowledge and the history of ideas; *excurses* on time and space that glance at the Marxist and structuralist writings of Gaston Bachelard, Gilles Deleuze, and Roland Barthes; orations to persuade the reader that 'contextuality' is everything and that the crucial period in the development of the asylum in France and England occurs between 1750 and 1800;[67] further arguments contending that the period before 1750, after 1800, or even after 1850, is not nearly so essential; persuasion through the use of arcane materials (names, places, events) - so arcane that the reader grows distracted from the main subject that there is a language of 'medical spaces' in the Enlightenment; discourses on discoursing in the impersonal self-reflexive mode, in which the effective subject disappears; a political ideology that is explicitly anti-authoritarian; the 'clinic', 'asylum', or 'hospital' as essential imagery invoked in discourse aimed at demonstrating that no subject in history or philosophy is so crucial (i.e. superlatively) as the announced 'principle of verbalization'.

These 'inventories' do not answer the question of whether or not Foucault is concerned with the so-called 'straight' history of medicine: that is, history of medicine conceived in traditional terms and tra-

[65] Originally published as *Naissance de la clinique* (Paris, 1963) and then under the above title in 1973, trans. A. M. Sheridan Smith. This approach to the 'clinic' or 'hospital' as a special kind of space needs to be compared to that of J. D. Thompson and Grace Goldin in *The Hospital: A Social and Architectural History* (New Haven, Conn., 1975).

[66] *Birth of the Clinic* (see note 65), xvii–xviii.

[67] The 'national histories' discussed above consider the chronological distinction important, but Foucault does not, nor does he comment on the matter.

ditional rhetorical moulds, and without preoccupation with the historian's metaphors, degree of self-reflectiveness, or other 'principle of verbalization'. This is why the inclusion in the book of a passage such as the following, that appears on the fourth page of Foucault's preface, is so ambiguous:

> In 1764, J. F. Meckel set out to study the alterations brought about in the brain by certain disorders (apoplexy, mania, phthisi); he used the rational method of weighing equal volumes and comparing them to determine which parts of the brain had been dehydrated, which parts had been swollen, and by which diseases. Modern medicine has made hardly any use of this research. Brain pathology achieved its 'positive' form when Bichat, and above all Récamier and Lallemand, used the celebrated 'hammer, with a bread thin end' . . .[68]

It may have been true in the early 1960s, when Foucault was presumably writing these passages, that 'modern medicine [had] made hardly any use of this research', and it may have been equally true that brain pathology achieved the 'positive forms' Foucault affirms it did after 1800 under the conditions he describes (i.e. Bichat, who died in 1802 and performed most of his important work at the end of his very short life of thirty-one years). But all this substance notwithstanding, it is Foucault's empirical accuracy, or lack of, that constitutes the *least* reliable and interesting aspect of his method; the aspect, also, that has least attracted students to his books. The matter seems simple but should not be casually dismissed, for analysis of historical accuracy reveals the truest clue to decoding the recent historiography of scholarship in our field of inquiry.

In the passage just cited, for instance, Foucault derives his date of 1764 from a journalistic account of Meckel's experiment published in a popular Paris weekly.[69] Foucault may or may not know the difference between the significant and insignificant work of the elder Johann Friedrich Meckel, or at least the less significant work of his son of the same name. But *The Birth of the Clinic* offers no evidence that Foucault has actually read any of the elder Meckel's works – a charge that speaks for itself. Moreover, the sentence about 'modern medicine', translated literally from the same two French words, is ambiguous. Does Foucault mean 'the practitioners of modern medicine', or recent scholarship in the history of medicine, or something else? Surely the basis of his claim here depends upon clarification of his ambiguity.

[68] *Birth of the Clinic* (see note 65), xii.

[69] *Ibid.*, 20, n.18, for the account in the *Gazette salutaire*, XXI, 2 August 1764.

After all, why should modern medicine rehabilitate Meckel's medicine?[70] The point about the 'celebrated hammer' is less controversial; every American, and probably European, medical student learns about it. Historians of medicine concur about its significance and usually there is little to discuss. But why are Récamier and Lallemand given more credit ('above all') than Bichat? Does Foucault know something medical historians do not? Is he withholding some information? The real issue in deciphering this prose lies in 'moving from sentence to sentence'; after sufficient decoding it becomes clear that any ostensible subject of the discourse is ancillary to Foucault's verbal pyrotechnics. But Foucault must be trusted when he discloses in the preface, already quoted, that he is most concerned 'with the analysis of a type of discourse'.[71] Such qualification ('a type') would seem to absolve him.

Further study of *Naissance de la clinique* and similar Foucaultian books would demonstrate the reason for their success. These texts do not isolate the 'psychiatric history' or 'medical history' as a constitutive subject. Rather, they transform the old 'text' into new 'discourse' and the old 'medical history' into new 'contextuality'. Therefore they represent, as the Swiss scholar Jean Starobinski has noticed, 'an experiment in a new way of writing the history of science, a testing

[70] Most histories of medicine written since 1950 do not mention Meckel, nor do histories of psychology. The only work about him appears in German medical journals; for an example see H. Schierhorn, 'J. F. Meckel', *Anatomische Anzeiger*, CXXXVII (1975), 221–56.

[71] *Birth of the Clinic* (see note 65), xvii. I have attempted to interpret Foucault's method in R. Allen (ed.), *The Eighteenth-Century Bibliography for 1974* (Iowa City, 1975), 790–4, and in 'Foucault on madness and civilization', *ECS*, IV (1970), 90–4. Hayden White has written admirably on similar matters in 'Foucault decoded: notes from under-ground', *History and Theory*, XII (1971), 52–7, but he focuses on Foucault's method of 'deconstruction' whereas I try to demonstrate that the rhetoric employed permits the categories found. Other secondary studies are listed in Meaghan Morris and Paul Patton (eds.), *Michel Foucault: Power, Truth, Strategy* (Sydney, 1979), 92–100. But further analysis is necessary, for Foucault's prose style has altered considerably since the 1960s, especially in one crucial aspect: before this time his implied reader was anonymous – everyone, every man. Afterwards, Foucault began to write with specific reader-scholars in mind, not merely in his short essays but in books as well. Thus, 'Theatrum philosophicum', originally published in *Critique*, CCLXXXII (1970), 885–908, and now translated into English and printed in D. Bouchard (ed.), *Language, Counter-memory, Practice* (Oxford, 1977), 165–96, specifically addresses Deleuze and Pierre Klossowski. Whole sections of this important essay are written with a specific reader in mind. How else can one interpret pp. 190–1 of the English version and Deleuze's incredible footnote on p. 191?

ground'.[72] I would add that the 'experiment' is rhetorically loaded: it contains just the right amount of obnubulation, to use T. S. Eliot's term,[73] to mystify and confuse the unsuspecting reader, and to cause the suspicious reader to pause from his own writing and investigate the new mode. But, further to account for the success of these historico-medical texts, Foucault also possesses what Garrison did not: a sufficiently brilliant imagination to connect the most disparate terrains. By so doing Foucault successfully constructs a rhetorical illusion of having broken ground – i.e. discovered a new epistemology – for the history of medicine. Furthermore, it is crucial to compare Foucault's passage quoted above with what Garrison wrote about the elder Meckel in his internalist and quasi-prosopographical study of the history of medicine:

> Johann Friedrich Meckel (1724–74), of Wetzlar, graduate at Göttingen in 1748, with a noteworthy inaugural dissertation on the fifth nerve (Meckel's ganglion), became professor of anatomy, botany, and obstetrics at Berlin in 1751, and was the first teacher of midwifery at the Charité. He was the first to describe the submaxillary ganglion (1748), and made important investigations of the nerve-supply of the face (1751) and the terminal visceral filaments of the veins and lymphatics (1772).[74]

Foucault, by way of contrast, does not provide as much verifiable information but inflates Meckel's position by rhetorical hyperbole ('Modern medicine has hardly made any use of this research', etc.) and cajoles the reader into believing that a new area of late-eighteenth-century psychology has been tapped – even into thinking that the whole history of psychology and psychiatry can be rewritten in this new mode. It may be that it can and that it will be, for to study

[72] See *New York Review of Books*, XXII (22 January 1976), 18–22. It is not widely known that although Starobinski writes abundant literary criticism, he is a trained physician who holds a chair in the history of medicine at the University of Geneva. His career sheds light on the boundaries of knowledge – philosophy, psychology, physiology – studied in this chapter. The same Starobinski who writes impressionistically about Rousseau and Baudelaire writes such essays as 'The word reaction: from physics to psychiatry', *Psychological Medicine*, VII (1977), 373–86.

[73] In *The Function of Criticism at the Present Time* (1923), *Selected Essays of T. S. Eliot* (New York, 1960, new edn), 12–22, the pages where Eliot develops the concept. Eliot's analysis of Rêmy de Gourmont, the turn-of-the-century literary author who wrote on scientific topics, bears parallels with mine of Foucault: both are, in Eliot's phrase, 'master illusionists of fact' (p. 21).

[74] Garrison, *History of Medicine* (see note 58), 334.

Foucault's mode is to probe the insufficiencies of the internalist school, i.e. Garrison.

Historiographically speaking, several points are to be gathered. First and foremost is the fact that Garrison and his admirers or followers provide almost no analytic content whatsoever, while Foucault and his Franco-American epigoni probably include too much.[75] Yet an author such as Foucault is revered (he is already a cult figure everywhere except in America) because medicine, especially psychology, has enjoyed no analytical history for the period studied in this volume. The books by Garrison and others discussed above are contextual orphans and analytical surrogates; likewise, one could compose a bibliography of essays and articles that shed some light on facets of the subject. But no single book exists wherein one can learn about the evolution of psychology 1700–1800 from a trained, authoritative cultural historian who also writes clearly and perceptively. Richard Hunter and Ida Macalpine's book (see note 5 above) is the only attempt at a checklist, and it is an anthology, not a descriptive or analytic work in any sense of the term. Thus, when a shrewd and gifted stylist such as Foucault enters the vacant arena, he can ransack the primary sources in any manner he desires and still lend the impression that he is doing precisely what traditional medical and social historians should have done decades ago.

A second crucial observation needs to be made about historiography, this one in connection with the work of the late George Rosen, perhaps best known for his book *Madness in Society: Chapters in the Historical Sociology of Mental Illness*,[76] the middle sections of which deal with our period. Economic conditions and rational ideas form the two cornerstones of Rosen's explorations into mass mental illness, the first to correlate psychological movements with social status, the sec-

[75] In America, where Foucault's writing is taken much less seriously than it is in Europe, Foucault's 'epigoni' are not historians of medicine or science, but literary critics and general philosophers and historians. American institutes of higher criticism rarely engage in analysis of Foucault's books, whereas there are no professional historians of medicine to my knowledge who take Foucault seriously. His work has parallels in this sense with that of T. S. Kuhn, who merely diverted many historians of science when he published his anatomy of scientific revolutions but who has few followers today and virtually no students who would actually call themselves Kuhnians. Perhaps Theodore Brown, author of a study on the iatromechanical movement in England is an exception. Time alone will tell, both about Kuhn and Foucault, and the 'internalist-externalist' debates in the history of modern thought.

[76] Chicago, 1968. See also Rosen's 'Social attitudes to irrationality and madness in 17th and 18th century europe', *JHM*, XVIII (1963), 220–40.

ond, the ideational, to relate societal attitudes at a given time to philosophical movements and currents then apparent. The result is a work of lasting worth; but the diachronic period surveyed is so vast (from the ancient Greeks to the present day) that superficiality necessarily arises. Moreover, Rosen has not usually discriminated between the varieties of historical scholarship; trained as a physician, his sense of solid grounding in history was the compilation of an exhaustive number of index cards.[77] The result is fragmented discourse, to import Foucault's term, of the telephone-directory variety, 'discourse' that reads well only for those wishing to know what the terrain is, not for those searching to understand it.

Other materials available for the social history of psychology in our period require brief commentary: scattered in journals are many useful observations about the hospital movement, its conditions for patients and doctors, its facilities and conditions.[78] Much less accessible is information on the economics of this development, especially the fees paid to hospital physicians, their surgeons and apothecaries, and the costs to the public of all these services. After 1758 treatment was obtainable in England, usually without hope of total cure, but little is known about its precise nature, especially that in outlying areas, and we have failed to unearth even a few manuscripts that would significantly alter the picture of our knowledge. It may be that all this 'history of psychology' is of little importance in comparison with the type of analysis that has existed now for almost fifty years – that evident in the gulf that lies between a Garrison and a Foucault – but it is hard to pronounce on this matter too definitively. What is certain is that one can pick one's favourite historian, for example Perry Miller, author of *The New England Mind,* ponder his discussions about psychology in that book,[79] and then comprehend his discourse by measuring its proximity to or distance from the Garrison–Foucault poles, the extremes of internalism and externalism. Garrison never

[77] Rosen obtained his medical degree from Berlin University in 1935, never having obtained any degree in history. See pp. 199–202 below.

[78] See, for example, Richard Hunter and Ida Macalpine, who have written about the origins of the private madhouse system in England in *BMJ,* II (1972), 513–15. Eighteenth-century madhouses have been surveyed by: A. D. Morris, *The Hoxton Madhouses* (1968, privately printed); David J. Rothman, *The Discovery of the Asylum* (New York, 1971); W. Parry-Jones, *The Trade in Lunacy: A Study of Private Madhouses in England in the 18th and 19th Centuries* (1972); various biographies of William Cowper the poet which treat of Dr Nathaniel Cotton's 'Collegium Insanorum', a lunatic asylum in St Albans where Cowper was incarcerated.

[79] Disseminated through ch. 9, 'The nature of man', in I: *The Seventeenth Century* (New York, 1939).

speculates about any cultural or social development; even his value judgements - what is good or bad for medicine - are suppressed. Foucault proceeds by an opposite method, never recording fact *qua* fact but collapsing traditional subjects and their objects (e.g. a specific author, a given text, an episode or moment) into discourse about newly constituted subjects (e.g. the 'history of medicine', 'systems of medical classification'), and then setting this discourse into the broadest possible context. As a result, Foucault is the externalist *ne plus ultra*. He is, in fact, the externalist's externalist. That a thinker of his idiosyncratic brilliance and with his particular political matrix (i.e. Marxist *manqué*, anti-establishmentarian, sexual anti-hero, spokesman for all those who are sexually and mentally chained), should dwell repeatedly on the history of medicine as a primary category is endlessly fascinating in itself. That he began his career writing about discontinuities in the history of psychiatry must never be forgotten by readers interested in the historiography - especially the Enlightenment historiography - of psychology. But what perhaps ought to engage us more than the moment of Foucault or his perpetual medical interests is the formula for his literary blend. I dare to say 'ought to' because it is not easy to persuade internalist historians and internalist scientists that language, in any sense, is the crux of the matter.

VI. The social history of psychology

Measured against this rhetoric of history should be the stark facts of the social history of the period's mental life - otherwise no interpretation of whatever variety, no progress in scholarship, whether internalist or externalist, linear or dialectical, can be made. The chart that follows is admittedly reductionistic and pertains to one country only, Britain, but nevertheless epitomizes the germane developments as they evolved in the country most alive to humanitarian needs in the eighteenth century:

Year	Events
1674-	James Newton opens 'a Mad-house' in Wood's Close in the parish of St James's, Clerkenwell, which flourishes until the end of the seventeenth century.[80]

[80] Little biographical information is known about Newton, who was intimate with the Earls of Northampton and occupied their town house, the 'Manor House' at Clerkenwell. The private papers of the house are not extant, so far as is known, but a group of handbill advertisements remain in the British Library. See Add. MSS. C. 112, fol. 9. The 'madhouse' appears on the 1720 map of John Strype in Book 4 of Stow's *Survey of the Cities of London and Westminster* (1747).

1695– John Miles opens a madhouse at Hoxton in Islington, in a road just north of the present Old Street tube station.[81]

c. 1700– Lack of regulation permits relatives to incarcerate persons who are not deranged, and English law renders it almost impossible for the imprisoned to escape.

1713–14 A parliamentary effort to combine the numerous Poor Laws results in the passage of an act to shelter homeless lunatics (13 Anne, c. 26, Vagrancy). This is part of a continuing attempt to regularize and simplify laws related to vagrancy in Britain.

1715–43 Throughout the Walpole administration regulation continues; no attempt under Walpole to effect reform.[82]

1742 Parliament sets up a 'committee' to investigate solutions to problems that arise from the attempts to administer the existing vagrancy laws, including those sections concerning the nuisance of lunatics.[83]

1744 The committee makes recommendations leading to a Consolidation Act of 1713 (17 George II, c. 5), but these merely 'consolidate' the various laws enacted in 1713–14, and do not add new laws, and especially do not regulate private madhouses.

1751 St Luke's Hospital for Lunatics opens, with Dr William Battie, member of the College of Physicians, as its first physician.[84] This is the second public hospital for lunatics in England.

1751–2 Dr Battie opens a private facility for lunatic patients in the Islington Road.[85]

[81] See Morris, *The Hoxton Madhouses* (see note 78), who has studied Miles's house from the Restoration onwards. In 'The first psychiatric hospital of the Western world', *American Journal of Psychiatry*, CXXVIII (1972), 1305–9, R. D. Rumbant describes a lunatic asylum in colonial America during these same years.

[82] British social historians have not delved into the matter. See M. Dorothy George, *London Life in the xviiith century* (New York, 1925), who is as silent on the subject as Kathleen Jones, *Lunacy, Law and Conscience 1744–1845: The Social History of the Care of the Insane* (1955) and Parry-Jones, *The Trade in Lunacy* (see note 78).

[83] Jones, *Lunacy, Law and Conscience* (see note 82), sums up the committee as follows: 'In 1742, the Pretender and his son were still active on the continent, and the Hanoverian succession not yet fully established; yet the committee included both the King's friends and those who were potential enemies. Thus there is nothing to suggest that this was more than a routine investigation to find ways of mitigating a common nuisance.' The diversity of the large committee may support this generality but it would take more research than Miss Jones has performed, as well as more understanding of the politics of medicine (see section IX below), to substantiate it.

[84] No adequate biography of Battie exists, but some biographical material will be found in John Monro's, M.D., *Remarks on Dr Battie's Treatise on Madness* (1758; facsimile repr., ed. R. Hunter and I. Macalpine, 1964).

[85] The evidence is found in Hunter and Macalpine, *Three Hundred Years of Psychiatry* (see note 5), 402.

1753 Battie begins to admit medical students to St Luke's to permit them to observe the mentally ill firsthand.

1754 Battie assumes proprietorship of Newton's private madhouse in Wood's Close.[86]

1761 The case of Rex v. Turlington rocks England. In this case one John D'Vebre committed his allegedly insane wife to Turlington's Chelsea madhouse to rid himself of her. But her physician and relatives obtained a court order securing her release.[87]

1760–2 Tobias Smollett writes a popular novel, *Life and Adventures of Sir Launcelot Greaves,* the first to be serialized in a British magazine, satirizing the conditions in private madhouses. The heroine is Aurelia Darnel, a sane woman unlawfully confined. The conditions of Aurelia's imprisonment are nowhere mentioned; Sir Launcelot Greaves, however, is confined in the same madhouse as Aurelia and treated as badly.

1763 A new 'select committee' is appointed by parliament to investigate the regulation of private madhouses but does not investigate thoroughly. It produces an insubstantial report on which parliament fails to act.[88]

1766 The Manchester Hospital for Lunatics is opened.

1774 Parliament passes an 'Act for Regulating Private Madhouses' (14 George III, c. 9) which has greater legal than administrative significance. Kathleen Jones has admirably summed up its importance: 'Even if the visitation envisaged by the Act had been carried out systematically and conscientiously, the proprietors of

[86] Newton's son (see note 80 above), who had owned and managed the premises until his death in October or November 1750 (see *Gentleman's Magazine,* xx (1750), 525). It is not known who owned the house from 1750 to 1754, but Battie purchased it in 1754 (see *GM,* xxv (1754), 496). After Battie's death in 1776 the house was continued by John Monro, M.D., who bequeathed it to his son Thomas, who relinquished it in 1803. During this entire period preachers of whatever denomination were prevented from entering the premises: 'We [preachers] are forbid to go to Newton [i.e. the madhouse now owned by Battie] for fear of making them wicked; and to Bedlam, for fear of driving them mad.' See N. Curnock (ed.), *The Journal of the Reverend John Wesley* (1938; bicentenary edn, 8 vols.), III, 455.

[87] Jones, *Lunacy, Law and Conscience* (see note 82), discusses the case known as 'Rex v. Turlington' in the legal literature (see *The English Reports* (176 vols., Edinburgh, 1900–30), XCVII, 741).

[88] Some consideration of the petition of Sir Cordell Firebrace of Melford Hall in Suffolk to the College of Physicians attempting to regulate private madhouses may illuminate the social history of insanity in England during the 1760s. The College flatly turned it down. Sir George Clark's analysis of the rejection in *A History of the Royal College of Physicians* (Oxford, 1966), II, 582–4, is authoritarian and politically coloured. Jones's estimate in *Lunacy, Law and Conscience* (see note 82), is shrewder: 'The impression left on the reader is that, although the Committee was bound to investigate, it did not *want* to investigate too deeply' (italics mine).

private madhouses would have remained almost as free from the fear of legal penalties as before. The Act said nothing on the subject of the medical supervision of patients, diet, overcrowding, mechanical restraint, or deliberate brutality of treatment. Its primary purpose was to provide safeguards against illegal detention, but it failed even in this simple object, since there was no means of forcing the proprietor to comply with the orders of the Commissioners.'[89]

1777 The York Hospital for Lunatics is opened.

1788 The King's second bout of illness opens legal questions about regency and, once again, about madness.[90]

1790 The Liverpool Hospital for Lunatics is opened.

After 1790 the pace of change, at least in England, was so accelerated that one wonders if it should be considered in this survey. It may be that its nature, inspired in part by certain reforms occurring in France and reported in England almost daily, belongs to the activities of the next half century. Be this as it may, the above reduction may demonstrate that while things were getting better, they were nevertheless improving at a slow pace; even one predisposed to anti-Whiggish judgements would have to be blind not to concede that, in view of the number of years involved, progress was slow.

In contrast, however, to this basically optimistic picture is the country's dark side - dark so far as madness and sanity are concerned. This aspect of the Enlightenment embraces everything other than so-called light and reason, and has only recently grasped the attention of some of the best students of the period: E. P. Thompson writing on crime in *Albion's Fatal Tree* (1976), Lawrence Stone on problematical sexuality in *The Family, Sex and Marriage in England 1500–1800* (1796), and Foucault on repressed sexuality in *Discipline and Punish: The Birth of the Prison* (1977). The improvements discussed above must be balanced against this very different state of affairs if one hopes to formulate an accurate picture of the times.

VII. Mysticism and witchcraft: the dark underbelly of the eighteenth century

Every epoch enjoys or suffers - depending on one's point of view - secret cults and psychic epidemics, ranging from the Orphic rites of

[89] Jones, *Law, Lunacy and Conscience* (see note 82), 39. Of chronological interest here is George Mora, 'The 1774 Ordinance for the hospitalization of the mentally ill in Tuscany: a reassessment', *JHBS*, XI (1975), 246–56.

[90] All the bouts have been studied by Hunter and Macalpine in *George III and the Mad-Business* (see note 25), a controversial book that stirred up further controversy.

the ancient Greeks to the massacre in 1979 in Guyana. Oddly enough, the eighteenth century also witnessed a mass suicide in Guyana. This fact is not recorded to reawaken interest in Guyana or claim modernity for the eighteenth century, but rather to suggest that in some essential ways, for those that cut through the superficies of historical events, centuries resemble each other. George Rosen says this about the activities of our period:

> The eighteenth century, like most other periods of history, is difficult if not impossible to characterize with a few limited phrases. It was not all enlightenment, rationalism, urbanity, and good sense. At the very time when religious enthusiasm was considered with suspicion and reproach, there appeared in several parts of Europe and America religious groups and movements which found a characteristic expression not only in their doctrines but equally in strange bodily agitations and extravagant behaviour.[91]

Rosen's point is valid, especially the quip about 'all enlightenment, rationalism . . .'. But his analysis of the behaviour and appearance of these 'doctrines' and 'agitations' must engage us more:

> These include the Camisard prophets of the Cévennes in Southern France; the Jansenist convulsionaries at the cemetery of St Médard in Paris; the sect called Shakers; the Great Awakening initiated by Jonathan Edwards in New England; the early methodist movement; the mystical, enthusiastic Russian sects, particularly the group known as Chlysti; and the Jewish sects, specifically the Frankist group and the Hasidism of the Baal-Shem Tov, that developed in the wake of the messianic movement of Sabbatai Zevi. To discuss these developments in detail would exceed the bounds of this presentation, and I shall limit myself to the essential features of some of them.

Rosen's 'essential features' are those noticed by every historian: the mass rituals, occult ceremonies, secret divinations practised in dens of sorcerers and conducted in gothic surroundings; but one wonders why no one has followed his lead and fathomed the whole story of these movements of 'great awakening' in England and France.

As early as 1677 John Webster, the radical reformer of the English universities and a practising physician, connected witchcraft with mental illness and magic with derangement. In *The Displaying of*

[91] *Madness in Society*, 209. The following passage is also found on this page. I intentionally avoid discussion of Mesmer and 'Mesmerism' in this section, as it has little to do *au fond* with my subject.

Witchcraft Webster affirmed 'that there are many sorts of deceivers and imposters' who are 'but divers persons under a passive delusion of melancholy and fancy'.[92] But the medical doctors did not take up the cudgels for Webster: only preachers and other clerical types did, pouring forth a literature *contra* enthusiasm on grounds that it was but insanity in disguise. These ephemera, as Christopher Hill has shown in various of his books,[93] were directed primarily against 'radical madness', the insanity of a carefree nation overrun by political rebels. William Erbery's *The Mad Mans* [*sic*] *Plea* (1653) is an example of the genre in the century before the one of interest here. After 1688 and the Revolution, political circumstances change 'the island of Great Bedlam': there is less enthusiasm and, consequently, less radical madness. From roughly 1700 onward this variety of insanity is replaced by the derangement that results from wretched environment, ill-starred love, and, of course, in Dr Battie's words, the 'ill-conditioned state of the nerve'.

One brief interlude, however, rocked the island at the beginning of the century, the history of which remains to be told. This is the episode of the French Camisard prophets. They had revolted in 1704 and virtually disappeared into England after that time.[94] Some Camisards committed suicide and others migrated into the hinterlands of central Europe; but those who fled to Britain gained quite a few followers, not the least vocal among whom were Sir Richard Bulkeley, the Irish baronet who established a 'cult' at Dunlavan in county Wicklow,[95] and John Lacy, the fanatical Englishman who was known to many of the Wits including Addison, Steele, Defoe, Swift, Whiston, and D'Urfey.

Both Bulkeley and Lacy spoke 'in tongues', prophesied hard days

[92] The passage quoted is from the subtitle of Webster's tract. Information about Webster and witchcraft in the Restoration is found in Christopher Hill, 'Mechanic preachers and the mechanical philosophy', in *The World Turned Upside Down: Radical Ideas during the English Revolution* (1972), 287–305.

[93] *Ibid.*, 277–86 especially, in the chapter entitled 'The island of Great Bedlam.'

[94] See Charles Almeras, *La révolte des camisards* (Paris, 1959), 41–50, for the Camisards in England after 1704, and Ronald Knox, *Enthusiasm: A Chapter in the History of Religion with Special Reference to the Seventeenth and Eighteenth Centuries* (Oxford, 1950), 356–71. See M. C. Jacob, 'Newton and the French prophets', *HS*, XVI (1978), 134–42, and my study, in preparation, 'The French prophets and the English establishment: a study in cult and ritual experience'.

[95] Illness seems to have been the cause of Bulkeley's conversion, especially his wish that the Camisards could cure his ailments: by 1707 he began to hold occult meetings in Ewell in Surrey, where he ministered over a church; in 1708 he wrote *An Impartial Account of the Prophets of the Cévennes in a letter to a Friend*, one of several such treatises.

for England, and gained many hundreds of English converts to the 'French religion of the Camisards'.[96] For ten or twelve years their influence was considerable: witness, for example, the sheer size of the British satirical literature directed against them in the years 1707–9.[97] But after the death of Queen Anne they seem to have become extinct, never to be reinvigorated in a similar shape or form, not even in the fanatical excesses of George Whitefield. This is not the place to speculate about the reasons for their temporary success in Britain, yet Rosen is surely correct to notice that the French prophets 'did not allow the tradition of ecstatic trance and violent transport to die out' and that these are the psychic phenomena that interested the British. But it is impossible, without the performed research, to know precisely what region of the mind of the average person in England and France they occupied. Historians of the Quaker movement in eighteenth-century England, for example, continue to suggest that the country 'practised' far more agitation and ecstasy than we usually think and that much of this supernatural experience has literally been obliterated by prolific 'Whig historians' such as George Saintsbury and George Trevelyan who have a vested interest in propagating myths about the peaceful eighteenth century as an Age of Reason; but it is impossible to generalize adequately, let alone authoritatively, about this aspect without a thorough revaluation of the religion of the century.[98]

[96] Lacy was converted shortly after the arrival of the prophets in 1705. In 1707 he translated Maximilian Misson's *Théâtre sacre des Cevennes as A Cry from the Desert, or Testimonials of the Miraculous Things lately come to pass in the Cevennes verified upon Oath and by other Proofs*. Daniel Defoe, William Whiston, Richard Steele – all well-known persons – tried to reason him out of his belief but to no avail.

[97] Written by many of the most popular Wits of the period. The silence of Swift is a mystery: he spent the winter of 1707–8 in London, just at the time of the prophets' notoriety, and he was wholly opposed to their brand of enthusiasm, as he had made clear in *A Tale of a Tub*.

[98] This has not been undertaken, and the little scholarship that exists is fragmented. Basically the situation in the country relative to the town is that of a gap beginning to appear between the administration of folk remedies used to cure mental derangements – or at least assuage them – and the remedial practices of physicians, a widening gap abundantly documented for England by Parson Woodforde's diaries; see N. C. Hultin, 'Medicine and magic in the eighteenth century: the diaries of John Woodforde', *JHM*, xxx (1975), 349–66. But even this is merely a facet of the entire picture: as English and French society grew increasingly secular, the question of what it meant to stand apart from so-called 'normal' and conventional men and women became more difficult than ever to answer and, furthermore, as the process of secularization marched forward, and as technology changed the social fabric of daily life,

In France the darkness regarding madness and sanity was somewhat less intense. Eighteenth-century France was more centrally organized than England in the regulation of its asylums. Its national programme built public hospitals for the mentally deranged and administered them more adequately than the English did theirs. France did not have 'private madhouses' *per se,* and there are no equivalents in France of the English legislation of 1713–14, 1744 and 1774; what private asylums there were existed at the discretion of the landlord or manager of the property. These distinctions, however, may not do justice to the differences between the two countries. The clue is the legal action, described above, of 1774: regulation of the insane, especially the type of treatment they ought to receive, was frequently discussed in England before this date, but until 1774 lunatics were merely considered as a subgroup of vagrants.[99] Not until then did the British government do anything genuine to help lunatics *qua* lunatics. All these differences may combine to explain why the activities of the Jansenist convulsionaries at the Parisian cemetery of St Médard, mentioned *en passant* by Rosen, are to be expected, and why one can readily conceive that the common people of France were far more capable than the English of embracing such religious enthusiasm. Robert Favre, the contemporary French social historian, has recently

some of the young especially found their religious needs unfulfilled. Some folk turned to religion with a zeal unknown to their parents and grandparents, and were consequently labelled 'mad' – lumped often together with incurables and those whose grief had clearly deranged their wits. Some background to this development is found in T. J. Schoeneman, 'The role of mental illness in the European witch hunts of the 16th and 17th centuries: an assessment', *JHBS,* XIII (1977), 337–51, and R. E. Hemphill, 'Historical witchcraft and psychiatric illness in Western Europe', *Proceedings of the Royal Society of Medicine,* LIX (1966), 891–902, but no one except George Rosen has extended the study beyond 1700. Then there is the matter of the behaviour of enthusiastic devotees of dissenting religions. Some eighteenth-century physicians believed these groups to be mentally ill, suffering from physiological 'mania' or demonically inspired 'deformity of the mind' (a phrase that needs scrutiny), but the manner in which religious belief coloured the opinion of 'mad doctors', as they were then called, has not, to my knowledge, been studied.

[99] See the chart beginning on p. 190. No study such as Jones's *Lunacy, Law and Conscience* (see note 82) exists for France. Instead, one discovers short pieces dealing with mental hospitals or medical treatment in a region. See, for example, E. H. Ackerknecht, 'Political prisoners in French mental institutions during the Revolution, and under Napoleon I', *MH,* XIX (1975), 250–5, and J. Alliez and J. P. Huber, 'L'assistance aux malades mentaux au xviiie siecle à Marseille', *Histoire de sciences médicales,* x (1976), 60–71.

demonstrated that French conceptions of death in the period had deep-rooted medical implications, especially in the area of 'God-willed illness'.[100] Here, again, is an aspect in which the two countries differed significantly: the secularizing effects of English medicine had occurred before 1700 and influenced, in turn, popular conceptions of disease, dying, and the life after death. Such differences between French and English medicine and religion, and the differing governmental policies towards each, cause one to wonder whether a 'national approach' to the historiography of psychology in the eighteenth century is not sensible after all.

Considering England again, far more knowledge about the undercurrents of popular psychology can be gained by microscopic examination of the 'enthusiastic' aspects of methodism, a field barely tapped, and the rites of secret societies such as the Medmenham Monks and the Beef-Eaters, than by cross-cultural comparisons of England and France, France and Holland, Holland and Spain, and so forth. It is true that the clubs were not religious cults as such; but they reached out to the contemporary religions, especially radical dissenting ones, to propagate their private rites and ceremonies - for example in the religious ecstasies and erotic activities of the 'Monks'. The methodological difficulty here is that so much of the evidence needed to reconstruct these so-called 'cults' has disappeared, or was never written in the first place.[101] Furthermore, sociological factors involving class structure, political position, and psychological environment (i.e. urban planning, architecture, available space) also influence the reasons why these various clubs existed in the first place. Then there is the zone of the erotic imagination, especially the fantasy life, of the period, a territory illuminated in erotica, in controversial tracts on *Onania,* and, of course, in the imaginative literature of the day. The imaginative literature in England is illustrated by the fictions of Defoe, Richardson, Cleland and Sterne, but even more lucidly by the

[100] R. Favre, *La mort au siècle des lumières* (Lyon, 1978): 'La médecine: du rêve à l'espoir', 221–43; 'Vers un politique de la santé publique', 244–72. See also Philippe Ariès, *Western Attitudes toward Death from the Middle Ages to the Present* (Baltimore, 1974), who notes that 'in the second half of the eighteenth century, things [i.e. attitudes to death] changed'. Such changes of attitude reflected a new concept not only of society but of personal psychology within that society.

[101] This may be why there has not been an adequate description of their activities from the eighteenth century onwards. Existing works include L. C. Jones, *The Clubs of the Georgian Rakes* (New York, 1942), and Iwan Bloch's *Sexual Life in England* (1938).

works of those female novelists who gazed into the world of 'male sexual fantasy': Mary Manley's *The Adventures of Rivella* (1714); Mary Davys' *The Accomplish'd Rake* (1727); Charlotte Charke's *The History of Henry Dumont esq* (1755); and, most notably, Eliza Heywood's *Love in Excess* (1719), *The Secret History of the Present Intrigues of the Court at Caramania* (1727) and *Life's Progress through the Passions* (1748).

Methodism, on the other hand, is a less difficult phenomenon to get hold of: from the differences between a Wesley and Whitefield one can gauge, even at this removal of time, what factors must have influenced the people, and in what ways they were being affected by social progress. But all this, perhaps surprisingly, has not been done, and remains an open territory for the historian of culture, especially the historian with a broad gaze who can integrate various areas of human concern.

Nevertheless, it is one thing to study the 'radical fringes' of a culture and another to determine norms and middle grounds for that culture at large – for its degree of superstition and its submission to a popular mythology. One can summarize, as George Rosen did, the mystical traditions of the century, which ranged in scope from the activities of those just discussed to the various Masonic orders, Rosicrucian sects, and other heterodox cults. But bibliographical summary, as all will agree, is one thing and exploration with a view to microscopic analysis another; and unless these in-depth studies are correlated with the conservative and middle zones of eighteenth-century psychological behaviour they will lack significance and expose themselves to charges of triviality. This, *au fond*, is the inadequacy of books such as Knox's *Enthusiasm*. It is excellent in what it attempts – the charting of the evolution of religious enthusiasm throughout English culture – but it hardly synthesizes its terrain, synthesis being the only attribute that would give it a context meaningful enough for one to know how to interpret it. And then the late author, a prominent member of the Catholic Church, was so confident about his own point of view of 'the true God' that his bias is soon evident. It is true that Peter Gay, the German-born Yale historian, has performed this act of synthesis for the 'century of illumination',[102] but he has, alas, so very little to say about psychic epidemics, extravagant behaviour and strange bodily agitations; and, furthermore, he has such a simplistic teleology of improvement, that one cannot imagine him seriously perplexed, as was Rosen, with the dark underbelly or radical fringe of his period. At the very least, these topics remain desiderata.

[102] In *The Enlightenment: An Interpretation* (2 vols., New York, 1966–9), II, 'The science of man'.

VIII. Saints and hagiographers: the 'great scholars'

The number of pioneering scholars in the field of psychology is few. If the criterion is seminal books, then this assertion gathers strength. There have been two kinds of metaphorical saints: those who have discovered significant new information, and those who have surveyed the whole field. The former group, as I shall attempt to demonstrate, sometimes interprets as well and sometime does not; the latter spends all its time assimilating and interpreting.

Precisely how our 'saints' interpret is another matter: what values, assumptions, even political predispositions do they bring to the act of interpretation? With the exception of one case (i.e. Hunter and Malcalpine), which I have already discussed (see p. 151 above), the group that has discovered significant new information has strangely not been accorded greater scholarly kudos than the broad surveyors. Indeed there is every reason to believe that many students of the subject cannot tell the difference between the two groups. So little attention is given to historiography in the history of medicine that most practitioners do not even know what it is, or how it differs from history.[103] Scholars since 1950 who properly ought to be considered in the first group - those who 'discover' - include, as already mentioned, Richard Hunter and Ida Macalpine, a team which has compiled the single most important anthology of primary documents with critical commentary.[104] Also to be included are Kathleen Grange, who studied British psychological theory in the 'age of Johnson',[105] C. A. Moore, the American literary historian, who first pointed out the significance of the condition melancholy in imaginative literature from the time of Dryden to that of Sterne,[106] and George Rosen, whose important work has already been discussed. All five followed on the heels of Gregory Zilboorg, M.D., discussed above, who also engaged in primary research. But most of Zilboorg's work was com-

[103] This point is implicit in the essays collected by E. Clarke and is discussed further in an essay of mine entitled 'Ephebi, Epigoni, and Fornacalia,' *Theory and Interpretation: The Eighteenth Century*, xx (Autumn 1979), 203–26.

[104] See note 5 above. I have not found similar anthologies for French, German or Italian psychiatry.

[105] See 'Dr Johnson and the passions' (Ph.D. thesis, University of California, 1960), a pioneering but still unpublished work written before Hunter and Macalpine compiled the bibliographies now used.

[106] C. A. Moore, 'The English malady', the longest section of *Backgrounds of English Literature 1700–1760* (Minneapolis, 1950). The defective aspect of Moore's work is his simplistic notion of the concept of 'disease'; he demonstrates little of the analytic ability valued in historians of ideas. The work of Ilza Veith on this subject is, in my estimate, less important.

pleted by 1950, the agreed-upon point of departure for this appraisal of the historiography of psychology, and whereas it is admittedly an arbitrary date, it nevertheless compels omission of one scholar, one saint, without whose work the landscape of this field would look altogether different.[107] This is not because Zilboorg discovered a multitude of facts unknown before, but rather because he established the categories and the framework in which Enlightenment psychology has been viewed ever since: rational and irrational, normative and aberrant, accepted and unaccepted, prevalent and scarce, in fact all the categories and assumptions underlying the essay the reader now reads, with the exception of those dealing with the politics of knowledge (my concluding section) and the territory of historiography.

The second group – those who survey – has had somewhat less impact on the subject and includes those who have written histories such as those discussed above in sections I and II above. With the possible exception of George Mora's *oeuvre* and Franz Alexander's, which are eclectic and refine their points by assessing a variety of contributing factors, most similar 'histories' are derivative.[108] For example, if one were to isolate a theme such as the condition of madhouses or the theory of the mentally ill, and were to examine books where these subjects are treated, it would soon become evident that each 'history' had consulted its predecessors – in most cases plundered its ideas from those works. It may even be true that some recent structuralist obsessions with 'origins' (although it would be nonsense to imply that all the structuralists have been interested in this question) are applicable here:[109] namely, that unless one goes back to the earliest 'histories' of insanity, such as those written in the 1790s, one cannot comprehend why the categories of our post-1950 surveys exist.

Among the various 'saints' Foucault is the exception, an anomaly in both classes, perhaps because he is a philosopher of sorts rather than

[107] See *A History of Medical Psychology* (New York, 1941); *Mind, Medicine and Man* (New York, 1943).

[108] Such histories of psychology and psychiatry are many, and too lengthy to be listed here. Another exception may be S. Selesnick's *The History of Psychiatry* (New York, 1966), which appears to be much less derivative than most others.

[109] See Edward Said, *Beginnings* (New York, 1975); G. Canguilhem, *La connaisance de la vie* (Paris, 1952); G. Bachelard, *La formation de l'esprit scientifique: contribution à une psychanalyse de la connaissance objective* (Paris, 1938); *idem*, 'La double illusion du continuisme', in M. Fichant and M. Pecheux (ed.), *Sur l'histoire des sciences* (Paris, 1974), 155; M. Foucault, *Les mots et les choses* (Paris, 1966); *idem*, *L'ordre du discours* (Paris, 1971); Dominique Lecourt, *Marxism and Epistemology: Bachelard, Canguilhem, Foucault* (1975).

an historian. He has performed in both groups: presented new information, especially in the resuscitation of obscure eighteenth-century French medical theory, and he has written many 'histories', such as the *Histoire de la folie*[110] and the multi-volume series dealing with the history of sexuality.[111] He stands alone in our hagiography because his point of view is oblique. His politics shapes his 'histories' to such a degree, and his language is so rhetorically charged, that his contribution to each category – both the new information offered and the resulting survey – must be viewed sceptically. Politics and rhetoric do not render him a lesser 'saint', indeed there is reason to believe that he will be read when others discussed in this essay have been long forgotten, but he will be read as 'literature' rather than 'scientific history'. Moreover, I believe that Foucault's works, especially his 'histories' (I use this term cautiously) of madness and sexuality, will be more appreciated when he is dead and his biography made known – when all the parts of his life, as it were, can be viewed organically. Only then will it be clear what emphasis Foucault has given to the period of the Enlightenment in all his medico-historical works. Fortunately, Foucault's style is growing less rather than more obscure. But even in his most recent volumes the language is dense and opaque, and his blend of considered elements consistently elusive. All this notwithstanding, he is the one writer, if one had to choose, who seems to know how to revolutionize the field: first by selecting it and elevating it to new significance;[112] then by discovering arcane primary materials and integrating them (whether he reads them or not) with his political and linguistic beliefs; and finally by constructing a matrix of thought in such compelling prose language that others hypnotically forget their own work under the compulsion to read and assimilate his. For no one else writing about the life sciences in the Enlightenment can the same be said, and that may be why he is the only thinker discussed in every chapter in this part (i.e. 'Life and its environment') of *The Ferment of Knowledge*. One must grant that an exception is found in the work of Gilles Deleuze. His massive project on schizophrenia and society is as important as Foucault's various histories. One volume, written in collaboration with Félix Guattari, has already appeared: *L'Anti-Oedipe* (Paris, 1972), and a second entitled *Schizo-analyse* is in preparation. But Deleuze is not read very much outside France, and that is why I have omitted discussion of his works here.

[110] See note 65 above.

[111] One volume has appeared to date: *La volonté de savoir* (Paris, 1976).

[112] Gilles Deleuze, alone among recent scholars, seems to have understood this.

The *raison d'être* for treating of supposed 'saints' in the first place is to weigh their comparative merits. It is the single moment in a topical survey such as this one when one can stand away and compare, place substance beside substance and imagine its worth in eternity. In this connection it may not be unprofitable to compare the work of Foucault with that of George Rosen.[113] One is French; the other American. Foucault's training was humanistic, largely philosophical; Rosen's was medical – he came rather late to medical history. Foucault reads mostly French, the language of 95 percent of his citations; Rosen read and cited in all languages, even Polish, Russian, Czech, Japanese. Foucault uses the footnote sparingly; Rosen used it profusely. Foucault's language is dense, grammatically paratactic, highly metaphoric; Rosen's prose antithetical to Foucault's. Foucault's academic posts have been in the 'history of systems of thought' – although the American universities have recently offered him visiting professorships in literature; Rosen, at least for the last ten years of his life, was professor of medical history and epidemiology at Yale. Foucault has no professional medical training; Rosen was both M.D. and Master of Public Health. Foucault and Rosen dwell on the same crucial period: the end of the eighteenth century. Each covers the very same chronological years in some of his books, but no one reading *Histoire de la folie* and *Madness in Society* side by side could possibly contend that the two books have anything in common except their historical period.

This is not the place to extend the survey. Suffice it to say that while Rosen thought and wrote as an empiricist – finding his material, finding out what others had said about it, and then reporting it without an axe to grind – Foucault never worked or works in this way. The last thinker to have empirical tendencies, he is a thoroughgoing rationalist, a system-builder no less than Descartes, Hegel or Kant. An original thinker with an original mind, his medico-historical terrain is merely a touchstone from which to spring to his own hypotheses. Whether these hypotheses are capable of verification; whether two independent researchers would confirm any of his findings; whether his notion of the evolution of 'Enlightenment madness' is 'scientifically' valid is far less important to Foucault than the fact – the blatant fact – that they are *his* hypotheses. Among recent 'saints' he is

[113] Whether their work *and life* should be compared is another matter, depending very much on one's notion of the relationship. I am reasonably certain that our sense of Foucault will shift rather noticeably when his biography is written and it is fully realized that all along he has been a member of the political-radical fringe. I would be surprised if anyone writes a biography of the late George Rosen, as fine a scholar as he was.

our thinker, Rosen our historian. A study in historiography such as this one makes the distinction palpable, hopefully even concrete.

IX. The politics of knowledge: future ferments

I have already suggested that many projects in the history of Enlightenment psychology are crying out for students, and I have named the main areas. It remains for me to suggest that many of these are politically determined, by which I mean that both the project (e.g. a reconsideration of the types of therapy offered lunatics in England 1750–1800) and the finished artefact (i.e. a book on this subject) will have political colouring. This is not a notion readily grasped by English-speaking scholars, who have tended, especially in America, to view all scholarship *qua* scholarship and as distinct from political considerations; but even a cursory glance at the history of European scholarship, particularly on the continent, makes it evident that the matter is not so simple. For our period, the politics of Newtonian science is today undergoing radical revision,[114] and I am going to suggest, albeit briefly, that the same is true of 'psychological science'. I would even go as far as to contend that this aspect – the politics of knowledge – has been the least understood, the least researched, and commanded the least amount of attention in all the secondary literature I have surveyed. It ought to be the theme of my essay, and is the one territory I would explore if I were writing a history rather than an historiography of psychology.

If we consider, for example, the regulation of private madhouses in eighteenth-century England, a vexing issue about which many Britons had something to say from the start to the very end of the century, it is evident that the matter is anything but straight-forward. Scholars have demonstrated that it was not in the best interests of the various 'appointed committees' (who 'appointed' them is all-important) to search too hard for irregularities,[115] but no one has bothered to determine why. Those 'select committees' should be reconstructed, as Namier reconstructed parliamentary committees. For we need to know the background of the members: their monied interests, party affiliations, private ideologies. One wonders whether it can have been economically advantageous or politically shrewd for the committee member to overlook the irregularity of the asylums. And one asks whether there is any evidence to show that

[114] Others in this volume have discussed the work of M. Jacob and J. Jacob, who reinterpret the reasons for Newton's instantaneous success in England.

[115] See Jones, *Lunacy, Law and Conscience* (see note 82).

eighteenth-century 'lunatic physicians' also had vested interests; if they were in collusion with asylum owners who gave them rake-offs for incarcerated but healthy patients about whom the physician had lied. The Batties and Pinels, men of standards and character surely, are exceptions – humanitarians in an age often barbarically insensitive to the needs of the mentally ill – but one learns almost nothing about the others in recent literature.

Explanations of the melancholy and hysteria endemic to women and certain segments of the upper classes in the eighteenth century have also suffered somewhat from a low threshold of explanation. It is true that much has been written in the last twenty years about 'the English malady'.[116] Yet, curiously, no one has gazed at the politics of melancholy: at the *social* fibre of the organism, at its economic, sociological, political dimensions, especially the political allegiances of patient and physician in a relationship that was clearly understood by both. It is now widely accepted that melancholy was then the most fashionable of diseases; that it afflicted the rich far more than the poor (although the literature boasts some bizarre cases where it touched the poor);[117] that it especially attacked high-class women; and that the flood of remedies, not merely pills and potions, that were administered to cure it often made millionaires out of those dispensing them.[118]

[116] See Thomas Chaplin, *Medicine in England during the Reign of George III* (1919); R. H. Gillespie, *Hypochondria* (1928); William K. Richmond, *The English Disease* (1958); Charles Trench, *The Royal Malady* (New York, 1964); Ilza Veith, *Hysteria: The History of a Disease* (Chicago, 1965); G. S. Rousseau, *Introduction to Sir John Hill 'Hypochondriasis: A Practical Treatise'* (Los Angeles, 1969); T. H. Jobe, 'Medical theories of melancholia in the seventeenth and early eighteenth centuries', *Clio Medica*, XI (1976), 217–31. W. Lepenies, *Melancholie und Gesellschaft* (Frankfurt, 1969).

[117] The class character of melancholy is described in many eighteenth-century texts – scientific, medical, social, literary. See, for example, *The Letters of Doctor George Cheyne to Samuel Richardson (1733–1743)*, ed. C. F. Mullett (Columbia, Mo., 1943); Dr John Gregory, the Edinburgh medical ethicalist, *A Father's Legacy to his Daughters* (1774), 126; James Makittrick Adair, *Essays on Fashionable Diseases* (1790), ch. on 'Fashionable Maladies'. For France see Anne Charles Lorry, the physician of the fashionable and author of *De melancholia et morbis melancholicis* (2 vols. Paris, 1762), which recognizes two classes of melancholics, one nervous the other humoural, both of which have class origins. Modern students (see note 116 above) have often noticed this class distinction, but have not explained its dynamic.

[118] The list is too long to be discussed here but it would inclue 'therapy sessions' for rich ladies, pills and potions such as Dover's drop, James's powders, Berkeley's tar water and Hill's Hungary water – manufactured remedies that made these men rich.

But the politics, especially the economics, of the matter has not been explored. The hypothesis, for instance, that those patenting 'melancholic medicines' paid a commission to ministers who praised them and newspapers that advertised them has not been investigated – although it may be the only reason medicines were advertised so regularly in the main London newspapers. This is a case in which economic historians have not ventured into medical fields and vice versa. Nor has the notion that hypochondria is a socially explicable phenomenon been studied in any detail by historians. Surely it was fashionable to be melancholic; the imaginative literature of the period, both in England and France, assures us of this. It was a trademark of upper-class gentility. Thus all types of socially mobile women, such as Moll Flanders in Defoe's novel, Miss Harlowe in Richardson's *Clarissa Harlowe,* and Elizabeth Bennett in Austen's *Pride and Prejudice,* to mention the three best-known heroines among dozens of similar types, may have developed their melancholy for purely social reasons, as 'signs' in the semiotic sense of a certain class consciousness, unaware that they were 'contracting' the awful condition. But the dynamics of contraction are not so facile as this: money-seeking doctors and apothecaries are involved; so-called 'patients' are involved; theorists spinning out large tomes are involved, as are their all-too-obscure publishers; resorts and spas, such as Bath and Tunbridge Wells, are involved; newspapers and magazines – the media – are involved; a host of appurtenances including 'melancholic gowns' and alleged 'pharmaceutical panaceas' are involved; and all these exist in the realm of the concrete. Nothing as yet has been said about these in the 'realms of gold', the imagination: literature, art, even music, where a myth begins to develop about the melancholic composer. To relate the whole story dynamically would further understanding of our period,[119] but it requires extensive research as yet not undertaken. What it reveals may change our whole sense of the politics of madness.

Another direction research ought to take is toward the relationship of psychology to other arts. This is, admittedly, a more nebulous field than most medical historians like to elect but nevertheless one whose

[119] A 'dynamic' explanation includes the socio-economic factors as well as psychological assumptions of the age. It does not consider an explanation adequate unless it deals with the interaction of these assumptions as a bare minimum, nor does it view with approbation surface-deep descriptive accounts that attempt to pass for genuine explanation. Therefore, it would not suffice to explain the history of melancholy in our period merely as a 'history of science'. To do that would tell only a fragment of the story.

history is very much needed at the present time. The Enlightenment is replete with 'mad' poets, 'mad' painters, 'mad' musicians; yet their alleged madness has generally been construed in narrow terms, viewed as a black-and-white issue. The existing biographies of some of these figures make this evident,[120] yet the advances of social history today render the approach defective and incomplete. Continuing in this suggestive vein is the representation of dementia and other mental states in the arts: insanity portrayed in painting and etching, especially in the self-analyses of such possessed lunatics as Christoph Haizmann (the Austrian painter) and in the caricature art enjoyed so much in the Enlightenment. We direly need an iconography of mental illnesses and mental states; nothing at present even remotely approximates this.[121]

Futher along lines of social history, it would be surprising if politi-

[120] Cowper, Smart, Collins, Samuel Johnson, and other British writers of the period are the exceptions because they have generally enjoyed adequate biographies, as have their physicians. E.g. see the essay on Smart's physician, Nathaniel Cotton, by B. Hill, '"My Little physician at St Albans"', *Practitioner*, CLXXXXIX (1967), 363–7. This fact notwithstanding, few modern students differentiate between eighteenth- and twentieth-century self-reflectiveness about insanity: i.e. it is one thing for something to happen to one, and quite another to explain away that occurrence. In this connection, Lady Mary, the noted female-traveller, is instructive. In a long letter about Richardson's novels written to her daughter-in-law, Lady Bute, she comments on the sexual attractiveness of demented women: 'He [Richardson] is not a Man Midwife, for he would be better skill'd in Physic than to think Fits and Madness any Ornament to the Characters of his Heroines, tho his Sir Charles [Grandison] had no thoughts of marrying Clementina till she had lost her Wits, and the Divine Clarissa never acted prudently till she was in the same Condition, and then very wisely desir'd to be carry'd to Bedlam, which is realy all that is to be done in that Case. Madness is as much a corporal Distemper as the Gout or Asthma, *never occasion'd by affliction*, or to be cur'd by the Enjoyment of their Extravagant wishes. Passion may indeed bring on a Fit, but the Disease is lodg'd in the Blood, and it is not more ridiculous to attempt to releive [*sic*] the Gout by an embroider'd Slipper than to restore Reason by the Gratification of wild Desires' (italics mine). See R. Halsband (ed.), *The Letters of Lady Mary Wortley Montagu* (3 vols., Oxford, 1967), III, 96. This position should be compared with other popular notions such as that found in the French traveller Pierre Jean Grosley's *A Tour to London; or, New Observations on England* (1772), I, 243: 'all the people here [in Bedlam] were here because it was occasioned either by love or religious enthusiasm'.

[121] See notes 60 and 65 above. Historians of architecture have dissected the plans for mental asylums and prisons, but there is no iconography of its paintings and caricatures along the lines studied by Grace Goldin, the historian of architecture, in 'A painting in Gheel', *JHM*, XXI (1971), 12–23. We need one that extends from Cuyp's turn-of-the-century caricatures of 'Bedlam' to J. C. Lavater's engravings of 'an idiot' in his 1789 *Essays on Physiognomy*. R. Herrlinger's *History of Medical Illustration* (1969) does not include psychiatry.

cal factors did not influence artists in the representation of mental conditions and mental states, but one ought not to pronounce before sufficient research has been undertaken. Likewise for the so-called 'sensibility movement' and for the 'melancholic conditions' discussed: its victims, if they may be called such, and symptoms are patent, as is the jargon of melancholy, by now an almost foreign tongue including a vernacular of words about black and white bile and neologisms barely known today. 'When I first dabbled in this art', Nicholas Robinson, the popular London 'hyp doctor', writes in a rare confessional mood just after George II had been crowned, 'the old distemper call'd *Melancholy* was exchang'd for *Vapours*, and afterwards for the *Hypp*, and at last took up the now current [1728] appellation of the *Spleen*, which it still retains, tho' a learned doctor of the west, in a little tract he hath written, divides the *Spleen* and the *Vapours*, not only into the *Hypp*, the *Hyppos*, and the *Hyppocons;* but subdivides these divisions into the *Markambles*, the *Moonpalls*, the *Strong-Fiacs*, and the *Hockogrokles*.'[122] None of these words is found in medical dictionaries, of the present or of the eighteenth century. Yet it is important to determine the extent and diversity of this jargon, particularly as a clue to the political, economic and social aspects of the condition. Even when that is accomplished more remains to be explained: namely the ways in which a medical or pseudo-medical condition, validated by the entire medical profession, was enlarged into a national myth, at least in Britain. Such mythology, as it deserves to be called, has still not been interpreted, neither its sociology nor its politics. Yet many of those who were afflicted continued successful, in clinical terms enjoying the secondary gain of the illness. Some of the 'gain' reflected the fact that, although suffering, they were succeeding professionally; some the old belief - as old as Burton's *Anatomy* - that melancholy of whatever type of Burton's eighty-eight varieties, had beneficial as well as negative aspects.

It is, therefore, not remarkable that James Adair, the astute Scottish physician, would write:

> After the death of Anne and the demise of the 'Wits' there arose a fashionable disease the likes of which had never been seen in England. Some call'd it melancholy, others by names too various to be enlisted here; but one thing about its development was clear: it enjoyed a 'progress' as distinct as that of any medical condition, and it is known that it became a fashionable disease only at that time, and that persons of all

[122] See Nicholas Robinson, M.D., *A New System of the Spleen, Vapours, and Hypochondriack Melancholy* (1729), 34–5.

manner wanting to advance themselves suddenly developed it.[123]

This is a fine passage, revelatory for the historian, but it probably will not suffice for decoding the complex phenomenon that C. A. Moore has called 'an age of melancholy'.[124] For that, its politics, the ways persons 'advanced themselves' professionally, must first be decoded, then its semiology interpreted.[125] This is why the politics of knowledge is so crucial to any study of the historiography of psychiatry and psychology in every age, not merely in the Enlightenment.

There are other urgent areas too. No one has asked why there is no adequate history of 'sensibility', a concept without understanding of which the psychology of the age dissolves. No one, it seems, has wondered how idiosyncratic was Dr Johnson's notorious fear of 'diseased imagination' – i.e. madness – and whether it had cultural determinants without which he would not have dreaded it so direly.[126] Recently there have been several studies of lunatics in literature, but drawn without enough sense of the cultural landscape to provide an interpretation that explains anything.[127] No one has related this social background to Smollett's penultimate novel, *Life and Adventures of Launcelot Greaves*, the work of imaginative fiction 'in the manner of Cervantes' already mentioned, whose main theme is madness.[128] This prose treatise of over 100,000 words needs to be analysed for its references to madness, as well as dissected for the 'mythology' of madness it provides. And if one inquires why, the answer will be found in Freud's essay on 'Mourning and melancholia': 'The description of the human mind is indeed the domain which is not his [the

123 See Adair, *Philosophic and Medical Sketch* (1787), 234.

124 See notes 29 and 106 above.

125 Medical historians have been so inattentive to the history of hysteria as the malady of the century, even in the latest clinical research on hysteria (see Mardi Horwitz, *The Hysterical Personality* (New York, 1977), that it may not be unreasonable to demand a 'semiotics of melancholy' that isolates the chief signs.

126 See W. J. Bate, *Samuel Johnson* (New York, 1977), ch. 21, 'Approaching breakdown: religious struggles; fear of insanity', and E. Verbeek, M.D., *The Measure and the Choice: A Pathographic Essay on Samuel Johnson* (Ghent, 1971). Various scholars and physicians have argued that Johnson depicts a 'classic schizophrenic' in *Rasselas* (1759).

127 See Max Byrd, *Visits to Bedlam: Madness and Literature in the Eighteenth Century* (Columbia, S.C., 1974), and M. V. DePorte, *Nightmares and Hobbyhorses: Swfit, Sterne, and Augustan Ideas of Madness* (San Marino, Cal., 1974). Both studies survey the literary aspects but their medico-scientific and socio-economic backgrounds are thin.

128 See P. G. Boucé, *Les romans de Smollett: étude critique* (Paris, 1971), 230–42.

creative writer's] own; he has from time immemorial been the precursor of science, and so too of scientific psychology.'[129] No one has concerned himself, moreover, with the interpretations that arise when sheer 'teleological history' is written, or when national boundaries are not crossed.[130] Nor has anyone conjectured about the relation of sexual repression in the Enlightenment to consequential madness, a correlation that physicians in the twentieth century believe to be practically infallible.[131]

These and other questions are large, and they ask for much – perhaps too much for a young field, medical history, still thought by some to be in infancy.[132] Nevertheless, there is a common denominator among these desiderata that includes first the need to explore the social history of the period and then the necessity of interpreting it more imaginatively and daringly than has been done. The question, of course, is whether it ought to be interpreted as inventively and idiosyncratically as Foucault (and now Deleuze) has done: that is to say, to a degree that places 'the threshold of possible explanation', as Foucault has commented in the epigraph at the beginning of this essay, 'too high'.

In conclusion, this threshold may be too complicated. Certainly it is too high for the internalists who still view such a modest and cautious externalist as George Rosen – let alone the likes of a Foucault or Deleuze – with suspicion. These French thinkers stand apart from a Rosen and any number of other externalists by maintaining that at the moment when one decides to write about a state of affairs in human history one assumes the burden of coping with 'discourse' as opposed to the mere confrontation with 'texts', the latter construed simply as the word on the page. This is not the place to enter into discussion about the complicated differences between texts and discourses, and the implications of these differences for the history of science, but it is ultimately the reason why Foucault and Deleuze place their thresholds of explanation so high.

What remains to be charted by future students, then, is the course

129 See James Strachey and Alan Tyson (eds.), *The Standard Edition of the Complete Psychological Works of Sigmund Freud* (24 vols., 1953–), XIV, 239–40.

130 In this connection see Toby Gelfand, *The Training of Surgeons in Eighteenth-Century Paris...* (Baltimore, 1973).

131 Stone, *The Family, Sex and Marriage in England* (see note 21), 385ff., speculates about the matter, but not in relation to developmental insanity.

132 Edwin Clarke's collection, *Modern Methods in the History of Medicine* (see note 5), lends the impression that there is an historiography of medicine; see especially p. ix, n.12, and O. Temkin's essay, 'The historiography of ideas in medicine'.

of externalism: not the simple history of its survival but the nature of its future life. Now, "at the beginning of a new decade," there is every reason to believe that the internalist camp is beginning to lose ground. One sees the loss in fields other than the history of science and medicine, even in the so-called pure sciences. Certainly the internalists have lost some of the brightest young minds of the sixties and seventies to the externalists. Furthermore, the massive surveys included in this volume provide abundant signs of trends to come. Finally, if Foucault is indeed correct about psychiatry now enjoying such 'a low epistemological profile', then psychiatry – among the various sciences – must continue to stand in the foreground of externalist treatments; it must necessarily occupy the labours of scholars in the future who are as concerned with the social, political and economic institutions of past eras as they are with the history of psychiatry, psychology or any other science. The key issues, then, are those social institutions: their precise regulations, exigencies, manipulations, deceptions – every aspect of their nature. Only the externalist approach can begin to cope with a synthesis of the sciences of man in relation to these institutions. And it is hard to imagine in decades to come a return to internalism after the externalists have just delivered so much and promised even more.

Acknowledgements

I am grateful to Drs Gloria Gross, William Bynum, Norland Berk, Roger Hambridge and Roy Porter for commenting, sometimes harshly, on various drafts of this essay and for suggesting means by which I could focus better on the very large terrains I attempt to cover. I am also indebted to Stephen Morris of Selwyn College and Simon Schaffer of St John's College, for discussions in Cambridge during the Michelmas term 1979 about the internalist–externalist debates. Without the unrelenting questions of the former and the enlightened provocation of the latter, I would not have known as well as I now do why I had constructed the essay as I did, and why I had to include discussion of 'the politics of knowledge'.

5

Health, disease and medical care

W. F. BYNUM

Contents

I.	Introduction	211
II.	Theories	215
III.	Actions	226
IV.	Interactions	234
V.	Living and dying	244
VI.	Conclusion	252

I. Introduction

Ours is not the first generation to be preoccupied with modernity, nor the first to search for clues as to when 'our' kind of medicine began. Conventional wisdom generally locates that origin in the Revolutionary Paris of the 1790s. Thus, R. H. Shryock's chapter on early-nineteenth-century French medicine is entitled 'The emergence of modern medicine, 1800–1850', and the 'clinic' described by Michel Foucault does not simply concern the activities of Bichat and Laennec, but also the 'conditions of possibility of medical experience in modern times'.[1] This French hospital medicine coincided with political and social upheaval and with 'the end of the Enlightenment', a fact which deepens the apparent chasm separating our medicine from that of the Enlightenment and before.[2]

If our own medicine derives in some important sense from the hospitals of Revolutionary France, the medical forms, patterns, and practices of the Enlightenment can be viewed in two antithetical ways: as having created the possibility of modernity; or, by way of reaction, the ultimate victim of it. The latter has been the more common historical attitude. Thus Shryock, while applauding eighteenth-

[1] R. H. Shryock, *The Development of Modern Medicine* (1948); Michel Foucault, *The Birth of the Clinic* (English trans., 1973), xix.

[2] David M. Vess, *Medical Revolution in France, 1789–1796* (Gainesville, Florida, 1975); Robert Darnton, *Mesmerism and the End of the Enlightenment in France* (Cambridge, Mass., 1968); E. H. Ackerknecht, *Medicine at the Paris Hospital, 1794–1848* (Baltimore, 1967); Pierre Huard, *Sciences, médecine, pharmacie de la Révolution à l'Empire (1789–1819)* (Paris, 1970).

century advances in mathematics, physics, chemistry, and the social sciences, discovered a variety of social and intellectual reasons why medicine lagged behind. The recent survey of eighteenth-century medicine by Guy Williams repeatedly uses the image of darkness before the dawn, the Enlightenment thereby being rather curiously cast in the role of shadow to the glorious triumphs that followed.[3] Foucault's books on insanity and on medicine were written without explicit reference to his concept of the *'episteme'* (developed subsequently in *The Order of Things*), but all his writings actually feed into a grand scheme of historical periodization, his synthetic vision of seemingly disparate bodies of knowledge undergoing regular transdisciplinary epistemological switches. According to Foucault, not only medicine and psychiatry, but biology, economics, linguistics, criminology - indeed, man himself - were born around 1800.[4] *The Birth of the Clinic* is really too gentle a title, for his metaphors are those of mutation rather than birth, violence rather than fruition. For him, Baroque, Classical (or Enlightenment) and Modern denote sharply defined epochs, though his rigorously detached writing betrays no hint of nostalgia for any particular period, our own included.[5]

A similar accent on the contrasts between eighteenth- and nineteenth-century medicine informs the work of several University of Leicester medical sociologists. This group has looked at hospitals, medical education, disease concepts, the medical profession, and the doctor–patient relationship, primarily in terms of broad-scale patterns. Thus, according to Jewson, eighteenth-century physicians employed holistic disease concepts in a client-dominated setting, whereas their nineteenth-century counterparts used the authority of the hospitals, localistic pathology, and esoteric vocabulary (crucial characteristics of modern medicine) to dominate their patients.[6] During the Enlightenment it was still possible for physicians to diagnose by post, since

[3] Guy Williams, *The Age of Agony* (1975).

[4] Michel Foucault, *Madness and Civilization* (English trans., 1967); *idem, The Order of Things* (English trans., 1970).

[5] Foucault himself is now the subject of a mini-industry. Cf. H. White, 'Foucault decoded: notes from underground', *History and Theory*, XII (1973), 23–54; J. C. Guédon, 'Michel Foucault: the knowledge of power and the power of knowledge', *BHM*, LI (1977), 245–77; Angèle Kremer-Marietti, *Foucault et l'archéologie du savoir* (Paris, 1974).

[6] N. D. Jewson, 'Medical knowledge and the patronage system in 18th century England', *Sociology*, VIII (1974), 369–85; *idem*, 'The disappearance of the sick-man from medical cosmology', *Sociology*, X (1976), 225–44; Ivan Waddington, 'The role of the hospital in the development of modern medicine: a sociological analysis', *Sociology*, VII (1973), 211–24; *idem*, 'General practitioners and consultants in early nineteenth-century England: the sociology of an intra-professional conflict', in John Woodward and David Richards (eds.), *Health*

the patient's own description of his illness was the pivotal element in the diagnostic process.[7] After the 'clinic' was born, physicians no longer passively listened to accounts of illness, they actively probed the patient's body, looking for disease. 'Where does it hurt?' replaced 'What is the matter with you?' as the question doctors wished to have answered.[8]

There is sufficient validity in these contrasts to make the official report calling for the reorganization of the French medical schools in 1794 a beginning of post-Enlightenment medicine. I shall so consider it in this essay, even though I shall suggest below that the middle decades of the century were just as pregnant as the decade which later gave birth to the 'clinic'. Foucault's metaphor of mutation is simply too stark, and while commentators closer to the scene often used a revolutionary vocabulary to describe early-nineteenth-century French hospital medicine, they were aware of its deep historical roots. That medical philosopher of revolution P. J. G. Cabanis (1757–1808) viewed the medicine he helped create as simply the latest in a series of 'revolutions' which had punctuated the history of medicine and, for him, there had been at least three other of these medical transformations in the eighteenth century, above all that achieved by G. E. Stahl, 'one of those extraordinary men, whom nature seems to produce, from time to time, for the purpose of effecting the reform of the sciences'.[9] Furthermore, for different reasons, both English- and German-speaking doctors were detached from French events, and while some striking parallels can be drawn between France and Britain – particularly after the close of the Napoleonic Wars – local traditions in Britain and the rest of Europe remained strong even after Paris became the Mecca of the medical world. Thus, in the 1830s John Bostock bypassed the French and suggested that 'modern' medicine dated from Haller's work of the 1740s.[10] In our own century, D'Arcy Power was convinced that Samuel Johnson (1709–1784) had witnessed

Care and Popular Medicine in Nineteenth Century England (1977). The work of the other member of the group, S. W. F. Holloway, has been principally concerned with the nineteenth century.

7 Stanley Joel Reiser, *Medicine and the Reign of Technology* (Cambridge, 1978), ch. 1; Guenter Risse, 'Doctor William Cullen, Physician, Edinburgh: a consultation practice in the eighteenth century', *BHM*, XLVIII (1974), 330–51.

8 Foucault, *Birth of the Clinic* (see note 1), xviii; Karl Figlio, 'The historiography of scientific medicine: an invitation to the human sciences', *Comparative Studies in Society and History*, XIX (1977), 262–86.

9 P. J. G. Cabanis, *Sketch of the Revolutions of Medical Science, and Views Relating to its Reform* (English trans., 1806), 132.

10 J. Bostock, *Sketch of the History of Medicine* (1835), 198–9: 'we may regard the publication of [Haller's] Elements of Physiology as having introduced a new era into medical science'.

the replacement of the old medical order by the new.[11] Other historians have used Harvey and other seventeenth-century doctors as the symbols of the great transformation, dismissing Enlightenment medicine as sterile, speculative, and riddled with perniciously 'heroic' therapeutics: a fallow century sandwiched between two periods of intellectual excitement.[12]

The medicine of the eighteenth century still elicits a variety of attitudes among historians. Some absorb contemporary critiques of high-technology medicine and discover in the Enlightenment the positive virtues of an easy commerce between doctor and patient.[13] Others note the technological inadequacy of Enlightenment medicine, yet see in the values of that century – its secularization, urbanity, and wordly concern for health – the 'rise of modern paganism'.[14] Other historians have held that the medical caricatures of Hogarth and Rowlandson embody the true spirit of that age of brutality, pompous doctors, quacks, and treatments far worse than diseases.[15] The Enlightenment is rich enough to accommodate a plethora of interpretations.

We flirt with historical teleology even to gauge Enlightenment medicine in terms of modernity,[16] and my own bias favours the tighter if more modest persuasiveness of local textures, piecemeal development, and firm empirical placement. Grand periodization and sweep-

[11] D'Arcy Power, 'Medicine', in A. S. Turberville (ed.), *Johnson's England* (2 vols., Oxford, 1933), II, 265.

[12] E.g. Victor Robinson, *The Story of Medicine* (New York, 1944), 332: 'Facts discovered in the seventeenth century frequently produced theories in the eighteenth'; or Fielding H. Garrison, *An Introduction to the History of Medicine*, 4th edn (Philadelphia, 1929), 310: 'For medicine, aside from the work of a few original spirits like Morgagni, Hales, the Hunters, Wolff, and Jenner, the age was essentially one of theorists and system-makers.'

[13] A view implicit in much of the literature cited in note 6 above. Despite Illich's polemics against modern 'medicalization of life', he is ambivalent towards the Enlightenment, which 'attributed a new power to the doctor, without being able to verify whether or not he had acquired any new influence over the outcome of dangerous sickness'. Ivan Illich, *Limits to Medicine* (Harmondsworth, 1977), 198–9.

[14] This is the thrust of Peter Gay's view, in his *The Enlightenment: An Interpretation* (2 vols., New York, 1966–9). The phrase in the text is the sub-title of Gay's first volume.

[15] Cf. Williams, *Age of Agony* (see note 3).

[16] For general considerations of the 'modernity' of the Enlightenment, see A. M. Wilson, 'The *Philosophes* in the light of present-day theories of modernization', *Studies on Voltaire and the Eighteenth Century*, LVIII (1967), 1893–913; and H. B. Applewhite and D. G. Levy, 'The concept of modernization and the French Enlightenment', *Studies on Voltaire and the Eighteenth Century*, LXXXIV (1971), 53–98.

ing sociological generalization can be goads to research. But there are still so many dark patches in our vista of eighteenth-century medicine that synthesis seems premature. Besides, medicine is much more than simply applied knowledge, and the more manageable history of ideas approach has less applicability than it does in physics or chemistry. If synthesis eludes, though, there is still more sound historical scholarship than one essay (or one historian) can cover.

For convenience, I have adopted topical headings, though there is inevitable overlap among themes, individuals, and historical debates. In the following section I survey the literature on the principal Enlightenment medical theorists. Section III examines more practical aspects of medical activity, while the fourth section looks at the social and professional dimensions of medical institutions, corporations, and societies. A fifth section deals with historical demography and the literature on diseases and social values. I make no apology that these headings invite historical rather than exclusively historiographical reflections, for no historiographical genre has a monopoly on verisimilitude. Having already suggested that Enlightenment medicine is not strictly synonymous with the eighteenth century, I shall assume that 1688, 1689, or 1690 began it: the Glorious Revolution, the death of Thomas Sydenham, or the publication of John Locke's *Essay concerning Human Understanding*. Somewhat arbitrarily, the last date and the *magnum opus* of an ageing physician-philosopher can be taken as the beginning of Enlightenment.

II. Theories

Every one nowadays pretends to neglect theory, and to stick to observation. But the first is in talk only, for every man has his theory, good or bad, which he occasionally employs; and the only difference is, that weak men who have little extent of ability for, or have had little experience in reasoning, are most liable to be attached to frivolous theories. (William Cullen)[17]

Despite a common portrait of Enlightenment medicine as theoretical and speculative, only a handful of theoreticians have been examined in any detail. The major figures tended to be associated with the universities, where systems were developed for teaching purposes, and while there was – particularly in Britain – a perceptible move towards avowedly 'theory-free' medical writing from the 1750s, Cullen's comment, quoted above, goes closer to the heart of the matter. Earlier, Sydenham had counted himself on the side of the theory-free

[17] William Cullen, *Works*, ed. J. Thomson (2 vols., Edinburgh, 1827), I, 402.

angels, though it could be argued that his most specific legacy for the Enlightenment can be found in the preface to the third edition (1675) of his work on acute diseases. There he affirmed his conviction that diseases can be classified in the way that plants can, and while this nosological ideal jostled uneasily with the strong Hippocratic elements in Sydenham's writings, nosology became an Enlightenment enterprise of no small moment. Recent historical studies of the subject have complemented and extended, rather than superseded, Knud Faber's outstanding monograph of 1930.[18] The principal nosologists – Linnaeus, Sauvages, Cullen, Macbride, Pinel – date primarily from the second half of the century, influenced by the prestige of Linnean taxonomy. The three most important medical theorists of the early Enlightenment – Boerhaave, Hoffmann, and Stahl – did not pick up Sydenham's gage, though in his old age Boerhaave did commend the young Sauvages' initial efforts in nosology.

Of the three, Stahl developed the 'purest' system, though each of them typifies (even if in rejection) the lingering shadow of Descartes on philosophy, physiology, and medical work in the period. Stahl's ideas on phlogiston have found their way into histories of chemistry, but many aspects of his complicated and frequently obscure medical writings remain unexplored. Pagel's essay on 'Van Helmont, Leibnitz, und Stahl' elucidates Stahl's general intellectual inheritance, and Rather, Frerichs, and King have placed his animism into philosophical perspective.[19] Like Descartes, Stahl divided the world into two basic categories, though his concept of anima served rather more diverse functions than did the Cartesian soul. For Stahl, the anima was simultaneously the agent of consciousness, physiological regulation, and defence against disease. Only the anima held back the natural tendency of bodily corruption and death, a notion which led Stahl perhaps more than any other Enlightenment physician to stress the healing power of nature.[20] Stahl's metaphysical foundations were

[18] Knud Faber, *Nosography, the Evolution of Clinical Medicine in Modern Times* (New York, 1930); Foucault, *Birth of the Clinic* (see note 1); Lester King, *The Medical World of the 18th Century* (Chicago, 1958), ch. 7; *idem*, 'Boissier de Sauvages and 18th century nosology', *BHM*, XL (1960), 43–51; F. Dagonet, *Le catalogue de la vie* (Paris, 1970).

[19] Hélène Metzger, *Newton, Stahl, Boerhaave et la doctrine chimique* (Paris, 1930); Walter Pagel, 'Helmont, Leibniz, Stahl', *Sudhoffs Archiv*, XXIV (1931), 19–59; L. J. Rather, 'G. E. Stahl's psychological physiology', *BHM*, XLV (1961), 37–49; *idem* and J. B. Frerichs, 'The Leibniz-Stahl controversy', *Clio Medica*, III (1968), 21–40, and V (1970), 53–67; Lester King. *The Philosophy of Medicine: The Early Eighteenth Century* (Cambridge, Mass., 1978), esp. 143–51. King's article on Stahl in the *Dictionary of Scientific Biography* is particularly full.

[20] Max Neuberger, *The Doctrine of the Healing Power of Nature throughout the Course of Time* (English trans., New York, 1943), 61–9.

devoutly religious, and he is most clearly seen in relation to pietism.[21]

Stahl and Hoffmann both taught at the University of Halle, where the latter eventually became the dominant medical figure. King and Rothschuh have recently rescued Hoffmann from relative neglect through a series of chapters and essays, and a translation of Hoffmann's short précis of medical knowledge, the *Fundamenta medicinae* (1695), a work which bears fruitful comparison with Boerhaave's *Institutiones medicinae* (1708).[22] Hoffmann's basic mechanistic proclivities are evident ('Medicine is the art of properly utilizing physico-mechanical principles, in order to conserve the health of man or to restore it if lost', reads the first sentence of *Fundamenta medicinae*), but Rothschuh has stressed his relationship to Leibniz, and King finds some striking structural parallels between Hoffmann and Stahl, despite their apparent differences and the fact that after Stahl's death Hoffmann directed a polemical work against his former colleague.[23] Hoffmann's interpretive medical framework was based on a somewhat abstract physiological base which King has summarized:

> In his physiology Hoffmann explained all phenomena by particles having essentially hypothetical qualities and undergoing hypothetical motion, combined in hypothetical proportions, under the control of a special kind of hypothetical force, that acted largely through hypothetical animal spirts, controlled by a hypothetical *anima*, with which, in man, was associated a conscious mind.[24]

But Hoffmann was also a clinical observer of considerable power, who left behind solid descriptions of a number of diseases and actively promoted various therapeutic measures, including massage and bathing. Interestingly his major systematic work was never translated from the Latin, but an English edition of its practical portions appeared in 1783, forty years after Hoffmann's death, aimed, according to one of its translators, at those who prefer 'useful facts to fanciful speculation'.[25]

Though Hoffmann had British contacts, including Robert Boyle,

[21] A theme currently being investigated by Dr J. Geyer-Kordesch in Oxford.

[22] Friedrich Hoffman, *Fundamenta medicinae*, trans. Lester King (1971); Lester King, *The Road to Medical Enlightenment, 1650–1695* (1970), ch. 5; Karl E. Rothschuh, 'Studien zu Friedrich Hoffmann', *Sudhoffs Archiv*, LX (1976), 163–93, 235–70.

[23] Rothschuh, 'Friedrich Hoffmann' (see note 22); Lester King, 'Stahl and Hoffmann: a study in eighteenth century animism', *JHM*, XIX (1964), 118–30.

[24] King, *Medical Enlightenment* (see note 22), 193.

[25] William Lewis and Andrew Duncan, *A System of the Practice of Medicine* (2 vols., 1783), I, v.

whom he met during a trip to England early in his life, Halle never attracted British students in large numbers.[26] Nevertheless, Hoffmann made a considerable impact on British medicine through Cullen, who preferred him to Boerhaave, particularly because of Hoffmann's stress on the role of the nervous system in health and disease. Hoffmann conceived the aether as a mediator between mind and body; French's recent essay on the aether in eighteenth-century physiology complements Putscher's study of *Pneuma, Spiritus, Geist* in earlier medicine.[27]

There is a sound history of the medical faculty at Halle, but Stahl and Hoffmann remain shadowy personalities. On the other hand, Boerhaave has become a relatively full-blooded historical figure, largely through the efforts of his biographer, bibliographer, editor, and promoter, G. A. Lindeboom.[28] Indeed, as a teacher, Boerhaave is unique in the whole history of medicine. It is a commonplace that, since early modern times a number of universities or medical schools have attracted student bodies of rich international composition: Padua, Paris, Berlin, Vienna, and, to a certain extent, London and Edinburgh. But in the sheer magnitude of the operation (relative to the number of physicians produced in Europe), and in the way it coalesced around Boerhaave, early Enlightenment Leyden was exceptional, even in the above select group. Boerhaave's writings are more impressive in their range and breadth than in their depth, and his fluent eclecticism may help explain his wide appeal. We have recently been well advised against applying terms such as iatrochemist and iatromechanist too confidently,[29] though like Hoffmann, Boerhaave favoured mechanical explanations. Newton was a direct influence on Boerhaave, and Schofield's monograph sheds light on medical and physiological 'Newtonianism', even if his distinction between

[26] Simon Adolphus (1676–1753), listed as coming from London, took a degree at Halle in 1730, but British and French students are exceptionally rare in the lists published in Wolfram Kaiser and Karl-Heinz Krosch, *Zur Geschichte der medizinischen Fakultät der Universität Halle im 18. Jahrhundert* (2 vols., Halle-Wittenberg, 1964–7); *idem, 250 Jahre Collegium Clinicum Halense 1717–1967* (Halle, 1967).

[27] Roger French, 'The ether and dynamic physiology', in G. Cantor and M. J. S. Hodge (eds.), *The Subtler Forms of Matter* (Cambridge, 1980); Marielene Putscher, *Pneuma, Spiritus, Geist* (Wiesbaden, 1973).

[28] G. A. Lindeboom: *Hermann Boerhaave* (1968); *Bibliographia Boerhaaviana* (Leyden, 1959); *Boerhaave and Great Britain* (Leyden, 1974); *Boerhaave's Correspondence* (3 vols., Leyden, 1962–79), contain much of Lindeboom's work on Boerhaave.

[29] King, *Philosophy of Medicine* (see note 19), chs. 4 and 5 (though, ironically, the chapters are entitled 'Iatrochemistry' and 'Iatromechanism' respectively).

mechanism and materialism seems somewhat artificial. Jacob's recent study of the political, social, and natural theological context of Newtonianism provides a richer framework for understanding Newton's appeal among certain groups.[30] Hall, Brown, Thackray, Coleman, Duchesneau, and others have examined additional aspects of early Enlightenment physiology, and its natural philosophical base which through the dominant physiological content of the 'Institutes of medicine' had direct import on medical teaching and theorizing.[31] Brown in particular has stressed the professional and ideological uses of mechanical physiology in the English context, and while the experimental basis of this physiology was sometimes meagre, Stephen Hales, longtime vicar of Teddington, was simultaneously an ardent exponent of mechanical explanations and one of the most competent experimental physiologists of the century.[32] Finally, Vartanian's illuminating studies of the continuing impact of Descartes on early-eighteenth-century French physiology and philosophical biology suggest that the road from beast-machine to man-machine was fairly straight.[33]

A number of factors reinforce the view that the middle decades of the century were just as pregnant with change as those surrounding its end. Stahl, Boerhaave, and Hoffmann died between 1734 and 1742, and the men who replaced them were of a different stamp. It is well known that Boerhaave had taught all of the original professors of the Edinburgh Medical School, but by mid-century Cullen, Whytt, and Monro *secundus* had reached maturity. Cullen replaced the vascular with the nervous system as the key to understanding health and disease, and modified Boerhaave's order of teaching medical subjects.

30 R. E. Schofield, *Mechanism and Materialism: British Natural Philosophy in an Age of Reason* (Princeton, N. J., 1970); M. C. Jacob, *The Newtonians and the English Revolution, 1689-1720* (Ithaca, 1976).

31 T. S. Hall, *Ideas of Life and Matter* (2 vols., Chicago, 1969); Theodore Brown, 'From mechanism to vitalism in eighteenth-century English physiology', *Journal of the History of Biology,* VII (1974), 179-216; William Coleman, 'Mechanical philosophy and hypothetical physiology', in Robert Palter (ed.), *The Annus Mirabilis of Sir Isaac Newton 1666-1966* (Cambridge, Mass., 1970); François Duchesneau, 'G. E. Stahl: antimécanisme et physiologie', *Archives internationales d'histoire des sciences,* XXVI (1976), 3-26.

32 A. E. Clark-Kennedy, *Stephen Hales* (Cambridge, 1929); H. Guerlac, 'The Continental reputation of Stephen Hales', *Archives internationales d'histoire des sciences,* XV (1951), 393-404. Theodore Brown's work is discussed in the contribution of Steven Shapin to this volume.

33 Aram Vartanian, *Diderot and Descartes: A Study of Scientific Naturalism in the Enlightenment* (Princeton, N.J., 1953); *idem, La Mettrie's* L'homme machine (Princeton, N.J., 1960); L. C. Rosenfield, *From Beast-machine to Man-machine* (New York, 1941).

In addition, the second generation of Edinburgh professors emphasized local pathology.[34] English medicine was emerging from what Le Fanu has called its 'lost half-century'.[35] In Vienna, van Swieten and de Haen continued to revere the Leyden Master, but van Swieten in particular cultivated a broad interest in public health, which was to flower in the work of Johann Peter Frank.[36] Gaubius replaced Boerhaave at Leyden, but in may ways he bears closer comparison with Cullen than with his teacher.[37] In addition, La Mettrie's notion of the man-machine and Haller's experimental work on irritability and sensibility both appeared around mid-century and, from different vantages, went some way towards circumventing the problems inherent in Cartesian dualism.[38]

These and other aspects of medical thought, teaching, and experimentation during the 1740s and 1750s mark off these decades rather distinctly from the preceding ones. Just as the intellectual origins of the French Revolution are sometimes traced from 1749, many of the concerns and attitudes of late Enlightenment doctors seem to find a resting place around this date.[39] Certainly the patterns of medi-

[34] C. J. Lawrence, 'Early Edinburgh medicine: theory and practice', in R. G. W. Anderson and A. D. C. Simpson (eds.), *The Early Years of the Edinburgh Medical School* (Edinburgh, 1976); *idem* 'The nervous system and society in the Scottish Enlightenment', in B. Barnes and S. Shapin (eds.), *Natural Order: Historical Studies of Scientific Culture* (1979); R. E. Wright-St Clair, *Doctors Monro* (1964); R. K. French, 'Sauvages, Whytt and the motion of the heart: aspects of eighteenth century animism', *Clio Medica,* VII (1972), 35–54; *idem, Robert Whytt, the Soul and Medicine* (1966). The forthcoming Ph.D. theses of C. J. Lawrence (University College London) and Rosalie Stott (University of Edinburgh) should do much to clarify the Edinburgh medical scene from the second half of the eighteenth century.

[35] W. R. Le Fanu, 'The lost half-century in English medicine, 1700–1750', *BHM,* XLVI (1972), 319–48.

[36] Erna Lesky and Adam Wandruska (eds.), *Gerard van Swieten und seine Zeit* (Vienna, 1973); Frank T. Brechka, *Gerald van Swieten and his World* (The Hague, 1970); Erna Lesky (ed.), *Wien und die Weltmedizin* (Vienna, 1974); Max Neuburger, *British Medicine and The Vienna School* (1943); J. P. Frank, *A System of Complete Medical Police*, trans. Erna Lesky (Baltimore, 1976).

[37] L. J. Rather, *Mind and Body in Eighteenth Century Medicine* (1965); John Thomson, *An Account of the Life, Lectures, and Writings of William Cullen* (2 vols., Edinburgh, 1859), I, 220–1.

[38] Cf. note 33 above and Aram Vartanian, 'Trembley's polyp, La Mettrie, and eighteenth-century French materialism', in P. P. Wiener and A. Noland (eds.), *Roots of Scientific Thought* (New York, 1957); Albrecht von Haller, *A Dissertation on the Sensible and Irritable Parts of Animals,* introd. Owsei Temkin (Baltimore, 1936); K. M. Figlio, 'Theories of perception and the physiology of mind in the late eighteenth century', *HS,* XIII (1975), 177–212.

[39] The years 1748 and 1749 saw the publication of the first volume of Buffon's *Histoire naturelle;* Montesquieu's *De l'esprit des loix;* Rousseau's *Discours sur les arts et sciences;* Diderot's *Lettre sur les aveugles,* etc.

cal theorizing seem more firmly rooted in local or national contexts later in the century. Boerhaave was the last *communis Europae praeceptor.* This is partly due to the increasing use of the vernacular in medical teaching and writing, the growth of private, hospital-based, and university medical teaching, and, strikingly in the case of British medicine, a strain of individualism with a concomitant decline in appeal to authority.[40]

In so far as there were trans-national models for medical theorists during the later Enlightenment, they were provided by Haller's work, and the ideal of Newton. Sensibility, the capacity to feel, and irritability, the capacity to react to stimuli, were presented as the basic biological properties, and Haller's conceptual framework was assimilated by many concerned with all aspects of the life sciences.[41] In medicine, John Brown believed that he had bettered Haller in reducing physiology, pathology and therapy to a series of variations around the mean of irritability.[42] In America, Benjamin Rush unified medicine and pathology even further, though with opposite therapeutic consequences, for while Brown stimulated his patients Rush depleted them. For Rush, all disease could be reduced to a result of what he called *hypertension,* or praeternatural vascular action.[43]

As the 'American Hippocrates', Rush has attracted considerable historical attention, and it is sometimes through him that Cullen, Brown, and other influences on him are viewed. In fact, Brown's theory is relatively simple to grasp, and since his central thoughts are contained in his *Elementa medicinae* (1780), the only book he published under his name, he is not accurately described as a neglected thinker. Risse's forthcoming book should shed further light on the phenomenon of 'Brunonianism', apparently richer in Italy and Germany than

[40] General aspects of the theme are discussed in John Roach, *Social Reform in England 1780–1880* (1978). Cf. Alan MacFarlane's interesting contribution to the debate about the singularity of English history, *The Origins of English Individualism* (Oxford, 1978); and E. P. Thompson's brilliant defence of 'The peculiarities of the English', in his *The Poverty of Theory* (1978).

[41] Gerd Buchdahl, *The Image of Newton and Locke in the Age of Reason* (1961); H. Guerlac, 'Where the statue stood: divergent loyalties to Newton in the eighteenth century', in Earl R. Wasserman, *Aspects of the Eighteenth Century* (Baltimore, 1965); R. H. Shryock, *Medicine in America* (Baltimore, 1966), esp. essay 1; and the works of Shryock and King cited in notes 1, 18, and 19 above.

[42] W. R. Trotter, 'John Brown and the nonspecific component of human sickness', *Perspectives in Biology and Medicine,* XXI (1978), 256–64.

[43] R. H. Shryock, 'The medical reputation of Benjamin Rush: contrasts over two centuries', *BHM,* XLV (1971), 507–52; Chris Holmes, 'Benjamin Rush and the yellow fever', *BHM,* XL (1966), 246–63; N. G. Goodman, *Benjamin Rush* (Philadelphia, 1934), esp. ch. 10.

in Britain.[44] But a social study of British Brunonians would be illuminating, particularly since Brown himself had a reputation as something of a freethinker and some of the British doctors indebted to him, such as G. H. Toulmin and Thomas Beddoes, held radical political opinions.[45] Erasmus Darwin can probably be included in this group, and his writings reflect a number of fundamental medical, biological, and social themes of the late Enlightenment.[46]

But for British medicine of the period, Cullen was undoubtedly the most significant figure. His nineteenth-century biographer has not yet been bettered, though recent selective studies have extended our knowledge of his chemistry and his relation to colleagues such as John Gregory, and work now underway should clarify other features of medicine in the Scottish Enlightenment.[47] At the moment, the specific channels of his influence must be pieced together from studies of his individual pupils, and our knowledge of Cullen's reputation in France, where many of his works were translated, is confined mostly to the case of Philippe Pinel who, apparently through forgetfulness, translated one of Cullen's papers on two separate occasions.[48]

Medical theory in France during the late *ancien régime* is usually identified with the vitalism dominating the medical school at Montpellier.[49] Stahl was a background influence there and certain aspects of Bichat's work were in the Montpellier tradition. But

[44] Risse's book is based on his University of Chicago thesis; B. Hirschel, *Geschichte d. Brown'schen Systems u.d. Erregungstheorie* (Leipzig, 1850) is still indispensable, as is Thomson, *William Cullen* (see note 37), II, 222–486.

[45] R. S. Porter, 'George Hoggart Toulmin's theory of man and the earth in the light of the development of British geology', *AS*, XXXV (1978), 339–52; *idem*, 'Philosophy and politics of a geologist: G. H. Toulmin (1754–1817)', *JHI*, XXXIX (1978), 435–50; T. H. Levere, 'Dr Thomas Beddoes and the establishment of his Pneumatic Institution: a tale of three presidents', *NR*, XXXII (1977), 41–9; F. F. Cartwright, *The English Pioneers of Anaesthesia* (Bristol, 1952); J. E. Stock, *Memoirs of the Life of Thomas Beddoes* (1811). There is also interesting material on Beddoes in James E. Cronin (ed.), *The Diary of Elihu Hubbard Smith 1771–1798* (Philadelphia, 1973).

[46] D. King-Hele, *Doctor of Revolution* (1977), a recent biographical study, does not fully exploit the inherent richness of Darwin's writings.

[47] Thomson, *William Cullen* (see note 37); Lawrence and Stott (see note 34), plus the literature cited in Shapin's essay in this volume.

[48] Aubrey Lewis, 'Philippe Pinel and the English', in his *The State of Psychiatry* (1967).

[49] Elizabeth Haigh, 'Vitalism, the soul and sensibility: the physiology of Theophile Bordeu', *JHM*, XXXI (1976), 30–41; *idem*, 'The vital principle of Paul Joseph Barthez: the clash between monism and dualism', *MH*, XXI (1977), 1–14; French, 'Sauvages, Whytt and the motion of the heart', and *Robert Whytt* (see note 34).

'vitalism' is a loose term conveying only a limited amount of information, and historically the vitalistic tradition has embraced a large number of keen observers and ardent experimentalists. John Hunter, possibly the most gifted experimental biologist of the century, was an avowed vitalist;[50] so were Réaumur, Spallanzani, Haller, Hewson, and Whytt. Indeed, if, as Schofield suggests, vitalism replaced mechanism as the principal physiological metaphysic around mid-century, it might be argued that the former was the more productive starting-point.[51]

The relationship of this experimental physiology to medical theories might be clarified by a study of holders of chairs in the Institutes of Medicine at selected universities. Haller's physiological work, too, is still insufficiently explored, both on its own terms and in reference to later physiologists and medical theorists.[52]

Various key topics offer convenient foci for comparative studies of both details and wider uses of theory. Fevers, for instance, generated an enormous literature which frequently reveals much about persistent fundamental principles and the tensions created by attempts to fit observations into theoretical structures.[53] Inflammation is another subject the analysis of which could yield insights into the world of medical controversy, and non-febrile conditions such as gout,

[50] June Goodfield-Toulmin, 'Some aspects of English physiology: 1780–1840', *Journal of the History of Biology,* II (1969), 283–320.

[51] Hilda Heine, 'The endurance of the mechanism-vitalism controversy', *Journal of the History of Biology,* V (1972), 159–88; and, for a later period, June Goodfield, 'Changing strategies: a comparison of reductionist attitudes in biological and medical research in the nineteenth and twentieth centuries', in F. J. Ayala and T. Dobzhansky (eds.), *Studies in the Philosophy of Biology* (1974).

[52] From a steady stream of books and papers on Haller, the following are representative of the most active Haller scholars: Erich Hintzsche (ed.), *Albrecht von Hallers Briefe an Auguste Tissot 1754–1777* (Bern, 1977); Urs Boschung, 'Albrecht von Haller als Arzt', *Gesnerus,* XXXIV (1977), 267–93; R. Toellner, 'Staatsidee, aufgeklärter Absolutismus und Wissenschaft bei Albrecht von Haller', *Medizinhistorisches Journal,* XI (1970), 206–19; *idem, Albrecht von Haller über die Einheit im Denken des letzten Universalgelehrten, Sudhoffs Archiv Suppl. 10* (1971); Shirley A. Roe, 'The development of Albrecht von Haller's views on embryology', *Journal of the History of Biology,* VIII (1975), 167–90; R. G. Mazzolini, 'Sugli studi embriologici di Albrecht von Haller negli anni 1755–1758', *Annali internationali storico italo-germanico di Trento,* III (1977), 183–242; Otto Sonntag, 'The motivations of the scientist: the self-image of Albrecht von Haller', *Isis,* LXV (1974), 336–51; and the ten commemorative addresses in *Albrecht von Haller, 1708–1777* (Bern, 1977).

[53] King, *Medical World* (see note 18), ch. 5; Thomson, *William Cullen* (see note 37), esp. II, 47ff., 107–73, etc.; and J. H. Powell, *Bring out your Dead* (Philadelphia, 1949), give a cross-section of both theoretical and practical discussions.

'tumours', dropsy, 'diabetes', and asthma provide interesting examples for investigating changing medical perceptions.[54] The social and ideological roots of theoretical positions may be uncovered, as several promising studies have shown. The spread of tea drinking created a polemical literature which touched a number of medical and social issues.[55] Models of physiological function may embody attitudes to children and the aged, to men and women, class and race, refinement and civilization, and to existing and desired social systems.[56] Notions of 'sympathy' and 'sensitivity' may cut across 'scientific' and 'ideological' divisions and be used simultaneously as units of medical analysis and of social legitimation.[57] More generally, the nervous system and solidist pathology should be closely examined, not only for what this could tell us about the relations between Enlightenment and early-nineteenth-century French medicine, but also as a way into the medicine of Morgagni, Cullen, John Hunter and Matthew Baillie.[58] What, for instance, is the practical significance of Cullen's remark that 'the dissection of morbid bodies is one of the best means of improving us in the distinction of diseases', or Cullen's insistence that even Sauvages, for all his overt use of external signs and symptoms in his nosology, actually tacitly employed autopsies 'in an hundred instances'.[59] Nosology itself is a useful starting-point for examining a number of theoretical and practical issues, and many doctors who did

[54] For an introduction to the literature of inflammation, cf. Peter H. Niebyl, 'Venesection and the concept of the foreign body' (Ph.D. dissertation, Yale University, 1969), ch. 12; L. J. Rather, *Addison and the White Corpuscles* (1972), Introduction; *idem, The Genesis of Cancer* (Baltimore, 1978); W. S. C. Copeman, *A Short History of the Gout and the Rheumatic Diseases* (Berkeley, 1964); S. Jarcho, 'An eighteenth century treatise on dropsy of the chest (Buchner, 1742),' *Bulletin of the New York Academy of Medicine,* XLV (1969), 799–806.

[55] Henry E. Sigerist, 'A literary controversy over tea in eighteenth century England', *BHM,* XXII (1943), 185–99.

[56] Figlio, 'Theories of perception' (see note 38); *idem,* 'The metaphor of organization: an historiographical perspective on the biomedical sciences of the early nineteenth century', *HS,* XIV (1976), 17–53; and the literature cited in Shapin's article in this volume.

[57] C. J. Lawrence, in Barnes and Shapin (eds.), *Natural Order* (see note 34); G. S. Rousseau, 'Nerves, spirits and fibres: towards the origins of sensibility', in R. F. Brissenden (ed.), *Studies in the Eighteenth Century,* III, *Proceedings of the David Nichol Smith Conference* (Canberra, 1975).

[58] E. R. Long, *A History of Pathology* (New York, 1965), ch. 5; A. E. Rodin, *The Influence of Matthew Baillie's Morbid Anatomy* (Springfield, Ill., 1973); P. Lain Entralgo, 'Clinica y pathologie de la ilustracion', in *Historia universal de la medicina* (Barcelona, 1973), V, 63–97.

[59] William Cullen, *Works,* (see note 17), I, 423.

not publish complete nosologies discuss the subject in monographs on acute diseases such as fevers or malignant sore throat, or chronic ones such as gout or arthritis.[60] Some of the outstanding clinicians of the period, such as William Heberden, Caleb Parry, Theodore Tronchin, or G. B. Borsieri remain relatively unexplored; even Morgagni awaits a full modern interpretation, and many descriptive but few analytical studies have been made of John Hunter.[61]

Not unreasonably, pure 'influence' studies are less popular than they once were, as historians have turned away from assumptions which made them seem credible. But Hippocratic medicine and Lockean philosophy presented two perennial positions frequently invoked by Enlightenment doctors. The history of Galenism, with its seventeenth-century decline, has been scrutinized; Hippocratism has not been subjected to the same examination, though one symptom of the real continuity between Enlightenment and French hospital medicine is the fact that Hippocrates continued to inspire Cabanis, Laennec, and their colleagues.[62] Indeed, in our own century, Hippocrates has been claimed as the source of much innovation in medicine (and in the social sciences as well), and certainly the Hippocratic writings provided the foundation for many Enlightenment medical enterprises, ranging from justifications of method to discussions of environmental factors in the aetiology of disease.[63] Bacon and Locke were also invoked on various title pages and programmatic introductions to medical works, and while modern historians are rightly suspicious of claims to naked empiricism, the assumptions underlying those claims may be fruitfully exposed. Not uncommonly, Enlightenment doctors drew fairly hard lines between fact and specula-

[60] E.g. William Falconer, *Observations on Dr Cadogan's Dissertation on the Gout and all Chronic Diseases*, 2nd edn (Bath, 1772).

[61] Brian Livesley, 'The resolution of the Heberden–Parry controversy', *MH*, xix (1975), 158–71; G. A. Lindeboom. 'Tronchin and Boerhaave', *Gesnerus*, xv (1958), 141–50; L. Belloni, 'G. B. Morgagni', in Lesky and Wandruska (eds.), *Gerard van Swieten* (see note 36); *idem*, Contributo all'epistolario Boerhaave-Morgagni', *Physis*, xviii (1971), 81–109; Jessie Dobson, *John Hunter* (Edinburgh, 1969).

[62] Owsei Temkin, *Galenism: Rise and Decline of a Medical Philosophy* (Ithaca, 1973); Ackerknecht, *Paris Hospital* (see note 2), 4–5, 9, 129–30, etc.

[63] For the twentieth century, see Bernard Barber (ed.), *L. J. Henderson on the Social System* (Chicago, 1970). For environmentalism, cf. the essay by Roy Porter in this volume; C. J. Glacken, *Traces on the Rhodian Shore* (Berkeley, 1967); and L. J. Jordanova, 'Earth science and environmental medicine: the synthesis of the late Enlightenment', in L. J. Jordanova and R. S. Porter (eds.), *Images of the Earth* (Chalfont St Giles, Bucks., 1979).

tion; they were prepared to abandon any particular speculation but were convinced that facts were sacred. 'A fact', wrote Benjamin Waterhouse, 'is worth a thousand arguments.'[64] Empiricism, however, carries its own theoretical baggage.

III. Actions

Actions speak louder than words. (Traditional proverb)

Historians have been more attracted to medical theories than they have to medical practices. This is partially, though not entirely, due to the nature of the literary evidence with which they are most comfortable. Theories about physiological function, ideas of health and disease, metaphysical presuppositions: these are amenable to intellectual analysis and sometimes seem to possess some progressive thread. On the other hand, the same basic tools of medical practice – blood-letting, cupping, stimulants, emetics, purgatives, sudorifics, mineral baths, modifications of diet or life-style, and alterations in climate or environmental circumstances – can be, and were, rationalized from a variety of theoretical perspectives. Practice seems to be relatively timeless and conservative when compared with theory: this is one of the themes of Ackerknecht's *History of Therapeutics* and it lies behind Lain Entralgo's perceptive comment that 'Medicine has always been, and has always had to be, in one way or another "psychosomatic"; this has not always been true in the case of pathology'.[65]

Therapeutics, of course, is not identical with pharmaceutics. Doctors in the Enlightenment (and earlier) aimed at 'management' or 'regimen', words with broad connotations, and while theories might differ, the goal of management was generally the Hippocratic ideal of balanced health.[66] Blood-letting could be claimed by proponents of differing systems to perform a multitude of tasks: it could remove the peccant or excess humour; reduce the tension in the arteries; relieve local inflammation or by sympathy ease congestion at some distant site; it could open up the body's pores or relax the body's fibres. In short, blood-letting, like so many remedies, could do almost anything, it seems. At the same time, the work of Lind with citrus fruit, Withering with digitalis, or Haygarth with a placebo, reminds us that

[64] R. Hingston Fox, *Dr John Fothergill and his Friends* (1919), 375.

[65] E. H. Ackerknecht, *Therapeutics from the Primitives to the 20th Century* (New York, 1973); Pedro Lain Entralgo, *Mind and Body* (1955), xii.

[66] Owsei Temkin, 'Health and disease', in P. P. Wiener (ed.), *Dictionary of the History of Ideas* (4 vols., New York, 1973), II, 395–407.

controlled observation was possible.[67] Though historians have found therapeutics embarrassing, or a subject fit for amused condescension, it was of real moment to Enlightenment doctors. Even if the classes of drugs and the kinds of advice and procedures were limited; even if individual doctors (as now) seemed to have used a few favourite remedies from the available armamentarium, therapeutics was reckoned as no simple affair, since the use of a remedy depended on clinical experience and a shrewd knowledge of the patient's individuality. Sir Hans Sloane owed part of his medical reputation to his constant use of the cinchona bark for fever and his special milk chocolate drink for consumptive patients.[68] But it was assumed that a remedy given at the wrong juncture of an illness could spell disaster, and a person used to high living might have to be treated differently from one from a lower social class whose diagnosis was identical. This meant that discussions of treatment were sometimes rambling, complicated, and contingent, though usually the poor could be lumped together as a group without too much individual variation, a fact which greatly simplified hospital and charity practice and permitted the busy doctor to see large numbers of the poor in rapid succession. In fact, Lawrence's work on the Edinburgh Royal Infirmary suggests that, even at the time that Cullen and Gregory were developing their respective medical systems, they were both using reasonably pure forms of Hippocratic medicine on their impoverished Infirmary patients. Although eighteenth-century case records survive much less frequently than do the minute books setting down the deliberations and activities of hospital governors, the sketchy primary clinical material could be examined with 'medical behaviour' in mind.[69]

Popular health treatises such as Wesley's *Primitive Physic,* Buchan's *Domestic Medicine,* or Tissot's *Advice to the People* were aimed at a buyer in a higher economic stratum than the average hospital patient, but a study of these and other medical handbooks written for laymen could be enormously revealing, not only in terms of the theory/practice dichotomy, but also with respect to patient expectations.[70]

[67] L. H. Roddis, *James Lind* (1951); T. W. Peck and K. D. Wilkinson, *William Withering of Birmingham* (Bristol, 1950); G. H. Weaver, 'John Haygarth: clinician, investigator, apostle of sanitation, 1740–1827', *Bulletin of the Society of the History of Medicine of Chicago,* IV (1928–35), 156–200.

[68] Cited by Le Fanu, 'The lost half-century' (see note 35).

[69] E. H. Ackerknecht, 'A plea for a "behaviorist" approach in writing of medicine', *JHM,* XXII (1967), 211–14.

[70] A. W. Hill, *John Wesley among the Physicians* (1958); G. S. Rousseau, 'John Wesley's "Primitive Physic" (1747)', *Harvard Library Bulletin,* XVI (1968), 242–56; C. J. Lawrence, 'William Buchan: medicine laid open', *MH,* XIX (1975), 20–35; Charles Eynard, *Essai sur la vie de Tissot* (Laussanne, 1839).

Indeed, it is not always obvious at whom medical treatises were aimed. Richard Blackmore wrote a series of specialized monographs on plague, smallpox, consumption, the spleen, gout, and dropsy, of which Dr Johnson commented: 'There is scarcely any distemper of dreadful name which [Blackmore] has not taught the reader how to oppose'. Wesley, himself a layman, assumed that his readers could instantly recognize specific conditions such as scurvy and consumption, and distinguish between ordinary and dry asthma, or slow and intermittent fevers. Buchan, too, presupposed that his fellow-physicians had needlessly complicated medical theory and practice and that his book contained virtually all that was necessary for most situations.[71] The continuing market for older works such as *Aristotle's Master-piece: or the Secrets of Generation* and Cornaro's *Sure and Certain Methods of Attaining a Long and Healthful Life* has greater significance than the unsurprising lay interest in reproduction, sex, and longevity. At the century's end A. F. M. Willich delivered a long series of public medical lectures in Bath, repeating them shortly afterwards in Bristol.[72] Itinerant science lecturers are a well-known group in the eighteenth century, and public medical lectures are an established feature of the Victorian landscape.[73] But virtually nothing is known of the extent to which Enlightenment doctors took to the platform, though entrepreneurial teaching for medical students grew apace in London and the major provincial centres.

The practical activities of doctors, in hospitals, or through lecturing or popular medical publishing, reveal more about doctor–patient relationships than do the systematic medical treatises. Much more detail about upper-class patients is needed before generalizations concerning client domination can be substantiated. Domestic correspondence of aristocratic and wealthy individuals could be revealing in this context, as could analyses of watering-places and health resorts much frequented by these classes.[74] Personal, social, and economic aspects

[71] C. J. Lawrence and M. R. Neve, 'Sudden death: an essay in the historiography of eighteenth century medicine', paper delivered at the conference of the Society for the Social History of Medicine, September 1978. For France, cf. William Coleman, 'Health and hygiene in the *Encyclopédie:* a medical doctrine for the Bourgeoisie', *JHM*, XXIX (1974), 399–421; *idem*, 'The people's health: medical themes in 18th century French popular literature', *BHM*, LI (1977), 55–74.

[72] A. F. M. Willich, *Lectures on Diet and Regimen* (1799). On Cornaro, cf. G. J. Gruman, *A History of Ideas about the Prolongation of Life* (Philadelphia, 1966).

[73] Nicholas Hans, *New Trends in Education in the Eighteenth Century* (1951); J. R. Millburn, *Benjamin Martin* (Leyden, 1976).

[74] The client domination point is primarily from Jewson, 'Medical knowledge' (see note 6); there are some interesting relevant suggestions in L. W. B.

of setting up a practice have never been looked at systematically, though some information for select doctors is available through biographies and preserved diaries and correspondence.[75] Physicians' use of the London coffee house as a consulting room, to see both patients and puzzled apothecaries, is well known, though the belief that this was a common practice may have arisen from the fame of two or three physicians of note who habitually did it.[76] Secret remedies for conditions like the stone and venereal disease were common, and though some doctors complained that secrecy smacked of quackery, many famous physicians, including James, Hoffmann, Fowler, and Gregory, increased their reputations through remedies bearing their names. A work on eighteenth-century patent medicines comparable to Young's monograph on the nineteenth-century American scene is a desideratum; such a study with a comparable Anglo-French dimension could reveal much about differing roles of the state in the two countries, for in France the sale of such remedies was more strictly controlled.[77] Throughout Europe, though, formal professional ties were much weaker than they became in the next century, and the pursuit of wealthy and influential patrons by individual doctors seems to have been a common avenue for establishing one's reputation. Particularly after mid-century, doctor identification with one or

Brockliss, 'Medical teaching at the University of Paris, 1600–1720', *AS*, xxxv (1978), 221–51 (pp. 239ff.). Representative studies from the extensive balneological literature include Marcel Bolotte, *Alise-Sainte-Reine aux xviie et xviie siècles* (Dijon, 1970); E. S. Turner, *Taking the Cure* (1967); and M. Michler, 'Hufelands Beitrag zur Bäderheilkunde', *Gesnerus*, xxvii (1970), 191–228. For Bath, see M. R. Neve, 'The medical profession in eighteenth century Bath', *MH*, xxi (1980), in press. Domestic correspondence has been used in illuminating ways by Lawrence Stone, *The Family, Sex, and Marriage in England 1500–1800* (1977); and Randolph Trumbach, *The Rise of the Egalitarian Family* (New York, 1978).

75 Among relevant studies of eighteenth-century doctors are Fox, *Dr John Fothergill* (see note 64); C. C. Booth and B. C. Corner (eds.), *Chain of Friendship, Selected Letters of Dr John Fothergill of London, 1735–1780* (Cambridge, Mass., 1971); J. J. Abraham, *Lettsom: His Life, Times, Friends and Descendants* (1933); G. C. Peachey, *A Memoir of William and John Hunter* (Plymouth, 1924).

76 Richard Mead spent his afternoons from about 1740 at Tom's Coffee House, Russell Street, Covent Garden, in consultation. More commonly, coffee houses were used as outlets for patent medicine and 'quack' remedies. Cf. the remarkable reference volume by Bryant Lillywhite, *London Coffee Houses* (1963), 73, 132, 592, etc.

77 James Harvey Young, *The Toadstool Millionaires* (Princeton, N.J., 1961). For the regulation of French patent medicines, see Caroline Hannaway, 'Medicine, public welfare and the state of eighteenth century France: the *Société Royale de Médecine* of Paris (1776–1793)' (Ph.D. thesis, the Johns Hopkins University, 1974), ch. 2.

more of the many medical organizations and societies which sprang up was also an important means of publicly establishing one's credentials (cf. section IV below). Doctors undoubtedly did treat their aristocratic patients gingerly, but that these patients really dictated the specific terms of therapy remains more an interesting suggestion than a proved fact.

What is certain is that the language of disease was more widely understood than it was to become in the nineteenth century. This was partially because the aims, if not all the nuances, of therapeutics, and most of the cultural images of health and disease, were part of a received and shared Classical inheritance. Hippocrates and Celsus remained vital figures during the Enlightenment. John Huxham's *Essay on Fever* is redolent of crises and concoctions, though it was the wisdom rather than the *a priori* authority of ancient writings which appealed. As William Heberden, himself one of the best classical scholars of his time, wrote, 'We can never therefore safely practise upon any of [the Hippocratic] aphorisms that have not been confirmed and made probable by later experience'.[78] In one sense, the overthrow of ancient authority represented by the seventeenth-century demise of Galenism was unalterable.

What Tröhler has called the 'quantification of experience' was, from the 1760s, an increasing preoccupation with a number of doctors, many of them Scottish or Edinburgh trained.[79] His recent study of the use of numerical reporting in the evaluation of therapy between 1750 and 1830 traces the development of what Pierre Louis was to name the *méthode numérique* to the Enlightenment ideals of empiricism, objectivity, and progress. In medical and surgical therapy, diagnosis and nosology, quantitative reporting, based on multiple cases, became a desired norm with reformist-minded doctors in the army and navy, and in newly founded hospitals and dispensaries. Not surprisingly, several of these, such as William Black, John Coakley Lettsom, and John Millar were associated with the Medical Society of London, an organization whose founders sought to break down some of the

[78] For instance, the British Library possesses nine eighteenth-century Latin editions of Celsus' *De medicina,* plus an English translation, and six Latin, two German, and two English editions of extracts. For language, though, cf. Jean-Pierre Goubert 'L'art de guérir: médecine savante et médecine popullaire dans la France de 1790', *Annales: Economies, sociétés, civilisations,* XXXII (1977), 908–26. The quotation from Heberden is from Le Roy Crummer, 'An introduction to the study of physic, by William Heberden', *Annals of Medical History,* X (1928), 226–41, 349–67 (p. 364). See also R. M. S. McConaghey, 'John Huxham', *MH,* XIII (1969), 280–7.

[79] Ulrich Tröhler, 'Quantification in British medicine and surgery 1750–1830, with special reference to its introduction into therapeutics' (Ph.D. thesis, University of London, 1978).

barriers between physicians, surgeons, and apothecaries.[80] Surgical procedures were particularly amenable to numerical reporting, since the outcome was usually precise (recovery or death), and various operations, such as alternative approaches to bladder stones, or the timing and method of amputations, were debated on the basis of figures.

Surgeons were known, even to their contemporaries, as men of action, and though anaesthesia and Listerian techniques were in the future, Enlightenment surgery did not deserve the label of barbaric butchery that is sometimes pinned on it. Most surgeons treated a wide variety of complaints, including venereal disease and skin disorders, and even within the relatively limited range of feasible major operations (e.g. amputations, lithotomy) there was room for debate over indications, timing, and preferred technique. Technical innovation was common in the surgery of the century, sometimes, as in Douglas's work in lithotomy, or Hunter's surgical treatment of aneurysms, the direct result of anatomical or physiological research.[81] Trauma and war wounds provided a principal source of surgical experience in major operations; unsurprisingly, many young surgeons spent some time in the army or navy. The monographs of Cantlie and of Keevil, Lloyd and Coulter have set out the main details of medical practice in the British army and navy, respectively, and various continental military medical services have been studied.[82] But much more systematic work should be done, particularly because better records often survive for the military than for comparable civilian situations.[83] Owsei Temkin's classic paper pointed out the role of surgery in the development of modern medicine and its disease concepts, and Ackerknecht and Vess have stressed the crucial role of the military in early-nineteenth-century French medicine.[84] But the wider impact of

[80] Thomas Hunt (ed.), *The Medical Society of London 1773–1973* (1972).

[81] Tröhler, 'Quantification in British Medicine' (see note 79) chs. 6,7; K. B. Thomas, *James Douglas of the Pouch and his Pupil William Hunter* (1964); Z. Cope, *William Cheselden, 1688–1752* (Edinburgh, 1953); L. A. Wells, 'Aneurysms and physiologic surgery', *BHM*, XLIV (1970), 411–24; Philip G. Ritterbush, *Overtures to Biology* (New Haven, 1964).

[82] Neil Cantlie, *A History of the Army Medical Department* (2 vols., Edinburgh, 1974); J. J. Keevil, C. Lloyd and J. L. S. Coulter, *Medicine and the Navy* (4 vols., Edinburgh, 1957–63); Friedrich Ring, *Zur Geschichte der Militärmedizin in Deutschland* (Berlin, 1962); J. Rieux and J. Hassenforder, *Histoire du Service de Santé Militaire et du Val-de Grâce* (Paris, 1951).

[83] Peter Mathias, 'Swords and ploughshares: the armed forces, medicine and public health in the late eighteenth century', in J. M. Winter (ed.), *War and Economic Development* (Cambridge, 1975).

[84] Owsei Temkin, 'The role of surgery in the rise of modern medical thought', *BHM*, XXV (1951), 248–59; Ackerknecht, *Paris Hospital* (see note 2); Vess, *Medical Revolution* (see note 2).

war experience on Enlightenment civilian medical theory and practice has yet to be explored, even though 'modern' military medicine is conventionally attributed to the activities of Pringle, Lind, van Swieten, Blane, and Heister.[85] A study of the upward mobility of the army or ship's surgeon turned physician (Cullen, Lind, Robert Robertson) could be illuminating: as a group, they do not fit the stereotyped image of the eighteenth-century physician as simply a gentleman, tending towards pomposity and essentially uninterested in the physical aspects of medical practice.

Midwifery was another sphere of male activity which has recently been of considerable interest, especially to those concerned with the history of women.[86] In terms of the number of deliveries, the male midwife was relatively insignificant throughout the century, and the vast majority of children continued to come into the world surrounded by women. But attitudes to the presence of men in the confinement room involved a rich range of questions, including the 'disease concept' of pregnancy, sexuality, doctor–patient relationships, female physiology and social roles, and patterns of influence from the aristocratic and upper-middle classes, where the desire to have male attendance apparently originated.[87] Lying-in hospitals were one tangible legacy of the phenomenon of the male midwife.[88] Midwives as a group are less visible than their relatively few male counterparts, but local studies might reveal more about the women, and a prosopographical examination of the doctors themselves might be possible, though the extent of occasional obstetrical practice by surgeons or provincial physicians remains obscure. Comparative international studies of the regulation and social positions of midwives would also be of interest.[89]

[85] Though see Tröhler, 'Quantification in British medicine' (see note 79); Mathias, 'Swords and ploughshares' (see note 83); C. G. Lorber, 'Die Bedeutung Lorenz Heisters in der Hasenscharten-chirurgie', *Medizin-historisches Journal*, x (1965), 81–93.

[86] Jean Donnison, *Midwives and Medical Men* (1977); Jane Donegan, *Women and Men Midwives* (Westport, Conn., 1978); T. R. Forbes, *The Midwife and the Witch* (New Haven, 1966).

[87] Donnison, *Midwives and Medical Men* (see note 86); David Hunt, *Parents and Children in History* (New York, 1970), 84–5; Trumbach, *Rise of the Egalitarian Family* (see note 74), 180ff.

[88] E.g. Philip Rhodes, *Doctor John Leake's Hospital* (1977).

[89] R. L. Petrelli, 'The regulation of French midwifery during the Ancien Régime', *JHM*, xxvi (1971), 276–92; E. H. Ackerknecht, 'Zur Geschichte der Hebammen', *Gesnerus*, xxxi (1974), 181–91; Jacques Gélis, 'Sages-femmes et accoucheurs: l'obstétrique populaire aux XVIIe et XVIIIe siècles', *Annales: économies, sociétés, civilisations*, xxxii (1977), 927–57.

If one of the characteristics of Enlightenment was enhanced concern with the business of this world, it is not surprising that health and disease were conscious preoccupations of laymen and doctors alike.[90] A number of afflictions, such as venereal disease, presented more practical than theoretical problems, but even 'fever' – a most theory-laden concept of eighteenth-century doctors – generated an enormous practical literature, particularly after mid-century.[91] Theory/practice studies can be more valuable than mere expositions of theoretical positions, if for no other reason than they require the historian to pay closer attention to the nuances of past medical thought. And historians sometimes overlook the fact that what we take to be 'common-sense' practices, such as emphasis on cleanliness and fresh air, had their theoretical underpinnings. A medical history of cleanliness could be fascinating and might be one point where Mary Douglas's anthropological concepts could bear fruit, and where literary and iconographical evidence would be germane.[92] Lind, Pringle and some of their colleagues do seem to be apostles of a new way of thinking about cleanliness, which, after all, could be seen as next to godliness.[93] In Lind's case, at least, much of his thinking about health and preventative medicine came directly from his experience in the alien climates of Africa and the West Indies, and in the claustophobic confines below deck. Environmentalism could be easily assimilated into a Lockean epistemology and of course squared well with much Hippocratic writing.[94] Indeed, 'neo-Hippocratism' was a major formative stance in late Enlightenment medicine, French and British; Sydenham too, continued to enjoy an enviable reputation at home and abroad.[95]

[90] Gay, *The Enlightenment* (see note 14), II, ch. 1; John Passmore, *The Perfectibility of Man* (1970), ch. 10.

[91] For venereal disease, see Theodor Roseburg, *Microbes and Morals* (1971); Stone, *The Family, Sex and Marriage* (see note 74), 572–99; W. B. Ober, *Boswell's Clap and other Essays* (Carbondale, Ill., 1979).

[92] Mary Douglas, *Purity and Danger* (1966).

[93] The *aperçu* comes from John Wesley, for whom cf. Hill, *John Wesley* (see note 70), 117–19, and Rousseau, 'John Wesley's "Primitive Physic"' (see note 70). For illuminating comments, cf. Owsei Temkin, 'An historical analysis of the concept of infection', in *Studies in Intellectual History* (Baltimore, 1953), 139ff.; Dorothy George, *London Life in the Eighteenth Century* (Harmondsworth, 1965), 69–72, 110–13; and Stone, *The Family, Sex and Marriage* (see note 74), 485ff.

[94] Philip Curtin, *The Image of Africa: British Ideas and Action, 1780–1850* (1965); Glacken, *Traces on the Rhodian Shore* (see note 63), esp. ch. 12; and the essay by Roy S. Porter in this volume.

[95] F. N. L. Poynter, 'Sydenham's influence abroad', *MH*, XVII (1973), 223–34; Erna Lesky, 'Vom Hippokratismus Boerhaaves und de Haens', in G. A. Lindeboom (ed.), *Boerhaave and his Time* (Leyden, 1970).

If nineteenth-century professionalization was to confer positional status on doctors, individual and personal characteristics were of greater consequence in the medical world of the eighteenth century. But informal networks and associations can also operate effectively, and at many points the interaction of Enlightenment medicine and society was not simply an individual doctor tending his patient. The absence of formal licensing laws is sometimes seen as blurring the boundary between 'doctor' and 'quack'. But in specific local settings social perceptions were probably rather precise, as witnessed by the regulation passed by the lay governors of the Royal Salop Infirmary in 1790, denying treatment to any patient known to have visited a quack.[96]

IV. Interactions

The united judgments . . . of men correct and confirm each other by communication, their frequent intercourse exites emulation, and from this comparison of different *phaenomena,* remarked by different persons, there often result general truths, of which, from one of these *phaenomenon,* no man of the greatest sagacity could entertain any suspicion.

(*Essays and Observations,* 1754)[97]

It is a fact of some curiosity that a century for which individualism seems so appropriate a unit of analysis should have produced three important kinds of medical institutions: hospitals, medical societies, and medical schools. The Enlightenment did not, of course, create them *de novo,* and these manifestations of corporate activity were transformed in the following century. But medicine was the focus of a considerable amount of associative behaviour, and these institutions have been subjected to a good deal of historical attention, usually in terms of narrower issues of medical life, less often as starting-points for analysing the interactions of medicine and society.

Hospitals and infirmaries became a particularly striking feature of the eighteenth-century landscape, and to the four medieval foundations (St Bartholomew's, St Thomas's, Bethlem, Christ's) which Henry VIII turned over to the city of London, Enlightenment philanthropists added literally dozens more, starting with the Westminster in 1719.[98] By mid-century there were four more general hospitals in

[96] W. B. Howie, 'The administration of an eighteenth-century provincial hospital: the Royal Salop Infirmary, 1747–1830', *MH,* v (1961), 34–55.

[97] 'Preface' to *Essays and Observations, Physical and Literary* (3 vols., Edinburgh, 1754–71), I, v–vi.

[98] John Woodward, *To do the Sick no Harm* (1974); J. G. Humble and Peter Han-

London, and most provincial towns with any sense of civic pride established a hospital or infirmary sometime during the period.[99] These voluntary hospitals were financed, staffed, and regulated along lines similar to those laid down at the Westminster. Subscribers paid varying annual or once-off sums for a variety of privileges including the election of medical staff, establishment of regulations, and nomination of patients judged worthy of receiving the services of the charity. Although these institutions courted aristocratic subscribers to add prestige to the enterprise, subscription lists substantiate the extent to which Enlightenment philanthropy was preponderantly financed by middle-class individuals, some of them landed, of course, but others engaged in business, commerce, manufacturing, medicine, law, and other worldly activities. Medicine was a major outlet of philanthropy, and concern with health surely one manifestation of Enlightenment secularism.[100] On the other hand, Enlightenment hospitals also possessed specific moral and religious (though generally not narrowly doctrinal) functions, and the successful hospital stay would find the discharged patient improved in body, morals, and spirit. Architectural details often reveal fascinating hierarchies of values (chapels were a prominent feature, separate operating theatres did not appear with regularity until late in the century), and it would probably be unwise to make too sharp a divorce between philanthropy and religion during the period.[101] Indeed, it may be that one reason for the success of the hospital as a charitable institution was the extent to which Anglicans and Dissenters could actually cooperate (something more difficult in educational establishments, for example), and in some localities all the local ministers took it in turn to act as chaplain.[102]

sell, *Westminster Hospital* (1966); F. N. L. Poynter (ed.), *The Evolution of Hospitals in Britain* (1964).

[99] Some of the more useful examples from this extensive literature include A. E. Clark-Kennedy, *The London: A Study of the Voluntary Hospital System* (2 vols., 1962); A. C. Cameron, *Mr Guy's Hospital 1726–1948* (1954); William Broadbank, *Portrait of a Hospital 1752–1948* [Manchester Royal Infirmary] (1952); A. Logan Turner, *Story of a Great Hospital: The Royal Infirmary of Edinburgh 1729–1929* (Edinburgh, 1937).

[100] David Owen, *English Philanthropy 1660–1960* (Cambridge, Mass. 1965); Coleman, 'Health and hygiene' and 'The people's health' (see note 71).

[101] John D. Thompson and Grace Goldin, *The Hospital: A Social and Architectural History* (New Haven, 1975); Adrian Forty, 'The modern hospital: the social and medical uses of architecture', in A. King (ed.), *Buildings and Society* (1980); Rod Morgan, 'Divine philanthropy: John Howard reconsidered', *History*, LXII (1977), 388–410.

[102] Howie, 'The Royal Salop Infirmary' (see note 96), 45.

There is still much to be learned about the financing of hospitals: who gave and why? This is a particularly intriguing problem since most of the books and articles describing these institutions have been vaguely meliorist in outlook, and have accepted as unproblematic the literature generated by the sponsors themselves. More recently, however, historians have been interested in the underside of philanthropy: in individual motives of self-aggrandisement and social legitimation, and covert hospital functions bluntly described as social control.[103] Certainly individual subscribers did not object to having their names published each year, and the atmosphere in hospital was aimed at increasing the deference and gratitude of the patient, thereby reinforcing class differences. But ambiguity of motive is not unique to the Enlightenment, and the enriched historical texture which such modes of analysis provide should not obscure the fact that patient care was after all the primary aim of these institutions.

Accordingly, doctors - physicians, surgeons, apothecaries - are key figures, even though in many ways they remained the servants of the lay subscribers or governors who elected them and to whom they were responsible. Apothecaries were generally full-time and resident, and therefore paid a salary. Physicians and surgeons donated their services gratuitously, in return for the considerable indirect benefits which the appointments carried. To be elected to an honorary post at the local hospital was a public vote of confidence from some of the locality's leading citizens; it offered opportunities for personal contact with potential private patients, and for surgeons in particular the hospital provided clinical material useful for the training of their own apprentices. And if routine hospital-based medical education was still in the future, its value was certainly widely recognized in the Enlightenment. Gelfand's work has demonstrated precisely how hospital-orientated the Paris Company of Surgeons was, particularly from the 1750s, and John Hunter made considerable pedagogical use of his appointment at St George's.[104] Nor was hospital-based clinical research unheard of: Lind took advantage of his position at the massive naval hospital at Portsmouth (Haslar); Blane adapted methods of

[103] E.g. Michel Foucault *et al.*, *Les machines à guérir: aux origines de l'hôpital moderne* (Paris, 1976); Forty, 'The modern hospital' (see note 101).

[104] Toby Gelfand, 'The training of surgeons in eighteenth-century Paris and its influence on medical education' (Ph.D. thesis, Johns Hopkins University, 1973); *idem*, 'The hospice of the Paris College of Surgery (1774–1793)', *BHM*, XLVII (1973), 375–93; G. C. Peachey, *The History of St George's Hospital* (1910–14); F. N. L. Poynter (ed.), *The Evolution of Medical Education in Britain* (1966). A revised version of Gelfand's thesis is to be published with the title *Professionalizing Modern Surgery: Medical Science and Institutions in Eighteenth Century Paris* (Westport, Conn., 1980).

clinical reporting learned in the navy when he became physician to the Westminster; and John Coakley Lettsom founded the Public Dispensary with (among other things) clinical research in mind.[105] In short, much of the contrast between eighteenth- and nineteenth-century hospitals is quantitative rather than qualitative. Prosopographical studies of the physicians and surgeons elected to the honorary positions would help specify the use such appointments played in career patterns, though examples like John Fothergill, who grossed more than £5,000 per annum at the height of his practice, remind us that hospital attachments may have been desirable but were not absolutely necessary for success.[106] There is nothing for Enlightenment Britain comparable to Jeanne Peterson's recent study of mid-Victorian London doctors, though Jean-Pierre Goubert's monograph on *Malades et médecine en Bretagne, 1770–1790,* is an example of the kind of integrative work that can be done.[107]

The organization of medical care for the poor varied considerably in different national settings, and workhouses, bridewells and poorhouses contained many of the sick poor, while much of the population undoubtedly passed from the cradle to the grave (many of them in double-quick time) without coming into contact with a medical man or institution. In Britain, medical services have generally had short shrift in studies of the Old Poor Law, but surviving overseers' account books could help fill in this picture.[108] The Mineral Water Hospital, Bath, apparently catered for the sick poor from all over the country, inhabitants of Bath being excluded; more generally parish guardians subscribed to the local hospital and recommended sick paupers, though significant differences in *per diem* costs between workhouse and hospital discouraged rate-conscious guardians from sending their charges to the latter very often. Friendly and benefit societies reflect active involvement by the workers themselves.[109]

105 Tröhler, 'Quantification in British medicine' (see note 79), 103–6; Abraham, *Lettsom* (see note 75).

106 Fox, *Dr John Fothergill* (see note 64), 20.

107 Jeanne Peterson, *The Medical Profession in mid-Victorian London* (Berkeley, 1978); Jean-Pierre Goubert, *Malades et médecine en Bretagne, 1770–1790* (Rennes, 1974).

108 Dr Joan Lane of Lanchester Polytechnic, Coventry, is currently looking at this issue in Warwickshire. See Dorothy Marshall, *The English Poor in the Eighteenth Century* (1926); G. W. Oxley, *Poor Relief in England and Wales 1601–1834* (Newton Abbot, 1974), 65–73; G. B. Hindle, *Provision for the Relief of the Poor in Manchester, 1754–1826* (Manchester, 1975).

109 Neve, 'The medical profession' (see note 74); S. C. McIntyre, 'Towns as health and holiday resorts: the development of Bath, Scarborough and Weymouth 1708-1915' (D. Phil. thesis, University of Oxford, 1974); E. Posner,

An analogous distinction between workhouse and hospital existed in France in the *Hôpital general* and the *Hôtel Dieu*. Foucault has insisted that the *Hôpital general,* each with its assortment of beggars, orphans, vagabonds, prostitutes and thieves jostling with the sick and the mad, represented a major repressive institution in Enlightenment France: epoch of 'the great confinement'.[110] The specifically medical connotation was greater in the *Hôtel Dieu,* more generally run by religious orders, but of the 2,005 inmates of the *Hôpital general* at Rouen in 1777 more than half were there as the result of some physical or mental infirmity.[111] Private and formal Church philanthropy was well developed in France during the period, but throughout the continent methods of financing public institutions varied, with the state playing a more visible role in many areas of Europe than obtains for Britain: as early as 1783 the Serafimer Hospital in Stockholm was partially financed from customs duty on tobacco, and elsewhere state funds supplemented income from tithes, offerings, and charitable collections.[112]

Though medical involvement with institutions serving medical functions varied, Enlightenment doctors recognized the educational value of hospitals. There were earlier Italian precedents, but Boerhaave's reputation rested partly on his use of twelve beds for clinical instruction, and this specifically Boerhaavian legacy may be found in Vienna, Edinburgh, Göttingen, and elsewhere.[113] There are particularly full analyses of the foundations of the Edinburgh Medical

'Eighteenth-century health and social service in the pottery industry of North Staffordshire', *MH,* XVII (1974), 138–45; P. H. J. Gosden, *The Friendly Societies of England* (Manchester, 1961); E. P. Thompson, *The Making of the English Working Class* (Harmondsworth, 1968), esp. 456ff.

110 Foucault, *Madness and Civilization* (see note 4); Olwen H. Hufton, *The Poor of Eighteenth-Century France 1750–1789* (Oxford, 1974); J.-P. Gutton, *La société et les pauvres: l'exemple de la généralité de Lyon, 1534–1789* (Paris, 1970); Cissie C. Fairchilds, *Poverty and Charity in Aix-en-Provence, 1640–1789* (Baltimore, 1976).

111 Hufton, *The Poor of Eighteenth-Century France* (see note 110), 150–1; Foucault et. al., *Les machines à guérir* (see note 103).

112 Wolfram Koch, 'Swedish medicine in the Gustavian period (1772–1809)', in *Jahrbuch der Universität Düsseldorf 1973–75* (Düsseldorf), 309–31; Muriel Jeorger, 'La structure hospitalière de la France sous l'Ancien Régime', *Annales: économie, société, civilisation,* XXXII (1977), 1025–51; Dieter Jetter, *Geschichte des Hospitals* (3 vols., Wiesbaden, 1966–72); Charles Coury, *L'Hôtel-Dieu de Paris* (Paris, 1969); Bernhard Grois, *Das Allgemeine Krankenhaus in Wien und seine Geschichte* (Vienna, 1965); G. Jaeckel, *Die Charité: die Geschichte des berühmtesten deutschen Krankenhauses* (Bayreuth, 1963).

113 Lesky and Wandruska, *Gerard van Swieten* (see note 36); E. A. Underwood, *Boerhaave's Men at Leyden and After* (Edinburgh, 1977); R. Rumsteller, *Die Aufange der medizinischen Poliklinik zu Göttingen* (Göttingen, 1958); W. H. Williams, *America's First Hospital: the Pennsylvania Hospital, 1751–1841* (Wayne, Pa., 1976).

School, and many though not all European and American universities with medical faculties established in the Enlightenment or before have had some historical description.[114] But the historiography of formal medical education is still in a rather primitive state except for the famous centres, and much more could be done on the extent to which hospitals not attached to universities were used in teaching.[115] Theses are another largely neglected indicator of the kinds of problems and methods deemed valuable in particular institutions, as Diethelm's patient work on theses of psychiatric interest documents.[116]

Although the records of private anatomy and medical schools are often frustratingly meagre, the importance of this characteristic Enlightenment enterprise is established, most firmly through the Hunters in London, and men like the Sues and Jacob Winslow in Paris.[117] Newspaper advertisements reveal that private anatomy and chemistry lecturing was surprisingly common in early-eighteenth-century London and elsewhere, and close examination of even Oxford and Cambridge can turn up a considerable amount of less formal educational activity.[118] Medical corporations, particularly surgeons' and apothecaries' companies, were concerned with education, and though

[114] J. B. Morrell, 'The Edinburgh Town Council and its University, 1717–1766', in Anderson and Simpson (eds.), *Edinburgh Medical School* (see note 34); A. R. Cunningham, 'Aspects of the history of medical education in Britain in the 17th and early 18th centuries' (Ph.D. thesis, University of London, 1974); J. R. R. Christie, 'The origins and development of the Scottish scientific community, 1680–1760', *HS*, xii (1974), 122–41. Dr Cunningham is presently writing a book on the prehistory and foundation of the Edinburgh Medical School. Other histories of eighteenth-century medical faculties not cited elsewhere in this paper include: E. Geist and B. von Hagen, *Geschichte der medizinischer Fakultät der Friedrich-Schiller-Universität Jena* (Jena, 1958); Leon Binet and Pierre Vallery-Radot, *La faculté de médecine de Paris* (Paris, 1952); H. K. Beecher and M. D. Altschule, *Medicine at Harvard: The First Three Hundred Years* (Hanover, N.H., 1977).

[115] W. B. Howie, 'Medical education in 18th century hospitals', *Scottish Society for the History of Medicine, Report Proceedings* (1969–70), 27–46.

[116] Oskar Diethelm, *Medical Dissertations of Psychiatric Interest Printed before 1750* (Basel, 1971).

[117] Peachey, *William and John Hunter* (see note 75); Toby Gelfand, 'The "Paris manner" of dissection: student anatomical dissection in early eighteenth-century Paris', *BHM*, xlvi (1972), 99–130; Sir Charles Illingsworth, *The Story of William Hunter* (Edinburgh, 1967); Th. Vetter, 'La vie active de Jacques-Benigne Winslow à la lumière des "Commentaires de la Faculté de Paris'", *Nordisk Medicin historik Årsbok* (1971), 107–29.

[118] Kenneth Dewhurst (ed.), *Oxford Medicine* (Oxford, 1970); A. H. T. Robb-Smith, *A Short History of the Radcliffe Infirmary* (Oxford, 1970); R. Hunter and I. MacAlpine, 'William Pargeter and the Medical Society of Oxford, 1780–3', *MH*, ix (1965), 181–3; Arthur Rook, 'Medicine at Cambridge 1660–1760', *MH*, xiii (1969), 107–22; Arthur Rook (ed.), *Cambridge and its Contribution to Medicine* (1971).

formal curricula were not laid down outside the universities, less-official educational establishments were often sources of innovation. To the public eye, anatomy teaching appeared as the be-all of medical education, and the public found dissecting unsavoury and obscene. Since doctors defended it, even to the point of tacitly condoning body snatching, grave robbing, and the dissecting of condemned criminals, the issue was a chronic source of conflict between the public and the doctors, particularly surgeons.[119]

We may assume that the occasionally reiterated stereotype of the medical student as an atheistic, immoral libertine was patently exaggerated, but little is known of the nuances of the medical student sub-cultures during this period.[120] Most provincial surgeons and apothecaries were trained individually by more or less formal apprenticeships, some details of the economic and practical aspects of which are beginning to be uncovered.[121] Not all medical students were as pious and industrious as was the young Albrecht von Haller,[122] but students did form a number of debating societies, several of which had long and vigorous lives. Enlightenment Edinburgh was a particularly clubable place, but student societies in London also were active from the 1770s.[123] Both formal and informal medical societies also catered for the needs of many more senior doctors in London, Dublin, Bristol, Edinburgh, Warrington, Colchester, Plymouth, and elsewhere in Britain.[124] There were usually strong medical elements

119 Peter Linebaugh, 'The Tyburn riots against the surgeons', in D. Hay *et al.* (eds.), *Albion's Fatal Tree* (Harmondsworth, 1977); Gelfand, 'The "Paris manner" of dissection' (see note 117).

120 General aspects of student life can be gleaned from the essays by Sheldon Rothblatt, Nicholas Phillipson and James McLachlan in Lawrence Stone (ed.), *The University in Society* (2 vols., (Princeton, N.J., 1975), but the history of the Enlightenment medical student remains to be written.

121 J. G. L. Burnby, 'Apprenticeship records', *Transactions of the British Society for the History of Pharmacy* (1977), 145–94; *idem*, 'A study of the English apothecary from 1660 to 1760, with special reference to the provinces' (Ph.D. thesis, University of London, 1979).

122 E. Hintzsche and H. Balmer (eds.), *Albrecht Hallers Tagebücher seiner Reisen nach Deutschland, Holland und England, 1723–1727* (Bern, 1971); and E. Hintzsche (ed.), *Albrecht Hallers Tagebuch seiner Studienreise nach London, Paris, Strasburg und Basel, 1727–1728* (Bern, 1968).

123 D. D. McElroy, 'The literary clubs and societies of eighteenth century Scotland' (Ph.D. thesis, University of Edinburgh, 1952); James Gray, *History of the Royal Medical Society, 1737–1837* (Edinburgh, 1952); J. R. Wall, 'The Guy's Hospital Physical Society', *Guy's Hospital Report*, CXXIII (1974), 159–70.

124 E.g. Ian Inkster, 'Science and society in the metropolis: the Askesian Society of London, 1796–1807', *AS*, XXXIV (1977), 1–32; Walter Radcliffe, 'The Colchester Medical Society, 1774', *MH*, XX (1976), 394–401; J. F. Fulton, 'The Warrington Academy (1757–1786) and its influence upon medicine and science',

in the many formal continental and American academies, e.g. the American Philosophical Society, the Berlin Akademie der Wissenschaften, the Academy of Sciences in Paris, the Royal Society of Göttingen, or the academies in such places as St Petersburg, Madrid, Uppsala, Bordeaux, and Haarlem.[125] The importance of doctors in general associations, such as the Lunar Society of Birmingham and the Manchester Literary and Philosophical Society, is also well established.[126] Most of these organizations tried to consolidate their wider presence through the publication of proceedings or transactions; many were successful. In addition, joint works, like the five-volumed *Medical Essays and Observations* (1733–1744) of the Society for the Improvement of Medical Knowledge (Edinburgh), or the later London-based *Medical Observations and Inquiries* (6 vols., 1757–84), encouraged a new opennesss in clinical reporting. By the 1780s medical periodicals were relatively common and, as titles such as the *Journal de médicine, chirurgie et pharmacie* (1754) and the *Medical and Chirurgical Review* (1794) indicate, many were aimed at a wide spectrum of medical opinion. General periodicals such as *Gentleman's Magazine* and the *Monthly Review* also published articles and reviews of medical interest, and British and American medical works were frequently extracted in the *Bibliothèque britannique*.[127] The growth in the use of the vernacular increased the market for translations: much remains to be learned about the kinds of books selected for translation, the choice of translators, and the economics of medical publishing in the Enlightenment.

It is significant that many of these group activities – new societies, periodicals, or collective works – cut across the traditional medical hierarchies of physicians, surgeons, and apothecaries. Much of the literature examining the eighteenth-century medical profession has been conflict-oriented, and certainly formal relations between the

BHM, I (1933), 50–80. More generally, cf. Fox, *Dr John Fothergill* (see note 64); Tröhler, 'Quantification in British medicine' (see note 79).

125 See, for example, Werner H. Kopf, *Die Akademie der Wissenschaften der DDR* (Berlin, 1975); *Die Berliner Akademie der Wissenschaften in der Zeit des Imperialismus* (Berlin, 1975); Roger Hahn, *The Anatomy of a Scientific Institution: The Paris Academy of Sciences, 1666–1803* (Berkeley, 1971); Alexander Vucinich, *Science in Russian Culture* (1965); Daniel Roche, *Le siècle des lumières en province, Académies et académiciens provinciaux, 1680–1789* (2 vols., Paris, 1978).

126 R. E. Schofield, *The Lunar Society of Birmingham* (Oxford, 1963); Arnold Thackray, 'Natural knowledge in cultural context', *American Historical Review*, LXXIX (1974), 672–709.

127 Marc A. Barblan, 'Journalisme médical et échanges intellectuels au tournant du XVIIIe siècle: le cas de la "Bibliothèque Britannique" (1796–1815)', *Archives des sciences, Genève*, XXX (1977), 283–398.

Royal College of Physicians and the Society of Apothecaries over the Rose case in 1704, between the licentiates and fellows of the College of Physicians in the 1760s, or between the various Paris colleges, academies, and faculties of medicine and surgery expose rampant petty jealousies and tenacious guarding of privileges, earned or unearned. Even the commissioned historians of the London medical corporations are frequently driven to apologetics when describing this aspect of Enlightenment medical activity.[128] But the genuine differences in background, education, and social standing which marked off physicians, surgeons, and apothecaries should not be permitted to obscure important qualifications which could be elucidated through local studies and prosopography. Waddington has recently drawn attention to the fact that conventional wisdom about the early-nineteenth-century medical profession is somewhat biased by excessive preoccupation with the London scene; in provincial towns distinctions between physicians and surgeons were frequently less clear, and this pattern probably obtained during the previous century. Certainly individuals and groups bent on reform share characteristics which cut across ordinary allegiances to familiar medical orders. Licentiates of the College of Physicians wanted a voice in the running of the College, to be sure, but the 'medical militants' of the 1760s included ex-surgeons, male midwives, and others whose careers and outlooks were hardly in any standard mould.[129]

The concept of 'marginality' has increasingly been used in analysing social action during the period.[130] If employed sparingly, the term is meaningful, though almost anybody, even the leader of the political party in opposition, can in some sense be seen as marginal. Nevertheless, it is significant that many of those active in creating new societies, or interested in reforming existing ones, expressed disaffec-

[128] E.g. Sir George Clark, *A History of the Royal College of Physicians of London* (2 vols., Oxford, 1964–6); Sir V. Z. Cope, *The Royal College of Surgeons of England* (1959); C. Wall, *The History of the Surgeons' Company 1745–1800* (1937); C. Wall, H. C. Cameron and E. A. Underwood, *A History of the Worshipful Society of Apothecaries of London* (1963); W. S. Craig, *History of the Royal College of Physicians of Edinburgh* (Oxford, 1976). For Paris, see Hannaway, 'Medicine, public welfare and the state' (see note 77); Gelfand, 'The training of surgeons in eighteenth-century Paris' and 'The hospice of the Paris College of Surgery' (see note 104); and Emile Coornaert, *Les corporations en France avant 1789*, 2nd edn (Paris, 1968).

[129] I. Waddington, 'The struggle to reform the Royal College of Physicians, 1767–1771: a sociological analysis', *MH*, XVII (1973); 107–26; Bernice Hamilton, 'The medical professions in the eighteenth century', *Economic History Review*, 2nd ser., IV (1951), 141–69, (pp. 145ff.).

[130] Ian Inkster, 'Marginal men: aspects of the social role of the medical community in Sheffield, 1790–1850', in Woodward and Richards (eds.), *Health Care* (see note 6); Roy Wallis and Peter Morley (eds.), *Marginal Medicine* (1976).

tion at their preceived exclusion from power, prestige, or social position. Calls for meritocracies frequently emanate from individuals who feel that their own merits or sense of values have been improperly acknowledged or rewarded. Dissenters were excluded from Oxford and Cambridge and consequently, except under special circumstances, from fellowship in the College of Physicians. Leyden graduates in the first half of the century, and Edinburgh graduates in the second half, clearly leavened the British medical scene. Underwood's recent monograph on 'Boerhaave's men', though by no means definitive, makes a beginning for a study of the earlier group, but there is no systematic assessment of the impact of Edinburgh on British medicine. Among dissenting groups Quakers and Unitarians were especially good at producing doctors, and shared values often led to cooperation on many levels. A biographical guide to eighteenth-century medical men, even an incomplete one, is a desideratum. Webster and Pelling's work on sixteenth-century Norwich, and Inkster's on nineteenth-century Sheffield, provide examples of the local results which can be obtained.[131]

Detailed local digging will provide a much richer picture of medical care; of the extent to which unlicensed practitioners, midwives, bone-setters, wise women, clergymen, part-time apothecaries and others were part of the texture of this medical world. Quackery, too, is a subject which has too often attracted a casual historiography, though a generation of historians dismissive of the claims of the 'regular' doctors should take a more sympathetic view of nostrum pedlars like Joshua Ward or showmen like James Graham.[132] The word

[131] Underwood, *Boerhaave's Men* (see note 113); R. W. Innes Smith, *English Speaking Students of Medicine at the University of Leyden* (Edinburgh, 1932); Margaret Pelling, 'Medical practice in Norwich 1550–1640', *Bulletin of the Society for the Social History of Medicine,* xxiii (1978), 30–2, and the papers in C. Webster (ed.), *Health, Medicine and Mortality in the Sixteenth Century* (Cambridge, 1979); Arthur Raistrick, *Quakers in Science and Industry* (Newton Abbot, 1968); Michael Watts, *The Dissenters from the Reformation to the French Revolution* (Oxford, 1978).

[132] Eric Maple, *Magic, Medicine and Quackery* (1968); Marcelle Bouteiller, *Médecine populaire d'hier et d'aujourd'hui* (Paris, 1966); C. J. S. Thompson, *The Quacks of Old London* (1928); M. H. Nicholson, 'Ward's "pill and drop" and men of letters', *JHI,* xxix (1968), 177–96; Toby Gelfand, 'Medical professionals and charlatans: the *Comité de Salubrité enquête* of 1790–91', *Histoire Sociale/Social History,* xi (1978), 62–97. For one individual sometimes viewed by his contemporaries as a quack, cf. G. S. Rousseau, 'John Hill, universal genius manqué', in J. A. Leo Lemay and G. S. Rousseau (eds.), *The Renaissance Man in the Eighteenth century* (Los Angeles, 1978), 45–129. Rousseau has also demonstrated recently that the image of the quack was often used by imaginative writers in France and England as a metaphor for political corruption: see G. S. Rousseau, 'On ministers and measures', *Etudes anglaises,* xxxii (1979), 185–91.

'quack' was sometimes used by allopathic doctors as a general term of abuse, and undoubtedly competent but unlicensed doctors were on occasion included. In Britain, at least, there seems to have been remarkably little systematic challenge to allopathic medical theories, though homeopathy originated in Enlightenment Germany, and it has been noted how many different medical systems there were by the century's end in the German-speaking lands.[133] Quacks are individualists, and the absence of systematic challenge to traditional medical men may simply reflect the extent to which informal networks and associations were an Enlightenment characteristic, even among the regulars. And while it is correct to insist that there was no unified medical profession – certainly not in Britain or America and probably not anywhere on the continent – it is clear that by the end of the Enlightenment there were many sets of allegiances and patterns of corporate activity which were new.

V. Living and dying

> Birth, and Copulation, and Death.
> That's all the facts when you come to
> brass tacks.
>
> (T. S. Eliot, 'Sweeney Agonistes')

Most scholars are now agreed that the medicine we have been considering was passed over by Europe's demographic transition: doctors, so it would appear, played little or no part in the widespread European population increases which, with some regional variation, began around the middle of the eighteenth century. Indeed, many historians place the advent of demographically significant medicine much later – the beginning of this century for preventative medicine, within the past generation or so for curative.[134] Even if we accept this view, medical history still has considerable relevance for current debates on the causes and consequences of the modern rise in population.

The phenomenon itself is reasonably clear: throughout much of Enlightenment Europe populations, which had been virtually static during the first half of the century, started to climb somewhere around mid-century. Though not a uniquely British event, the British

[133] Werner Weibbrand, *Romantische Medizin* (Hamburg, 1937).

[134] Thomas McKeown, *The Role of Medicine: Dream, Mirage, or Nemesis?* (1976); John Powles, 'On the limitations of modern medicine', *Science, Medicine and Man*, I (1973), 1–30.

historical debate has been particularly lively, since the population increase roughly coincided with the beginnings of the industrial revolution.[135] Populations rise only because more people are being born, fewer people are dying, or from net immigration into an area. With the last possibility discounted, there is left only increased fertility, diminished mortality, or a combination of the two. A half century ago it was assumed that pre-industrial Europe had a traditional population structure, with early marriage, high fertility, and a high death rate, particularly among infants and children. Traditional non-Western societies with this population structure have been changed by Western medicine and technology, and two influential monographs written by Griffith and Buer in the 1920s claimed that much the same thing had happened in eighteenth- and early-nineteenth-century Britain.[136] Industrialization meant more and cheaper goods, and this, combined with medical improvements such as more hospitals, male midwifery, better medical education, local concern for public health, and discoveries in anatomy, physiology, and pathology, resulted in a reduced mortality and a population increase. Although Miss Buer admitted that eighteenth-century anatomy and physiology did not result in therapeutic advances save in obstetrics, she decried a tendency, evident even then, for medical historians to dismiss Enlightenment medicine as sterile and ineffective. She suggested that crude London death rates had fallen from 50 per thousand in the 1770s to 29.2 per thousand in the first decade of the nineteenth century and concluded:

> By the beginning of the 19th century, the general advance in society and in particular the efforts of the medical profession, had resulted in some approach to modern health conditions as contrasted with medieval ones. Plague, leprosy, and scurvy were extinct; rickets, smallpox and typhus were scotched, with a definite hope of ultimate elimination.[137]

The disappearance of leprosy and plague was of course completely irrelevant to Buer's case, and typhus was still a major threat in 1800; nevertheless, a generation of historians more or less subscribed to its medical meliorism.

[135] Michael Drake (ed.), *Population in Industrialization* (1969), reprints key articles and has an excellent introduction; cf. Neil Tranter, *Population since the Industrial Revolution, the Case of England and Wales* (1973); E. A. Wrigley, *Population and History* (1973).

[136] G. T. Griffith, *Population Problems of the Age of Malthus*, 2nd edn (1967); M. C. Buer, *Health, Wealth, and Population in the Early Days of the Industrial Revolution* (1926).

[137] Buer, *Health, Wealth, and Population* (see note 136), 210.

Then, in the early 1950s, the data of the Griffith/Buer thesis were challenged by K. H. Connell's work on rising eighteenth-century Irish birth rates, and H. J. Habakkuk's examination of the claim that English mortality fell in the century. In 1955 Thomas McKeown published the first of a series of articles and books in which he and his colleagues have developed an alternative causative framework to explain the modern rise in population.[138] Although McKeown has refined his own position since 1955, its main features have remained intact despite an accumulation of conflicting data. While accepting that falling death rates were primarily responsible for the eighteenth-century population increase, McKeown has consistently denied that medicine had anything to do with it. Hospitals, he contends, probably did more to spread diseases than to either cure or contain them; male midwifery was fraught with danger through needless instrumentation with dirty forceps; medical science was irrelevant for medical practice; and local public health activities, if well-intentioned, were ineffective. Rather, in his opinion, death rates fell because of a generalized improvement in nutrition: an agricultural rather than an industrial revolution. As evidence, he cites numerous agricultural innovations introduced since the seventeenth century, including crop rotation, new foodstuffs such as potatoes, maize, and turnips, and improved fertilizers.[139] While it is true that British agriculture *permitted* a doubling of the population between 1760 and 1830 without significant food imports, the causative role of nutrition remains an unproved case. For one thing, McKeown's hypothesis has difficulty accommodating T. H. Hollingsworth's careful demographic study of British ducal families between 1330 and 1954, since the aristocracy would not have been directly affected by improvements in nutrition. Hollingsworth's work demonstrated, in this small, select segment of the population, a striking decline in mortality during the mid-eighteenth century.[140] Peter Razzell has used this as evidence favouring one further medical acti-

[138] K. H. Connell, *The Population of Ireland, 1750–1845* (Oxford, 1950); H. J. Habakkuk, 'English population in the eighteenth century', reprinted in D. V. Glass and D. E. C. Eversley (eds.), *Population in History* (1965). McKeown and Brown's 1955 article is also reprinted there, as well as in Drake (ed.), *Population in Industrialization* (see note 135).

[139] T. McKeown, *The Modern Rise in Population* (1976); W. L. Langer, 'American foods and Europe's population growth, 1750–1850', *Journal of Social History* (Winter 1975), 51–66; T. McKeown, R. G. Brown and R. G. Record, 'An interpretation of the modern rise of population in Europe', *Population Studies*, XXVI (1972), 345–82.

[140] T. H. Hollingsworth, 'A demographic study of the British ducal families', reprinted in Glass and Eversley (eds.), *Population in History* (see note 138), and in Drake (ed.), *Population in Industrialization* (see note 135); cf. Trumbach, *Rise of the Egalitarian Family* (see note 74), 193–7.

vity which would have reduced mortality: smallpox inoculation.[141]

Smallpox was the first disease for which specific preventative measures existed, and while Edward Jenner's vaccination did not appear before the end of the century, protection in Europe was possible from the 1720s through inoculation with the actual virus.[142] The practice early received a royal imprimatur when George II (as he became) had two daughters inoculated by Charles Maitland, but until the late 1750s doctors chose to make the procedure elaborate, expensive, and dangerous. Then, a Suffolk surgeon, Robert Sutton, his sons (particularly Daniel) and disciples transformed inoculation by replacing the older method of multiple deep incisions with a much quicker, safer, and cheaper technique involving a light scratch on the arm and the introduction of a minimum amount of material taken from a smallpox pustule. Any inoculation carried the risk of spreading natural smallpox, but the Suttonians specialized in the wholesale inoculation of entire villages at risk from an epidemic in the vicinity. By 1776 the Suttonians claimed to have inoculated more than 300,000 people; Daniel Sutton performed the simple operation 487 times on a single day at Maldon, Essex. Direct fatalities were apparently very rare, since the passage of the virus from donor to recipient may well have attenuated it, and the light scratch favoured mild cases. The Suttons were never particularly effective in the larger towns and cities, where smallpox remained endemic, but in the countryside, with the disease appearing only periodically in epidemics, Razzell believes that the Suttonian method was demographically significant. His conclusion is based not simply on the various claims published by the Suttonians, but on the examination of numerous parish records where local clerks listed cause of death. At Maidstone, Kent, for instance, 252 died of smallpox between 1752 and 1763; 91 between 1782 and 1791; and only 2 between 1792 and 1801. Since smallpox had been a virtually universal disease with a particularly high mortality in pregnant women and children, the demographic consequences of inoculation could have been considerable. In addition, smallpox can cause infertility in males.[143]

[141] Peter Razzell, 'Population change in eighteenth century England: a reappraisal', in Drake (ed.), *Population in Industrialization* (see note 135); *idem*, *The Conquest of Smallpox* (Firle, Sussex, 1977); *idem*, *Edward Jenner's Cowpox Vaccine: The History of a Medical Myth* (Firle, Sussex, 1977).

[142] Genevieve Miller, *The Adoption of Inoculation for Smallpox in England and France* (Philadelphia, 1957); John B. Blake, *Benjamin Waterhouse and the Introduction of Vaccination: A Reappraisal* (Philadelphia, 1957); O. E. Winslow, *A Destroying Angel: The Conquest of Smallpox in Colonial Boston* (Boston, 1974).

[143] Razzell (works cited in note 141); W. F. Bynum, 'Introduction' to Symposium on 'Medicine and Industrialization in history', in E. G. Forbes (ed.), *Human*

This latter fact – that smallpox causes infertility in males – accords rather better with much of the available direct evidence for eighteenth-century English populations, primarily recovered through family reconstitutions, a method pioneered by French historical demographers. The fact itself is not in Razzell's original argument, but deserves to be explored. Although only a limited number of English parishes contain records complete enough to permit long-term analyses of parameters such as age at marriage, completed family size, age-specific mortalities, and other aspects of human biology, these limited data have yielded remarkable insights into the realities of pre-industrial England. E. A. Wrigley's analysis of Colyton, Devon, and Gautier and Henry's work on Crulai have been supplemented by the investigation of other parishes and villages, in Britain largely through the project of the Cambridge Group for the History of Population and Social Structure.[144] This Group has uncovered material suggesting that increased fertility rather than diminished mortality was a principal cause of the eighteenth-century rise in population.[145] In Colyton and other places a significant drop in the average age at marriage between 1700 and 1800 resulted in an increased average completed family size and a shorter generational span. Furthermore, deliberate family limitation was apparently more common in the later seventeenth century than it was a hundred years afterwards. Work by Dorothy McLaren on patterns of lactation, wet-nursing, and artificial infant feeding points to a possible decrease in nursing mothers in the second half of the eighteenth century, perhaps a result of new habits forced on women by the early industrial revolution. She provides impressive evidence that children who were nursed by their own mothers stood a better chance of surviving than those who were wet-nursed or fed by hand.[146] Certainly the factory system would not encourage working mothers to nurse their

Implications of Scientific Advance (Edinburgh, 1978), 197–9, describing remarks made by Razzell at the Symposium.

144 E. Gautier and L. Henry, *La population de Crulai: paroisse normande* (Paris, 1958); E. A. Wrigley, 'Mortality in pre-industrial England: the example of Colyton, Devon, over three centuries', in *Daedalus* (Spring 1968): 546–80; *idem* (ed.), *An Introduction to English Historical Demography* (1966).

145 E. A. Wrigley, 'Family limitation in pre-industrial England', reprinted in various collections, including Drake (ed.), *Population in Industrialization* (see note 135); Peter Laslett (ed.), *Household and Family in Past Time* (Cambridge, 1972); *idem, The World we have Lost*, 2nd edn (1971); *idem, Family Life and Illicit Love in Earlier Generations* (Cambridge, 1977).

146 Dorothy McLaren, 'Fertility, infant mortality, and breast feeding in the seventeenth century', *MH*, XXII (1978), 378–96; *idem*, forthcoming paper on wet-nursing in *MH*.

own children, something which they well might do if engaged in cottage industry. In villages experiencing what he calls protoindustrialization, Levine has discerned an increased infant and child mortality, and a decreased adult mortality, in working-class families during the period.[147] At the other end of the spectrum, however, Trumbach's recent study suggests that, between about 1765 and 1780, aristocratic women began routinely nursing their own babies. It is reasonable to assume that, other things being equal, infants nursed by their own mothers would have a greater chance of surviving than those farmed out to wet-nurses or artificially fed. The dating of this aristocratic move towards mother's milk makes it unlikely to be the principal cause of the rather sharp increase in ducal life-expectancy in the cohort born about 1750, and Trumbach's work further indicates that inoculation was rarely performed before children reached the age of five, leaving unprotected the youngest group, whose enhanced survival increased aristocratic life-expectancy. Diminished infant exposure to smallpox might have been partially protective, but Trumbach sees the growth - around 1750 - of romantic love, domesticity, and caring aristocratic mothers and fathers as the primary reason why these infants thrived more than their counterparts a generation earlier. He furnishes impressive epistolary documentation of the domestic values and mores of his thirty families, and his book is based on an interesting use of John Bowlby's fundamentally non-Freudian interpretations of the roles of attachment, separation, and loss in infant development. Though intriguing, Trambach's monograph is exposed to the inevitable difficulty of counter-examples of companionate marriages with high infant mortalities, and the demographic sharpness of Hollingsworth's data sits somewhat uneasily with what must have been the more gradual development of psychological attitudes and domestic arrangements.[148]

The Enlightenment was waning before the first British census (1801) permits some general perspective on these local or class extrapolations, and a degree of mortuarial accuracy is possible only with the advent of civil registration in 1837. Consequently, there is still disagreement on basic eighteenth-century issues, such as the relationships between industrialization and working-class standards of living, or the relative importance of fertility and mortality in the population rise. But while inoculation may have been the only specific medical activity with claim to demographic import, medical historians

[147] David Levine, *Family Formation in an Age of Nascent Capitalism* (New York, 1977).

[148] Trumbach, *Rise of the Egalitarian Family* (see note 74).

have much to learn from - and contribute to - these challenging debates. At the very least, historical demography has produced a heightened appreciation of the impact of disease in history.[149] Sixteenth- and seventeenth-century plague in particular has been investigated from the local perspective which the absence of national figures has forced on the historian, and the study of parish documents since the early modern period has accentuated the relative absence of periodic 'crises of mortality' during the later Enlightenment. The inadequacy of the London Bills of Mortality was recognized during the eighteenth century, and satisfactory disease patterns are treacherously difficult to construct, even for smallpox, not merely because of the quality of the records but because modern nosological categories only rarely conform to those used in earlier times.[150] However, Jean-Pierre Peter's examination of the massive epidemiological survey carried out between 1774 and 1795 by the Société Royale de Médecine promises much, and his approach could be employed on other, more limited Enlightenment caches.[151]

Moreover, local or regional studies are ideal for placing health, disease, and organized medical care into their genuine contexts. J. D. Chambers's study of the Vale of Trent is a model in its demographic and economic sensitivity, and Jean-Pierre Goubert gave a specifically medical focus to his outstanding work on disease and doctors in Brittany between 1770 and 1790.[152] Manchester is currently under

149 W. H. McNeill, *Plagues and Peoples* (Oxford, 1977); Leslie Clarkson, *Death, Disease and Famine in Pre-industrial England* (Dublin, 1975); François Lebrun, *Les hommes et la mort en Anjou aux 17e et 18e siècles* (Paris, 1971); Jean-Noël Biraben, *Les hommes et la peste en France et dans les pays européens et mediterranéens* (2 vols., Paris, 1975–6).

150 For aspects of London, cf. R. A. P. Finlay, 'The accuracy of the London parish registers, 1580–1653', *Population Studies*, XXXII (1978), 95–112; E. A. Wrigley, 'A simple model of London's importance in changing English society and economy 1650–1750', *Past and Present*, XXXVII (1967), 44–70; A. B. Appleby, 'Nutrition and disease: the case of London, 1550–1750', *Journal of Interdisciplinary History*, VI (1975), 1–22; George, *London Life* (see note 93). More generally, cf. J. D. Post, *The Last Great Subsistence Crisis in the Western World* (Baltimore, 1977), esp. ch. 4.

151 Jean-Pierre Peter, 'Malades et maladies à la fin du XVIIIe siècle', *Annales: économie, société, civilisation*, (1967), 711–51, reprinted, with other interesting articles, largely from *Annales*, in J.-P. Desgive *et al.*, *Médecins, climat et épidémies à la fin du XVIIIe siècle* (Paris, 1972). Peter's article has been translated in R. Forster and O. Ranum (eds.), *Biology of Man in History* (Baltimore, 1975); and into German in A. E. Imhof (ed.), *Biologie des Menschen in der Geschichte* (Stuttgart, 1978). See, also, Hannaway's excellent thesis, 'Medicine, public welfare and the state' (see note 77).

152 J. D. Chambers, *The Vale of Trent*, Suppl. No. 3, *Economic History Review* (1957); *idem*, *Population, Economy, and Society in Pre-industrial England* (1972); Goubert, *Malades et medécines* (see note 107).

investigation, and other localities with adequate local records and active medical communities should be investigated.[153] A broad perspective for northern Europe can be gleaned from A. E. Imhof's two stout volumes, and more generally, Imhof and Larsen's recent *Sozialgeschichte und Medizin* documents how many historical issues in medicine are viewed in sharper focus when they have a quantitative dimension.[154] German and Scandinavian archives are frequently rich.

Much of the inspiration for the various approaches mentioned above derives from what is affectionately known as the '*Annales* School', after the French journal whose founders were particularly concerned to extend the boundaries of historical investigation beyond its traditional political and diplomatic spheres. Bookish historians sometimes flinch at pages of numbers and charts, and rigorous historical demographers are sometimes disdainful of social historians who seem to intuit sweeping generalizations from meagre qualitative sources. But a grasp of quantitative biological parameters surely provides a richer context for many questions of medical interest. Illegitimacy, infanticide, and infant mortality are constitutive to our understanding of Enlightenment foundling hospitals;[155] child-rearing practices cannot be separated from issues of family structure, and from the perceptions and realities of death;[156] sexual mores form part of the history of venereal disease, lock and Magdalen hospitals, doctor–patient relationships, and physical examinations, as well as of fertility.[157] Nutrition, affection, love, dress, architecture, bathing, liv-

153 Dr J. V. Pickstone is in charge of a research project devoted to the history of hospitals and disease in the Manchester area; cf. Lebrun, *Les hommes et la mort* (see note 149).

154 A. E. Imhof, *Aspekte der Bevölkerungsentwicklung in den nordischen Ländern 1720–1750* (2 vols., Bern, 1976); *idem* and O. Larsen, *Sozialgeschichte und Medizin* (Oslo and Stuttgart, 1975); Imhof (ed.), *Biologie des Menschen* (see note 151).

155 Laslett, *Family Life and Illicit Love* (see note 145); J. H. Hutchins, *Jonas Hanway 1712–1786* (1940); W. L. Langer, 'Infanticide: a historical survey', *History of Childhood Quarterly*, I (1974), 353–65; A. Dupoux, *Sur les pas de Monsieur Vincent: trois cents ans d'histoire parisienne de l'enfance abandonnée* (Paris, 1958).

156 Lloyd de Mause (ed.), *The History of Childhood* (New York, 1974); T. H. Rabb and R. I. Rotberg (eds.), *The Family in History* (New York, 1971); P. Ariès, *L'enfant et la vie familiale sous l'ancien régime*, 2nd edn (Paris, 1973), trans. as *Centuries of Childhood* (Harmondsworth, 1973); *idem, Western Attitudes towards Death* (1974); *idem, L'homme devant la mort* (Paris, 1977); George Rosen, 'A slaughter of innocents: aspects of child health in the eighteenth century city', in H. Pagliaro (ed.), *Studies in Eighteenth Century Culture*, V (1976), 293–316; Ernest Caulfield, *The Infant Welfare Movement in the Eighteenth Century* (New York, 1931); Stone, *The Family, Sex and Marriage* (see note 74).

157 M. Foucault, *Histoire de la sexualité*, I (Paris, 1976); A. Leibbrand and W. Leibbrand, *Formen des Eros* (2 vols., Freiburg, 1972); E. Trudgill, *Madonnas and Magdalens* (1976).

ing standards, and social customs are all relevant to the history of medicine;[158] so are charity and philanthropy, crime and punishment, values and beliefs. Extending the boundaries of history has done the same thing for medical history, for Marc Bloch's call for a historiography in search of *mentalité* was complementary to Henry Sigerist's insistence that systems of medical care and ideas of disease cannot be divorced from the societies which have produced them.[159]

VI. Conclusion

As the literature cited in this essay has documented, scholars from many disciplines have concerned themselves with the history of health, disease, and medicine. Within the past generation, social and economic historians, sociologists, anthropologists, historians of literature, and historical demographers have explored new techniques and opened new areas of historical investigation which provide a broadened vista of medicine's past. The history of medicine will never again be the province of a few specialists working primarily in medical schools and writing chiefly for the medical community. These developments are to be welcomed, though this expanded and, in general, more sophisticated medical history must not cut itself off from its medical audience or from the genuine contribution which doctors and other health professionals can make to the subject.

I have already indicated a number of topics which need further study before a more comprehensive picture of Enlightenment medicine can be drawn. Many of these topics cluster around four interconnected themes.

(*a*) *The physical conditions of life.* Questions of historical epidemiology, nutrition, family structure, dress, social customs, and living standards are all integral to this widened perspective of medical history. The requisite research frequently necessitates a local or regional perspective, and the exploitation of records which traditional medical historicans have too often neglected. But only with a firm grasp of these physical realities can we understand the historical role of medicine in the community.

(*b*) *The history of the patient.* These physical realities also enable us to appreciate the silent partner in the history of medicine: the patient. Much more needs to be uncovered about lay perceptions of disease,

[158] Relevant works not previously cited include: J. C. Drummond and W. Wilbraham, *The Englishman's Food* (1939); and Arthur J. Taylor (ed.), *The Standard of Living in Britain in the Industrial Revolution* (1975).

[159] Marc Bloch, *Historian's Craft* (Manchester, 1977); Henry Sigerist, *A History of Medicine*, I (New York, 1967), ch. 1; *idem*, *Civilization and Disease* (Chicago, 1962).

pain, and death during the Enlightenment. Medical economics - neglected even from the vantage of the doctors themselves - is a subject which should be approached with the consumer in mind. Diaries, sermons, will, letters, and literary works can all be fruitfully conscripted into service, for there are important links between cultural values and systems of medical care.[160]

(*c*) *The structure of the total medical community.* We know too little about many medical elites; we know even less about the many kinds of individuals - ordinary practitioners, midwives, clergymen, druggists, nostrum pedlars, the local squire - who were part of the medical world of the Enlightenment. Most of history's silent majority have disappeared without trace, but until more effort is made to appreciate the mundane and ordinary, our understanding of the extraordinary - the famous doctor or the gifted medical scientist - is impoverished.

(*d*) *The role of the state.* Throughout Enlightenment Europe and America significant regional and national variations existed in the organization of medical education, the sources and aims of medical philanthropy, the sale of drugs, and the provision of medical care. Many of these differences were intimately linked to widely varying patterns of state involvement (or lack thereof), and a genuinely comparative grasp of Enlightenment medicine can be achieved only by studies which set medicine within its political context.

None of these themes was deemed worthy of serious consideration in the standard histories of medicine which Garrison, Castiglioni, and others produced earlier this century. Modern scholarship was extended, though not necessarily invalidated, the perspectives of an earlier generation. As Paracelsus once remarked, 'What was true for the Greeks is not true for us. Truth must be born in its own land'[161]: and, he might have added, in its own time.

Acknowledgements

For useful comments on earlier versions of this essay, I am grateful to the following: Diana Long Hall, Christopher Lawrence, Michael Neve, Roy Porter, G. S. Rousseau, Steven Shapin, and Ulrich Tröhler. Research expenses have been generously provided by the Wellcome Trust.

[160] Roy S. Porter, 'Medicina e illuminismo nell'Inghilterra del settecento', *Quaderni Storici*, XL (1979), 155–80, examines the possibility of a patient-orientated history of medicine for the period.

[161] The quotation comes from a conversation with Professor Henri Ellenberger, and I have been unable to find the precise passage. However, Dr Walter Pagel has pointed out to me that similar sentiments are expressed by Paracelsus in *Sämtliche Werke*, ed. K. Sudhoff (14 vols., Munich, 1929–33), I, 228; IV, 72; and XI, 135, 169.

6

The living world

JACQUES ROGER

Contents

I. Who studied what? 258
II. The triumph of natural history 263
III. What is life? 270
IV. From natural history to the history of nature 278
V. Towards a conclusion 282

It is somewhat difficult to describe and understand an intellectual revolution correctly when one has lived it day after day. Only by recalling how things were in the early 1950s may we take the measure of the change that has occurred in the last twenty-five years and has affected almost every field of research, including history in general and the history of science in particular.

We do not have to undertake such a general description here, and fortunately so: we must limit ourselves to the precise problems of the history of eighteenth-century life science. But because we are historians, we know how difficult it is to isolate a problem from its context and, as historians of science, we know that changes in our discipline are closely linked to the opinions prevailing in the philosophy of science and history. It will therefore be necessary to refer, even if briefly, to a more general situation if we want to understand what has happened in our field of inquiry. Eighteenth-century scientists, particularly naturalists, may be looked at in two different ways. If we read Charles Bonnet's observations of aphis parthenogenesis, Haller's experiments on muscular excitability, or Spallanzani's notes on artificial fertilization, we may be tempted to consider those scientists as the first 'modern' experimentalists. But if we think that they accepted the fantastic theory of the 'pre-existing germs', we cannot help considering them as 'archaic' thinkers.

Great historians of the first half of the twentieth century clearly adopted the first attitude, even if they felt obliged to blame or deplore the 'inconceivable blindness' of their heroes when they did not see things that are obvious to us and stuck to 'old errors'. It may be useful to remark here that those historians were very often scientists themselves.[1] As such, they were mainly interested in the development of their science, from the past that they studied as historians to the present that they knew as scientists. That development was of course directed toward the present and so was its history. 'Useful' discoveries were emphasized whereas 'errors' were labelled as such and explained by the supposedly 'infant state' of science at the time. 'Archaic' features of reasoning in great scientists were ignored or surreptitiously corrected: William Harvey could not believe in spontaneous generation, no matter what. According to that perspective, eighteenth-century science was 'modern science in the cradle', and this made it possible to apply modern criteria or modern categories to scientists for whom they did not exist.

That 'vertical' or diachronic reading of history was supported by the historical and epistemological persuasion that science progresses continuously, from discovery to discovery, through a simple process of accumulation. This had been the accepted view since the time of Fontenelle.[2] That persuasion was supported by another accepted view, namely that facts were the only important thing in science, a view based in turn upon the ideal of inductive science, also generally accepted since the eighteenth century, before being theorized by nineteenth-century philosophers of science. Of course, there was in the background the ideology of progress, for which history of science, again from the time of Fontenelle, was the firmest support.

Thus, the way the history of science was written was but the tip of the iceberg, and the iceberg was nineteenth-century science and epistemology, already abandoned by modern physicists, but still surviving among biologists and historians of biology.

Things are very different now, and the old system of thought has been replaced by a new one. That change deserves a much closer analysis than, unfortunately, can be offered here. Progress is no longer the key-word of history; the ideal of inductive science has given way to a more realistic description of scientific reasoning as a hypothetico-deductive process. Interrelations between facts and

[1] This was particularly clear in France and Great Britain, with eminent personalities like Bernal, Caullery, Dobbel, Guyénot, Needham or Jean Rostand. The most typical representative of the attitude I am describing here probably was Emile Guyénot, who was a distinguished geneticist and wrote an important book on *Les sciences de la vie aux 17e et 18e siècles* (Paris, 1941). That book is full of useful information but is a good example of what we would now consider as a misconception of the nature of the history of science. See my article 'Réflexions sur l'histoire de la biologie', *RHS*, XVII (1964), 25–40.

[2] See his Preface to the *Histoire de l'Académie des Sciences* (1699).

theories have been underlined and the importance of a new theory for the advancement of science is now fully recognized.[3] To put it briefly, the originality of twentieth-century science, especially in biology, has been realized, and the phrase 'modern science' can no longer be applied to nineteenth-century science. This in turn pushes eighteenth-century scientists back to a remote pre-scientific stage.

This general change took on some more-specific aspects in the field of the history of science, as a new generation of scholars entered it, who were not scientists, but philosophers or historians of ideas. They had read or studied with Bachelard, Lovejoy and Alexandre Koyré. They were more interested in history for its own sake, in understanding the past in its own terms, than in tracing modern scientific developments back to their alleged origins. Instead of a vertical or diachronic history, they tried to write a horizontal or synchronic one, which would emphasize intellectual links between science and other activities such as philosophy or theology. As epistemologists, they tried to reconstruct the internal consistency of ancient thought and understand how what we now call error and truth could coexist without contradicting each other: logical consistency was no longer denied to old science or to the 'savage mind'.[4]

Actually, what appeared was a new philosophy of history, in which continuity was replaced by discontinuity. Two books, written from very different points of view, underlined this point: Thomas Kuhn's *Structure of Scientific Revolutions* and Michel Foucault's *Les mots et les choses.*[5] Kuhn was mainly interested in the social aspects of the history of science, whereas Foucault was a philosopher and epistemologist. They had something in common however: they considered history, and particularly the history of science, as a discontinuous series of isolated episodes. Each episode was characterized by a 'paradigm' according to Kuhn, by an *'episteme'* according to Foucault, and these were separated by a 'scientific revolution' for the former and a

[3] Needless to say that Sir Karl Popper was particularly instrumental in that change, but we still have no study of his influence.

[4] It would be interesting to study the influence of 'structuralism' on this particular point, and especially that of Lévi-Strauss's 'structural anthropology'. But the study could be extended to the general problem of modern concerns with 'systems'. What we are now studying as historians of science is mainly 'systems of thought'.

[5] The first edition of *Structure of Scientific Revolutions* was published by the University of Chicago Press in 1962. The first edition of *Les mots et les choses* was published in Paris by Gallimard in 1966, and later translated into English as *The Order of Things* (1970).

discontinuity' for the latter.[6] Otherwise the two concepts of 'paradigm' and *'episteme'* had little in common, especially insofar as the reality they wanted to describe was concerned, but both defined lasting frameworks within which human thought had to work for a given time. They gave that thought its consistency and delimited a space where controversies were possible between scientists who spoke the same language.

This analysis, though much too brief, makes it at least possible to understand why we can no longer look at the eighteenth-century life sciences in the same way as our predecessors did. This is not to say that we are not greatly indebted to them, but we have to ask new questions and search for new answers.

I. Who studied what?

Michel Foucault was probably the first thinker to point out that we cannot use the word 'biology' when we speak of the eighteenth century, and for two reasons : first, because the word itself was not coined before the very end of the century;[7] second, because the very concept of 'life' could not exist in the mechanistic philosophy that was then prevalent.[8] We will return to this problem later but, if there was no biology, there also were no biologists. Thus, who studied the living world in the eighteenth century? Two questions are intertwined here : the first is epistemological and deals with the divisions of knowledge at the time we are studying; the second is sociological and related to the social status of those who studied living beings at that time, and the way they could make a living out of their scientific work. These questions have not been thoroughly studied so far, and here I can only indicate some directions research might take and make remarks of a general nature.

The study of living beings was traditionally reserved for medical doctors. Medical studies had for a long time been the only regular scientific training available. Anatomy, physiology and botany were regularly taught in medical schools, and a professorship in medicine was the only way to make a living out of those studies. It was there-

[6] According to Michel Foucault the end of the eighteenth century was precisely the time of such a 'discontinuity', especially for life sciences and medicine. See his *Naissance de la clinique* (Paris, 1963) translated into English as *The Birth of the Clinic* (1973).

[7] Everybody now knows that the word appeared simultaneously in Lamarck's and Treviranus's writings in 1802. The first occurrence probably appeared some years earlier in Germany.

[8] *Les mots et les choses* (see note 5), ch. 5, 139.

fore accepted that a professor of medicine should be primarily interested in anatomy or botany, provided that it was human anatomy and a botany considered essentially for its medical uses.

This state of affairs lasted and even developed during the eighteenth century, especially in countries such as Germany, the Netherlands or Italy where universities remained active centers of intellectual life. It seems, however, that scientific research in an 'ancillary science' was allowed a greater part in the professional life of a professor of medicine. Hermann Boerhaave and Friedrich Hoffmann were primarily professors of medicine, even if their teaching included a physiology of their own. Carl Linnaeus and Albrecht von Haller were also professors of medicine, but the former was first of all a botanist and the latter an experimental physiologist.

This institutional change paralleled a slow but steady transformation in the intellectual content of botany and physiology, and their growing independence from medicine. Botany no longer was the science of vegetable drugs: plants were now studied for themselves, and the ultimate aim of that study was to name and classify them, that is, to discover the way God had organized the vegetable kingdom. On the other hand, as physiology became more experimental, animal physiology became more and more important since, despite some curious suggestions for scientific exploration,[9] experimentation on man was considered as immoral. As far at least as botany is concerned, this trend is particularly visible in France, where new institutions, not so closely related to medicine, could offer teaching or research positions: Tournefort and the Jussieus were doctors of medicine but taught in the Jardin du Roi or in the Collège Royal, and did their research under the general patronage of the Academy of Sciences. Nothing similar, however, happened to physiology: it remained a part of medical studies, but French medical schools were unable to develop experimental research, so that human physiology remained merely theoretical and animal physiology made no progress in that country. This development probably contributed a great deal to the peculiar development of medicine in France, which emphasized theoretical questions and transformed medicine into a general 'science of man'.

Because seventeenth-century universities had stubbornly stuck to the old Aristotelian physics, modern science had often been the work of amateurs, till the time universities reformed their outlook, as they did in Italy or Germany, or new institutions could harbour the new

[9] In his *Encyclopédie,* Diderot proposed experimentation upon criminals condemned to the death penalty.

scientists, as did the Royal Society of London or the Paris Academy of Sciences. The Royal Society was quite informally organized, but the Academy of Sciences, especially after 1699, was divided strictly into sections, which mirrored the accepted divisions of knowledge at that time. In this division, life sciences are represented by anatomy and botany only, and all the members of those two sections were doctors of medicine. We still are in the traditional framework, borrowed from medical studies. However, some work in plant or animal physiology was done in the Academy, especially by Claude Perrault and Edme Mariotte. Interestingly enough, both belonged to the section of 'Physics', a science which encompassed all the activities of nature, according to the old Aristotelian scheme. However, Perrault, who dissected some animals and proposed a physiological theory, was a medical doctor who had more or less quit medicine, whereas Mariotte, who studied the motion of the sap in plants and also discovered the relationship between the pressure and the volume of gas about at the same time as Boyle, was a Catholic priest and, so to speak, the prototype of the eighteenth-century 'amateur' or 'virtuoso'.

Since the new professional scientists, in the universities and elsewhere, were not prepared to embark upon studies in the new fields of research, such as entomology or plant physiology, which became fashionable at the beginning of the eighteenth century, the category of amateur scientist emerged again. What these amateurs had in common was the fact that they were self-taught and did not belong to the traditional medical profession. They had learned natural history through reading and direct observation; but this did not prevent them from making some of the most striking discoveries of the century.

Some of them remained 'amateurs' for all their life, as did Stephen Hales, a clergyman, Charles Bonnet, who had independent means, and Abraham Trembley, who was pensioned by his former pupil, the Duke of Richmond. In France, they could become professional scientists by entering the Academy of Sciences, as did Réaumur or Duhamel du Monceau, and, eventually, they could get a position in other institutions, such as the Jardin du Roi: this was the case with such important figures as Buffon, who had been bred to law and became the Administrator of the Jardin, and Lamarck, a former army officer who was also attached to the Jardin, where he eventually taught invertebrate zoology. Sometimes, because natural history, owing to the pressure of public opinion, became more and more usually taught in teaching institutions, such 'amateurs' could get a position in an Academy, as Saussure did in Geneva, in a College, or even in a university, as Spallanzani did in Reggio d'Emilia, in Modena and, finally, in Pavia.

The names I have just quoted are famous enough to show how important such amateurs were in the development of life sciences in the eighteenth century but, if we know those particular cases, we do not know the social dimensions of the phenomenon or the extent to which it influenced the very nature of the science and its ideological and epistemological character. Is there a link between the fact that, in order to become an 'amateur', not only 'taste' and 'wit' but also money was required, and the religious and moral aspects of a literature which, with few exceptions, was written by bourgeois or clergymen? Is not the minute description of an isolated phenomenon easier for an amateur than the discussion of a general theory? To give but one example, is it a matter of chance that the greatest contribution to plant and animal physiology at the end of the eighteenth century is due to Priestley (a divine); to Lavoisier (a tax-collector) and to Saussure (a teacher-amateur), while medical doctors mainly concerned themselves with on-going theoretical debates about sensibility and irritability?

Another important social aspect of the life sciences in the eighteenth century, and closely related to the role of amateurs, was their success among the public. This general phenomenon, which was first studied for France by Daniel Mornet at the beginning of this century,[10] is particularly evident in the many collections that kings, noblemen and even rich bourgeois established throughout Europe. Natural curios, especially shells, insects and fossils, were sold at high prices in public auctions or by specialized merchants, like the famous Gersaint in Paris.[11] This fashion has often been pictured as frivolous and as having nothing to do with science. However, it contributed a great deal to the development of natural history. Without it, the very existence of the amateurs we spoke about would have been impossible, not only because many people would simply have ignored natural sciences, but also because fashion made it possible for some freelance scientists to make a living out of their writings or their public lectures. New journals were published, devoted to natural history. It might well be that even the style of natural history was

[10] *Les sciences de la nature en France au XVIIIe siècle* (Paris, 1911). See also Lamy, *Les cabinets d'histoire naturelle en France au XVIIIe siècle* (Paris, 1930, privately printed).

[11] See Y. Laissus, 'Les cabinets d'histoire naturelle' in R. Taton (ed.), *Enseignement et diffusion des sciences en France au XVIIIe siècle* (Paris, 1964). For England, see J. M. Chalmers-Hunt, *Natural History Auctions 1700–1792* (1976).

transformed by that fashion, because books were being written for the general reader more than for the specialized scientist: Buffon is a good example of this.

The general interest in natural history also played a role in the scientific discovery of the world. Even before the English or French governments sent scientific expeditions to the most remote parts of the planet, merchants, sailors and missionaries collected exotic plants and animals and sent them to Europe, for mercantile or scientific purposes. At first, these would-be naturalists were poorly trained and ineffective but, in time, scientists understood how useful they could be, gave them precise instructions and officially recognized their role: Buffon created an honorary title of 'Correspondent of the Jardin du Roi' and published their names in his *Natural History* – as a result of which he received a goldmine of information from such people. Unfortunately, we know very few of them.[12] The impact of newly discovered species has often been described as important,[13] but few attempts have been made to measure it precisely,[14] and the role of their discoverers has unfortunately not been fully estimated so far.

We are still lacking a sociological and institutional history of the life sciences in the eighteenth century. Physicists and mathematicians have been more accurately studied.[15] Such a study is probably a tremendously difficult one but it would shed plenty of light on not

[12] In my edition of the *Epoques de la nature* (Paris, 1962), I had tried unsuccessfully to discover the source of Buffon's very precise description of Guyana. This source was eventually identified by Elizabeth Anderson as Sonnini de Manoncourt, a young naturalist who also contributed to the *Natural History of Birds*. See E. Anderson, 'Some possible sources of the passages on Guiana in Buffon's *Epoques de la Nature*', *Trivium*, v (1970), 72–84, and vi (1971), 81–91; *idem*, 'More about some possible sources of the passages on Guiana in Buffon's *Epoques de la nature*', *Trivium*, viii (1973), 83–94, and ix (1974), 70–80; *idem*, 'La collaboration de Sonnini de Manoncourt à *L'Histoire naturelle* de Buffon', *Studies on Voltaire and the Eighteenth Century*, cxx (1974), 329–58.

[13] See F. Dagognet, *Le catalogue du vivant* (Paris, 1970), 29 and 42ff.

[14] For exotic birds, see Erwin Stresemann, *Die Entwicklung der Ornithologie von Aristoteles bis zur Gegenwart* (Berlin, 1951), translated into English as *Ornithology from Aristotle to the Present* (Cambridge, Mass., and London, 1975). See especially chapter 2, where the role of the Dutch merchants is particularly emphasized, and chapter 4, which deals more directly with the eighteenth century.

[15] There is relatively little about life sciences in Roger Hahn's brilliant book *The Anatomy of a Scientific Institution: The Paris Academy of Sciences, 1666–1803* (Berkeley, Cal., 1971) which, in any case, concerns itself only with the French institution. For the role of science in the French teaching institutions, see R. Taton (ed.), *Enseignement et diffusion des sciences en France au XVIIIe siècle* (Paris, 1964), but there is very little in this book about students and studies in natural history.

only the place of life scientists in scientific institutions and in society at large, but also the way they worked and on the kind of science they built. It is impossible to separate sociology and epistemology, even if their relationships are not easy to define. It might well be, for instance, that there were 'biologists' even before there was a 'biology'.[16] If we want to understand how the phenomenon known as 'physicks' in 1670 became what was known as 'biology' in 1802, we must not overlook the evolving status of the scientists themselves.

II. The triumph of natural history

Natural history had blossomed at the time of the Renaissance, but the seventeenth century was much less favourable. Mathematics, mechanics and physics - in the modern sense of the word - and even chemistry attracted the greatest number of scientists, whether professional or amateur. The sudden and general revival of natural history in the 1670s is thus all the more interesting, but it poses many problems. It is of course impossible here to give a detailed description of that revival, which apparently started in England with some 'amateurs' like Francis Willughby, John Ray and Robert Hooke, and eventually spread throughout Europe. In fact, the history of this revival is still to be written.

Michel Foucault has pointed out that, simply in order to exist, 'natural history' had to be 'natural'. This raises the first question: was natural history indeed something 'natural' for scientists who considered the world as the work of God, and studied it precisely in order to demonstrate the wisdom of God? In the first half of the eighteenth century natural history was written mainly for religious purposes, as is witnessed by the works of John Ray, George Cheyne, William Derham and, on the continent, of Pluche, Fabricius, Lesser and many others. That whole corpus of literature, whose success was enormous, has not yet been studied in its own right,[17] perhaps because it has been considered as an accident in the history of science. In fact it probably was an essential feature of the revival of natural history, a science that could teach man that he was but one being among all those created by God, and probably not the most admirable at that. In

[16] On this particular point, see Claire Salomon-Bayet, *L'institution de la science et l'expérience du vivant* (Paris, 1978), where the links between the institution of the Academy and the emergence of a new definition of a science of life are convincingly emphasized.

[17] For the Abbé Pluche, see Caroline V. Doane, 'Un succès littéraire du XVIIIe siècle: le *Spectacle de la nature de l'abbé Pluche*' (Unpublished dissertation, University of Paris, 1957).

this sense, interest in natural history was a part of the theocentric atmosphere of the late seventeenth century. When nature was eventually substituted for God, the religious purpose disappeared, but the moral lesson remained.

However, natural history was much more 'natural' at the end of the seventeenth century than it used to be in the sixteenth, because it was only concerned with the 'natural' features of what it described and was no longer burdened by unnecessary information – that is information which was now considered unnecessary – on the culinary or medical use of plants or animals, the legends about them, etc.[18] Nature was now studied for itself, not because man could use it. It was consequently the work of man that was driven out of natural history which reinforced the moral lesson I alluded to, but also allowed natural history to emerge as a science.[19]

Science is but a part of the general relationship of man to nature. When this relationship changes, as it did at the end of the seventeenth century, science cannot remain the same. The success of natural history was a sign and a result of an epistemological change, which substituted 'experimental philosophy' for abstract rationalism. Nothing is more opposed to natural history than the Cartesian spirit and its belief that nature can be fully understood by the human mind. It is not therefore surprising that natural history revived precisely at the time when, and in the country where, Cartesian philosophy was first abandoned and Locke developed a much more modest philosophy of knowledge. What is more surprising, incidentally, is that the history of Locke's influence on the Enlightenment has not yet been written. As far as science is concerned, it would be interesting to see how that influence was favoured by a mixture of religious and scientific motives, not very dissimilar from those which inspired the first humanists when they rebelled against scholastic philosophy.

The history of the word 'history' itself would deserve an accurate study. In the 1670s it often means a plain description of a natural phenomenon. In that sense, and even in physics, a 'history of nature' is opposed to a 'natural philosophy', which is a search for the causes of the phenomena. According to many scientists of the time one should first write 'an ample history of nature' before thinking of building a 'philosophy' of it. Here, it is easy to recognize the primacy of fact over theory

[18] This remark has also been made by M. Foucault in various parts of his books.

[19] This was at least an ideal, but many writers, including the Abbé Pluche, or Diderot in his *Encyclopédie*, felt it necessary to provide the reader with all the information related to the human use of natural things: this also was a feature of the Enlightenment. In the case of the *Encyclopédie*, however, that part of the articles was not written by a naturalist.

to which I alluded at the start of this paper. In such a context natural history obviously becomes the most appropriate science.

But this means that natural history had to limit itself to the mere description of natural beings, or to their classification. It did not need to explain how some of those beings, namely the living ones, were functioning: this was left to 'physicks' or physiology. The phenomena of life, as such, were necessarily excluded from natural history: between the 'three kingdoms of nature' there was no real difference, and birds could be described and classified exactly as stones could be.

If this was the status of natural history at the beginning of the eighteenth century – and, once again, no detailed study has been made of this problem – how long did that status last? How was it modified by the ideological and epistemological changes that occurred in the middle of the century, by the debates about classification, by the appearance of a 'natural history of man' and the development of comparative anatomy? In summary, how 'natural history', as it was called at the time of John Ray, became 'natural sciences' as it was known at the time of Cuvier, remains the largest of questions. This has not yet been studied in the way it ought to be.

The most conspicuous achievement of natural history in the eighteenth century was the work done in the field of classification. Here we find the great names of the century: Linnaeus, Tournefort, Adanson, A.-L. de Jussieu, to list only a few. It would be an error, however, to forget, as Foucault seems to do, that for many important naturalists of the age the most necessary task was not to classify, but to describe living beings. It is even curious to discover to what a small extent 'observers of nature' were interested in classification. Réaumur, who was their master, did not hesitate to consider a crocodile as an insect! His disciples – Trembley, Lyonnet, de Geer, and even the famous Charles Bonnet, who styled himself 'the observer' – took very little interest in classification, but spent their time observing and describing insects and plants.[20] By the middle of the century they unexpectedly found an ally in Buffon, who went so far as to declare that classification was not science, and that only description was useful. The agreement of a Christian philosopher like Réaumur and a sceptical materialist like Buffon suggests a deeper agreement about the limits of human knowledge. Nature cannot be 'represented' by our words, nor can the multifarious relationships between living beings be made visible on a two-dimensional picture. It is also not a matter of chance that Buffon and Bonnet both insist that

[20] For Bonnet however, see Lorin Anderson, 'Charles Bonnet's taxonomy and chain of being', *JHI*, XXXVII, 1 (1976), 45–58.

no clear-cut division can be drawn in nature and that the divisions made by the classifiers are but the product of our mind – even if the two naturalists do not have the same idea of the 'great chain of being'. Between 'observers' and 'classifiers' the difference is not of style or personal taste: it is a complete disagreement as to the very nature of natural history.

However it is clear that Buffon and Réaumur did not have the same thing in mind when they spoke of description. But, although we have some interesting books and articles on Réaumur, Lyonnet, Bonnet and Buffon,[21] as far as I know none of these secondary works tries to define precisely their style of description.[22] The problem, of course, is not a literary one, but a scientific and, perhaps, even a philosophical one. When Réaumur describes an insect, the way it catches its prey or builds a nest to live in as a nymph, he fully satisfies the curiosity of the reader and wants to engender in him a feeling of amazement and admiration at the spectacle of the richness and diversity of God's creation. In contrast, when Buffon describes the habits of the dog or the wolf, it is in order to show us how two very similar animals really belong to different species. Moreover, Réaumur's minute descriptions of insects could endlessly be put one after the other in an infinite series of volumes; nothing but death can bring such a study to an end.[23] Buffon, on the contrary, is always trying to organize the descriptions and draw some general conclusions from them.[24] The

[21] On Réaumur, see Dr Jean Torlais, *Réaumur, un esprit encyclopédique en dehors de l'Encyclopédie* (Paris, 1937), and the special issue of the *RHS*, XI, no. 1 (1958). On Lyonnet, see W. H. van Seters, *Pierre Lyonnet (1706–1789)* (The Hague, 1962). On Trembley, see J. R. Baker, *Abraham Trembley of Geneva (1710–1784)* (1952). On Bonnet, see R. Savioz, *La philosophie de Charles Bonnet de Genève* (Paris, 1948). There are several books and articles on Buffon, but very little on his description of animals. See *Buffon*, Muséum National d'Histoire naturelle (Paris, 1952) (a good paper on the 'Histoire naturelle du chien') and, from a more literary point of view, O. E. Fellows and S. Milliken, *Buffon* (New York, 1972).

[22] However see Jacques Marx, 'L'art d'observer au XVIIIe siècle: Jean Senebier et Charles Bonnet', *Janus*, LXI (1974), 201–20.

[23] Actually Réaumur left a seventh volume of his *Mémoires pour servir à l'histoire des insectes* incomplete. A part of the manuscript was first published in an English translation: *The Natural History of Ants*, translated with an introduction and notes by William Morton Wheeler (New York and London, 1926). A French edition was published the following year (Paris) by L. Bouvier and Ch. Perez.

[24] It would be unfair to Réaumur not to say that he was perfectly able to draw general conclusions from the facts he observed. But he was more timid – or more cautious – than Buffon, and contented himself more easily with precise descriptions.

whole history of observation in natural history would deserve a general study, whose result would be, among other things, a better knowledge of the circumstances in which ethology first appeared.

A great deal of work has been done on eighteenth-century taxonomy since the time Henri Daudin published his two remarkable books on that problem,[25] and yet, these books are still useful and illuminating. A new and interesting feature of recent research is that the epistemological basis of taxonomy has been studied by philosophers who believed that aspect of natural history to be the most typical of the eighteenth century.[26] One knows how Michel Foucault tries to relate taxonomy, political economy and the theory of language in such a way as to define the '*episteme*' of the 'classic age'. Even if we disagree with his reduction of eighteenth-century natural history to taxonomy, his attempt is extremely stimulating and has led him to make important comments on the eighteenth-century sciences of life. Generally speaking, recent studies of eighteenth-century taxonomy have focused less on its scientific value than on its relationship with the main intellectual trends of the age. We must be very cautious, however, before we generalize the feelings we may experience while reading some texts. It is tempting, for example, to contrast the arrogance of taxonomists, who do not hesitate to claim that they are able to discover the very order of God's creation, with the humility of observers, who are content to describe some of the many beings created by God. The psychological vocabulary of course only signifies two opposite philosophies of nature and two theories of knowledge. However tempting it is, this contrast is somewhat artificial, essentially because it is not so easy to define a general philosophy of taxonomy: various scientists held various opinions and, as regards some of them, it is very difficult to understand what they really meant.[27] It is also tempting to contrast taxonomists who tried to build an artificial but useful 'system' with those who tried to discover the 'natural method'; but we must not forget that a clear-cut distinction between system and method was not accepted before the time of

[25] *De Linné à Lamarck. Méthodes de la classification et idée de série en botanique et en zoologie (1740–1790)* (Paris, n.d. [1926]). *Cuvier et Lamarck : Les classes zoologiques et l'idée de série animale (1790–1830)* (Paris, 1926).

[26] See Foucault, *Les mots et les choses* (see note 5), ch. 5, and F. Dagognet, *Le catalogue de la vie* (Paris, 1970).

[27] For example, see the interesting book on Tournefort, published by the Muséum National d'Histoire naturelle (Paris, 1957). The historians who contributed to the book do not agree on Tournefort's attitude toward Cartesianism. Also see Philip R. Sloan, 'John Locke, John Ray and the problem of the natural system', *Journal of the History of Biology*, v (1972), 1–53.

Lamarck, and was therefore ignored by eighteenth-century taxonomists.[28] What is less disputable is the 'essentialist' nature of that taxonomy. This is not surprising when the scientists concerned believed that the species were fixed and directly created by God. Linnaeus's beliefs about the fixity of the species and about creation are no longer criticized, as they used to be by previous historians: they are now considered simply as necessary elements of Linnaeus's thought, and the very basis of his taxonomic work. It is more surprising to find almost the same essentialism in the thought of scientists who, like Buffon, did not believe that living beings were directly created by God, nor that species were absolutely immutable. This 'essentialism' is all the more important, since it would eventually become a formidable obstacle to the development of evolutionary theories. On the whole, the interpretation of eighteenth-century taxonomy is still a field open to new research, and fortunately so.

On the other hand, some interesting studies have been devoted to the relationship between taxonomy and other problems in the life sciences, particularly theories of generation and the definition of species – two closely related issues.[29] This has led some historians to an interesting reappraisal of Buffon's ideas on taxonomy, which used to be considered as childish and based on mere ignorance. Buffon's 'Discours premier' has been translated into English,[30] and a sympathetic effort has been made to understand how his attacks upon Linnaeus's classification may be explained by his views on species and generation.[31]

In a more traditional way, several books have been devoted to the great taxonomists of the age and, above all, to Linnaeus.[32] The time of

[28] On that point, see Emile Callot, 'Système et méthode dans l'histoire de la botanique', *RHS*, XVIII (1965), 45–53.

[29] For instance, see H. Engel, 'The species concept of Linnaeus', *Archives internationales d'histoire des sciences,* VI (1953), 249ff.; and Peter J. Bowler, 'The impact of theories of generation upon the concept of a biological species in the last half of the 18th century', *Dissertation Abstracts International,* XXXIII (1972), 244A.

[30] John Lyon, The "Initial Discourse" to Buffon's *Histoire naturelle:* The first complete English translation, *Journal of the History of Biology,* IX, (1976), 133–81.

[31] See Peter J. Bowler, 'Bonnet and Buffon : theories of generation and the problem of species', *Journal of the History of Biology,* VI (1973), 259–81; and Philip R. Sloan, 'The Buffon–Linnaeus controversy', *Isis,* LXVII (1976), 356–75.

[32] The fundamental book by K. Hagberg was translated into English as *Carl Linnaeus* (1952) and into French as *Carl Linné, 'le Roi des Fleurs'* (Paris, 1944). See also Norah Gourlie, *The Prince of Botanists, Carl Linnaeus* (1953); Heinz Goerke, *Carl von Linné, Arzt, Naturforscher, Systematiker* (Stuttgart, 1966); and James L. Larson, *Reason and Experience. The Representation of Natural Order in the Work of*

blind hagiography seems to be over, as does that of anachronistic criticism: the emphasis now to be specific, is put either on the complex personality of the man and his multifarious activity, as in Goerke's book, or on the philosophical and technical basis of his taxonomy, as in Larson's interesting study. The interaction between religion and science has been more precisely described, and neglected texts, particularly the enigmatic *Nemesis divina* and the seven volumes of the *Amoenitates academicae,* have attracted the attention of modern historians. This has unexpectedly led to new ideas of Linnaeus's role in the history of biology: his concept of 'natural economy' (*'oeconomia naturae'*), which was both scientific and religious, is now considered as a first attempt at an ecological study of the living world, and a remote starting point for some of Charles Darwin's ideas.[33] Even his belief in the fixity of the species seems to have been overestimated. The image of Linnaeus that we now have is quite different from that given by previous historians.

Linnaeus's disciples and the spread of his ideas throughout Europe have been studied more accurately than before.[34] Among other eighteenth-century taxonomists, Tournefort and Adanson are also better known.[35] On the whole, our knowledge of that part of eighteenth-century natural history is deeper and wider than it was twenty-five years ago. This is not to say that nothing may be added to our understanding of the matter. Chronologically, John Ray belongs to the seventeenth century, but he was the father of *all* the taxonomists of the eighteenth. His work is often alluded to, but we still are without the accurate study it obviously deserves. The same could be said of the Jussieu family. In general, botanical classification, which was more successful and more interesting both practically and

Carl von Linné (Berkeley, Cal., 1971). The *Systema naturae* of 1735 was republished (Nieuwkoop, 1964) and the *Öland and Gottland Journey* of 1741 was translated into English (1974).

33 See the French translation of five Linnean works by Bernard Jasmin and the introduction by Camille Limoges : C. Linné, *L'équilibre de la nature* (Paris, 1972). On the relationship with Darwin, see F. Egerton, 'Observations and studies of animal populations before 1860' (Unpublished dissertation, University of Wisconsin, 1967).

34 See Frans A. Staffleu, *Linnaeus and Linneans* (Utrecht, 1971). On Frederic Hasselquist, see F. S. Bodenheimer, 'Le vie et l'oeuvre de F. Hasselquist', *RHS,* IV (1951), 60–77.

35 On Tournefort, see the book quoted in note 27 above. On Adanson, see the important work published by the Hunt Botanical Library: H. M. Lawrence (ed.), *Adanson : The Bicentennial of Michel Adanson's 'Familles des Plantes'* (2 vols., Pittsburgh, 1963); Michel Guédès, 'La méthode taxonomique d'Adanson', *RHS,* XX (1967), 361–86; and Jean-François Leroy, 'Adanson dans (l'histoire de la pensée scientifique', *RHS,* XX (1967), 349–60.

theoretically, has been much more thoroughly studied than zoological taxonomy. Even Linnaeus's zoological work is somewhat neglected. Taxonomists of lesser rank are almost entirely forgotten, especially German naturalists like Klein, Fabricius and Pallas, ornithologists like Brisson, and taxonomists of the end of the century, like Lacépède, who seems to be already overshadowed by the formidable figure of Cuvier. For all those scientists, detailed studies would be welcome.

III. What is life?

We saw that, at the end of the seventeenth century, the divisions of science in our domain were quite clear: natural history describes and classifies natural beings, living or not; medicine deals with the diseases of the human body, but includes the study of its normal functioning; what remains, that is, the phenomena of life in plants and animals, belongs to the province of physics. This answers the old Aristotelian definition of physics as the study of the changes that occur in nature, but also reflects the new mechanical philosophy, which considered living beings as mere machines, simply more complicated than the non-living ones. This philosophy had been initiated by Descartes, but was best exemplified by Harvey's discovery of the circulation of the blood. What it did was to provide a technological model for the functioning of natural beings, a model that was supposed to make them understandable.

An interesting feature of this model was the new relationship that it established between organs and functions. According to the old Galenic physiology an organ performed a function because it possessed a special power, a *virtus* or *vis insita,* that was really responsible for the functioning of the organ. There was no necessary relationship between the function and the structure of the organ, let alone its chemical nature, though, according to Aristotle, its *temperamentum,* or particular proportions of the four elements in its matter, should play a role in its performing a particular function. According to the new philosophy, the function could and should be accounted for by the mechanical construction of the organ, that is, by the physical configuration of its parts, which were considered as the gears of a machine.[36] Was this enough to give birth to the concept of 'organiza-

[36] This view had not been entirely ignored by Galen, but the first appearance of a technological model, fully and consciously developed, is probably to be found in the study of the valvules of the veins by Fabrizio d'Acquapendente. See my article 'La situation d'Aristote chez les anatomistes padouans', in *Pla-*

tion', which was to become so important at the beginning of the nineteenth century?[37] It is true that the word appears at that time, and that a mechanical interpretation of a living organism should lead to the idea of a necessary correlation of its parts. It is doubtful, however, whether the word and the concept already had the full meaning that they were to acquire eventually. What is less disputable is that physiology was transformed by the new philosophy and became that '*anatomia animata*' which Haller would make famous.

The great weakness of the mechanical model though was that, in fact, it was almost entirely imaginary. It could work, at least to some extent, for the heart and the vascular system,[38] or for the motion of the limbs. For digestion it grossly simplified the phenomena. But, above all, many functions had to be explained by supposing a particular structure of inframicroscopic elements, which no direct observation could reach. Such was the case with glandular secretion, for instance. The primacy of structure over function, which is a characteristic of the age and which answered the accepted view of nature as passive in the hands of God, necessarily led to the creation of imaginary structures. The same phenomenon occurred in chemistry.[39]

The problem of 'generation' is a good example of that process: since no natural process could give birth to a structure - and this was contrary to Descartes' philosophy - no embryogeny was conceivable, and the whole structure of the new organism had to be already present in the seed or, for animals, in the egg or in the newly discovered

ton et Aristote à la Renaissance (Paris, 1976), 222–4. It is known that Harvey worked with Fabrizio at Padua.

[37] In a recently published book, the late Dr Schiller answers the question affirmatively, and gives Descartes all the credit for the invention and the success of the concept. See J. Schiller, *La notion d'organisation dans l'histoire de la biologie* (Paris, 1978). Several books have been published on the general history of physiology and biology, and particularly P. C. Ritterbush, *Overtures to Biology : The speculations of 18th-century Naturalists* (New Haven and London, 1964); M. J. Sirks and Conway Zirkle, *The Evolution of Biology* (New York, 1964); C.U.M. Smith, *The Problem of Life. An Essay in the Origins of Biological Thought* (London, 1976). See also Theodor Ballauf, *Die Wissenschaft vom Leben. Eine Geschichte der Biologie, von Altertum bis zur Romantik* (Freiburg in Brisgau, 1954).

[38] See A. Doyon and L. Liaigre, 'Méthodologie comparée du biomécanisme et de la mécanique comparée', *Dialectica,* x (1956), 292–335.

[39] I do not have to speak of human anatomy here. However, on the positive aspects of this search for microstructures, see Luigi Belloni, 'De la théorie atomistico-mécaniste à l'anatomie subtile (de Borelli à Malpighi) et de l'anatomie subtile à l'anatomie pathologique (de Malpighi à Morgagni)', *Clio Medica,* VI (1971), 99–107.

spermatozoon. And, since 'machines cannot build machines', as Fontenelle had said, the pre-formed structure had to be directly created by God.[40] Another difficult point was vegetable life: since no significant structures could be found in plants, scientists were obliged either to decide that plants were not living beings, or to imagine vegetable structures after the model of animal anatomy.[41] Hence the constant use in plant physiology, by Malpighi, Grew and others, of analogies drawn from animal structures or animal physiology, as in the famous theory of the circulation of the sap.[42]

This mechanical model, however popular it was at the end of the seventeenth century, did not fully satisfy everybody. Some scientists, like Claude Perrault, were reluctant to reduce animal behaviour to mere results of mechanical processes. Thus, they stuck to the old idea of an animal soul, whose metaphysical status was very uncertain. In the 1740s Réaumur was still asking questions about the fate of the soul of the freshwater hydra, that little animal discovered by Abraham Trembley which, when cut into several pieces, was able to regenerate the same number of perfect animals. On different grounds, the German chemist and professor of medicine Georg-Ernst Stahl decided that a soul was a necessary element of life.[43] His animism found many supporters until after the middle of the century, particularly in Montpellier and Edinburgh, including the famous physiologist Robert Whytt.[44]

Another difficulty of the mechanical model was that it clearly belonged to the age of Descartes and *a priori* rationalism. The new experimental philosophy preferred to describe phenomena accurately, even if their causes remained unknown. This was what New-

[40] For the problem of the generation of animals and the theory of pre-existing germs, see my book *Les sciences de la vie dans la pensée française du XVIIIe siècle*, 2nd edn (Paris, 1971), part 2, and Carlo Castellani, *La storia della generazione* (Milan, 1965).

[41] On the general problem of plant physiology, see François Delaporte, *Le second règne de la nature* (Paris, 1979).

[42] Delaporte, *Le second règne* (see note 41), ch. 2.

[43] There is no recent study of Stahl's works, which are difficult to read and understand. However, see L. S. King, 'Stahl and Hoffmann, a study in 18th-century animism', *Journal of the History of Medicine*, XIX (1964), 118–30; and François Duchesneau, 'G. E. Stahl : antimécanisme et physiologie', *Archives internationales d'histoire des sciences*, XXVIII (1976), 3–26.

[44] This very interesting figure of the Scottish Enlightenment has not yet been studied as he should be. However, see G. Canguilhem, *La formation du concept de réflexe* (Paris, 1955), 101–7; Robert E. Schofield, *Mechanism and Materialism: British Natural Philosophy in An Age of Reason* (Princeton, N.J., 1970), 202–4; Roger K. French, 'Sauvages, Whytt and the motion of the heart', *Clio Medica*, VII (1972), 35–54; *idem*, *Robert Whytt, the Soul, and Medicine* (1969).

ton had done with gravitation and what Stephen Hales did with animal and plant physiology. Hales's success was important, and his work served as a model for a new generation of scientists.[45]

However, the most important development of the theory of life in the first half of the eighteenth century still belongs to the history of medicine, of which we do not have to speak here. The leading figures were Hermann Boerhaave and Friedrich Hoffmann, whose thought was still imbued with iatromechanicism.[46] But a new field of research developed very soon, that of vegetable physiology, in which naturalists were particularly active. Here, the experimental method was more and more applied, and the influence of Hales were clearly visible, especially with Buffon and Duhamel du Monceau. We have a general survey of these problems, and some particular studies,[47] but the detailed history of plant physiology at that time is still to be written.

On a more theoretical level, the problem of life and its mechanical interpretation was studied by an interesting scientist and philosopher, whose name lay forgotten for too long, Louis Bourguet, a disciple of Leibniz and a friend of Vallisneri.[48] The particularly distinctive feature of Bourguet's thought, as far as our problem is concerned, is his definition of an 'organic mechanism', clearly distinguished from the general mechanism not only because it could be found only in living beings and was made possible by their particular structure, but also because it worked in a different way. More precisely, Bourguet introduced, or revived, an important distinction between the growth of a crystal, which occurs as a result of mechanism of juxtaposition or apposition of new molecules on the external parts of the crystal, and the 'organic' mechanism of 'intussusception', as a

[45] It is known that the *Vegetable Staticks* was translated into French by Buffon : see Lesley Hanks, *Buffon avant 'l'Histoire naturelle'* (Paris, 1966). Also see Henry Guerlac, 'The continental reputation of Stephen Hales', *Archives internationales d'histoire des sciences*, IV (1951), 393–404.

[46] On the general history of physiology, see the books by Karl E. Rothschuh, and particularly *Geschichte der Physiologie* (Berlin, Göttingen and Heidelberg, 1953) and *Physiologie. Der Wandel ihrer Konzepte, Probleme und Methoden vom 16. bis 19. Jahrhundert* (Freiburg and Munich, 1968). Also see Thomas S. Hall, *History of General Physiology* (Chicago and London, 1969), especially chs. 19–35.

[47] For a general survey, see Delaporte, *Le second règne* (see note 41). Buffon's experiments were studied by Hanks in *Buffon* (see note 45). For Duhamel du Monceau, see Lucien Plantefol, 'Duhamel du Monceau', *Dix-huitième siècle*, I (1969), 123–37.

[48] On L. Bourguet, see my *Sciences de la vie* (see note 40), 376–8, and Schiller, *La notion d'organisation* (see note 37), 28–33. Bourguet was also influential in the history of the earth sciences.

result of which new molecules penetrate the living body, are 'moulded' by it, so to speak, and make it grow from within. This distinction between juxtaposition and intussusception was to last till the end of the nineteenth century, and the metaphor of the mould was eventually picked up by Buffon. A detailed study of Bourguet would be useful, because he played an interesting role in the building of a new theory of life.

However, the turning point of this history was obviously the middle of the century, when, on the one hand, new theoretical approaches were proposed by Maupertuis and Buffon and, on the other hand, the experimental research realized by Haller introduced new facts and concepts in the definition of life. We have already alluded to the discovery of the phenomenon of regeneration, first made by Trembley on the freshwater hydra in 1740, and later by other naturalists on other animals. This phenomenon had already obliged scientists to ask themselves difficult questions about their mechanical interpretation of life. The discovery by Haller of muscular excitability or, as he said, 'irritability', raised other difficulties. As a peculiar property of muscular fibres, irritability could not be explained mechanically. We still have little information on the intellectual process that led Haller to that piece of research,[49] though we may suppose that his starting point was the problem of the motion of the heart.[50] In any case, it is clear that his discovery raised the much more general question of the nature of life, and the definition of physiology as well. As an inexplicable property of living matter, irritability was clearly similar to gravitation, at least as gravitation was more and more generally interpreted in the middle of the century. And, if physiology had to take such properties of matter into account, it was no longer the '*anatomia animata*' that Haller himself had defined.[51]

[49] Many interesting studies and documents on Haller were published by the late Professor Hintzsche over the last thirty years. Also see Giorgio Tonelli, *Poesia e pensiero in Albrecht von Haller*, 2nd edn. (Turin, 1956) and *Albrecht von Haller. Zehn Vorträge . . .*, Sonderdruck aus den Verhandlungen der Schweizerischen Naturforschenden Gesellschaft; Wissenshaftlicher Teil, vol. 1977, with an interesting but short essay on 'Albrecht von Haller und die Physiologie' by Nikolaus Mani.

[50] For that problem, see Karl E. Rothschuh, 'Geschichtliches zur Lehre von der Automatie, Unterhaltung und Regelung der Herztätigkeit', *Gesnerus*, XXVII (1970), 1–19; Gerhard Rudolph, 'Reiz, Erregung, Kontraktion : Zur Frühgeschichte der experimentellen Nerven – und Muskel Physiologie', *Christiana Albertina*, IX (May 1970), 69–78; and H. J. Müller, *Die Begriffe 'Reizbarkeit' und 'Reiz'* (Stuttgart, 1975).

[51] On this interpretation of Haller's discovery, see G. Canguilhem, 'La constitution de la physiologie comme science', in his *Etudes d'histoire et de philosophie des sciences* (Paris, 1968), 226–73 (p. 227), and G. Rudolph, 'Haller's Lehre von

What is less clear, however, is the influence of the Newtonian model on Haller's interpretation of his discovery.

In contrast, the role of Newton is perfectly clear in the thought of Maupertuis and Buffon when they tried to explain reproduction by supposing the existence of living atoms or 'organic molecules', endowed with peculiar properties and subjected to 'penetrating forces'. Maupertuis shifted very soon to a more Leibnizian image of those living atoms, and did not hesitate to endow them with elementary psychic qualities.[52] Buffon was more cautious and maintained only that there was a 'living matter' different from the dead one, and whose existence had to be supposed if one wanted to understand vital phenomena.[53] This was not yet vitalism, but already a clear limitation to a mechanical theory of life. It is not therefore surprising that French vitalists took advantage of Buffon's 'organic molecules' when they tried to make their opinion respectable.[54]

Vitalism appeared very soon, that is in the 1760s, in nearly all fields of research, and almost simultaneously in England, France and Germany.[55] Unfortunately, we still have no comprehensive study of the phenomenon, whose importance was enormous, not only in the history of biology but in the general history of ideas as well. It is clear,

der Irritabilität und Sensibilität', in K. E. Rothschuh (ed.), *Von Boerhaave bis Berger* (Stuttgart, 1964). It may be argued that irritability is a property linked to a very particular anatomical structure, that of the muscular fibre, but even so it could not be accounted for by that structure.

52 For Leibniz's ideas on life and vital phenomena, see my paper 'Leibniz et les sciences de la vie' in *Akten des internationalen Leibniz-Kongresses* (Wiesbaden, 1969), II, 209–19, and François Duchesneau, 'Leibniz et la théorie physiologique', *Journal of the History of Philosophy*, XIV (1976), 281–300. The revival of Leibnizianism in the middle of the eighteenth century, especially in life sciences, has not yet been studied as it should be. On Maupertuis, see P. Brunet, *Maupertuis* (2 vols., Paris, 1929), and E. Callot, *Maupertuis, le savant et le philosophe* (Paris, 1964). On Maupertuis and evolution, section IV of this paper below.

53 On that aspect of Buffon's thought, see my *Sciences de la vie* (see note 40), 547–9. Eventually, Buffon explained that this living matter had been produced by chemical reactions, at the beginning of the history of life. See my edition of the *Epoques de la nature* (Paris, 1962), 67–68.

54 Cf. my *Sciences de la vie* (see note 40), 633–4.

55 For England, see Schofield, *Mechanism and Materialism* (see note 44), especially ch. 9, and Theodor M. Brown, 'From mechanism to vitalism in 18th-century English physiology', *Journal of the History of Biology*, VII (1974), 179–216. For the more particular problem of animal heat, see Everett Mendelsohn, *Heat and Life, the Development of the Theory of Animal Heat* (Cambridge, Mass., 1964). For France, see my *Sciences de la vie* (see note 40), 618–41, and my paper 'Méthodes et modèles dans la préhistoire du vitalisme français', in *Actes du XIIe Congrès International d'Histoire des Sciences* (Paris, 1971), III.B, 101–8.

however, that eighteenth-century vitalism does not resemble that of the following century, and this is obvious primarily through its philosophical meaning. In the particular case of Barthez it has been shown that vitalism could be more cautious and closer to Newtonian mechanics than one could have thought.[56] Even German vitalism was not as metaphysical as has often been said.[57] Nevertheless, the general appearance of vitalism throughout Europe should be explained, and it seems difficult to believe that such a phenomenon had no other causes than the difficulties of bio-mechanism.[58] Now that vitalism is no longer a threat to the modern biochemical interpretation of life, it should be possible to study its long history peacefully, and even to define the positive role it may have played at some stages of the history of biological research.[59]

Was vitalism ultimately responsible for the definition, at the end of the century, of a new science uniquely devoted to the study of life, and whose name, accordingly, should be 'biology'? It is difficult to answer such a large question and, generally speaking, the end of the eighteenth century is not easy to describe in such simple terms. Its

[56] See Wolf Lepenies, 'De l'histoire naturelle à l'histoire de la nature', *Dix-huitième siècle*, XI (1979), 179–81; Paul Hoffman, 'La théorie de l'âme dans la pensée médicale vers 1778', *ibid.*, 201–12; and Federico di Trocchio, 'Fisiologia e meccanica newtoniana in P.J. Barthez', *Medicina nei secoli*, I (1978), 93–108.

[57] See James L. Larson, 'Vital forces: regulative principles or constitutive agents? A strategy in German physiology, 1786–1802', *Isis*, LXX (1979), 235–49; and, on the particular case of Caspar Friedrich Wolff, Shirley A. Roe, 'Rationalism and embryology: Caspar Friedrich Wolff's theory of epigenesis', *Journal for the History of Biology*, XII (1979), 1–43.

[58] In an interesting but controversial paper, Richard Toellner maintains that the emergence of vitalism is a good example of a Kuhnian 'change of paradigm', and considers Haller's discovery of irritability as responsible for that change throughout Europe, though Haller himself stuck to the old mechanical interpretation of life. See R. Toellner, 'Mechanismus – Vitalismus: ein Paradigmwechsel? Testfall Haller', in Alwin Diemer (ed.), *Die Struktur wissenschaftlicher Revolutionen und die Geschichte der Wissenschaften* (Meisenheim am Glan, 1977), 61–72. I agree with the first thesis, though I do not think that the change occurred overnight, as suggested by Toellner. But I do not believe that Haller's discovery was as instrumental in the change as Toellner thinks it was. In any case, early vitalists like Bordeu in France or Wolff in Germany considered Haller as an enemy (which is in accord with Toellner's thesis) but did not consider irritability as a vital principle, because it was not a general property of the living being, but only a property of the muscular fibre. Bordeu's 'sensibility' or Wolff's *'vis essentialis'* had nothing to do with Haller's irritability.

[59] G. Canguilhem has pointed out some of those positive aspects of vitalism in his chapter on 'Aspects du vitalisme' in *La connaissance de la vie* (Paris, 1952), 101–23, as well as in his book *La formation du concept de réflexe* (see note 44), where he refers more precisely to the works of Thomas Willis.

complexity may be illustrated by the case of Lamarck, who has recently attracted the attention of historians once again.[60] I will speak a little later of his role as a founder of the evolutionary theory but, since he was one of the coiners of the word 'biology', a clear definition of life could be expected from him. Actually, we know that he was a vitalist for a long time, and to such an extent that he could not consider life as a 'natural' phenomenon;[61] then he changed his mind, accepted the idea of spontaneous generation and tried to explain life through a physico-chemistry of his own. And it was after that conversion, which apparently took place in the late 1790s, that the word 'biology' appears in his writings, which is quite logical since no science of life could have been imagined as long as life itself was considered a non-natural phenomenon. It is interesting to discover the same vitalist temptation in the works of Treviranus, the other coiner of 'biology', as well as the same final recourse to chemistry.[62]

In any case, it is clear that biology did not appear suddenly at the end of the eighteenth century, as Athena sprang from Zeus's forehead. On the other hand, from the existence of the word we cannot conclude that the science itself really existed: one might well say that biology did not exist before the time the cell theory was fully developed. What interests us here is the slow and complicated process which led from the Cartesian bio-mechanism and the Stahlian animism that prevailed at the beginning of the century, with all their metaphysical implications, to a vitalism that abandoned neither the mechanical nor the chemical explanations but recognized the originality of living beings and made possible a more phenomenological approach. As far as theories are concerned we may summarize that development by the sequence: Cartesian bio-mechanism, Newtonian bio-mechanism, and vitalism. But the succession of theories, interesting as it is, unduly simplifies the complexity of the process, and speaking of 'a change of paradigm' does not help very much. One

[60] See *Colloque international Lamarck* (Paris, 1971); Max Vachon, Georges Rousseau and Yves Laissus (eds.) *Inédits de Lamarck* (Paris, 1972); Ernst Mayr, 'Lamarck revisited', *Journal of the History of Biology*, v (1972), 55–94, repr. in Ernst Mayr, *Evolution and the Diversity of Life* (Cambridge, Mass., 1976), 222–50; Richard W. Burkhardt, Jr, *The Spirit of System: Lamarck and Evolutionary Biology* (Cambridge, Mass., 1977); Madeleine Barthélemy-Madaule, *Lamarck ou le mythe du précurseur* (Paris, 1979); Giulio Barsanti, *Dalla storia naturale alla storia della natura* (Milan, 1979). It now seems to be possible to study Lamarck for himself, without trying to demonstrate that he was or was not a 'precursor' of Darwin, or responsible for the neo-Lamarckism of the 1880s.

[61] See his *Recherches sur les causes des principaux faits physiques* (Paris, 1794).

[62] See the interesting article by Brigitte Hoppe, 'Le concept de biologie chez G. R. Treviranus', in *Colloque international Lamarck* (see note 60), 199–237.

must take into account such changes as the development of animal physiology from Harvey to Haller and its relative set-back at the end of the century in the face of a much more aggressive theoretical medicine (an evolution paralleled by the changing situations of experimentation and observation); the steady increase of plant physiology; the appearance of the notion of a 'living matter' or a 'matter of life', which favoured the introduction of chemistry into the science of living beings, and that was instrumental in the birth of biology;[63] and, in the background, the changes in the philosophy of knowledge and in the relation of man to language, society, nature and himself. The history of life science in the eighteenth century is but a part of the general history of the Enlightenment.

IV. From natural history to the history of nature

I borrow the heading of this section from Wolf Lepenies,[64] because it clearly summarizes an important aspect of the change that occurred in natural sciences in the second half of the eighteenth century: namely, the introduction of temporality.[65]

At the beginning of the century nature was immutable, and so was the order that natural history discovered in her. There were many reasons for thinking so, and very few for thinking otherwise. The universe was immutable because God had created it; and since God's creation was necessarily perfect, no improvement was conceivable.[66] God had designed everything with a clear purpose in mind, and consequently all the existing beings should have been there from the beginning of the world and would stay there till the end. They were necessary elements of the order of the world. So was every organ, every part of every being. Here, mechanist philosophy agreed with Christian faith: one cannot imagine a machine that keeps on working

[63] See my article 'Chimie et biologie: des "molécules organiques" de Buffon à la "physico-chimie" de Lamarck', *History and Philosophy of Life Sciences*, I (1979), 43–64.

[64] It is the heading of the ch. 6 of his book *Das Ende der Naturgeschichte* (Munich and Vienna, 1976). Giulio Barsanti has chosen the same title for his book on Lamarck, quoted in note 60 above.

[65] Also see Francis C. Haber, 'Fossils and the idea of a process of time in natural history', in Bentley Glass, Owsei Temkin and William L. Straus, Jr (eds.), *Forerunners of Darwin: 1745–1859* (Baltimore, 1959), 222–61.

[66] The only conceivable evolution was a destruction of the primaeval order, more or less linked with the original sin, or a slow erosion of the first creation. Those ideas were more often expressed for the history of the earth (see Burnet or the Abbé Pluche) but they found their counterpart in the history of the living forms: see on Buffon below.

when one takes some gears away, or adds some new ones. The essence of each living machine is a special type of construction without which it will not work. God had given every natural being its structure and the exact quantity of motion or force necessary for its working. This was true for the solar system, according to Newton, and for everything in nature, according to most scientists and philosophers. One could only dispute, as did Samuel Clarke and Leibniz, whether God had built the machine once and for all or was still intervening in its running; in any case, one could not doubt the passivity of nature. According to the theory of pre-existing germs there was not even a real succession of individuals, since all living beings had been created simultaneously, once and for all, at the time of the creation. Generation was only an appearance. One could even say, with Leibniz, that individual lives were only apparently subject to temporality: once created, a 'monad' lived for ever. Birth and death also were appearances: substances do not die. As long as He wanted her to last, nature was participating in God's eternity.[67]

Such a theological view could not last for ever, especially in the context of the eighteenth-century thought, in which nature was more and more substituted for God. As Buffon said in 1749, 'in physics one must avoid as far as possible having recourse to causes outside nature'. If nature was now considered, by some scientists at least, as in active power, this did not mean, however, that a historical point of view was to prevail. Such a view would have disagreed too much with the widely accepted ideal of a Newtonian science which studied phenomena as they might be observed here and now, and refused to speculate on the origin of things. Newton's world was a steady-state universe, for both religious and epistemological reasons. When religion disappeared, epistemology remained. Newton believed that God had to intervene from time to time and restore the solar system to its primitive order, but astronomers of the eighteenth century discovered that the system was self-regulating, in such a way that it did not need God's intervention – but neither was it evolving. At the end of the century the model of a self-regulating and cyclic system was widely accepted, in Adam Smith's political economy as well as in Laplace's astronomy, Hutton's *Theory of the Earth* or Lamarck's *Hydrogéologie*.[68] Applied to the living world, this became the 'balance of nature', or Linnaeus's '*oeconomia naturae*': since every living species is

[67] This is not the place to examine the problem of earth sciences at the beginning of the eighteenth century and the historical character of certain theories of the earth.

[68] For Lamarck's *Hydrogéologie*, see Albert V. Carozzi, 'Lamarck's theory of the earth: *Hydrogéologie*', *Isis*, LXV (1964), 293–307.

at the same time a prey and a predator, every change in the size of a population is automatically checked and kept at a proper level.[69] This was to become an important element of Malthus's *Essay on the Principle of Population,* and eventually of Darwin's theory of natural selection.

With nature again considered as an active power, the theory of pre-existing germs came under attack. Here again, one of the leading figures is Buffon who, significantly, often uses a new word: reproduction. The nuance is not insignificant, especially because, for Buffon as for John Ray, species are now defined by reproduction, by the uninterrupted succession of similar beings. A species is a self-reproducing system, so to speak, and the similarity of parents and offspring is a logical necessity as well as an observed fact. The metaphor of the 'inner mould', which in some pages of the *Histoire naturelle* sounds quite Platonic,[70] actually expresses that necessary permanence of living forms. In spite of his attacks upon taxonomy, Buffon still considers species as essential elements of the order of nature. This is not to say that no variation is possible. There are variations in living forms indeed, and Buffon studies them in the human race as early as 1749. He even tries to discover their mechanism and explains them through the pressure of the environment. But those variations can only be a degeneracy from the original type and cannot go beyond the borders of the species, as defined by the 'inner mould'.[71] No evolution is possible.

And yet, as soon as nature is no longer a plaything in the hands of God, as soon as one tries to explain everything by natural causes, one may be tempted to search in the past for the causes of what exists in the present. This is precisely what Buffon does in 1749 when he proposes a cosmogonic hypothesis, whose role avowedly is to replace the direct creation by God. Thirty years later the hypothesis has become a historical truth, and the *Epoques de la nature* present the reader with a detailed narrative of the history of the earth and life upon it. The beginning of the book fully develops the comparison between the history of nature and that of mankind. More importantly, Buffon does

[69] See Frank N. Egerton, 'Animal demography from Lamarck to Darwin', *Journal of the History of Biology,* I (1968), 225–59, and 'The concept of competition in nature before Darwin', in *Actes du XIIe Congrès International d'Histoire des Sciences* (Paris, 1971), VIII, 41–6. Also see Camille Limoges, 'Darwinisme et adaptation', *Revue des questions scientifiques,* CXXXXI (1970), 353–74.

[70] Especially in the *Histoire naturelle du cheval* (1753).

[71] See my *Sciences de la vie* (see note 40), part 3, ch. 2, and Philip R. Sloan, 'The idea of racial degeneracy in Buffon's *Histoire naturelle*', in H. Pagliaro (ed.), *Studies in Eighteenth-Century Culture* (Cleveland and London, 1973), III, 293–321.

not hesitate to compute the duration of this history and to pretend that the earth is 75,000 years old, an incredible age for people who used to believe that the world had been created in 4004 B.C. And yet, this was only a conservative estimate, according to Buffon.[72] No wonder that the old naturalist - he was seventy-one when he published the book - got into trouble with theologians.

An interesting feature of Buffon's work is that it made use of fossil remains not only in order to reconstruct the history of the earth - this had been done from the beginning of the century - but also to reconstruct the history of life. This was made possible by recent discoveries, especially in Siberia and North America, of remains of gigantic mammals, including the famous mammoth. Buffon was not able to interpret those remains correctly, but this was the beginning of palaeontology, the historical science that was to be so successful with Cuvier.

The *Epoques de la nature* should have been the proper place for Buffon to propose an evolutionary theory. Actually, he did not use the opportunity,[73] and his views of the history of life were nearer to Cuvier than to Lamarck: the great mammals of the past had disappeared because new climatic conditions had made it impossible for them to survive. They were by no means the ancestors of present species. In order to explain the diversity of life Buffon preferred to rely upon spontaneous generation rather than evolution.

Buffon's history of nature was very influential, especially in Germany where he had a great impact on Herder,[74] and this influence merged into the general development of historical thinking. But the divorce that we observe here, as in Cuvier's work, between history and palaeontology on the one hand, and evolution on the other hand, is all the more worth noticing in that for so long we had been taught to consider history as a necessary element of evolutionary theory, and palaeontology as its firmest support. Actually, this is probably an error, as witnessed by the history - or prehistory - of the theory of evolution in the eighteenth century.

Maupertuis is traditionally hailed as the first scientist to propose a theory of evolution.[75] Nevertheless it is remarkable that he was abso-

[72] See my edition of the *Epoques de la nature*, introduction, LX/LXVII.

[73] On the limits of Buffon's 'transformism', see my *Sciences de la vie* (see note 40), part 3, ch. 2. There is a history of life according to Buffon, but that history is not an evolution.

[74] See Max Rouché, *La philosophie de l'histoire de Herder* (Paris, 1940), esp. 207.

[75] Among many other historians, see Bentley Glass, 'Maupertuis, pioneer of genetics and evolution', in *Forerunners of Darwin* (see note 65), 51-83. It has been recently argued, however, that Maupertuis never had the idea of evolu-

lutely not interested in a history of life, except when he suggested that geological catastrophes could have destroyed some species and created some gaps in the 'great chain of being'. What interested him was the hereditary transmission of some abnormalities like polydactyly or mutilations, that is, phenomena that he could directly observe or experiment upon. This was real science: history was not. The same thing could be said about Lamarck, who indisputably proposed a theory of evolution. History plays no role in his thought. Spontaneous generation is not for him, as it was for Buffon, a phenomenon of the past, that occurred once and for all at the beginning of life: it is a phenomenon that happens everyday, even if we cannot observe it. Palaeontology is not alluded to either: Lamarck does not believe in extinct species. Evolution is not a historical phenomenon: it starts everyday anew, as new 'monads' are spontaneously generated. Paradoxically, Cuvier, the fixist, was more of a historian than Lamarck, the evolutionist. But the paradox was a lasting one, and, one might maintain, that neither Darwin nor the present synthetic theory of evolution nor modern genetics are interested in history. Time is a necessary element of any evolutionary theory, history is not. But this is not to say that the success of historical thinking throughout Europe in the first half of the nineteenth century played no role in the spreading of evolutionary theory in public opinion. In that sense, Buffon, Herder and German *Naturphilosophie* were instrumental in its success. But the philosophy of 'development' that they presented could lead to Chambers's *Vestiges of the Natural History of the Creation* or to Spencer's philosophy of progress, not to Darwin's *Origin of Species*.

V. Towards a conclusion

When did the eighteenth century end and the nineteenth begin? Such questions cannot be answered: there are no such clear-cut divisions in the history of science. But old scientists die and new ones enter the scene. Between the death of Haller (1777) and those of Spallanzani, Ingenhousz and Saussure (1799), almost all the great naturalists of the century disappeared: Linnaeus (1778), Trembley (1784), Buffon (1788), Lyonet (1789), Charles Bonnet (1793), Caspar-Friedrich Wolff (1794).

tion, even if he proposed some of the genetical mechanisms that are now accepted, *mutatis mutandis*, in the modern theory of evolution. See M. Foucault, *Les mots et les choses* (see note 5), 166–7; François Jacob, *La logique du vivant* (Paris, 1970); and Anne Fagot, 'Le "transformisme" de Maupertuis', in *Actes de la Journée Maupertuis* (Paris, 1975), 163–78.

What they left to their successors was significantly different from what existed at the beginning of the century. Most of the specialized disciplines that were to develop in the nineteenth century had now come into existence: Embryology and comparative anatomy, palaeontology, plant and animal physiology, not to speak of anthropology. But, if the fields of research were already defined, methods generally had to be made more precise, a task that was left to nineteenth-century professional scientists.

Generally speaking, religious and philosophical preconceptions were much less visible than at the beginning of the century. This is not to say that they had disappeared, but they were more incorporated, so to speak, in the scientific reasoning, being probably also less conscious. A good example would be the famous Spallanzani and his experiments on spermatozoa.[76] This might well mean that science was becoming an autonomous and respectable activity, which no longer needed to be justified by external reasons. At the beginning of the century everybody spoke of God; at the end everybody spoke of nature, and they probably thought that great progress had been made, without realizing that they were substituting a new ideology for the old one. The philosophical character of German *Naturphilosophie* was of course obvious, but the relationships between Lamarck and the 'Idéologues' should be more precisely studied.[77]

However, experimental method and scientific precision were steadily gaining ground. It is by no means certain that Spallanzani was a better observer and experimentalist than Réaumur, but it is pretty sure that bold generalizations like Buffon's *Epoques de la nature* were less and less considered scientific by scientists who were themselves becoming more and more professional. In natural history and biology, as in mathematics and physics one century earlier, the time of amateurs was over and a new scientific establishment was now reigning. A new era was beginning for life sciences.

[76] Interesting work has been done recently on Spallanzani by Professor Castellani. See his editions of Charles Bonnet, *Lettres à M. l'Abbé Spallanzani* (Milan, 1971); Lazzaro Spallanzani. *Opere scelte* (Turin, 1978); and Lazzaro Spallanzani, *I giornali delle sperienze e osservazioni relativi alla fisiologia della generazione e alla embriologia sperimentale* (Milan, 1978) with a important introduction.

[77] On the 'Idéologues' see Sergio Moravia, *Il tramonto dell' Illuminismo* (Bari, 1968), and *Il pensiero degli Idéologues* (Florence, 1974).

7

The terraqueous globe

ROY PORTER

Contents

I.	The globe as subject	285
II.	Image and understanding of the globe c. 1700	288
III.	European civilization changes the environment	294
IV.	Social and institutional dimensions	300
V.	The growth of knowledge	305
VI.	Environment, ideas and ideology	313
VII.	Disciplines	317
VIII.	Towards new views of the environment	321
IX.	Future prospects	324

I. The globe as subject

With few exceptions, historians of science have paid little attention to a major branch of natural philosophy in the eighteenth century: the science of the system of the earth and its products. Distinguished general surveys, such as those of Dampier, Singer, Hall, Gillispie and Taton, have no connected analysis of the sciences of the terraqueous globe. The volume *Natural Philosophy through the Eighteenth Century* barely mentions the physical environment.[1] Most historians of science are still writing (sometimes Whiggish) 'tunnel histories', tracing the progress of modern scientific disciplines, rather than exploring

[1] Sir William Dampier, *History of Science and its Relations with Philosophy and Religion* (Cambridge, 1929; 4th edn, 1948); C. J. Singer, *A Short History of Scientific Ideas to 1900* (Oxford, 1959); A. R. Hall, *The Scientific Revolution, 1500–1800*, 2nd edn, (1962); C. C. Gillispie, *The Edge of Objectivity* (Princeton, N.J., 1960); R. Taton (ed.), *A General History of the Sciences* (4 vols., English trans., 1964–6); A. Ferguson (ed.), *Natural Philosophy through the Eighteenth Century and Allied Topics* (*Philosophical Magazine* commemorative number, 1948; repr., 1972). The lack of any recent interpretative survey of eighteenth-century science compounds this neglect, as does the absence of surveys of *popular* attitudes towards the earth.

the cognitive landscapes of the past.[2] And much scholarship in this field remains stubbornly biographical rather than problem-oriented.[3]

Why has there been this neglect of the sciences of the physical environment, of the earth as such? It is partly a reflection of the focusing down of science itself, where formerly dominant all-embracing frameworks, like the Great Chain of Being, have splintered into specialties like palaeobotany and seismology.[4] It is partly also a response to persuasive philosophies of science. Ever since the influential classifications of natural knowledge by Comte, J. S. Mill and Jevons, leading philosophers have argued that the ideal science is one of universal categories, amenable to abstraction and quantification. Thus physics rather than the 'extensive' environmental sciences becomes their model. Also, the more introspective science has become about its own methods, the greater the indifference to the unity and integrity of the *object* of inquiry. Concern with how to practise sedimentology, oceanography, meteorology, has diverted attention from conceptualizing the terraqueous globe as a whole.[5]

Since the last century, the environmental sciences have always played second fiddle. Geography has hovered uncertainly between humanity and science. T. H. Huxley's 'physiography' failed; ecology has hardly become separately institutionalized. Victorian mathematical physicists such as Lord Kelvin presumed to teach geologists scientific method,[6] while Darwinian evolutionism downgraded the physical environment to, at most, the goad to interspecific competition, and steered towards genetics. Furthermore, nineteenth-century philosophies of nature (whether Hegelian idealist, or materialist like Marxism), perpetuated ingrained Classical and Christian contempt for the 'mere' nature of physico-material being – views whose chick-

[2] I should confess my own 'tunnel history': *The Making of Geology: Earth Science in Britain, 1660–1815* (Cambridge, 1977).

[3] This point is enlarged in Roy Porter and Kate Poulton, 'Research in British geology, 1660–1800: a survey and thematic bibliography', *AS*, xxxiv (1977), 33–42.

[4] A. O. Lovejoy, *The Great Chain of Being* (Cambridge, Mass., 1936). For its demise see W. F. Bynum, 'The Great Chain of Being after forty years', *HS*, xii (1975), 1–28. This fragmentation explains why C. J. Glacken stopped his monumental history of environmental ideas at *c.* 1800: *Traces on the Rhodian shore: Nature and Culture in Western Thought from Ancient Times to the End of the Eighteenth Century* (Berkeley, Cal., 1967).

[5] Excellent here is C. F. A. Pantin, *The Relations Between the Sciences* (Cambridge, 1968). Much recent literature from the environmentalist lobby makes the same point.

[6] J. D. Burchfield, *Lord Kelvin and the Age of the Earth* (1975).

ens come home to roost in environmental vandalism.[7] The globe was just an object ripe to be given meaning, dominated, and expropriated by mankind.

Moved partly by contemporary ecological anxiety, however, consciousness is stirring about the history of conceptions of the environment. Scholars like Yi-fu Tuan, for instance, have recently illuminated our ideas of space and place, and the human needs met by different representations of the globe.[8] Yet, predictably, our grasp of the environmental sciences owes its most penetrating insights not to professional historians of science but to a medley of scholars, including historians of ideas and literature, students of the Enlightenment, and historical geographers.[9] Archaeologists of knowledge have argued the historicity of the categories of earth description,[10] and philosophical historians have explored the earth as a sacred - or profane - object.[11] Investigation of the moral meanings of cosmogonies by cultural anthropologists, sociologists of knowledge, and historians of religion has also suggested new approaches.[12]

My aim in this essay is to pinpoint major problems for understanding what eighteenth-century inquirers construed as the terraqueous globe; to examine existing research and its implications; and to outline agenda for mapping *terra mentis nondum cognita*. I make no pretence of offering a connected history of environmental science, and trust that

[7] J. A. Passmore, *Man's Responsibility for Nature* (1974). Perhaps this explains why Marxist scholars ignore environmental sciences, an instance being J. D. Bernal's *Science in History* (1954). For Marx himself see A. Schmidt, *The Concept of Nature in Marx*, trans. B. Fowkes (1971).

[8] Yi-fu Tuan, *The Hydrologic Cycle and the Wisdom of God* (Toronto, 1968); *idem*, *Topophilia* (Englewood-Cliffs, N.J., 1974); *idem*, *Space and Place: The Perspective of Experience* (1977).

[9] Alongside Glacken and Tuan (see notes 4 and 8), see Lewis Mumford, *Technics and Civilization* (1934); *idem*, *The Condition of Man* (1944); Georges Gusdorf, *Les sciences humaines et la pensée occidentale* (to date, 7 vols., Paris, 1966–76); M. H. Nicolson, *Mountain Gloom and Mountain Glory* (Ithaca, N.Y., 1959); Ton Lemaire, *Filosofie van het landschap* (Bilthoven, 1970).

[10] M. Foucault, *The Order of Things* (English trans., 1970); *idem*. *The Archaeology of Knowledge* (English trans., 1972).

[11] Cf. Passmore, *Man's Responsibility* (see note 7).

[12] M. Eliade, *Myth and Reality* (1964); *idem*, *Images and Symbols* (1961); *idem*, *The Myth of the Eternal Return* (1955); Mary Douglas, *Natural Symbols* (1970); *idem*, 'Environments at risk', in her *Implicit Meanings* (1975), 230–48; *idem*, *Cultural Bias*, Royal Anthropological Institute of Great Britain and Ireland, Occasional Paper 35 (1978); S. B. Barnes, *Scientific Knowledge and Sociological Theory* (1974); *idem*, *Interests and the Growth of Knowledge* (1977); D. Bloor, *Knowledge and Social Imagery* (1976); E. Durkheim, *The Elementary Forms of the Religious Life* (English trans., 1976).

the speculative and provisional nature of this exploratory voyage needs no apology.[13]

II. Image and understanding of the globe *c.* 1700

By the dawn of the Enlightenment the terraqueous globe had acquired certain new meanings resulting from the intellectual transformations of the Renaissance, Reformation and the 'new science'. The heliocentric revolution, switching the closed world into an infinite universe, had dislodged the earth from its central (albeit lowly) place in cosmogony and cosmography.[14] The earth was now nowhere in particular in the universe, no longer possessing any privileged relation - of influence, teleology or correspondence - with other spheres of creation. Late-seventeenth-century cosmogonists, such as Woodward, Burnet, Whiston and Leibniz, had detached the earth's creation from the origins of the universe.[15] Philosophies which hinged on cosmic influences and analogies - such as the Aristotelian element theory, Hermetic natural magic, or astrology - were waning (though we are learning how historians have hitherto packed them off prematurely).[16]

The earth had traditionally been viewed as the sub-lunary sphere, taking its nature from its centrality to the celestial hierarchy. In Carpenter, Blancanus, and Heylyn, geography was still part of cosmography, and in the Ptolemaic sense a branch of mixed mathematics. 'Geography is that part of mixed Mathematics', wrote Varenius, 'which explains the State of the Earth, and of it's Parts, depending on Quantity, *viz,* it's Figure, Place, Magnitude, and Motion, with the

[13] I know of few historiographical surveys in this area. See, however, chiefly for the history of geology, R. Rappaport, 'Problems and sources in the history of geology, 1749-1810', *HS,* III (1964), 60-77; V. A. Eyles, 'The history of geology; suggestions for further research', *HS,* V (1966), 77-86; and Roy Porter and Kate Poulton, 'Geology in Britain, 1660-1800: a selective biographical bibliography', *JSBNH,* IX (1978), 74-84.

[14] A. Koyré, *From the Closed World to the Infinite Universe* (Baltimore, 1957); Tuan, *Topophilia* (see note 8), ch. 10, 'From cosmos to landscape'.

[15] K. Collier, *Cosmogonies of our Fathers* (New York, 1934); J. Roger, 'La théorie de la terre au XVIIe siècle', *RHS,* XXVI (1973), 23-48.

[16] See for instance D. C. Kubrin, 'Providence and the mechanical philosophy. The creation and dissolution of the world in Newtonian thought. A study of the relations of science and religion in seventeenth century England' (Ph.D. thesis, Cornell University, 1968); H. Leventhal, *In the Shadow of the Enlightenment* (New York, 1976); B. Capp, *Astrology and the Popular Press: English Almanacs, 1500-1800* (1979).

Celestial Appearances.'[17] Very gradually this view was to yield to a conception of the earth as a self-contained, autonomous globe. The 'vertical' cosmos yielded – as Tuan suggests – to the 'horizontal' landscape. The applied mathematical and astronomical emphases of Ptolemaic geography faded. Problematic features of the globe were coming to be understood with respect to the terrestrial system, rather than to the design of the heavens. And this trend was graphically reinforced from the fifteenth century by European navigators experiencing the globe in the round, above all through circumnavigations.

Such explorations yielded new truths: the tropics could be crossed without their fires frazzling sailors to cinders; there *were* habitable antipodes. They also bore new scientific insights, such as the zoning of the earth's magnetic field, or the regular distribution of winds and ocean currents. But from Columbus onwards the problems of relating *terra antiqua* to *terra nova,* or *terra cognita* to *terra incognita,* remained daunting. What were the historical and functional relations between continents? Was America absolutely a newer world, or merely new to Europeans? Why did the climate and fertility of landmasses differ, thus creating distinct environments for specific flora and fauna?[18] Relating fundamental and universal physical forces (such as central heat, or tides) to local configurations (like volcanic eruptions or coastlines) became key *explicanda* in theories of the physical environment such as De Maillet's *Telliamed* (written about 1700), Buffon's *Histoire naturelle* (1749–) and Lamarck's *Hydrogéologie* (1802).[19]

Severing the terraqueous globe from the cosmos at large urgently posed the problem of conceptualizing the planet as a body. Seeing the earth as a living animal still claimed a few supporters – like Thomas

[17] B. Varenius, *A Compleat System of General Geography,* trans. Dugdale, 3rd edn (2 vols., 1736), I, 2. Compare the similarly traditionalist statement in John Newton's *Cosmographia* (1679), title page: 'Geography is a science concerning the measure and distinction of the Earthly globe.'

[18] For excellent scholarship on the Old World looking at the New, see D. Echeverria, *Mirage in the West: A History of the French Image of American Society to 1815* (Princeton, N.J., 1957); E. O'Gorman, *The Invention of America* (Bloomington, 1961); H. Mumford Jones, *O Strange New World* (1965); E. H. P. Baudet, *Paradise on Earth* (New Haven, 1965); J. Elliott, *The Old World and the New* (Cambridge, 1970).

[19] A. V. Carozzi (ed.), *Telliamed, or Conversations between an Indian Philosopher and a French Missionary on the Diminution of the Sea, by Benoit de Maillet* (Urbana, 1968); J. Roger (ed.), *Buffon: Les époques de la nature,* Mémoires du Muséum National d'Histoire naturelle, ser. C, vol. x (Paris, 1962); A. V. Carozzi (ed.), *Lamarck: Hydrogeology* (Urbana, 1964).

Robinson or William Hobbs – at the close of the seventeenth century.[20] Such literal anthropomorphic animism, however, was in retreat before the 'new science's' mechanical philosophy – and with it went its bedfellows, such as *mundus senescens*, the earth in decay.[21] Yet conceptions of the environment as organism, a quasi-animal, with its own internal heat and energy, and with metabolic systems of change, restoration, circulation and equilibration, remained key metaphors through the eighteenth century, vying with other images of the earth, as a machine, a chemical laboratory, or simply a chance congealing lump.[22]

Interpretations of terrestrial phenomena are not hermetically sealed. Michel Foucault, above all, has affirmed that meanings of particular phenomena such as fossils and mountains lie deep within comprehensive 'discursive formations' – structures of thought which have construed nature as grid-like, or genetic, organic or mechanical. Thus earlier conceptions of figured stones ('fossils') as *lapides sui generis*, or of volcanoes as the earth's anus, were predicated on a wider philosophy of nature as an active, living organism. By contrast, for later naturalists who pictured the earth as a law-governed machine, figured stones became organic *reliquiae*, and volcanoes safety valves. What remain to be investigated are the lie and boundaries of such subterranean thought-grids. Foucault himself appears Janus-faced. *The Order of Things* tells us to seek the fundamental '*episteme*' which governs all that in any epoch it is possible to say, uniting superficially contradictory theories; whereas by contrast *The Archaeology of Knowledge* uncovers the discontinuities between speciously commensurable views.[23]

For all its posturing, Foucault's 'archaeology' is a sharper scalpel for skinning the physiology of geophilosophies than a traditional history of ideas which views truth as emerging from the clash of diametrical opposites. To chronicle the theorizing of the physical environment as battles between Christian and secular, Newtonian and Cartesian, or

[20] F. J. North, '"The anatomy of the earth" – a seventeenth century cosmogony', *Geological Magazine*, LXXI (1934), 541–7; Roy Porter, 'William Hobbs of Weymouth and his *The earth generated and anatomized* (?1715)', *JSBNH*, VII (1976), 333–41.

[21] G. L. Davies, *The Earth in Decay* (1969); B. Olsson, 'Mundus senescens', *Lychnos* (1954–5), 66–81.

[22] For organicism, see R. H. Dott, Jr, 'James Hutton and the concept of a dynamic earth', in C. J. Schneer (ed.), *Toward a History of Geology* (Cambridge, Mass., 1969), 122–41, and Porter, *Making of Geology* (see note 2), 184–98. For the earth as mechanism, see *ibid.*, 70–8; or as a body flung off the sun, see Buffon, in Roger, *Epoques* (see note 19).

[23] See note 10.

static and dynamic, concepts, with one destined for victory, is to oversimplify Whiggishly.[24] Historians are too-little sensitive to the architectonic images used to render the earth intelligible. They have charted the growth of positive knowledge and concepts about the earth, but have been deaf to the *meanings* with which these resonated.[25]

Despite the astounding geographical discoveries of the age of reconnaissance and of the revolution in astro-physics, scientific approaches to the *oikoumene* remained essentially traditional up to 1700. 'Cosmography', 'geography', and 'chorography' were still cast in Ptolemy's and Strabo's categories by leading geographers such as Heylyn.[26] This traditionalism is mirrored by Glacken in his concentration, in *Traces on the Rhodian Shore,* on the continuity from Classico-Christian philosophies of nature to the Enlightenment. Belief that the physical environment was divinely designed as a habitable earth to support vegetable and animal life, and finally man; that man has a right to dominion over the environment, to use and improve it (reciprocated by a duty of preserving its creatures according to God's law); that man was ontologically distinct from, and superior to, his physical surroundings through possessing reason and a soul, and that terrestrial existence was, ultimately, only a fleeting way-station - ideas common both to the Bible and to Greek philosophy - had hardly been challenged. Similarly, the Judaeo-Christian belief in the earth's recent

[24] Thus John C. Greene's *The Death of Adam* (Ames, Iowa, 1959), sees the mechanico-dynamic Cartesian world-view soldiering eventually through to Darwinian evolution. This account is effectively undermined by Foucault, *Order of Things* (see note 10), and Martin J. S. Rudwick, *The Meaning of Fossils* (1972), chs. 3 and 4.

[25] Insensitivity to the meaning-loaded, myth-bearing language which pervades theorizing about the terraqueous globe (for instance, images of wombs, revolutions, origins, extinctions, order and chaos, height and depth) is specially puzzling as such analysis has been triumphantly undertaken for 'primitive' cosmologies by cultural anthropologists (see note 12) - though see now the essays by G. N. Cantor and M. J. S. Rudwick in L. J. Jordanova and Roy Porter (eds.), *Images of the Earth* (Chalfont St Giles, 1979). Literary historians have applied such analysis to imaginative environment writing. Cf. Leo Marx, *The Machine in the Garden* (New York, 1964); A. Kolodny, *The Lay of the Land* (Chapel Hill, N. Carolina, 1975); and J. L. Lowes, *The Road to Xanadu* (1927).

[26] For seventeenth-century geography see E. G. R. Taylor, *Late Tudor and Early Stuart Geography, 1583-1650* (1934); R. E. Dickinson, *The Makers of Modern Geography* (1969); E. Fischer, R. D. Campbell and E. S. Miller, *A Question of Place; The Development of Geographic Thought* (Arlington, Va., 1967). For Varenius see esp. 'The geography of Bernhard Varenius', in J. N. L. Baker, *The History of Geography* (Oxford, 1963), 105-18. See also P. E. James, *All Possible Worlds: A History of Geographical Ideas* (Indianapolis, 1972).

creation, and in its eschatological transformation by God's future hand, even gained strength in cosmogonies towards the end of the seventeenth century.[27] Christian natural philosophers had ensured that 'neo-Aristotelian' and 'Spinozist' bids to prove the world eternal were nipped in the bud. Compendia such as Athanasius Kircher's *Mundus subterraneus* (1664–5) continued to convey traditional learning about the geography of the Ancients, Atlantis, the Abyss, the alembic theory of the origin of rivers, and so on, to the late-seventeenth-century reader, knowledge reproduced in popular format through the next century.[28]

By 1700 neither crushing weight of new evidence nor any urgent reorientations of interest and experience amongst naturalists had compelled a radical reformation in environmental visions. One feature of this was the undiminished ignorance and pseudodoxy about so many facets of the globe. The interiors of most continents were still dark to Europeans. The 'North-West Passage', and the 'Terra Australis nondum cognita' were confidently engraved onto maps – vindicating Swift's cynicism:

So geographers, in Afric maps,
With savage pictures, fill their gaps.

Heylyn and Varenius both plotted the Rhipaean Mountains, North of the Black Sea, about which even Strabo had been sceptical; and anthropological monstrosities, such as the Patagonian giants, strode their plates. 'There are many particular falsehoods', bemoaned Joseph Addison, 'suited to the particular climates and latitudes in which they are published, according as the situation of the place makes them less liable to discovery'.[29] Great tracts were still mapped – or rather unmapped – as *terra incognita*.[30] Even at the end of the seventeenth

[27] The pervasiveness of Chiliasm has been underscored in M. C. Jacob, *The Newtonians and the English Revolution, 1689–1720* (Hassocks, Sussex, 1976), and M. Farrell, 'The life and work of William Whiston' (Ph. D. thesis, University of Manchester, 1973).

[28] For how almanacs and encyclopaedias gave a new lease of life to traditional knowledge, see Leventhal, *Shadow of the Enlightenment* (see note 16); A. Hughes, 'Sciences in English encyclopaedias, 1704–1875', *AS*, VII (1951), 340–70, IX (1953), 233–64; and discussion in Elizabeth Eisenstein, *The Printing Press as an Agent of Change* (Cambridge, 1979).

[29] See P. G. Adams, *Travelers and Travel Liars, 1660–1800* (Berkeley, Cal., 1962) – Addison's rebuff is quoted at p. 44; N. Broc, *La géographie des philosophes: géographes et voyageurs français au XVIIIe siècle* (Paris, 1975); bk 4, 'Les enigmes persistantes'; and Sir Thomas Browne, *Pseudodoxia epidemica* (1646).

[30] E.g., A. L. Farley, 'Fact and fancy in mapping Northwest America to 1800', *Occasional Papers (Geographical, of the Canadian Association of Geographers, British Columbia Division)* III (1962), 27–35.

century the question of the globe's figure was still utterly open.[31] A theory of the earth's history – such as Thomas Burnet's *Telluris theoria sacra* (1681–) – might still make no mention of fossils. On other matters of fact – such as gauging the earth's heat, or charting the tides – accurate data were deficient. Small wonder there was no agreement as to a *theory* of (say) the nature of fossils, the origin of rivers, or the causes of the tides.[32] Lacking tight controls, theories of the earth mushroomed in tropical profusion.

Of course, dawning awareness of ignorance could become a light for dispelling darkness – though Robert Hooke rightly feared

> the number of Natural Histories, Observations, Experiments, Calculations, Comparisons, Deductions, and Demonstrations necessary thereunto seem to be incomprehensive and numberless . . .[33]

Natural philosophers did, indeed, set out to explore the world around them. Edmond Halley spent his life investigating tides, magnetic fields, winds, currents, air pressure, oceanic depths, terrestrial rotation, the earth's age, its figure, the column of the atmosphere, and many other dimensions of the environment. Hooke, Huygens, Mariotte and Marsigli – to name a few – were no slouches.[34] Yet what is perhaps most striking about their age – in contrast to that of such successors a century later as de Saussure, Cuvier, or Humboldt – is the halting, fragmentary condition of specialized research, stymied at bottom by lack of secure data and by limitations in experimental techniques. Thus physical obstacles baulked measurements of the compo-

[31] V. Bialas, *Der Streit um die Figur der Erde: Zur Begründung der Geodäsie im 17. and 18. Jahrhundert* (Munich, 1972); M. W. Makemson, 'Old and new beliefs about the earth's figure', *Proceedings of the 10th International Congress of the History of Science*, II (1962: publ. 1964), 1027–9; D. H. Hall, *History of the Earth Sciences During the Scientific and Industrial Revolutions* (Amsterdam, 1976), 121ff.

[32] On fossils, see Rudwick, *Meaning of Fossils* (see note 24); H. K. Butler, 'The study of fossils in the last half of the seventeenth century' (Ph.D. thesis, University of Oklahoma, 1968); H. Hölder, *Geologie und Paläontologie in Texten und ihren Geschichte* (Freiburg and Munich, 1960). On rivers see Tuan, *The Hydrologic Cycle* (see note 8), and A. K. Biswas, *History of Hydrology* (Amsterdam, 1972). On tides see M. Deacon, *Scientists and the Sea, 1500–1900* (1971), chs. 2–5.

[33] Robert Hooke, *Posthumous Works*, ed. R. Waller (1705), 279.

[34] E.g., for Halley, see A. Armitage, *Edmond Halley* (1966); N. J. W. Thrower, 'Edmond Halley and thematic geo-cartography', in C. J. Glacken and N. J. W. Thrower, *The Terraqueous Globe* (Los Angeles, 1969); Simon Schaffer, 'Halley's atheism and the end of the world', *NR*, XXXII (1977), 17–40; A. K. Biswas, 'Edmond Halley, F.R.S., hydrologist extraordinary', *NR*, XXV (1970), 47–57; C. A. Ronan, 'Edmond Halley and early geophysics', *Geophysical Journal of the Royal Astronomical Society*, XV (1968), 241–8.

sition and temperature of the deep-sea bed, while the technical and conceptual problems of representing three-dimensional qualities on paper impeded development of visual languages, such as maps and sections, in several fields of environmental science.[35]

III. European civilization changes the environment

The very concept of 'environment' already construes nature through the humans who inhabit, destroy, transform, improve, pollute and hymn it. 'Environment' is a human not a natural category. The very same age (from about 1470 onwards) which saw Europeans discover, explore, and dominate other civilizations, their territory and resources, is also that of the genesis of modern science, including the sciences of the *oikoumene*. These two developments were mutually – though often obliquely – reinforcing. 'The entire face of the Earth bears today the stamp of the power of man', adjudged Buffon.[36]

Yet remarkably few historians have probed the links between political and cognitive imperialism, between the physical conquest of the globe and its mental appropriation. This applies both to Europe's control of the wider world, and to the march of economic dominance over tracts of Europe hitherto wild. The leading political, economic and military historians of colonialism have disappointingly little to say on its scientific dimensions.[37] Intellectual and literary historians, for their part, have chiefly evaluated white man's contact with strange *cultures*.[38] And European historians of science have left the meaning

[35] For obstacles to sea science see Deacon, *Scientists and the Sea* (see note 32); on visual representations, cf. Martin J. S. Rudwick, 'The emergence of a visual language for geological science, 1760–1840', *HS*, xiv (1976), 149–95, and the literature there cited. For break-throughs in instrumentation see W. E. K. Middleton, *The History of the Barometer* (Baltimore, 1964); G. R. de Beer, 'The history of the altimetry of Mont Blanc', *AS*, xii (1956), 3–29; J. A. Wolter, 'The height of mountains and the length of rivers', *Quarterly Journal of the Library of Congress*, xxix (1972), 187–205.

[36] Quoted by C. J. Glacken in 'Count Buffon on cultural changes of the physical environment', *Annals of the Association of American Geographers*, l (1960), 1–21 (p. 10).

[37] The following deal admirably with Western expansion, but say little about transformed scientific perspectives: J. H. Parry, *Europe and a Wider World, 1415–1715* (London, 1949); F. Mauro, *L'Expansion européenne 1600–1870* (Paris, 1964); I. M. Wallerstein, *The Modern World System* (New York, 1974); F. Braudel, *Capitalism and Material Life, 1400–1800* (English trans., 1974). A partial exception is G. Williams, *The Expansion of Europe in the Eighteenth Century: Overseas Rivalry, Discovery and Exploitation* (London, 1966), which investigates exploration in the context of national rivalry.

[38] For instance, B. Smith, *European Vision and the South Pacific* (Oxford, 1960); B.

of exploration and discovery to others: chiefly to those nautical bibliophiles who so scrupulously edit voyages and biographize their captains.[39] Marxism has hitherto contributed little.[40] As so often, the most valuable insights have come from maverick scholars riding their own hobby-horses.[41]

Enlightenment society was unprecedentedly aggressive towards the environment. Science's mediating role in that surge of Europocentric expansion can be examined under four categories.

Exploration and appropriation

Under the Enlightenment's Baconian banner of *Plus Ultra*, physical and intellectual exploration complemented each other. A prime scientific goal was to discover and docket all that existed within the *orbis terrarum;* the tasks of locating, describing, and charting were to the fore. Thus, during the century, the etherial 'Terra Australis nondum cognita' turned into the Australia and New Zealand, precisely mapped by Captain Cook and his successors.[42] North America was crossed for the first time – by MacKenzie in 1790–1 and by the Lewis and Clark expedition (1804–6) – which opened an era of well-equipped scientific

Bissell, *The American Indian in English Literature of the Eighteenth Century* (New Haven, 1925); P. D. A, Curtin, *The Image of Africa: The British and Action, 1780–1850* (1965); K. George, 'The civilized West looks at primitive Africa', *Isis,* XLIX (1958), 62–72.

39 E.g., Hakluyt Society publications (for specific references, see below, section V). Grateful as one is for the immaculate scholarship of these editions, it must be said that few show any sustained interest in science, or in the perceptual problems which *have* been addressed in the 'What did Columbus find?' debate. The situation is worse still amongst Anglo-American histories of geography, which are still deplorably antiquarian and stereotyped, rarely showing the conceptual sophistication of historians of ideas, or the technical expertise of historians of science.

40 See note 7. Hall's *History of the Earth Sciences* (see note 31) has commendably viewed scientific exploration as a function of Western power, using the Marxist perspectives of G. J. Crowther. Unfortunately its scholarship is third-hand and frequently erroneous. K. A. G. Mendelssohn's *Science and Western Domination* (1976) has little on eighteenth-century expansion. A more penetrating Marxist account – though over-schematic – is W. Leiss, *The Domination of Nature* (New York, 1972).

41 As well as the works listed in notes 9–12, see Lewis Mumford, *The Culture of Cities* (1938), and Francis Klingender, *Art and the Industrial Revolution* (1947), which is highly suggestive about science as well.

42 J. C. Beaglehole, 'Pacific exploration before Cook', *Endeavour,* XXVII (1968), 18–22; *idem,* 'Eighteenth century science and the voyages of discovery', *New Zealand Journal of History,* III (1969), 107–23; A. Sharp, *The Discovery of Australia* (Oxford, 1963); J. D. Mack, *Matthew Flinders, 1774–1814* (Melbourne, 1966).

exploration.[43] Ever more detailed instructions were issued to travellers: La Peyrouse received sixteen pages of them from the Académie des Sciences.[44] And scientific aspects of exploration were accorded more respect. Cook's *Endeavour* was described as a cross between a fortress and a travelling laboratory.[45]

Within nations the story is similar. Mines bureaux were set up in nations such as France and Sweden (Bergskollegium) to compile inventories of natural resources.[46] Naturalists like Pallas were commissioned by enlightened absolutists to map terrain and discover mineral riches. The first mineralogical maps were compiled.[47]

Fresh knowledge needed new frameworks of intelligibility to contain it. Thus experience of near-simultaneous earthquakes and volcanic eruptions suggested theories of seismic wave propagation, and of central heat.[48] The proliferation of varieties of minerals and rocks prompted attempts to classify them according to genera and species as with flora and fauna; to order collections on geographical as well as physical and chemical principles; and – in the case of Werner – to posit the reality of quasi-universal geognostical formations.[49] A key attraction of the Universal Deluge lay in explaining geological similarities the world over.

European colonization demanded new scientific acumen in many departments. Thus exotic climates and tropical diseases had to be studied, and the adaptability of crops to soil conditions gauged;

[43] E.g., C. I. Wheat, *Mapping the Trans-Mississippi West 1540–1861* (2 vols., San Francisco, 1958); W. P. Cumming, *The Exploration of North America, 1630–1776* (1974).

[44] Smith, *European Vision* (see note 38), ch. 2; George W. Stocking, Jr, 'French anthropology in 1800', *Isis*, LV (1964), 13–150.

[45] See G. M. Badger (ed.), *Captain Cook, Navigator and Scientist* (1970).

[46] T. Frängsmyr, 'Swedish science in the eighteenth century', *HS*, XII (1974), 29–42; F. B. Artz, *The Development of Technical Education in France, 1500–1850* (Cambridge, Mass., 1966).

[47] Rudwick, 'Visual language' (see note 35), and the literature there cited; Rhoda Rappaport, 'The geological atlas of Guettard, Lavoisier and Monnet', in Schneer (ed.), *Toward a History of Geology* (see note 22), 272–87.

[48] A Geikie, *Memoir of John Michell* (Cambridge, 1918); A. V. Carozzi. 'Rudolf Erich Raspe and the basalt controversy', *Studies in Romanticism*, VIII (1969), 235–50.

[49] For the use of Linnaean mineral classification see H. W. Scott (ed.), *John Walker: Lectures on Geology, including Hydrography, Mineralogy and Meteorology* (Chicago, 1966). For Werner, see A. M. Ospovat, 'The place of the *Kurze Klassifikation* in the work of A. G. Werner', *Isis*, LVIII (1967), 90–5; *idem*, 'The work and influence of Abraham Gottlob Werner: a re-evaluation', *Actes du XIII congrès international d'histoire des sciences* (1971; publ. 1974), VIII, 123–30; *idem*, 'Abraham Gottlob Werner and his influence on mineralogy and geology' (Ph.D. thesis, University of Oklahoma, 1960).

cameralist politics required statistics of human geography – population, land tenure, wealth. And science left its stamp, sometimes in conspicuous ways: maps were Europocentric. Personal, dynastic and national names given to locations reinforced ownership and control: thus Georgia, Georgetown, South Georgia, King George's Sound (=Nootka Sound) – even the *Georgium sidus* (William Herschel's original proposal for Uranus).[50] Dutch decline was endorsed when New Holland and New Amsterdam became Australia and New York. Science also made its mark by more subtle shifts of attitude. Thus Enlightenment science plumed itself in rolling back the clouds of ignorance, superstition and error, which had always shrouded exotic parts. Subsuming terrestrial features under natural law established intellectual overlordship, and made men good Stoics when nature wrought nemesis in such catastrophes as the Lisbon earthquake.[51]

Exploitation

Intellectual control of course went hand in glove with physical mastery. As the eighteenth-century economy steamed from manufacture into machinofacture, and from a wood-and-water to a coal-and-iron technology, it had to exploit vaster mineral resources. To meet these needs, several absolutist princes established mining schools, such as those at Freiberg and Schemnitz, and official corps of technocratic mineralogists and metallurgists.[52] In other industrializing nations, principally Britain, production of technical expertise was entrusted to market forces, and thus to private engineers, like John Smeaton and William Smith, with their empiricist methods. Research has yet to evaluate the cost-effectiveness of applying theoretical earth science to mineral exploitation. It may not have mattered: bureaucratic control and fiscalism were themselves major functions of mines academies.[53]

[50] See the conceit along these lines in the verses dedicatory to King James II in Robert Plot's *Natural history of Staffordshire* (Oxford, 1686),
Described Provinces his Empires greet
And throw their plenty at his glorious feet. (etc.)
For medical cameralism, see G. Rosen, *From Medical Police to Social Medicine* (New York, 1974).

[51] T. Kendrick, *The Lisbon Earthquake* (1956).

[52] See Martin Guntau, 'The emergence of geology as a scientific discipline', *HS*, XVI (1978), 280–90, for important bibliography on mining establishments in Central and Eastern Europe; also *Abraham Gottlob Werner: Gedenkschrift aus Anlass der Wiederkehr seines Todestages* (Leipzig, 1967); D. M. Farrar, 'The Royal Hungarian Mining Academy, Schemnitz. Some aspects of technical education in the eighteenth century' (M.Sc. thesis, University of Manchester, 1971).

[53] Roy Porter, 'The industrial revolution and the rise of the science of geology', in M. Teich and R. M. Young (eds.), *Changing Perspectives in the History of Science* (1973), 320–43.

Mining and quarrying were vital stimuli to earth science. Such operations' exposures revealed the superimposition of strata down perpendicular shafts, and also the dip and faults along horizontal galleries underground. Coal mining, in particular, indicated consistent order amongst bedded rocks, suggestive to geologists of the conditions of their formation during sedimentation, and to prospectors of the resources of the bowels of the earth.[54] Advancing knowledge, and use, of artesian wells likewise benefited both science and exploitation.[55] Interpretative histories of the theory and use of mineral waters, building stones, and canal technology, are badly needed.

Improvement

Believing that art perfects nature, and that he had a right to tailor nature for his own benefit, eighteenth-century man set about improving his natural surroundings: by extending towns; by building roads, canals, harbours, country houses; by landscape gardening; and perhaps above all through agriculture. Because (as Francis Bacon had stipulated) nature could be commanded only if first known and obeyed, knowledge of tides, climate, rivers, water supply, drainage, soil fertility, erosion, forestry, and many other forms of technical expertise were at a premium.[56]

The links between earth science and land surveyors and improvers are ill-understood and would be well-worth pursuing, as regards both famed pioneers like William Smith, Robert Bakewell and Humphry Davy, and the great body of practitioners. As is well known, in addition to his *Theory of the Earth* (1795) James Hutton left a manuscript treatise on agriculture. Publication of this would prove highly illuminating, for the idea of the circulation of terrestrial materials in his geology suggests crop rotation on a well-managed farm, and, of course, reminds one of Hutton's own early career in farming.[57]

Particularly in France, schools of thought developed that aimed to found a naturalistic science of the environment in relation to man considered as a progressive being, stressing the improvement of

[54] Cf. Joan M. Eyles, 'William Smith: Some aspects of his life and work', in Schneer (ed.), *Toward a History of Geology* (see note 22), 142–58.

[55] D. King-Hele, *Doctor of Revolution: The Life and Genius of Erasmus Darwin* (1977), 18, 19, 157–9.

[56] For discussion of the relations between science and technology in the industrial revolution, see the essay by D. S. L. Cardwell in the volume.

[57] See Sir E. B. Bailey, *James Hutton, the Founder of Modern Geology* (Amsterdam, 1967), for extracts from the MS, which is in the possession of the Royal Society of Edinburgh; Peter Eden (ed.), *Dictionary of Land Surveyors and Local Cartographers of Great Britain and Ireland, 1550–1850* (3 vols., 1975–6); and Morris Berman, *Social Change and Scientific Organization* (1978).

virgin lands, soil productivity, and the grasp of how climate, habitat and living conditions were key determinants of health. French *agronomes*, physiocrats, neo-Hippocratic doctors and 'Idéologues' have fortunately been the subject of impressive research.[58] We need to know much more about similar aspirations elsewhere.

Fundamental to almost all visions of the environment was belief in a divine plan, wherein through applying knowledge man could perfect himself and nature. Both for Christians like Jean André de Luc and for sceptics like Buffon the earth itself had progressed from uninhabitable chaos to temperate habitat, from rudeness to refinement. The current (seventh) age, that of tranquillity, was the age of man. Savages – like the North American Indians – languished through failing to escape domination by the environment. But civilized man could shape the environment to his needs, to the benefit of both – as Europe's superiority to America proved.[59]

Aesthetics

Enlightenment man took fresh delight in viewing the face of nature. Topophilia lay not just in its traditionally agreeable aspects (pastoral; 'Nature improved'; bucolic figures in a civilized landscape) but in new modes – in the Sublime, the Picturesque and then the Romantic.[60]

58 M. W. Rossiter, *The Emergence of Agricultural Science* (New Haven, 1975); A. J. Bourde, *Agronomie et agronomes en France au XVIIIe siècle* (Paris, 1967); Elizabeth Fox-Genovese, *The Origins of Physiocracy* (New York, 1976). English agrarian science has been studied in E. J. Russell, *History of Agricultural Science in Britain, 1620–1954* (London, 1966); G. E. Fussell, *Science and Practice in Eighteenth Century British Agriculture* (Berkeley, 1969). For Neo-Hippocratic medicine, the Idéologues on their roots in Montesquieu, see Glacken, *Traces on the Rhodian Shore* (see note 4), part 4; S. Moravia, 'Philosophie et géographie à la fin du XVIIIe siècle', *Studies on Voltaire and the Eighteenth Century*, LVII (1967), 937–1011; *idem*, *Il pensiero degli Idéologues* (Florence, 1974).

59 For the massive debate comparing European and American environment and civilization, see G. Chinard, 'Eighteenth century theories of America as a human habitat', *Proceedings of the American Philosophical Society*, IXC (1941), 27–58; P. Boerner, 'The images of America in eighteenth century Europe', *Studies on Voltaire and the Eighteenth Century*, CLI (1976), 323–32; H. W. Church, 'Corneille de Pauw and the controversy over his *Recherches philosophiques sur les Américains*', *Proceedings of the Modern Language Association*, LI (1936), 178–206; A. Gerbi, *The Dispute of the New World* (English trans., Pittsburgh, 1973). The pivotal source is Buffon: see Glacken, *Traces on the Rhodian shore* (see note 4), ch. 14, and Roger, *Epoques* (see note 19).

60 The minimum introduction would be M. H. Nicolson, *Newton Demands the Muse* (Princeton, N.J., 1946), and *Mountain Gloom* (see note 9); S. H. Monk, *The Sublime: A Study of Critical Theories in XVIII Century England* (New York, 1935); W. J. Bate, *From Classic to Romantic: Premises of Taste in Eighteenth Century England* (Cambridge, Mass., 1946); W. J. Hipple, *The Beautiful, the Sublime*

Such tides of taste were in dialogue with intense scientific debate as to the essence of beauty, the psychology of form and colour, association and evocation.[61] Scientific poetry, buttressed with copious scholarly apparatus, fused all these concerns in one genre.[62]

It is hardly surprising, then, that naturalists unashamedly viewed the environment through aesthetic filters, such as order and disorder, regularity, symmetry, organic form. Furthermore, such value-laden concepts stood as yardsticks of *scientific* truth. Thus, for Lamarck or Hutton, any geological system which construed the earth as a jagged ruin, formless and decaying, was blind to the beauty of its deeper, lawful operations, and was shallow science. For them, as for Kant, science was the enterprise which drew out the intelligibility of the world. The beautiful was one mode of the rational; he who was blind to beauty would see no system in things.[63]

Criteria of beauty and order, economy and simplicity, proportion and perspective were integral to conceptualizing the environment. Thus Dennis Dean has dubbed geology the Romantic mode of scientific perception of the earth.[64] Rival aesthetics were heavy artillery in the war between Christian and Deist natural theologies, and in the pantheistic naturalism of *philosophes* such as d'Holbach, Helvétius, and Boulanger. Ernst Cassirer argued for the cardinality of the aesthetic dimension in his *The Philosophy of Symbolic Forms* (1953–7), and this sensitivity to aesthetic criteria informs his *Philosophy of the Enlightenment* (1955). It is a black hole in the scholarship that these leads have hardly been followed up.

IV. Social and institutional dimensions

The social and institutional basis of the growth of science has been – till recently – notoriously neglected. This applies more to the

and the Picturesque in Eighteenth Century British Aesthetic Theory (Carbondale, Ill., 1957); C. Hussey, *The Picturesque* (1967); E. Sewell, *The Orphic Voice: Poetry and Natural History* (New York, 1960); A. O. Lovejoy, 'Nature as an aesthetic norm', in his *Essays in the History of Ideas* (Baltimore, 1948), 69–77; G. S. Rousseau, 'Literature and science: the state of the field', *Isis*, LXIX (1979), 583–91.

61 Tuan, *Topophilia* (see note 8); D. Lowenthal, 'Past time, present place: Landscape and memory', *Geographical Review*, LXV (1975), 1–36; J. Hunt, *The Figure in the Landscape* (Baltimore, 1976).

62 W. P. Jones, *The Rhetoric of Science* (1966); *idem*, 'The vogue of natural history in England, 1750–70', *AS*, II (1937), 345–52.

63 See for example, Porter, *Making of Geology* (see note 2), 191ff.

64 D. R. Dean, 'Geology and English literature. Crosscurrents, 1770–1830' (Ph.D. thesis, University of Wisconsin, 1968).

eighteenth than to later centuries, and particularly to the amorphous body of investigators of the terrestrial environment.[65] More *information* is becoming available, through well-researched biographies and studies of institutions such as the Académie Royale des Sciences. But few scholars have as yet availed themselves of tools such as prosopography and addressed themselves to the question of naturalists *as a community*.[66]

What were the official channels through which public investigation of the terraqueous globe was organized? These were rarely the abodes of traditional humanistic learning: university sponsorship was a development of the following century.[67] Neither was the eighteenth century the hey-day of private clubs of specialist investigators: the particularly British surge in geographers', explorers', and climbers' clubs was a nineteenth-century development, triggered by the founding of the Geological Society of London in 1807. Rather, the typical eighteenth-century institution was attached to the state, and implemented royal policy. Some of these were general scientific bodies acting at a distance as offices of state, such as the Académie Royale des Sciences, the Berlin Academy, the St Petersburg Academy, the Bologna Academy. Some were highly subject-specific, such as the Ecole des Mines (1783). Some were government bureaux in the strict sense, such as the Ecole des Ponts et Chaussées (1775). The British government set up the Board of Longitude and the Ordnance Survey (formally constituted in 1791) to meet the needs of military surveying

[65] For samples of fruitful work done on equivalent nineteenth-century problems see (e.g.) J. G. O'Connor and A. J. Meadows, 'Specialization and professionalization in British geology', *SSS*, VI (1976), 77–89; P. D. Lowe, 'Amateurs and professionals: the institutional emergence of British plant ecology', *JSBNH*, VII (1976), 517–35.

[66] For head-counts, see Steven Shapin and Arnold Thackray, 'Prosopography as a research tool in history of science: the British scientific community, 1700–1900', *HS*, XII (1974), 1–28; L. Pyenson, '"Who the guys were": prosopography in the history of science', *HS*, XV (1977), 155–88. For social analysis see D. E. Allen, *The Naturalist in Britain* (1976), and *idem*, 'Natural history and social history', *JSBNH*, VII (1976), 509–16, which raises key issues.

[67] There were, of course, exceptions, stimulated by particular individuals. Kant was one of few eighteenth-century professors to lecture on geography (the first chair was founded in 1820, at Berlin, for Ritter). Linnaeus sent out his 'apostles' – such as Solander and Sparrman – from Uppsala: see S. Lindroth, *A History of Uppsala University, 1477–1977* (Uppsala, 1976), ch. 4. Even where universities flourished (e.g., Scotland) investigation of the *oikoumene* might be marginal: 'The two hundred year gap between Varenius and Humboldt remains a dark age in the history of academic geography.' Alan Downes, 'The bibliographic dinosaurs of Georgian geography (1714–1830)', *Geographic Journal*, CXXXVII (1971), 379–87 (p. 379).

– though in general it preferred offering prizes (e.g., for discovering the North West Passage) to establishing boards. From 1776 the United States urgently mapped its frontiers through the General Land Office. In all such institutions mercantilist projects such as surveying, mapping, compiling inventories of natural resources, evaluating performance in agriculture, forestry, mining, quarrying, canal and road building were priorities.[68] Oceanic voyages of discovery were utterly dependent – for ships, crews, equipment, finance and diplomatic clearance – upon state funding.[69]

Of course, national investment in science fell short of aspirations. Thus the great series of maps of France produced by Cassini de Thury was launched with state funds, but these were withdrawn, and it had to be completed through raising private finance. Yet funding was notably more generous than in previous centuries, indicating greater political awareness of the utility of scientific knowledge. John Ellis remarked of Cook's *Endeavour* voyage, 'No people ever went to sea better fitted out for the purpose of Natural History', and, looking back, Cuvier – always eagle-eyed for state patronage – judged, 'This voyage deserves to be noted as forming an epoch in the history of science.'[70]

Employment of men of science for national purposes constantly sparked tensions between savant and paymaster, between the ethos of individual science (private property within a free market in knowledge), and national advantage. Navigators like Cook travelled with secret agenda from their governments. To prevent intellectual privateering, officers had to yield up their logs for scrutiny. J. R. Forster was banned by the Admiralty from publishing his private account of Cook's voyage. On board ship, the goals of navigation and

[68] See especially R. Hahn, *The Anatomy of a Scientific Institution: The Paris Academy of Sciences, 1666–1803* (Berkeley, 1971), and for geography at the Académie, Broc, *La géographie* (see note 29), 16ff; R. Calinger, 'Frederick the Great and the Berlin Academy of Sciences (1740–1766)', *AS*, xxiv (1968), 239–50; Sir A. Day, *The Admiralty Hydrographic Service, 1795–1919* (1967); J. B. Harley, 'The Society of Arts and surveys of English counties, 1795–1809', *Journal of the Royal Society of Arts*, cxii (1964), 43–6, 119–24, 269–75, 538–43; W. R. Mead, 'The eighteenth century military reconnaissance of Finland. A neglected chapter in the history of Finnish cartography', *Acta Geographica Universatis Comenianae*, xx (1968), 255–71; R. A. Skelton, 'The military survey of Scotland 1747–55', *Scottish Geographical Magazine*, lxxxiii (1967), 5–16; R. Rappaport, 'Guettard, Lavoisier and Monnet: geologists in the service of the French monarchy' (Ph.D. thesis, Cornell University, 1964); L. Aguillon, 'L'ecole des Mines de Paris: notice historique', *Annales des mines*, 8th ser., xv (1889), 433–686.

[69] E.g., for the financing of Cook's second voyage, see M. Hoare, *The Tactless Philosopher: Johann Reinhold Forster, 1729–98* (Melbourne, 1976), 73f.

[70] Quoted in Smith, *European Vision* (see note 38), 36.

discovery clashed with natural history, as Forster endlessly bewailed. 'Had Cook paid the same attention', wrote Banks, 'to the Naturalists [as Flinders later did] we should have done more at the time'.[71]

Many issues are raised here. How did men of science see their missions? How serious were clashes over intellectual property? Did secrecy squeeze science? Did naturalists writhe in the Kameralwissenschaft straitjacket, or glory in bureaucratic status? These issues await systematic study.[72]

Such tensions came to the boil in trans-oceanic explorations, where national competition was endemic. The trumpeted 1761 and 1769 observations of the transits of Venus instance – Faivre's phrase, 'international Venusian year' notwithstanding[73] – national scientific coexistence rather than cooperation. Sir Gavin de Beer claimed that eighteenth-century governments elevated the freedom of science above narrow diplomatic squabbles, but this internationalist view has been challenged as idealizing both scientists and states. The pursuit of knowledge was never far isolated from national and commercial profit.[74]

For all that, was not the level of state investment in, and patronage of, science for mercantilist purposes extremely paltry – especially when contrasted with outlays on cultural luxuries like building and the arts, and with the greater open-handedness in the next century? Endowment of exploration of the physical environment tests the hypothesis that science was a key function of state policy under the *ancien régime*. In any case, the relations between science and absolutism need much further investigation.[75]

The vigour of individual, 'amateur' involvement in the environmental sciences at this time is impressive, as indeed it continued to be

[71] Hoare, *The Tactless Philosopher* (see note 69), 102.

[72] J. P. Carswell, *The Prospector: being the Life and Times of R. E. Raspe, 1737–94* (1950), shows how becoming a functionary could offer a leg-up to an aspiring young scholar, while eventually frustrating the desire for scientific fame.

[73] J.-P. Faivre's roseate 'Savants et navigateurs: un aspect de la co-opération internationale entre 1750 et 1840', *Journal of World History*, x (1966), 98–124; H. Woolf, *The Transits of Venus* (Princeton, N.J., 1959).

[74] Contrast G. R. de Beer, *The Sciences were Never at War* (1960), with A. Hunter Dupree, 'Nationalism and science, Sir Joseph Banks and the wars with France', in D. H. Pinkney and T. Ropp (eds.), *A Festschrift for Frederick B. Artz* (Durham, N.C., 1964), 37–51; S. L. Chapin, 'Scientific profit from the profit motive: the case of the Lapeyrouse expedition', *Actes du Congrès Internationale d'Histoire des Sciences* (1968; publ. 1971), xi, 45–9; J. Beaglehole, *The Exploration of the Pacific* (1966), 233ff.

[75] For timely warnings against proleptically inflating state investment in 'professional' careers in science see Roger Hahn, 'Scientific careers in eighteenth century France', in Maurice Crosland (ed.), *The Emergence of Science in Western Europe* (1975), 127–38.

through the next century.[76] The greatest explorers – from de Saussure and Dolomieu, through to von Humboldt and Charles Darwin – were all privately motivated and self-financing, and, needless to say, affluent. Joseph Banks, who lavished perhaps £10,000 on Cook's first expedition, and was lifelong a generous patron of science, was a landed gentleman of £6,000 per year.[77]

Gentlemanly science has not aroused the same attention as professionalization. That environmental and field sciences have tended to remain amateur, in fact and in ethos, till quite recent times, is well known. But we need studies of how and why significant numbers of leisured gentlemen came to set themselves lifelong scientific projects; how the career of science in many cases pre-dated the remunerated profession of it.[78]

Partial answers suggest themselves. Field sciences united the delights of the Grand Tour, the hunt, military campaigns and landed proprietorship.[79] Yet the natural history of the gentleman traveller must be pressed further. Did the deeply personal motivations of men like de Luc create idiosyncratic disciplines? How far did economic security enable such travellers as Sir James Hall, or speculators like Erasmus Darwin, to stand outside the intellectual controls of Church and academy? Did nobiliary naturalists produce a 'white man's science', naturalizing the political outlooks of the governing orders?[80]

[76] For doubts about use of the term 'amateur', see Susan F. Cannon, *Science in Culture* (Folkestone, 1978).

[77] See H. C. Cameron, *Sir Joseph Banks: the Autocrat of the Philosophers, 1744–1820* (1952); P. Tunbridge, 'Jean André de Luc, F.R.S. (1727–1817)', *NR*, XXVI (1971), 15–33; D. W. Freshfield, *The Life of Horace Benedict de Saussure* (1920); A. Lacroix, *Deodat Guy Silvain Tancrede Gratet de Dolomieu, 1750–1803* (2 vols., Paris, 1921); Sir A. Geikie (ed.), *Faujas de St Fond: A Journey through England and Scotland to the Hebrides in 1784* (2 vols., Glasgow, 1907); B. Fothergill, *Ambassador Extraordinary: a Life of Sir William Hamilton* (1969).

[78] But see O. Sonntag, 'The motivations of the scientist: the self-image of Albrecht von Haller', *Isis*, LXV (1974), 336–51; David Knight, 'The scientist as sage', *Studies in Romanticism*, VI (1966), 65–88; *idem*, 'Science and professionalism in England, 1770–1830', *Proceedings of the XIV International Congress of the History of Science* (Tokyo, 1971), I, 53–67; Dorinda Outram, 'The language of natural power: the "éloges" of Georges Cuvier and the public language of nineteenth century science', *HS*, XVI (1978), 153–78; Roy Porter, 'Gentleman and geology: the emergence of a scientific career, 1660–1920', *Historical Journal*, XXI (1978), 809–36.

[79] See the penetrating article by Morris Berman, '"Hegemony" and the amateur tradition in British science', *Journal of Social History* (Winter 1975), 30–50, and Allen, *The Naturalist* (see note 66), chs. 1–3.

[80] For such ideas of a 'white-man's science', see M. Duchet, *Anthropologie et histoire au siècle des Lumières* (Paris, 1971); J. Burrow, *Evolution and Society* (Cambridge, 1966); G. Weber, 'Science and society in nineteenth century anthropology', *HS*, XII (1974), 260–83.

Or perhaps the reverse: was aristocratic naturalism a disdainful declaration of independence, a critical pastoral idyll vindicating nature against commercial culture?

The other striking milieu shift of Enlightenment environmental science lies in its popularization. Far greater numbers than ever before were exploring local, national and exotic terrains, seeking science, piety and pleasure. The terraqueous globe was brought to the public through sumptuous folio picture books, encyclopaedias, prints, museums, collections, landscape painting, itinerant lecturers, topographical poetry, maps, children's books and the like. Collections of voyages and travels enjoyed a staggering vogue.[81] Popular interest in the globe had tangible consequences, one being the emergence of a new scientific rank-and-file, supporting science careers by plying their trade to leisured consumers. But we know little as yet about what Steven Shapin in this volume has termed the 'social uses' of this literature. For readers of the Abbé Pluche's *Spectacle de la nature* or Gilbert White's *Natural History of Selborne* the romance of nature was probably a modulation of religiosity. But how far did such relations put down new roots in the soil for a socially and geographically mobile society? Did natural history medicate the diseases of urban industrialism? Did travelogues appeal to specific social groups? Were distinct bodies and styles of environmental science produced by and for different layers of the community? Answers are lacking, and they will take on their full meaning only when placed in a comparative context of different circumstances and different nations, which will impose adequate controls upon the conclusions to be drawn.

V. The growth of knowledge

The eighteenth century unleashed unprecedented exploration of the face of the globe and the bowels of the earth. Continents like North

[81] For France, D. Mornet, *Les sciences de la nature en France au XVIIIe siècle* (Paris, 1911) is a classic; R. Taton (ed.), *Enseignement et diffusion des sciences en France au XVIIIe siècle* (Paris, 1964); R. Pomeau, 'Voyages et lumières dans la littérature française du XVIIIe siècle', *Studies on Voltaire and the Eighteenth Century*, LVII (1967), 1269–89. For Britain, and mainly from a geological angle, see Porter, *Making of Geology* (see note 2), ch. 4; V. A. Eyles, 'The extent of geological knowledge in the eighteenth century, and the methods by which it was diffused', in Schneer, *Toward a History of Geology* (see note 22), 159–83; A. Frost, 'The Pacific Ocean – the eighteenth century's "new world"', *Studies on Voltaire and the Eighteenth Century*, CLII (1976), 779–882 (p. 284f.); P. G. Adams, *Travelers* (see note 29); G. R. Crone and R. A. Skelton, 'English collections of voyages and travels, 1625–1846', in E. Lynam (ed.), *Richard Hakluyt and his Successors*, Hakluyt Society 2nd ser., XCIII (1946), 65–140; L. J. Cappon, 'Geographers and mapmakers, British and American from about 1750 and 1789', *Proceedings of the American Antiquarian Society*, LXXXI (1971), 243–71.

America were newly trudged, the Pacific Ocean scoured, the strata delved into. Minute regional investigation went side-by-side with heroic circumnavigation.[82]

Original types of exploration were devised, such as the systematic scaling of peaks.[83] Virgin areas of the globe were visited: Cook was the first European to break the Antarctic Circle. New kinds of recordings were taken with novel instruments. Humboldt was to take to South America forty different kinds of instruments.[84]

[82] It is impractical to provide an account here of scholarship upon *regional* exploration. The more important literature upon global exploration includes Numa Broc, 'Voyages et géographie au XVIIIe siècle', *RHS*, XXII (1969), 137–54; E. Hennig, *Terrae incognitae, Eine Zusammensfassung und kritische Bewertung der witchtigsten vorcolumbischen Entdeckungseisen an Hand der daruber vorliegenden Originalberichte* (Leiden, 1956); G. Williams, *The British Search for the North West Passage in the Eighteenth Century* (1962); L. P. Kirwan, *The White Road: A History of Polar Exploration* (New York, 1969); G. R. Deacon, *Oceans, an Atlas History of Man's Exploration of the Deep* (1962); J. Dunmore, *French Explorers in the Pacific* (2 vols. Oxford, 1965–9); H. Friis (ed.); *The Pacific Basin: A History of its Geographical Exploration* (New York, 1967); R. I. Rotberg (ed.), *Africa and its Explorers: Motives, Methods, Impact* (Cambridge, Mass., 1970); A. Sharp, *The Discovery of Australia* (Oxford, 1963). On individual travellers see J. C. Beaglehole, *The Journals of Captain James Cook on his Voyages of discovery*, Hakluyt Society, Extra Series, nos. 34–36 (Cambridge, 1955–67); *idem*, *The Life of Captain James Cook* (1974); *idem*, *The 'Endeavour' Journal of Joseph Banks, 1768–71* (2 vols., Sydney, 1963); R. E. Gallagher (ed), *Byron's Journal of his Circumnavigation*, Hakluyt Society, 2nd series, no. 122 (Cambridge, 1964); A. M. Lysacht, *Joseph Banks in Newfoundland and Labrador, 1766: his Diary, Manuscripts and Collections* (1971); Helen Wallis (ed.), *Philip Carteret: Voyage Round the World, 1766–1769*, Hakluyt Society, 2nd series, nos. 124 and 125 (Cambridge, 1965); B. Anderson, *Surveyor of the Sea: the Life and Voyages of Captain George Vancouver* (Seattle, 1960); F. A. Golder, *Bering's Voyages* (2 vols., New York, 1922–5); J.-E. Martin-Allanic, *Bougainville, navigateur, et les découvertes de son temps* (Paris, 1964). It is a pity that so much of the scholarship of voyages remains superior narrative. So many questions remain to be investigated both about the perceptions of the navigators and the impact of their writing (for example, James Hutton the geologist read travel literature above all else). For a brief introduction to geological exploration see G. S. Tikhomirov, *Bibliograficheskii ocherk istoriigeografii Russu XVIII veka* (Moscow, 1968); G. A. White, 'Early American geology', *Scientific Monthly*, LXXVI (1953), 134–41; *idem*, 'Early geological observations in the American Midwest', in Schneer (ed.), *Toward a History of Geology* (see note 22), 415–25; and see Porter and Poulton, 'Research in British geology' and 'Geology in Britain' (see notes 3 and 13).

[83] G. R. de Beer, *Early Travellers in the Alps* (1930).

[84] L. de Voysey, Jr, 'Hydrography: a note on the equipage of eighteenth century survey vessels', *Mariners' Mirror*, LVIII (1972), 173–77. See also M. Archinard, 'De Luc et la recherche barometrique', *Gesnerus*, XXXII (1975), 235–47; G. R. de Beer, 'The history of the altimetry of Mont Blanc', *AS*, XII (1956), 3–29. It was not until the eighteenth-century improvement in instrumentation and techniques that Mont Blanc was recognized as the highest mountain in Europe.

All aspects of the physical environment were subjected to scrutiny: climate,[85] running waters,[86] earthquakes,[87] volcanoes,[88] cave systems,[89] the strata,[90] the oceans,[91] glaciers,[92] mountains,[93] the figure of the earth and terrestrial magnetism,[94] minerals and crystals.[95] Vast collections were amassed.[96] And, perhaps most strikingly, navigation

W. E. Knowles Middleton, *The History of the Barometer* (Baltimore, 1964); *idem, The Invention of the Meteorological Instruments* (Baltimore, 1969).

85 See Aubert de la Rue, *Man and the Winds* (New York, 1955); W. E. Knowles Middleton, *A History of the Theories of the Rain and Other Forms of Precipitation* (1966); A. K. Khrgian, 'The history of meteorology in Russia', *Actes du VII Congrès Internationale d'Histoire des Sciences*, I, (1956), 445–8; *idem, Meteorology: A Historical Survey* (Jerusalem, 1970); W. E. Knowles Middleton, 'Chemistry and Meteorology (1700–1825)', *AS*, XX (1964), 125–41; W. J. Remillard, 'The history of thunder research', *Weather*, XVI (1961), 243–53. Interpretative histories of almost all aspects of eighteenth-century meteorology, taken in the widest sense, are badly needed.

86 A. K. Biswas, *History of Hydrology* (Amsterdam, 1972).

87 E.g., A. V. Carozzi, 'Robert Hooke, Rudolf Erich Raspe and the concept of "earthquakes"', *Isis*, LXI (1970), 85–91; J. Taylor, 'Eighteenth century earthquake theories: a case history investigation into the character of the study of the earth in the Enlightenment' (Ph.D. thesis, University of Oklahoma, 1975).

88 E.g., G. R. de Beer, 'The volcanoes of Auvergne', *AS*, XVIII (1962), 49–62; A. Rudel, *Les volcans d'Auvergne* (Clermont Ferrand, 1962).

89 T. R. Shaw, 'A history of the scientific investigation of limestone caves to 1900' (Ph.D. thesis, Leicester University, 1975).

90 See for instance R. Rappaport, 'Lavoisier's geologic activities 1763–1792', *Isis*, LVIII (1967), 375–84; A. V Carozzi, 'Lavoisier's fundamental contribution to stratigraphy', *Ohio Journal of Science*, LXV (1965), 71–85.

91 Deacon, *Scientists and the Sea* (see note 32); S. Schlee, *The Edge of an Unfamiliar World: A History of Oceanography* (New York, 1973); C. P. Idyll, *The Science of the Sea: A History of Oceanography* (1970).

92 Cf. S. Thorarinson, 'Glaciological knowledge in Iceland before 1800, a historical outline', *Jokull*, X (1960), 1–17.

93 N. Broc, *Les montagnes vues par les géographes et les naturalistes de langue française au XVIII*^e^ *siècle* (Paris, 1969); G. L. Davies, 'The eighteenth century denudation dilemma and the Huttonian theory of the earth', *AS*, XXII (1966), 129–38; Nicolson, *Mountain Gloom* (see note 9).

94 H. Balmer, *Beiträge zur Geschichte der Erkenntnis des Erdmagnetismus* (Aarau, 1956); S. Chapman, 'Alexander von Humboldt and geomagnetic science', *Archive for the History of Exact Science*, II (1962), 41–51; H. L. Burstyn, 'Early explanations of the role of the earth's rotation in the circulation of the atmosphere and the ocean', *Isis*, LVII (1966), 167–87. For the journeys of La Condamine (1735) and Maupertuis (1736) to measure the arc of the meridian at different latitudes, see Hall, *History of the Earth Sciences* (see note 31), 125f.

95 J. G. Burke, *Origins of the Science of Crystals* (Berkeley, 1966); P. H. Groth, *Entwicklungsgeschichte der mineralogischen Wissenschaft* (Wiesbaden, 1970).

96 N. Broc, 'Un musée de géographie en 1795', *RHS*, XXVII (1974), 37–43; M. E. Jahn, 'The Old Ashmolean Museum and the Lhwyd Collections', *JSBNH*, IV (1966), 244–8; S. Bedini, 'The evolution of science museums', *Technology and*

was perfected and cartography dramatically improved.[97] New territories were charted, specialized maps – mineralogical and tidal – developed, and national mapping surveys instituted. Hachuring, spot heights, topographical symbols, and isogenes appeared. Geodesy developed; triangulation was perfected. The cartographical ignorance which on late-seventeenth-century maps had joined Australia to New Guinea, New Zealand, Tasmania and the 'Southern Continent' had been dispelled by the end of the century – thanks largely to Cook.[98]

The eighteenth century saw La Condamine sail down the Amazon, Carsten Niebuhr traverse unknown Arabia, and Bruce reach the source of the Blue Nile; it vastly extended the European's knowledge of his planet. Interpreting these developments obviously poses historiographical problems. One lies in gauging the relations between

Culture, VI (1965), 1–29; W. Campbell Smith, 'A history of the first hundred years of the Mineral Collection in the British Museum', *Bulletin of the British Museum (Natural History) Historical Series,* III (1969), 235–59; E. Laury, *Les cabinets d'histoire naturelle en France au XVIII*[e] siècle (Paris, 1931); D. Murray, *Museums: Their History and Their Use* (3 vols., Glasgow, 1904); D. Hudson, *A Social History of Museums* (1975).

[97] E. G. R. Taylor, *The Haven Finding Art* (1956); A. H. W. Robinson, *Marine Cartography in Britain: A History of the Sea-Chart to 1855* (Leicester, 1962); W. E. May, *A History of Marine Navigation* (Henley, 1973); S. L. Chapin, 'A survey of the efforts to determine longitude at sea, 1660–1760', *Navigation,* III (1952–3), 188–91, 242–9, 296–303. See also the article by Eric G. Forbes in this volume.

[98] For brief introduction to the massive scholarship on the history of cartography, see L. A. Brown, *The Story of Maps* (Boston, 1950); G. R. Crone, *Maps and their Makers* (1953); W. W. Ristow, *Guide to the History of Cartography* (Washington, 1973); *idem, A la carte: Selected Papers on Maps and Atlases* (Washington, 1972); *idem* and C. E. LeGear, *A Guide to Historical Cartography* (Washington, 1961); R. V. Tooley, *A Dictionary of Mapmakers, including Cartographers, Geographers, Publishers, Engravers, etc., from the Earliest Times to 1900* (1965); *idem, Landmarks of Mapmaking: An Illustrated Survey of Maps and Mapmakers* (Oxford, 1976); L. Bagrow, *History of Cartography* (1964); E. and D. S. Berkeley, *Dr John Mitchell, the Man who Made the Map of North America* (Chapel Hill, 1974); Karol Buczek, *Drieje kartografii polskiej od XV do XVIII wieku. Zarys analityczno-syntetczny* (History of Polish cartography from the 15th to the 18th century. An analytic-synthetic outline), (Wroclaw, 1963); W. P. Cumming, *British Maps of Colonial America* (Chicago, 1974); H. G. Fordham, *Studies in Carto-bibliography, British and French* (1969); R. A. Skelton, *Explorers' Maps, Chapters in the Cartographic Record of Geographical Discovery* (1958); *idem, County Atlases of the British Isles* (1964); N. J. W. Thrower, *Maps and Man: An Examination of Cartography in Relation to Culture and Civilization* (Englewood-Cliffs, N.J., 1972); R. W. Tooley, *Maps and Mapmakers* (1949). For geological maps, see Rappaport, 'Geological atlas' (see note 47); Rudwick, 'Visual language' (see note 35); V. A. Eyles, 'Mineralogical maps as forerunners of modern geological maps', *Cartographic Journal,* IX (1972), 133–5; R. Bush, 'The development of geological mapping in Britain, 1795–1825' (Ph.D. thesis, University of London, 1974).

the explorers' pre-programming through their novitiate in traditional concepts and practices, the observations they then made, and the gelling of these experiences into a new world-picture – or, to adapt Broc's phrase, the transition 'de voyages à géographie'. For example, Werner's teaching at Freiberg led his pupils to anticipate a regular and predictable order of strata-formations, and to distinguish ('genuine') volcanic productions from (supposedly aqueous) trap (= basalt). But, as is well-known, certain of those pupils, such as d'Aubuisson and Humboldt, abandoned those interpretations when they undertook fieldwork distant from Werner's native Saxony.[99] Others, like Robert Jameson, did not. How do we explain this divergence? Was it because Jameson was primarily a teacher? On the same issue, the British geological fieldworker Alexander Catcott is a revealing instance of one who *privately* experienced tensions between his over-arching theoretical and religious expectations and his field findings, but *publicly* suppressed or overcame them.[100]

Furthermore, the sheer quantity of newly available information threatened chaos. Buffon wrote, 'Ce globe immense nous offre à la surface des hauteurs, des profondeurs, des prairies, des mers, des marais, des fleuves, des cavernes, des gouffres, des volcans, et à la première inspection nous ne découvrons en tout cela aucune regularité, aucune ordre.'[101] Diversity had to be pulled into order. One visual solution lay in the development of what Cannon has termed 'iso-maps', from Halley, via Buache, through to Humboldt. Similarly, faced with evidence of the unexplained distinct location of diverse kinds of rocks, John Woodward had argued that they lay in the stratigraphical column in order of specific gravity. Yet fresh evidence, in other people's hands, despatched that theory smartly enough. This stimulated later attempts – for instance by Linnaeus – to ground a rock typology on the basis of species and genera, and subsequently Werner's historical/local grid for aligning the diversity of rocks in a historical column.[102]

[99] W. Coleman, 'Abraham Gottlob Werner vu par Alexander von Humboldt', *Suddhoffs Archiv*, XLVII (1963), 465–78.

[100] M. Neve and R. Porter, 'Alexander Catcott: glory and geology', *BJHS*, X (1977), 37–60.

[101] G. le C. Buffon, *Histoire naturelle*, I (1749), 97.

[102] Cf. Porter, *Making of Geology* (see note 2), 78f., 170f.; H. D. Hedberg, 'The influence of Torbern Bergman (1735–1784) on stratigraphy', in Schneer (ed.), *Toward a History of Geology* (see note 22), 186–93; C. St Clair, 'The classification of minerals: some representative systems from Agricola to Werner' (Ph.D. thesis, University of Oklahoma, 1966); and W. R. Albury and D. R. Oldroyd, 'From Renaissance mineral studies to historical geology in the light of Michel Foucault's *The Order of Things*', *BJHS*, X (1977), 187–215. For 'iso-maps' see Cannon, *Science in Culture* (see note 76).

One crucial consequence of the new deluge of information was that the *diversity* of terrestrial phenomena indelibly etched upon naturalists' imaginations nature's genuine complexity. It compelled them to abandon specious attempts to comprehend the globe within monocausal reductionist physical models. A constant theme of Buffon, and of popularizers like Oliver Goldsmith in his *History of the Earth and Animated Nature* (1774), is the presumption of naturalists who attempt to box the kaleidoscope of nature within the narrow compass of human reason and artificial classifications. In a similar vein of empiricist scepticism, Rudolf Erich Raspe's *Specimen historiae naturalis globi terraquaei* (1763) opened with a lengthy stipulation of the problems which a natural history of the globe could *not* resolve:

> 1. Whether and at what time the world was created
> 2. Whether something was created from nothing
> 3. Whether the physical world, pre-existing from eternity, finally took form by chance or rather by order of some absolute, ordaining, purposeful rational being, taking form in this marvellous contrivance which is called earth... [and so forth].[103]

I should suggest three tests for gauging the impact of new data. The first is to probe areas where new knowledge compelled the conceptualization of aspects of nature hitherto inchoate, leading to new problem areas, and perhaps sub-disciplines. Thus, up to the late seventeenth century investigators had known or cared little about the different beds of rock which made up the earth's crust.[104] Strata had had little *meaning*. During the eighteenth century field investigation across broad terrains proved that a great part of the crust was indeed stratified. Observers such as Lehmann, Arduino, Füchsel, Michell, and a little later Bergman and Werner, having established that strata lay in a regular order, began to interpret this as an index of the gradual and successive accumulation of sediments in the history of the ocean floor.[105] Great debates over the meaning of strata were still raging at the end of the century. The point is that a new scientific

[103] R. E. Raspe, *An Introduction to the Natural History of the Terrestrial Sphere*, trans. and ed. A. N. Iversen and A. V. Carozzi (New York, 1970), 1.

[104] See C. J. Schneer, 'The rise of historical geology in the seventeenth century', *Isis*, XLV (1954), 256–68; for Britain, see Porter, *Making of Geology* (see note 2), 57–61, 170–83. For fossils as a parallel case, see Rudwick, *Meaning of Fossils* (see note 24), chs 1 and 2.

[105] See G. Stegagno, *Il Veronese Giovanni Arduino e il suo contributo al progresso della scienza geologica* (Verona, 1929); B. von Freyberg, *J. G. Lehmann, 1719–1767* (Erlangen, 1955); R. Möller, 'Mitteilungen zur Biographie George Christian Füchsels', *Freiberger Forschungshefte*, ser. D, XLIV (Leipzig, 1963); for Werner, see note 49.

'space', stratigraphy, had been colonized, opening up earth history, with ramifications for the interpretation of other terrestrial features, such as fossils, and thence the history of life.

A parallel is the 'basalt controversy' which flared from the 1760s.[106] Growing acquaintance with rock masses adjacent to volcanoes sparked bitter controversy over the origins of basalt. 'Aqueous' and 'igneous' camps formed, with waverers trailing between both. In this case, the immediate, substantive problem was effectively settled within a generation. Yet it inexorably opened up further debate – whether *all* rocks were igneous in origin – which divided geology for far longer, and posed profound questions of the earth's origins.[107] In other words, fresh data were an irritant, provoking new problems, and new domains of conceptualization, rather than simply confirming or refuting old theories. Data are more dynamic than either empiricists or anti-empiricists sometimes allow.[108]

The second case is where new knowledge challenged the essentially static terraqueous globe envisaged by seventeenth-century geography and natural theology. The data could better be fitted into a world of change, process and dynamism. 'Thus succeeds Revolution to Revolution', pronounced Richard Joseph Sulivan in 1794, pondering the century's new knowledge of the globe.[109] The discovery of tropical fossils near the poles popularized theories of a slow tracking of climatic zones across the latitudes. Fossil and strata evidence caused belief in the decay and renovation of continents to become universal by the end of the century. Physical geographers, from de Maillet, through le Cat and Buffon, to Lamarck, postulated the gradual procession of the continents round the globe, propelled by

106 K. L. Taylor, 'Nicholas Desmarest and geology in the eighteenth century', in Scheer (ed.), *Toward a History of Geology* (see note 22), 339–56; C. S. Smith, 'Porcelain and Plutonism', in *ibid.*, 317–38; Carozzi, 'Raspe and the basalt controversy' (see note 48).

107 I. R. Proctor, 'The concept of granite in the later eighteenth century and early nineteenth century in relation to theories of the Earth' (M. Litt. thesis, Cambridge, 1973); H. H. Read, *The Granite Controversy* (1957).

108 Further examples could be listed. Alpine exploration opened theorizing about glaciation: cf. D. W. Freshfield, *The Life of Horace Benedict de Saussure* (1920). Mapping the South Seas suggested to Philippe Buache and his successors the notion that the earth was divided into 'land' and 'sea' hemispheres. Experience with highlands indicated uniformities in the topography of mountains and river valleys – e.g., the regularity of salient and re-entering angles: see K. L. Taylor, 'Natural law in eighteenth century geology: the case of Louis Bourguet', *Actes du XIII Congrès Internationale d'Histoire des Sciences* (1971; publ. 1974), VIII, 72–80.

109 R. J. Sulivan, *A View of Nature* (6 vols., 1794), II, 169.

orbital spin and tidal action. Disruptive evidence could be accommodated if the dimension of change were embraced.

Thirdly, as more facts accumulated from all quarters of the earth, understanding of processes and structures was more than ever conceived in respect of the globe *as a whole*. In his *Essai de géographie physique* (1756), Philippe Buache, 'legislateur de la géographie', suggested integrating global relief through the conception of drainage basins defined by mountain ranges. Philosophical geologists like Hutton proposed that homeostatic balance governed the constant net global distribution of land and sea. Indeed, the dialectic of land and sea formed the central conundrum of eighteenth-century theories of the earth – a fact too often neglected by historians of geology who interest themselves in the story of landmasses almost to the exclusion of oceans.[110] Study of the physical distribution of land and sea throughout the globe gradually led away from the Ptolemaic emphasis on coordinates of location towards a new physical geography; from *klimata* to climates.

'Globalization' set the terms for natural history. Malthusian demography premised one single closed world. Similarly, for Buffon, the relations between the Old and the New Worlds made sense within the history of the globe as a whole. His *Epoques de la nature* sought to articulate the content of the three kingdoms of nature, the Great Chain of Being, within the progressively unfolding history of the earth as an organized planet.[111] Linnaeus, although Buffon's greatest rival, is similarly to be understood as an encyclopaedist grappling with data from the whole earth by proposing an adequately capacious classification, and at the same time grasping the nature of terrestrial equilibrium.[112] It is a mark of the ferment of knowledge that Linnaeus and Buffon stood poles apart, and that both shifted outlook during their life's work.

And the same need to understand the dynamics of the earth as a whole fired the lifelong labours of Alexander von Humboldt in the first part of the nineteenth century, becoming the agenda of modern physical geography. Humboldt's programme supposed the interconnection of all things terrestrial in a harmony of nature. He attempted a precisely measured 'physique du monde', focused on the areal associations of physical and organic phenomena, seeking 'unity in diversity'.[113]

[110] One honourable exception is the work of A. V. Carozzi (see note 19).
[111] See Roger, *Époques* (see note 19).
[112] Cf. James Larson, *Reason and Experience* (Berkeley, 1971); and C. Linné, *L'équilibre de la nature*, trans. B. Jasmin and introd. C. Limoges (Paris, 1972).
[113] See, for instance, his studies of the relations of altitude to flora, or his use of isothermic lines on maps. D. S. Botting, *Humboldt and the Cosmos* (1973); H.

VI. Environment, ideas and ideology

The most fruitful recent thrust in the historiography of scientific ideas has been to reveal that theories of nature embody ideas which do not merely run ahead of the facts, but which resonate with a staggering range of (non-scientific) associations and analogies.[114] Images of nature encode multiform interests: political, sectional, personal, sexual. Multivalent language is the carrier of such ideological projections.[115] Thus historians of geography have shown how the very conceptualization of environmental space fixes subjective associations, as Immanuel Kant's own construction of the nature and limits of geographical knowledge had made abundantly clear.[116] Similarly, the very concept of 'geology' is culture-relative, an historical construct.[117]

Because so much scholarship has been biography- or discovery-oriented, historians have rather neglected those *filters* through which the terraqueous globe is seen – filters inevitably potent in the language, even if not in the full consciousness of the theorists. Which are the chief ideological resources, to whose use eighteenth-century environmental historians need to be alert?

The first is surely the forging of pervasive naturalism.[118] The tenet

Beck, *Alexander von Humboldt* (2 vols, Wiesbaden, 1959); Hans Baumgärtel, 'Alexander von Humboldt: remarks on the meaning of hypothesis in his geological researches', in Schneer (ed.), *Toward a History of Geology* (see note 22), 19–35. For an excellent recent characterization of 'Humboldtian science' (as the 'science of measuring world-wide variables') see Cannon, *Science in Culture* (see note 76), ch. 3. The most interesting figure in many ways is Lamarck. Cf. R. W. Burkhardt, Jr, *The Spirit of System: Lamarck and Evolutionary Biology* (Cambridge, Mass., 1977). Many naturalists of the second half of the eighteenth century sought to encompass all the sciences of the physical environment, aiming to integrate all features of the globe. Torbern Bergman, James Hutton, Jean André de Luc, H. B. de Saussure, J. A. Pictet, and Richard Kirwan come readily to mind.

114 For theoretical dimensions see the works of Barnes and Bloor (see note 12), Shapin's essay in this volume, and Barry Barnes and Steven Shapin (eds.), *Natural Order: Historical Studies of Scientific Culture* (Beverley Hills and London, 1979).

115 For a relevant study of language as a value carrier, see Martin J. S. Rudwick, 'Poulett Scrope on the volcanoes of Auvergne', *BJHS*, VII (1974), 205–42.

116 J. A. May, *Kant's Concept of Geography and its Relation to Recent Geographical Thought* (Toronto, 1970); D. Lowenthal and M. J. Bowden (eds.), *Geographies of the Mind* (1976).

117 Porter, *Making of Geology* (see note 7).

118 For general discussions of Englightenment naturalism, see C. Kiernan, *The Enlightenment and Science in Eighteenth Century France*, rev. edn (Banbury, 1973); A. O. Lovejoy, 'The parallel of Deism and Classicism', in his *Essays in the History of Ideas* (Baltimore, 1948), 78–98; *idem*, 'Nature as an aesthetic norm', *Modern Language Notes*, XLII (1927), 44–50; J. Ehrard, 'L'evolution de

that the system of the earth is ordered by nature is not necessarily antipathetic to religious providentialism, teleology, catastrophism, or anthropocentrism. Indeed, eighteenth-century naturalism endorsed many of these commitments.[119] But for most *philosophes,* such commitments had to be mediated through, and sanctioned by, an overriding regulative conception of nature: a nature which worked through laws; which was simple, economical and made no waste; which could do no wrong; which engineered the good of the whole through the subordination of the parts; a nature which was, above all, uniform, organized and integrative. As John Greene has implied, despite all their ambiguities and tendentiousness, Enlightenment conceptions of nature as a self-sufficient system, 'capable de tout', endorsed a vision of the earth's constitution as autonomous, a self-sustaining organization of matter in motion, of process and change.[120] What historians now need to explore are precisely these ambiguities: that is to say, the values being mediated by naturalism. Is it correct – as has been suggested – that catastrophic visions of nature have encoded both Christain commitments, and, in cases such as Diderot and Boulanger, radical values, whereas uniformitarian ideas have embodied deist affirmation of gradual progress?[121] Was anti-anthropocentric naturalism a strategy for securing the independence of man from kings and priestcraft?[122]

Once the notion of the self-contained *systema naturae* became established, it exerted wide sway. It encouraged concepts of cycles, symmetry and equilibrium.[123] What Beaglehole has called 'a classical ar-

'idée de nature du XVIe au XVIIIe siècle', *Revue metaphysique et morale* (1953), 108–29; E. R. Wasserman, 'Nature moralized: the divine analogy in the eighteenth century', *English Literary History,* xx (1953), 39–76; J. Arthos, *The Language of Natural Description in Eighteenth Century Poetry* (Ann Arbor, 1949); A. Vartanian, *Diderot and Descartes* (Princeton, N.J., 1953).

[119] Though for tensions, see R. Hooykaas, *Natural Law and Divine Miracle* (Leiden, 1963).

[120] Greene, *The Death of Adam* (see note 24).

[121] Cf. Kiernan, *The Enlightenment and Science* (see note 118); Roy Porter, 'Charles Lyell, uniformitarismo, e l'attegiamento del secolo XIX verso la geologia dell'illuminismo', in A. Santucci (ed.), *Eredità dell'illuminismo* (Bologna, 1979), 395–433.

[122] Roy Porter, 'Philosophy and politics of a geologist: George Hoggart Toulmin (1754–1817)', *JHI,* xxxix (1978), 435–50.

[123] For such ideas in Hutton, see Dott, 'James Hutton' (see note 22). For the deeper levels of cyclicalism see Eliade, *Eternal Return* (see note 12). For quasi-mystical ideas of nature in Linnaeus, see K. Rob. V. Wikman, *Lachesis and Nemesis: Four Chapters on the Human Condition in the Writings of Carl Linnaeus* (Stockholm, 1970), esp. ch. 4.

chitecture of geography' sprang to mind, in which the oceans were presumed as deep as the mountains were high, the New World balanced the Old, and (in the hopes of Buache, des Brosses and Dalrymple) a Southern Continent must counterpoise the Northern. Viewing nature as a self-recycling system tended to eclipse interest in the original creation of the earth, or its future destruction. The millennium ceased to be a topic for mainstream earth science, though notions of gradual progress were, of course, linguistically and conceptually integrable with naturalism.[124]

Naturalism established the earth as an independent object, valid in its own right. On the one hand, naturalists became predisposed to construe the globe as relatively independent of God. Transcendental and miraculous explanations lost favour, bowing to natural law. Even Christian geologists abandoned the view that the earth had been *recently* created by God and was facing imminent destruction. Likewise, seating Hell at the earth's core entirely died out among educated opinion – a facet of the general 'disenchantment' of nature.[125]

At the same time, naturalism meant that the terraqueous globe came to be regarded as relatively independent of men. Acceptance of its high antiquity gave the earth an aeons-long existence before ever being trodden by humans. Similarly, environmental determinists out of Montesquieu's stable showed how man – in complex and oblique ways – was nature's child as well as her master.[126] More radical Enlightenment pantheists and atheists such as d'Holbach, Helvétius and Toulmin asserted the earth's eternity and deemed man no more than a 'small but necessary part' of the terrestrial economy.[127] For Buffon, man was locked in a struggle against the independent forces of the globe, which only the civilized races could actually win. The earth was still man's habitat, but it had to be tamed.

Why this naturalistic ideology of the globe viewed as an independent entity gained momentum, remains curiously unstudied. It is

124 Visions of the earth's future have hardly been studied. But see Roy Porter, 'Creation and credence: the career of theories of the Earth, *c.* 1660–1820', in Shapin and Barnes (eds.), *Natural Order* (see note 114), 97–124. For Buache, see N. Broc, 'Un géographe dans son siècle: Philippe Buache (1700–1773)', *Dix-huitième siècle*, III (1971), 223–35. See also E. L. Tuveson, *Millennium and Utopia* (Berkeley, 1949).

125 Cf. D. P. Walker, *The Decline of Hell* (1964); and, more generally, K. Thomas, *Religion and the Decline of Magic* (1971).

126 Glacken, *Traces on the Rhodian Shore* (see note 4), esp. ch. 12.

127 Roy Porter, 'George Hoggart Toulmin's theory of man and the earth in the light of the development of British geology', *AS*, XXXV (1978), 339–52.

obviously part of the secularizing of science's rendering of nature (a subject which itself still awaits its historian).[128] It must at one level register awe at the overwhelming extent – in time and space – and diversity of nature, which were impressing themselves ever more forcibly during the century. Such feelings were strengthened by the growing perception of landforms as aesthetic laws unto themselves, not to be judged by geometry or the conventions of taste.[129] The exploration of regions of the globe uninhabited by man, and resistant to civilization, also underlined man's telluric marginality.

But desacralization, or demystification – the other face of the 'objectification' – of the globe also doubtless reflects a growing determination amongst Europeans to use, exploit and dominate it. As the earth became less of a mother, and turned into an *object*, it was less revered, and lay more open to violation.[130] The eighteenth-century scientist's gaze lay on the terraqueous globe, as it currently was. The actual order was what concerned geological uniformitarians. They wished to show that the earth here-and-now expressed its natural, quintessential condition. The earth no longer directed the eye upwards to Heaven, or inwards to man's soul. It was very tangible clay, for man to use as he would.

In short, grasping the actual order of nature became the key to Enlightenment environmental thought. If man were to be lord of creation it would have to be through his own self-creation – a process which, men were confident, would enhance rather than desolate the terraqueous globe. There was little fear that man was wrecking an ecological balance, or polluting his planet.[131] Rather, the bleak frontierist vision of puny man, clinging onto an inhospitable globe, haunted the *philosophes* even as they preached 'il faut cultiver notre jardin'. Numa Broc has pertinently asked: 'Peut-on parler de géographie humaine au XVIIIe siècle en France?'[132] Suitably reinterpreted this question urgently needs further investigation.

[128] Though Glacken, *Traces on the Rhodian Shore* (see note 4), ch. 11, has some astute observations.

[129] See the works cited in note 60.

[130] It was during the eighteenth century that animals finally disappeared from maps. Other aspects of the stripping of human associations from the earth, like the history of holy places, would repay study.

[131] Lamarck is one exception.

[132] Numa Broc, 'Peut-on parler de géographie humaine au XVIIIe siècle en France?', *Annales de géographie*, LXXVIII (1969), 57–75. See also his *La géographie des philosophes: géographes et voyageurs français au XVIIIe siècle* (Paris, 1975); S. Moravia, 'Philosophie et géographie à la fin du XVIIIe siècle', *Studies on Voltaire and the Eighteenth Century*, LVII (1967), 937–1011.

VII. Disciplines

The sciences of the environment - like all others - were deep in methodological and epistemological self-reflection in the eighteenth century. Baconian-Lockean empiricism had utterly distanced what existed in the world through the spectacles of what men might experience. God's transcendental order was being reconceptualized as nature's. And, as was insisted from Hume to Kant, Linnaeus to Buffon, Hutton to Humboldt, 'nature' might well be no more than a principle invented by the human mind to order the mysteries of the phenomena. But were there any guarantees that the mind of man and the nature of things were congruent? What attempts were made to establish ontologies, epistemologies, and methodologies?[133]

Few naturalists wished to see, or felt resigned to seeing, one possible extreme: the theatre of the globe as meaningless chance flux.[134] There might not be strict speciation among minerals and rocks, or a globally identical superposition of the strata, but approximate constancies were not hard to plot.[135] Local eccentricities might grossly disturb the timing of tides, but natural philosophers agreed that they were regimented by their lunar mistress.

There was fierce dissent, however, on the status of the presiding law and order. Some saw regularities which others thought were phantoms. Thus to some geographers the earth's mountain chains were patently ranged along the ordinal points of the compass; others denied any such arrangement. More generally, some theorists thought that nature, to be orderly, must be uniform, and operate slowly, repetitively, by nuances. For others (both *philosophes* like Maupertuis and *dévots* like de Luc) there was nothing paradoxical about a system of nature which wrought cataclysms, acted rapidly and with violence, and which evolved with time.[136] Yet *some* concep-

[133] Cf. the excellent discussion in Glacken, *Traces on the Rhodian Shore* (see note 4), 504–50.

[134] See Lester G. Crocker, 'Diderot and eighteenth century French transformism', in B. Glass, O. Temkin and W. L. Straus, Jr (eds.), *Forerunners of Darwin, 1745–1859* (Baltimore, 1959), 114–44.

[135] R. Hooykaas, 'The species concept in eighteenth century mineralogy', *Actes du VIe congrès international d'histoire de sciences*, II (1950; publ. 1953), 458–68; Albury and Oldroyd, 'Renaissance mineral studies' (see note 102).

[136] Hooykaas, *Natural Law* (see note 119); *idem*, 'Catastrophism in geology, its scientific character in relation to Actualism and Uniformitarianism', *Kon. Ned. Akad. Wet. Lett. Meded., N.S.*, XXXIII (1970), 271–316; M. Guntau, *Der Aktualismus in den geologischen Wissenschaften*, Freiberger Forschungshefte 55 (Leipzig, 1967); Martin J. S. Rudwick, 'Uniformity and progression: reflections on

tion of the uniformity of natural processes was vital to facilitate induction and inference. Constant correlation between surface symptoms and stratigraphical structure had to be supposed for mine-prospecting to proceed. The regularity of the earth's magnetic field was a precondition for navigation. And, perhaps above all, the fidelity – in some form – of nature's laws over time had to be presumed to validate inferences about earth *history*, inaccessible except by analogy with the present.

The vitriolic debate which still rages within geology over the principle of uniformitarianism has clouded the historical record with polemic.[137] Preconceptions in the terms used to describe geo-philosophies – particularly when so many, like 'catastrophism', are emotionally loaded – still prevent dispassionate assessment of the schemes of terrestrial order envisaged by naturalists like de Saussure, Dolomieu, de Luc and Cuvier. And the distinctions need to be teased out between those who embraced the uniformity of nature solely as a heuristic principle, and those who seized it as a full-blooded theological and metaphysical truth.[138]

Fundamental, of course, to all attempts to propose methodologies for environmental sciences was the issue of what kinds of sciences they should be. Yet questions of the constitution and reconstitution, the 'space' and boundaries, of sciences remain sadly neglected. It has been all too common to take for granted the nature and distribution of present scientific disciplines, their contents and their parameters, as if they timelessly reflected nature. Yet, the reverse is true. That disciplines are, on the contrary, historical and cultural products is evident from the fact that terms such as 'biology' and 'geology', that are

the structure of geological theory in the age of Lyell', in D. H. D. Roller (ed.), *Perspectives in the History of Science and Technology* (Norman, Oklahoma, 1972), which applies equally to the late eighteenth century.

137 Cf. C. C. Albritton (ed.), *The Fabric of Geology* (Stanford, 1963); *idem*, (ed.), *Philosophy of Geohistory, 1785–1970* (Stroudsberg, Penn., 1975); *idem*, *Uniformity and Simplicity* (New York, 1967); for mystifications in the history of uniformitarianism, see Porter, 'Charles Lyell' (see note 121).

138 We are endebted to Hooykaas for 'demystifying' uniformitarianism, revealing that the principle is not a necessary methodological presupposition internal to geology but rather part of a tendentious ideology of nature. Over the last decade Hutton's uniformitarianism, in particular, has ceased to be seen as an innocent, heuristic device, and has been reinterpreted as integral to his deism, natural philosophy and epistemology. Cf. Porter, *Making of Geology* (see note 2), 185–96, for this literature. Kiernan, *Enlightenment and Science* (see note 118) has lively suggestions about the wider philosophical underpinnings of 'uniformitarianism', understood in a slightly different sense.

currently taken for granted, were coined scarcely two centuries ago.[139]

Indeed, in many studies of the terraqueous globe, disciplinary boundaries, methods and nomenclatures were notably chaotic in the second half of the century. Thus investigation of the crust passed under a multitude of heads – subterraneous geography, oryktology, geognosy, cosmogony, cosmography – whereas by the early decades of the nineteenth century terms such as geology and stratigraphy were becoming standard.

The constitution of the environmental sciences became problematic, for two reasons above all. Firstly, the traditional pantechnicon discipline, natural history, articulated by the metaphysics of the *échelle des êtres,* could no longer contain the expanding, complex multi-dimensional grid of natural objects known to the late Enlightenment, particularly as nature became viewed as ever more dynamic. Disciplinary sub-cultures, and the specialization of scientific labour, were emerging.[140] Secondly, the over-arching Biblical creationist account of the order, history and meaning of the earth was – at least in its literal form – losing its scientific credibility. It could not cope easily with new theories, such as extinction. It was challenged by powerful philosophical lobbies, such as the deists. And the scriptural, patriarchal, transcendental myth of the two cities was losing its ideological appeal as theodicy in a society ever more confident about men's powers to shape their progressive destiny for themselves. Pressures to demarcate the (supposedly) distinct disciplines of theology and science also reinforced the need to construct a separate, more secular, natural history of the globe.[141] Whatever the precise causes, the all-inclusive theories of the earth of Burnet, Whiston, or J. J. Scheuchzer in the age of Newton had been discredited fifty years on. Cosmogonical issues – questions of *origins* – became deeply problematic. A diplomatic Christian like Cuvier was remarkably oblique and opaque in his tracing of God's finger.

[139] F. D. Adams, 'The earliest use of the term geology', *Bulletin of the Geological Society of America,* XLIII (1933), 121–3; *idem,* 'Further note on the use of the term geology', *ibid.,* XLIV (1933), 821–6. D. R. Dean, 'The word "geology"' *AS,* XXXVI (1979), 35–44.

[140] See the works cited in note 1.

[141] See Porter, *Making of Geology* (see note 2), ch. 8; J. Roger, 'La théorie de la terre au XVIIe siècle', *RHS,* XXVI (1973), 23–48; Rhoda Rappaport, 'Geology and orthodoxy: the case of Noah's Flood in eighteenth century thought', *BJHS,* XI (1978), 1–18; T. L. Frängsmyr, *Geologi och skapel setro* (Stockholm, 1969); C. C. Gillispie, *Genesis and Geology* (Cambridge, Mass., 1951).

Although many naturalists, like de Luc and Lamarck, were ambitious to construct comprehensive theories of the environment, integrating elements and processes past, present and future, no hegemonic paradigm was actually forged.[142] The earth's complexity defied them. No naturalist could singly know enough. Buffon and Hutton were both slated for spinning over-ambitious phrases out of their heads which ill-matched reality. And the conjectural natural histories of *philosophes* such as Vico, Boulanger, d'Holbach, Toulmin and Herder, cut little ice with hard-nosed empirical naturalists.

Finding the right terms for explicating the environment was troublesome. Quantification (e.g., of the earth's age, or of its rate of denudation) was dubious. Explanations of the earth's features in terms of primary physical forces such as gravity, magnetism, heat-loss, or chemical theories of the successive precipitation of the strata, too readily succumbed to the complex diversity of the material surroundings and to fashions in physics and chemistry.[143] Attempts to nurse up the earth from chaotic origins, in the manner of Descartes or Leibniz, foundered for lack of primordial evidence. But *a priori* instructions to restrict understanding of the environment strictly to its present order, as issued by Hutton or Toulmin, grated as a gratuitous human imposition. Furthermore, prestigious Newtonian forces of attraction and repulsion, gravitation and the aether, could not on their own satisfactorily explain topography, the distribution of the continents, the vagaries of coastlines. In any case, Newtonianism became too identified with essentially static conceptions of order to be readily assimilable to the more dynamic philosophies emerging from mid-century and was itself (as many contributors to this book point out) fraught with internal contradictions.

The major consequence was that, out of the chaos of the dissolving old order, new, restricted sub-disciplines were grouping towards the end of the century. Buttressed by their own new specialist nomenclatures, geologists separated themselves from natural historians; stratigraphers from mineralogists. Specialist palaeontologists began to appear. And comprehensive sciences divided into subdepartments: thus general geography divided from special, and physical from human. In the nineteenth century these specialisms hatched their own societies, publications, university chairs and courses, re-

[142] Some evidence of this is contained in the welter of geological schemes considered by Desmarest in his four volumes of *Géographie physique* published in the *Encyclopédie méthodique* (Paris, 1794/5–1828).

[143] Thus see Porter, *Making of Geology* (see note 2), 170–82.

placing the general scientific societies, and the undifferentiated chairs of natural philosophy and natural history earlier typical.[144]

Yet we must not ante-date the clear-cut specialties of the nineteenth century. Certainly the Enlightenment closed with the old certainties of Scripture and the three kingdoms of nature disintegrating. Yet hopes remained of creating a new universal science of the environment (man included) to replace them.[145] In the event, no new intellectual framework succeeded (till perhaps evolutionism) in weaving together the sub-disciplines. Robert Young has argued that – in Britain at least – natural theology continued to provide that kind of hub of consensus, though John Brooke has suggested the fragmentation was running just as deeply through natural theology itself.[146]

Under what circumstances are new disciplines formed? Do intellectual coalescences always mirror institutional regroupings (as, say, in England, where the coming of geology was marked by the secession of the Geological Society of London from the Royal Society)?[147] How far are discipline boundaries differently drawn in different nations, or in different tongues? Do deep, epistemic changes – e.g., as posited by Foucault – underlie the reformations of scientific disciplines? Research must be addressed to these questions before we can construct adequate maps of the shifting configurations of knowledge.

VIII. Towards new views of the environment

Towards the end of the eighteenth century naturalists were grappling with the mounds of evidence which had newly become available,

144 Guntau, 'The emergence of geology' (see note 52); K. L. Taylor, 'Geology in 1776: some notes on the character of an incipient science', in C. J. Schneer (ed.), *Two Hundred Years of Geology in America* (Hanover, New Hampshire, 1979), 75–90.

145 L. J. Jordanova, 'Earth science and environmental medicine: the synthesis of the late Enlightenment', in Jordanova and Porter (eds.) *Images of the Earth* (see note 25), 119–46.

146 R. M. Young, 'Natural theology, Victorian periodicals and the fragmentation of the common context' (unpublished paper); John Brooke, 'Natural theology and the plurality of worlds: observations on the Brewster–Whewell debate', *AS*, xxxiv (1977), 221–86.

147 See Martin J. S. Rudwick, 'The foundation of the Geological Society of London', *BJHS*, i (1963), 325–55. But also P. Weindling, 'Geological controversy and its historiography: the pre-history of the Geological Society of London', in Jordanova and Porter, *Images of the Earth* (see note 25), 248–71; and Rachel Laudan, 'Ideas and organizations in British geology: a case study in institutional history', *Isis*, lxviii (1977), 527–38.

seeking to define disciplines adequate to articulate it. One sees the formulation of heroic strategies, and the building of ambitious – precarious – structures. But most were failures. Lamarck's dynamic, 'materialist', naturalistic, evolutionary plan of nature convinced few. Ardent spirits such as Goethe and Humphry Davy (in his *Consolations in Travel, or the Last Days of a Philosopher,* written in 1829) had to use visionary forms in order to express their Romantic sense of nature's wholeness. Humboldt devoted his life to the ideal of articulating the harmony of the cosmos, unity in diversity. By contrast, other naturalists, like Cuvier and Blumenbach, practised their science by accepting the fragmentation of nature, and by soft-pedalling certain basic issues.

Yet, late in the century, some new vantage points were emerging to provide new perspectives on the terrain. All could now agree that the terraqueous globe was dynamic, an infinitely complex interlocking economy of materials and processes, continually making and unmaking itself. Land and sea, organic and inorganic, the natural and the cultural, continuously interpenetrated.

Alongside process, *time* had also become crucial to understanding the *systema naturae* by late century. In part, this meant acceptance that the earth had existed for ages unimaginable – or unacceptable – a century earlier.[148] Further research is needed to ascertain naturalists' computations of the earth's antiquity, and to understand why it was *now* that Christian dogma on the earth's novity was first broken down. In some ways the liberalization of the time dimension followed on from the infinite space of the earlier astronomical revolution. The more relaxed Christianity of the Enlightenment – and deistic and pantheistic pressure – eased the passage. And invoking vast drafts of time proved the only way to accommodate the grand revolutions of Nature – the denuding and rebuilding of continents, and the migration of rocks, flora and fauna round the globe – increasingly suggested by evidence of fossils and the strata column. In particular, we need sensitive readings of how time was construed by certain eighteenth-century geologists such as Buffon, Hutton and Playfair: as not merely an objective, metaphysical parameter, but a force in its own right, nature's 'grand oeuvrier', to use Buffon's phrase.[149]

This revolution is part of an emergent integrating vision of the

[148] See F. C. Haber, *The Age of the World: Moses to Darwin* (Baltimore, 1959); S. E. Toulmin and J. Goodfield, *The Discovery of Time* (Harmondsworth, 1967); H. Meyer, *The Age of the World* (Allentown, Pa, 1951).

[149] Dr N. Rupke, of Wolfson College, Oxford, has a history of geological time nearing completion.

environment: the growth of a *historical* conception of the earth, and hence of genetic explanations of its phenomena. Temporalizing nature promised to settle many problems. Before its demise, the Great Chain of Being was temporalized. Palaeontology, the new science of fossils, gave life a history, and thereby rationalized extinction.[150] Evolutionary theories, in Lamarck and Erasmus Darwin, suggested a succession of life-forms. Natural history was becoming the history of nature. And, integrating these different strands, the earth itself assumed a defined past, through the rise of a historical geology to interpret the strata archives.[151]

Understanding late-Enlightenment notions of nature's history still poses difficulties. Foucault has argued – with Oldroyd's blessing – that some apparently historical philosophies, such as Buffon's and Lamarck's, are not truly 'historical' at all: no more than moving staircases, lacking genuine genesis. Oldroyd has argued that we should distinguish between theories of the globe undergoing merely a preprogrammed sequence of developments in time, and, on the other hand, the new and genuinely 'historicist' commitment of geologists to discover, by conscientious fieldwork, the actual historical succession of the ecosystem in particular localities.[152]

For early-nineteenth-century geologists, the earth's meaning lay in its coming-into-being. If Oldroyd's distinction holds, we must inquire what led certain eighteenth-century stratigraphers, and later theorists like Lyell, to subscribe to such a 'historicist' standpoint. How much did the historicizing of the earth owe to the more general emergence of a historical consciousness, so apparent in the flourishing of historical scholarship, Romantic antiquarianism, evolutionary social theory, and ideas of progress in the late eighteenth century? Were the new historical views of the environment transferred from social experience? Or was the influence the other way round? Were historical views of the environment closely connected with the historicization of other fields of natural philosophy, such as the rise of embryology or

[150] See Rudwick, *Meaning of Fossils* (see note 32), ch. 3.

[151] The fullest argument, with detailed references, is D. R. Oldroyd, 'Historicism and the rise of historical geology', *HS*, XVII (1979), 191–213, 214–43. For the emergence of a 'genetic' *episteme* see Foucault, *Order of Things* (see note 10), ch. 6, 'The age of history'. William Whewell's *History of the Inductive Sciences from the Earliest to the Present Times* (3 vols., 1837) gave one of the earliest accounts of the rise of the 'palaetiological sciences'. See also W. Lepenies, *Das Ende der Naturgeschichte* (Munich, 1978).

[152] Cf. J. S. Wilkie, 'The idea of evolution in the writings of Buffon', *AS*, XII (1956), 48–62, 212–27, 255–66.

nebular theory?[153] These issues are central, not just for understanding the emergence of the most important new framework for understanding the globe to appear in the last two centuries, but for enhancing our understanding of the generation and transfer of scientific ideas.

IX. Future prospects

To draw together the threads of this paper, I should like to suggest four pathways for future scholarship:

(1) Historians still have much work to do in thinking themselves out of 'presentist' disciplinary boundaries, like 'history of geology' and 'history of oceanography', and back into the far more inclusive vision of the terraqueous globe so common in the eighteenth century.

(2) Recapturing that sense of the wholeness and interconnectedness of the earth's economy will in turn make us pose with clarity the question: how far were eighteenth-century concepts of the terrestrial order created by that century's unprecedented programme of scientific exploration?

(3) The social history and the ideological meaning of exploration and explanation of the globe are both still in their infancy. We need to know who the naturalists were, and the relation of naturalists to 'naturalism'.

(4) The present generation has produced much technically expert, rigorous scholarship of the eighteenth-century environmental sciences, and also stimulating perspectives in intellectual, cultural and literary history. Can we hope for a closer walk between these in the next generation?

[153] For a good modern survey, see R. L. Numbers, *Creation by Natural Law: Laplace's Nebular Hypothesis in American Thought* (Seattle, 1977). Compartmentalization in the history of science has hitherto precluded adequate consideration of the relations of geology with cosmology.

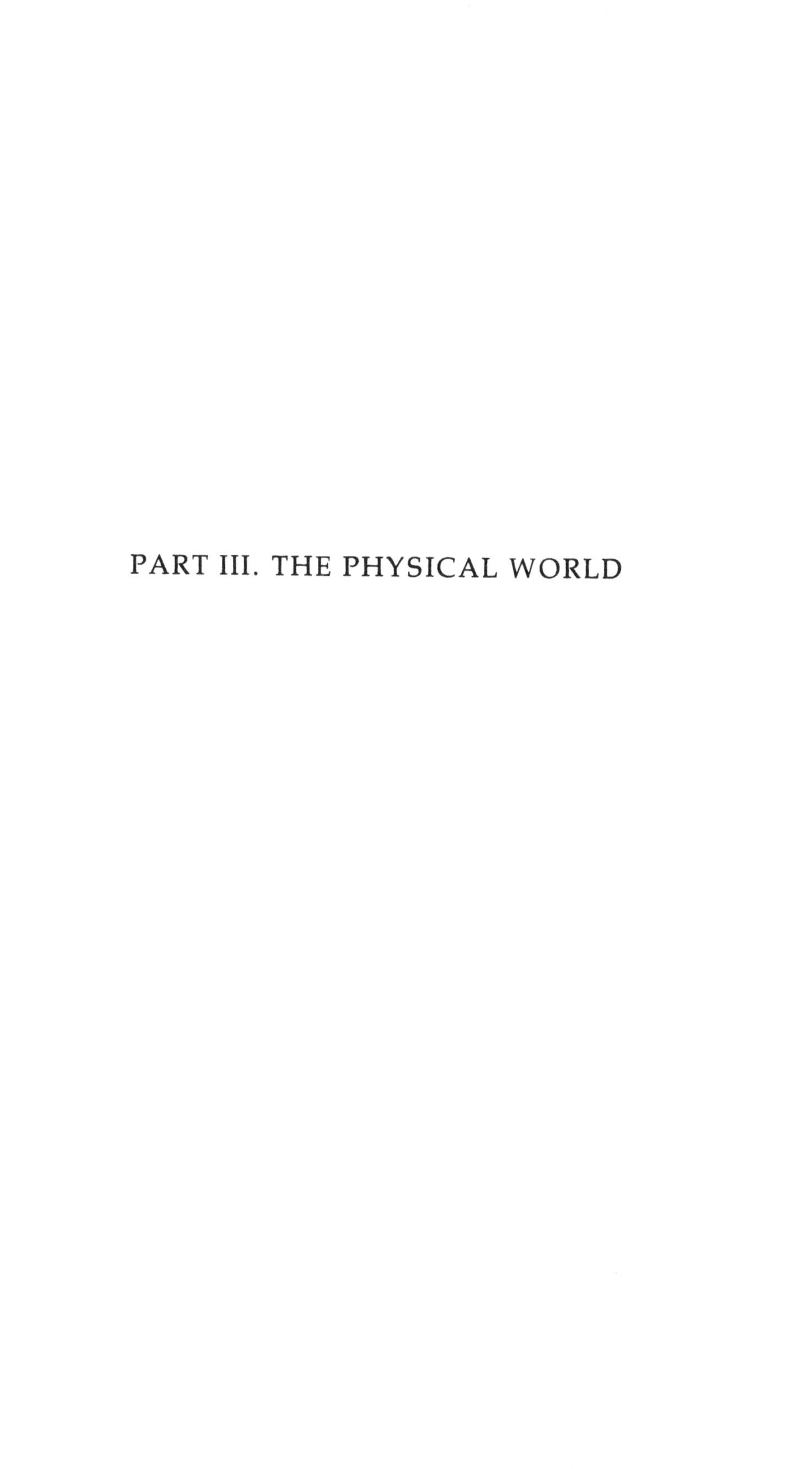

PART III. THE PHYSICAL WORLD

8

Mathematics and rational mechanics

H. J. M. BOS

Contents

I.	Introduction	327
II.	The shaping of a picture	328
III.	A new view of rational mechanics	333
IV.	Analysis	341
V.	The context of the mathematical sciences	351
VI.	Conclusion	355

I. Introduction

In most historical accounts of mathematics and mechanics, the eighteenth century is recognized as a separate period. But that recognition seems conditioned less by the characteristics of eighteenth-century mathematical science itself, than by the fact that the eighteenth century straddles two exciting periods in the development of mathematics and mechanics: the seventeenth and the nineteenth centuries. The seventeenth century saw the creation of algebraical symbolism by Viète and Descartes, of analytical geometry by Descartes and Fermat, the beginning of a mathematical theory of chance through Pascal, Fermat and Huygens, the foundation of classical mechanics by Galileo, Huygens and Newton, and the invention of the calculus by Newton and Leibniz. And the nineteenth century witnessed the foundation of rigorous analysis through Cauchy and Weierstrass, the creation of complex function theory by Cauchy, Riemann and Weierstrass, the emergence of new schools in geometry, projective geometry and non-Euclidean geometry in the work of Poncelet, von Staudt and others, new extensions of mathematical physics in relation to the theory of heat and electricity, the creation of modern algebra through Galois, Dedekind and Kronecker, and the beginnings of set theory and foundational studies of mathematics in the works of Cantor.

Available historical accounts create distinct pictures of both these periods, with persons and achievements clearly placed in the foreground – images well structured in the common memory of the community of mathematical scientists. The eighteenth century does not enjoy such clarity. From the distance of one and a half centuries the achievement of the eighteenth century seems to be merely to have brought the mathematical sciences from the peaks of the seventeenth century to those of the nineteenth. There were great scientists – the Bernoullis, Euler, Lagrange – and an enormous amount of work was done, but it seems impossible to characterize that work in such simple and clear-cut formulations as we have for the achievements of the previous and succeeding centuries.

This lack of structure in the vision of eighteenth-century mathematical science epitomizes the challenge which that period offers to historians of science. It is a twofold challenge. Historians of mathematics and the mathematical sciences shape the common memory of the community of mathematical scientists. Such a common memory is not detailed: it consists of general pictures like the ones I stated above for the seventeenth and nineteenth centuries. But a structureless picture, a history without highlights, does not remain in the memory. I believe that the community of mathematical scientists is ill-served by its amnesia of the eighteenth century. Hence the challenge to historians to provide such a picture must be taken seriously.

Previous lack of success in structuring the period suggests that the approach to history which provided well-defined visions of the mathematical sciences in the seventeenth and nineteenth centuries is not appropriate for the intermediary period. Thus the second challenge of the eighteenth century is that it requires historians to reconsider the questions and interests which guided successful studies of other periods, and to explore other issues and methods which may yield a better-structured and more instructive view of eighteenth-century mathematical science.

In this essay I shall report on some new trends in the historiography of the mathematical sciences in the eighteenth century. In line with the remarks above I shall be interested primarily in trends which have been restructuring the common vague picture of the mathematical sciences in the period, or which may be expected to do so. Before that, however, I shall discuss how it has happened that a series of general histories of mathematics and mechanics have left the picture of the eighteenth century as vague and structureless as it has been till quite recently.

II. The shaping of a picture

The first general history of mathematics to include the eighteenth century was Montucla's *Histoire des mathématiques*.[1] The two last vol-

[1] J. E. Montucla, *Histoire des mathématiques* (4 vols., Paris, 1799–1802).

umes, which deal with the eighteenth century, were completed by Lalande after the author's death in 1799. Here we have a view of eighteenth-century mathematics as it appeared to a contemporary. Like any history of recent events, it lacks the structure which only a time perspective can give, but it certainly is alive. The first thing that strikes the modern reader is the range of Montucla's mathematics. In his vision – one generally shared in the eighteenth century – mathematics consists of pure mathematics and mixed mathematics, the latter including among other things optics; theoretical or analytical mechanics; machines; planetary, observational and physical astronomy; navigation of ships; and shipbuilding. All these subjects are treated in Montucla's work.

The object of mathematics (here also Montucla formulates the eighteenth-century viewpoint) is the mutual relations of magnitude and number of any objects which are capable of increase or decrease.[2] This explains the terminology 'pure' and 'mixed'. Pure mathematics treats the relations between (variable or constant) quantities irrespective of the objects they measure or count; mixed mathematics deals with quantities, and their relations, as they occur in natural objects which can be counted or measured.[3] The terminology is indeed an appropriate one, better than the division into 'pure' and 'applied' now in use, which overlooks the dialectical nature of the use of mathematics and suggests that one either practises pure mathematics or takes a ready parcel of mathematics and applies it elsewhere.

The picture of eighteenth-century pure mathematics and analytical mechanics that emerges from Montucla's narrative is as follows. In pure mathematics there is stagnation as well as progress. Geometry in the ancient manner (which means the non-analytical study of conics and higher curves) is a worthy, dignified but unproductive field, treated only in England. In algebra the complexity of the calculations sets unsurmountable barriers to progress. This applies to the classification of algebraic curves, problems of elimination and, above all, the solution of equations of degree higher than four. But perhaps, says Montucla, in the last-mentioned subject Lagrange's new approach may point in a more fruitful direction. There is progress in analysis, the field is rich and open, and subfields within it are being

[2] 'La science des rapports de grandeur ou de nombre, que peuvent avoir entr' elles toutes les choses qui sont susceptibles d'augmentation ou de diminuation', Montucla, *Histoire des mathématiques* (see note 1), I, 3.

[3] *Ibid.* Compare: 'Mathematicks is commonly distinguished into pure and speculative, which consider quantity abstractly; and mixed, which treat of magnitude as subsisting in material bodies, and consequently are interwoven every where with physical considerations.' *Encyclopaedia Britannica* (3 vols., Edinburgh, 1768–71), *s.v.* 'Mathematicks'.

formed: the theory of partial differential equations through d'Alembert and Euler, and the variational calculus through Euler and, on a completely analytic footing, by Lagrange. Probability is also a budding field; it is less extensive than analysis but its applications. as in voting procedures, insurance and statistics, are more diverse.[4]

Montucla's story of theoretical mechanics in the eighteenth century[5] highlights the central problems (tautochrones, the vibrating string, projectile movement under air resistance, hydrodynamics) and the various principles (of live forces, of least action, and d'Alembert's principle) which were developed in the attempt to find a general mathematical science of mechanics. In his view, this attempt was crowned by Lagrange, who 'reduced all the problems to general formulae whose development provides all the equations necessary for the solution of each problem'.[6]

What strikes one in going through Montucla's narrative is the absence of historical distance. The mathematicians whose work he discusses all seem to be contemporary with each other: they speak the same language and work at the same mathematics. Perhaps when Montucla wrote his history he had more of the ancient sense of that word in mind: history as organized knowledge. He gives an organized survey of the mathematics as it is alive in his day and it seems almost accidental that the order he provides in the survey is the chronological order.[7]

And so one might say that the first truly historical survey of eighteenth-century mathematics is not in Montucla's work but in Moritz Cantor's *Vorlesungen*[8] of a century later. Indeed Cantor writes history, and this in a 'wissenschaftlich' manner, which means that he attempts to write the full and reliable chronicle of the mathematical writings of the past and their authors. Mechanics, astronomy and other neighbouring fields are not taken into account. His history is basically a report on mathematical literature, with remarks on persons and influences interspersed between résumés of books and articles. Within this conception of his task Cantor produced impressive volumes, still very useful as a source, but hardly creating a memorable picture of the past – especially not for the eighteenth century. In fact Cantor's approach barely reached that century; there was too much

[4] Montucla, *Histoire des mathématiques* (see note 1), III, 4–426.

[5] *Ibid.*, III, 606–719.

[6] *Ibid.*, III, 606.

[7] Montucla mentions as the most important objective of his book to give the 'idée distincte et les veritables principes de toutes les théories de quelque considération qui composent le système des mathématiques'. *Ibid.*, I, iii.

[8] M. Cantor, *Vorlesungen über Geschichte der Mathematik* (4 vols., Leipzig, 1880–1908).

material to be covered. The year 1800 was reached only with the help of nine historians of mathematics who composed the fourth volume of the *Vorlesungen* as a collection of long surveys of the literature for nine subfields of mathematics in the second half of the eighteenth century.[9]

No picture of eighteenth-century mathematics emerges from the mass of detail in Cantor's *Vorlesungen.* It seems that he sees three highlights in the period: the priority question between Leibniz and Newton over the invention of the calculus, and Euler's *Introduction to the Analysis of Infinites* (1748)[10] and his *Differential Calculus* (1755).[11] But the method of presentation hardly provides a focus, and the chapters on Euler's books are mere excerpts of their contents.

Thus the view of eighteenth-century mathematics lost the liveliness it still had in Montucla's vision without gaining a structure. In that respect nineteenth-century mathematics was more fortunate; Klein, in his *Vorlesungen* (1926),[12] did give a clear picture of that period. Compared with the mathematics of the centuries before and after, eighteenth-century mathematics became pale and vague. In Boyer's *History of Mathematics* (1968) we read about the 'misfortune' of the eighteenth century to be between the seventeenth and the nineteenth.[13] Hofmann, who in his *Geschichte der Mathematik* (1953) tried to characterize epochs in mathematics through the periodizations of cultural history (Humanism, Baroque, Enlightenment, etc.), can persuade the reader that seventeenth-century mathematics, with geniuses like Descartes, Newton, Leibniz, who transformed whole areas of mathematics, fits into the Baroque age of genius. But he cannot analogously relate eighteenth-century mathematics to the characteristics of the Enlightenment, and he characterizes the period merely by spelling out what it lacked (and what the nineteenth century provided): secure foundations.[14]

The most recent effort to picture eighteenth-century mathematics

9 The fourth volume of Cantor's *Vorlesungen* (see note 8) was published in 1908; it covers the period 1759–99. The historians of mathematics who contributed to it were S. Günther, F. Cajori, E. Netto, V. Bobynin, A. von Braunmühl, V. Kommerell, G. Loria, G. Vivanti and C. R. Wallner. Cantor himself only contributed a short 'Überblick'.

10 L. Euler, *Introductio in analysin infinitorum* (2 vols., Lausanne, 1748), in *Opera omnia*, ser. 1, VIII, IX.

11 L. Euler, *Institutiones calculi differentialis* (St Petersburg, 1755), in *Opera omnia*, ser. 1, X.

12 F. Klein, *Vorlesungen über die Entwicklung der Mathematik im 19. Jahrhundert* (2 vols., Berlin, 1926–7).

13 C. B. Boyer, *A History of Mathematics* (New York, 1968), 510.

14 J. E. Hofmann, *Geschichte der Mathematik* (3 vols., Berlin, 1953–7), cf. I, 7, and III, 35.

as a whole is Kline's *Mathematical Thought from Ancient to Modern Times* (1972), which is both more detailed and more intent on characterizing broad lines in the development of mathematics than were previous histories of the subject. Kline notes the receding of geometrical methods and concepts in eighteenth-century mathematics, the loosening of rigour in proofs, the interest in and reliance on analytical formalisms. He mentions the strong interpenetration of mathematics and science (in particular mechanics), the increase in the means of scientific communication and the importance of academies as institutional bases for the development of mathematics. New fields were developed in the century but they were fields of problems rather than theories. General methods were what eighteenth-century mathematicians were looking for, but did not find.[15]

This is the picture of eighteenth-century mathematics which is now common. It is not very graphic. It is noteworthy that it does not stress persons, achievements and theories, but rather styles, problems and circumstances. One of the reasons for the lack of structure in this view of eighteenth-century mathematics may well be that styles, problems and circumstances are more difficult to study historically than persons, achievements and theories.

As to eighteenth-century mechanics, the picture given by Montucla, who in fact followed Lagrange's view of the matter,[16] remained for a long time essentially unchanged. Moreover his vision of mechanics fitted that articulated by Mach in his influential book *Die Mechanik* (1883).[17] Mach argued that a science develops through three phases: one of gathering observations of phenomena, one 'deductive' phase in which the scientists learn to deduce logically and theoretically certain phenomena from others, and finally a phase of formal development in which certain observations are recognized as the basic principles of the science and in which the deductions are organized in a theoretical unity. In this scheme the 'economy of thought' demands that the deduction of phenomena from basic principles should be as easy and straightforward as possible. Mach told the history of mechanics to support this view. In this vision, Newton concluded the first phase in the history of mechanics, Newton's laws, which Mach considered to be based on observations, were the principles from which 'without the help of any new princi-

[15] M. Kline, *Mathematical Thought from Ancient to Modern Times* (New York, 1972), cf. 614–25.

[16] As expounded in the historical preface of J. L. Lagrange, *Mécanique analitique* (Paris, 1788), in *Oeuvres*, XI–XII.

[17] E. Mach, *Die Mechanik in ihrer Entwicklung historisch-kritisch dargestellt*, 7th edn (Leipzig, 1912).

ples all mechanical cases occurring in practice, either statical or dynamical, can be understood'.[18]

Eighteenth-century mechanics, in Mach's view, belongs to the second and third phases. The various principles (d'Alembert's, *vis viva*, least action, etc.) are all implied in Newton's laws, being attempts at economic organization of deductions within mechanics. The phase of formalization consisted in the creation of an analytic mechanics, begun by Euler and completed by Lagrange. Mach is less interested in the ultimate structure of this analytical mechanics than in the question of whether it involves any metaphysics; he argues forcefully that it does not. Thus Mach's two main interest – the conclusion of the deductive phase in mechanics through Newton's laws, and the absence of metaphysics in the whole structure – have the effect of making the eighteenth century in his vision an uneventful and uninspiring period in mechanics.

Dugas' *Histoire de la méchanique* (1950), though more historical and less partisanly committed to one particular view of science, gives the same impression of eighteenth-century mechanics as Mach: 'Classical mechanics was formed in the seventeenth century. The task remained of organizing it and of developing its general principles: that was to be the work of the eighteenth century.'[19] The creation of this mechanics was done by the great three of Galileo, Huygens and Newton; Dugas orders his presentation of the seventeenth century according to persons. The eighteenth century receives a different treatment: Dugas tells its history ordered by principles (virtual work, live forces, least action) and subfields (such as hydrostatics), closing the narrative with a section on Lagrange. Through Lagrange the collective work of the eighteenth century finally attains 'an ordered science whose form approaches perfection'.[20]

In this way historians have shaped our view of eighteenth-century mathematics and mechanics. But pictures are not static, and in the following sections I shall discuss a number of recent trends which are bringing essential changes, or may do so in the near future.

III. A new vision of rational mechanics

The common picture of eighteenth-century mechanics rests on the view that all mechanics till Einstein is implicit in Newton's laws. Since the 1950s this simplistic view has been effectively challenged. We can

[18] Mach, *Mechanik* (see note 17), 272.

[19] R. Dugas, *Histoire de la méchanique* (Neuchâtel, 1950), 221.

[20] *Ibid.*

no longer pass over the eighteenth century as the period which merely exploited the rich mine whose entrance had been opened by Newton's *Principia*. On the contrary, we now realize that Newton's laws do *not* imply all of mechanics. We have come to see that it was in the eighteenth, rather than the seventeenth, century that the foundations and the main concepts of mechanics were formulated. Indeed the very programme of the science of mechanics appears as a creation of the eighteenth century. This deeper insight into the history of mechanics has been opened to us through the work of one historian of science, Clifford Ambrose Truesdell III, and the core of his contribution to the knowledge of eighteenth-century mechanics is to be found in his three magnificent introductions to volumes of Euler's *Opera omnia.*[21]

Truesdell's picture of eighteenth-century mechanics is more structured, richer and more exciting than the previous accounts. But it also has a fundamental simplicity. This simplicity stems from Truesdell's view of his subject. He writes the history of *rational mechanics,* that is, the axiomatic mathematical science of mechanics. This is not the mechanics to which physicists are used, but rather it is a part of pure mathematics, treated by 'mechanicists'.[22]

Only fairly recently - about twenty years ago - has rational mechanics reached the stage of a truely axiomatic science. Hence it is also a fairly recent insight that the two basic axioms of rational mechanics are the principle of linear momentum and the principle of moment of momentum.[23] From these axioms rational mechanics derives the differential equations governing the motion of mechanical systems. These systems themselves are characterized by constitutive equations.[24] The systems may be composed of point masses or rigid bodies; they may involve fluids, in which case the constitutive equations require a well-defined concept of *pressure,* and they may

[21] L. Euler, *Opera omnia* (3 series, 67 vols. to date; Leipzig, 1911–). Truesdell's introductions are:
1954: 'Rational fluid mechanics, 1687–1765', in L. Euler, *Opera omnia,* ser. 2, XII (1954), l–cxxv.
1956: 'I. The first three sections of Euler's treatise on fluid mechanics (1766); II. The theory of aerial sound (1687–1788); III. Rational fluid mechanics (1765–1788)', in L. Euler, *Opera omnia,* ser. 2, XIII (1956), vii–cxviii.
1960: 'The rational mechanics of flexible or elastic bodies 1638–1788', in L. Euler, *Opera omnia,* ser. 2, XI, sect. 2 (1960).
A number of shorter articles by Truesdell on the history of mechanics are collected in C. Truesdell, *Essays in the History of Mechanics* (New York, 1968).

[22] Truesdell, *Essays* (see note 21), 334–40.

[23] *Ibid.,* 268–71.

[24] *Ibid.,* 173.

involve flexible or elastic bodies, in which case a well-defined concept of *stress* is needed. The differential equations derived from the axioms, combined with the constitutive equations, fully imply the motion of the system. The further determination of that motion is a question of mathematics (but certainly not despised by the mechanicists), namely the solving of the differential equations.

This style of rational mechanics was the creation of the eighteenth, not of the seventeenth, century. Its history, as Truesdell tells it, has a clear simple structure and obvious highpoints: the formulation of the principles of linear momentum and moment of momentum (both by Euler, in 1750 and 1775 respectively), the correct definition of the concept of pressure (by Euler in the early 1750s), and the 'creation and unfolding of the concept of stress',[25] crowned by Cauchy's introduction of the stress tensor in 1822. Thus the main traits of Truesdell's new picture of eighteenth-century mechanics are as follows.

In 1750 Euler for the first time formulated the principle of linear momentum in its full generality.[26] The principle states that, for force F (with coordinates F_x, F_y, F_z) and mass M, with position (x,y,z) in space:

$$M\,\frac{d^2x}{dt^2} = F_x,\ M\,\frac{d^2y}{dt^2} = F_y,\ M\,\frac{d^2z}{dt^2} = F_z.$$

Euler made clear that the equations apply to all mechanical systems, discrete (in which case one can sum over the masses) and continuous (in which case M and F must be taken as differentials dM and dF and integration performed over the whole system or over the subsystem one wants to consider).

These equations are known as 'Newton's equations', but Truesdell argues at length[27] that it is inadmissible hindsight to read these equations in Newton's second axiom or law of motion, which states that 'the change of motion is proportional to the motive force impressed, and is made in the direction of the right line in which that force is impressed'.[28] Newton's formulation remained far from the generality provided by the mathematical formula; neither did Newton use his second law to set up differential equations of motion for systems of more than two bodies (let alone for continuous media).

[25] Cf. the article of that title, *ibid.*, 184–238.

[26] L. Euler, 'Découverte d'un nouveau principe de mécanique', *Histoire de L'Academie des Science, Berlin* (year 1750, publ. 1752), 185–217 in *Opera omnia*, ser. 2, v, 81–108.

[27] Truesdell, *Essays* (see note 21), 114–17, 167–71.

[28] I. Newton, *Philosophiae naturalis principia mathematica* (1687), lex. II.

Euler, Truesdell argues, gave the principle its full generality, which is twofold. The formulation in terms of differential equations in rectangular coordinates allows the principle to be applied directly to any configuration of bodies in three-dimensional space. This gain of mathematical generality and simplicity is illustrated by a comparison with Euler's *Mechanics* of 1736,[29] in which, for want of a general principle incorporating Newton's second law, Euler had had to work out a whole apparatus of differential geometry for skew curves.

Moreover, Euler realized that the principle equally applies to each subsystem of a given system, and that through this it yields the equations of motion for many systems. In particular its applicability to infinitesimal subsystems (as volume-differentials in fluids) gives it its function as the fundamental principle in continuum mechanics. Indeed in the article in which Euler published the 'new principle', he used it to derive 'Euler's equations' for the motion of rigid bodies.

The second axiom of rational mechanics relates the moment of momentum H, defined for a system of point masses as

$$H = \sum_k \boldsymbol{r}_k \times m_k \dot{\boldsymbol{r}}_k ,$$

to the total torque L of the external forces F^{e}:

$$L = \sum_k \boldsymbol{r}_k \times \boldsymbol{F}_k^{\mathrm{e}} .$$

The axiom, or the principle of moment of momentum, states that

$$\dot{H} = L.$$

(In the case of continuous bodies, H and L are defined as the corresponding integrals with respect to differentials $\mathrm{d}M$ and $\mathrm{d}F$.) This is an independent axiom of mechanics, though in certain restricted cases it can be derived from the principles of linear momentum; Truesdell notes that many physicists are not aware of this independence.[30] The principle was formulated as an independent axiom of mechanics by Euler in 1775;[31] his studies on the form of bent elastic beams had led him to realize its independence. An early form of the principle occurs already in Jakob Bernoulli's work and Truesdell

[29] L. Euler, *Mechanica sive motus scientia analytice exposita* (St Petersburg, 1736), in *Opera omnia*, ser. 2, I–II.

[30] Truesdell, *Essays* (see note 21), 239–43.

[31] L. Euler, 'Nova methodus motum corporum rigidorum determinandi' *Novi Commentarii Academiae Scientarum Imperialis Petropolitanae*, XX (year 1775, publ. 1776), 208–38, in *Opera omnia*, ser. 2 IX, 99–125. Cf. Truesdell, *Essays* (see note 21), 263, 172–3.

traces the further development of the principle and of the awareness of its independence from the principle of linear momentum.[32]

Thus far the principles; the story of the concepts of pressure and stress, necessary for the constitutive equations, provides further structure to the picture. The concept of internal pressure of a fluid has its prehistory in the works of Archimedes, Stevin, Pascal, Newton and Clairaut. The concept occurs explicitly in Johann Bernoulli's *Hydraulica,*[33] for fluids in tubes. Euler extended it to the general case of fluids occupying any part of three-dimensional space. With this general concept of pressure and the principle of linear momentum he worked out a full theory of hydrodynamics in the years 1753–5.[34] The understanding of stress, however, notably the recognition that it is a tensor, did not come in the eighteenth century. Internal pressure is a special form of stress as it occurs in fluids. Other special deformable continuous bodies were studied in the eighteenth century, such as the catenary and the elastic band, and the concept of stress occurs in these studides, but they involved only special cases and therefore did not lead to the full understanding of the concept. It was not until 1822 that Cauchy formulated the general concept of stress.[35]

Truesdell's new approach to the history of mechanics is of great importance for the history of mathematics, because rational mechanics formed the context in which the most exciting parts of eighteenth-century mathematics were developed. All the great eighteenth-century mathematicians – the Bernoullis, Clairaut, d'Alembert, Euler, Lagrange, and others – worked both in mathematics and in the mathematical sciences, and did not think these fields distinct. Neither does Truesdell himself make such a separation and his historical studies show, better than any separate history of mathematics or of mechanics could do, the impressive richness and fruit-

[32] See the article 'Whence the law of moment of momentum?', in Truesdell, *Essays* (see note 21), 239–71.

[33] Johann Bernoulli, *Hydraulica nunc primum detecta ac demonstrata ex fundamentis pure mechanicis,* in J. Bernoulli, *Opera omnia* (4 vols., Lausanne and Geneva, 1743), IV, 387–493.

[34] L. Euler, 'Principes généraux de l'état d'équilibre des fluides', *Mémoires de L'Académie des Sciences,* XI (written 1753, year 1755 publ. 1757) 27–73, in *Opera omnia,* ser. 2, XII, 2–53; 'Principes généraux du mouvement des fluides', *Mémoires de L'Académie des Sciences, Berlin,* XI (year 1755, publ. 1757) 274–315, in *Opera omnia,* ser. 2, XII, 54–91; 'Continuation des recherches sur la théorie du mouvement des fluides', *Memoires de L'Academie des Sciences, Berlin,* XI (year 1755, publ. 1757) 316–61, in *Opera omnia,* ser. 2, XII, 92–132. Cf. Truesdell, *1954* (see note 21), lxxv–c, and *Essays* (see note 21), 219–32.

[35] Cf. note 25.

fulness of the interplay of mathematics and mechanics in the eighteenth century.[36]

The two fields that developed within analysis were closely bound up with mechanics. The theory of partial differential equations was 'the great gift of continuum mechanics to analysis'.[37] The problem of the vibrating string led to the first full solution of a partial differential equation (the wave equation) by d'Alembert in 1746; and Euler worked out much of the theory of partial differentiation and the solution of partial differential equations in his hydrodynamical researches. Variational calculus, originating partly in the study of mechanical problems such as the brachistochrone around 1700, was worked out by Euler and Lagrange as a purely analytical theory, being soon afterwards applied to many problems in mechanics.

In details as well, Truesdell's work shows the interplay between mechanics and mathematics. He claims that from the 1730s 'continuum mechanics gave rise to all the major new problems of analysis'.[38] Elliptic functions originated in Jakob Bernoulli's studies on the elastica. Euler's general solution of the linear differential equation with constant coefficients was occasioned by his study of the oscillations of an elastic band. Daniel Bernoulli's study on the vibrations of a free hanging rope involved Bessel functions. Related problems, such as vibrations of membranes, studied by Euler, led to the study of proper frequencies and proper functions.

Truesdell's framework for the history of mechanics in the eighteenth century is a very fruitful one. It brings clarity and order, and this not in a constricting way – the richness of developments can unfold within it. Moreover, this order and clarity make it possible to ask and answer meaningful questions. Thus there is the question of Newton's influence – so liable to be overrated – and the role of the *Principia* in the eighteenth-century development of mechanics. In discussing this question Truesdell points to parts of mechanics that Newton did not touch at all: continuum mechanics and hydromechanics. He shows that many traditions, concepts and problems in these fields entered the eighteenth century bypassing the *Principia*.[39]

Another question is why it took so long for the mechanics of elastic bodies to reach the coherent treatment which Cauchy finally gave it

[36] See especially the summary in Truesdell, *1960* (see note 21), 416–20.

[37] Truesdell, *1960* (see note 21), 418.

[38] *Ibid.*, 417.

[39] See the article 'Reactions of late baroque mechanics to success, conjecture, error, and failure in Newton's *Principia*', in Truesdell, *Essays* (see note 21), 138–83.

through the introduction of the stress tensor. As Truesdell shows,[40] the concepts required for this unification all lay in the hands of Euler by the 1770s. Why then the delay? Because, Truesdell argues, the successes of the theory in the restricted cases were so great that they hindered unification; 'nothing is harder to surmount than a corpus of true but too special knowledge'.[41]

Finally it becomes possible to reappraise the merits of the individual contributors to mechanics. One who has greatly benefited from this is Jakob Bernoulli. His achievements in mechanics were overshadowed by Newton's, and in mathematics the work of his impetuous and brilliant brother attracted more attention. He left no *magnum opus,* but hid his insights in private notebooks and a few articles. Thus he has not received much attention from historians of mechanics. But Truesdell, seeking the origins of rational mechanics as he sees it, finds many and deep insights in Bernoulli's work and brings these out.[42]

But the greatest change of focus concerns Euler. It seems indeed, that Truesdell's history is a game in which winner takes all, and the winner most certainly is Euler. The giants of previous histories, such as d'Alembert and Lagrange, are reduced to normal size (and, in the case of d'Alembert, belittled and sneered at,[43] but Euler towers above all. The reader reacts with wonder: who sets the rules of this game, Truesdell or history?

Certainly for a great part Truesdell keeps to the rules of history: his case for attributing the fundamental concepts and principles of rational mechanics to Euler rather than to others is well defended. But it is partly Truesdell who sets the rules. He has a definite view on what the mechanics is whose history he writes. It is a mathematical rational mechanics, built on as few axioms as possible. There is basically only one such mechanics, and so things are either right or wrong within this framework (and Truesdell does not refrain from telling right from wrong in history[44]). Of all eighteenth-century writers on mechanics it

[40] Cf. note 25.

[41] Truesdell, *Essays* (see note 21), 238.

[42] *Ibid.*, 101–5, 156–60, 204–11, 248–52.

[43] 'At the age of twenty-four, there enters our scene now a talented but sinister personality who is to make in six years a sequence of brilliant discoveries but thereafter will write endlessly in what seems today no more than a dogged attempt to confine the capacities of mathematics and to belittle the work of others . . . This is d'Alembert.' Truesdell, *1960* (see note 21), 186.

[44] 'In writings on the history of science today, as in all aspects of social intercourse, it becomes increasingly bad taste to call a spade a spade. In the particular application to the history of science, the compulsion to euphemy assumes the form of a solemn refusal to admit that there is such a thing as

is Euler whose work comes closest to this style of mechanics, and that is why he dominates the scene in Truesdell's story.

So it is Truesdell's vision of what mechanics really is which, at least in part, determines his picture. Truesdell himself repeatedly explains from which point of view he considers the history of mechanics. He also claims that it is the only sensible point of view a historian of science can take. This conviction is evident in almost every page of his work and it comes over very forcefully. Some readers will regret this strong *parti pris* and the preferences and dislikes connected with it. But such a regret has little sense. It is hardly imaginable that without these strong views a corpus of historical research so coherent and unified in style as Truesdell's work could be produced.

On the other hand, the critical reader and user of this material will do well to be aware of its idiosyncrasies. He should keep in mind that, from Truesdell's point of view, a number of aspects of eighteenth-century mechanics recede into the background. Mention should be made here of some of the aspects which, because of the special character of rational mechanics in Truesdell's view, receive less attention in his treatment of the history of mechanics.

For instance: in axiomatic rational mechanics principles must be formulated clearly and as generally as possible. Hence d'Alembert's principle receives no important role in Truesdell's record; it is restricted, and d'Alembert did not formulate it well.[45]

In rational mechanics the principles are axioms, or derivable from axioms; their metaphysical foundation does not concern the mechanicist. Truesdell, therefore, devotes little attention to, for instance, the *vis viva* controversy and its philosophical background, or the controversy over the principle of least action.

Again, in rational mechanics the fundamental undefined entities are force and torque,[46] the basic principles are differential equations involving these entities, and the special features of the systems studied are expressed in the constitutive equations. Most of the principles and formalisms worked out in the eighteenth century to unify the treatment of mechanical problems, such as d'Alembert's principle, the principle of virtual work, and Lagrange's formalism, were not, however, about force and torque, did not clearly separate the princi-

wrong in science. . . . However admirable this philosophy may be in promoting peace and mutual love among historians of science, it disregards one aspect of science that is not altogether negligible, namely, that scientists seek *the* truth, not *a* truth.' Truesdell, *Essays* (see note 21), 145–6; see also the revealing footnote *ibid.*

[45] Cf. Truesdell, *1960* (see note 21), 192.

[46] Truesdell, *Essays* (see note 21), 321–3.

ples from the constitutive equations, and were not completely general. Hence these principles and methods, which in previous histories were the central issue of the story of eighteenth-century mechanics, receive little emphasis in Truesdell's account. This explains the dramatic fall in appreciation for Lagrange. His formalism, acclaimed by Dugas and almost all previous writers on the subject as the final unification of the science of mechanics, is not a principle in the above sense and is not general. Hence in Truesdell's version of the story Lagrange is definitely in Euler's shadow.[47]

Finally, rational mechanics is an axiomatic, not an experimental science, and its relation to experience is the same as the relation of geometry to experience. Truesdell argues forcefully that practice did not have any positive effect on rational mechanics in the eighteenth century, and that the mechanicists of that century were very wise to disregard practice. The historian interested in the (direct or indirect) relations of science with technology and other practices, therefore, will find only negative evidence in Truesdell's work.[48]

Despite these points, however, Truesdell's studies are a great asset to the history of mathematics and mechanics. They are based on extensive and thorough study of the sources. They provide a new, clearly structured picture of mechanics in the eighteenth century in which the neglected fields of continuum mechanics receive their proper attention. They also bring out more clearly than ever before the interplay between mathematics and mechanics. Historians of science and mathematical scientists interested in the history of their field will greatly benefit from his contributions.

IV. Analysis

The most important developments within eighteenth-century pure mathematics occurred in analysis. These developments have formed a central theme for many historians of mathematics. In this field a number of new changes of interest and direction of research have occurred in recent years, which must be mentioned here.

If the effects of new trends in historiography are to be judged by the changes they bring in the common picture of historical events or periods, then there is one new approach to the history of analysis which above all deserves attention here, because it claims explicitly that the common picture of the development of analysis must change. This new approach is provided by a new mathematical discipline,

[47] *Ibid.*, 131–5.

[48] Cf. Truesdell, *Essays* (see note 21), 135–6, and *1960* (see note 21), 13–14.

non-standard analysis. Non-standard analysis gives a rigorous mathematical theory of infinitely small and infinitely large numbers. In terms of that theory the calculus can be reformulated as a theory based on the concept of the infinitely small differential. For instance, non-standard analysis enables us to define the derivative of a function as the quotient of the corresponding differentials of the dependent and the independent variable.

Non-standard analysis, which has provided at last a rigorous theory involving infinitely small differentials, has a great appeal, because for more than two centuries these differentials were the classic example of non-rigorously defined concepts. The creator of non-standard analysis, Abraham Robinson, claimed that, precisely because this new theory has made the infinitely small mathematically respectable, it is now necessary to supplement and redraw the historical picture of analysis. Robinson himself gave a sketch of such a redrawing of the picture in his book on non-standard analysis.[49]

Thus this approach is presented as a new trend. Before discussing its implications, though, something must be said about the picture of the history of analysis which non-standard analysis might supplement and redraw. That picture was drawn almost forty years ago by Carl B. Boyer in his book *The History of the Calculus and its Conceptual Development.*[50]

In Boyer's view the eighteenth century is 'the period of indecision' as to the foundations of the calculus.[51] Leibniz had taken the infinitely small *differential* as the basic concept of his calculus. Newton used the *fluxion,* a velocity or rate of change. Both concepts were insufficiently founded; Leibniz did not define 'infinitely small' and Newton took velocity as an undefined prime concept. The 'indecision' which Boyer senses in the eighteenth century concerns this lack of foundations. The lack was felt from the beginning of the century and the problem became more pressing after the brilliantly sharp critique by Bishop Berkeley in 1734. Berkeley attacked the calculus, showing that it was founded on notions much less evident than those commonly used in theological arguments. In Britain a number of mathematicians defended Newtonian fluxions against this critique; Maclaurin, in particular, presented the theory in rigorous classical geometrical style, but he could not solve the basic problem of defining consistently what fluxions are.

[49] A. Robinson, *Non-Standard Analysis* (Amsterdam, 1966), 260–81.

[50] C. B. Boyer, *The History of the Calculus and its Conceptual Development* (New York, 1949; formerly publ. under the title *The Concepts of the Calculus,* New York, 1939).

On the continent the reaction of the mathematicians of the Leibnizian school of analysis was mixed. Euler, like most working mathematicians, did not care much about the question. He suggested that differentials are mere symbols for 0, useful for distinguishing between the many values that the quotient 0/0 can assume. D'Alembert suggested in 1754 that the solution of the problem lies in considering dy/dx as the limit of the quotient $(y_1 - y)/(x_1 - x)$. Later Lagrange hoped to eliminate the question altogether by a purely formal approach based on the assumption that functions have power-series expansions and that the derivative functions can therefore be defined as the coefficients of the terms in these expansions.

These were the main lines of thought on the fundamental question in the 'period of indecision' which, in Boyer's view, lasted until Cauchy, in the early nineteenth century, took the final step and founded analysis by defining the derivative function by means of the concept of limit (which he used in a more precise way than d'Alembert).

It is not an exciting story. It is also rather unstructured. All the opinions on the fundamental question seem to be of a similar nature, namely wrong, and unaccountably obstructing the only correct solution: the limit concept as Cauchy was to use it. In fact Boyer thought that if Newton had spent more time over the fundamental problem the indecision of the eighteenth century would have been unnecessary: 'the calculus might have been established upon the concept of derivative a century before the time of Cauchy'.[52]

At first sight non-standard analysis does indeed completely change this picture. It explains why, on seemingly so insecure a basis as infinitesimals, such a rich and lasting theory could be built. The eighteenth century appears not as a period of indecision but as an age in which analysis worked informally on a basis which in principle could be justified. And it is tempting to hope that a better understanding of eighteenth-century analysis may result from reinterpreting its arguments in terms of non-standard analysis.[53]

Thus non-standard analysis provides a reappraisal of eighteenth-century analysis and a new tool for its study. As a tool it is certainly useful. For instance it now becomes possible to understand more

[51] The chapter on the eighteenth century in Boyer, *History of the Calculus* (see note 50) is called 'The period of indecision' (224–66).

[52] Boyer, *History of the Calculus* (see note 50), 196.

[53] See for instance D. Laugwitz, 'Euler und das "Cauchysche" Konvergenzkriterium', *Abhandlungen aus den Mathematischen Seminar der Hamburgischer Universität*, XLV (1976), 91–5.

deeply Euler's dealing with infinitesimals and infinites in his studies on series. Further research using this tool will most likely clarify a number of aspects of the techniques of eighteenth-century analysis.

But the role of non-standard analysis as a ground for reappraising earlier analysis, and the enthusiasm with which some acclaim this reappraisal,[54] must be approached with scepticism. One would think, indeed, that as rich and immediately successful a theory as eighteenth-century analysis needs no 'rehabilitation' to be recognized as deserving historical interest in its own right. There are signs that in the enthusiasm for this idea of a rehabilitation, mere translations of old texts in non-standard analysis terms are considered as historical research.[55]

The danger here is that this new inspiration from a modern mathematical theory leads to the same errors which caused the lack of structure in Boyer's picture, namely judging the mathematics of the eighteenth century in terms of modern concepts, and concentrating only on differentials, derivatives and infinitesimals.[56] It will therefore be useful to point out here that in a number of respects eighteenth-century analysis was decidedly different from non-standard analysis.

First, the impact of non-standard analysis on the 'rehabilitation' of infinitesimals rests on its proof that there is a mathematically consistent theory which incorporates these infinitesimals. This proof of the existence of infinitesimals could never have been given in the late seventeenth or eighteenth century. Neither do the arguments used in that period in defence of the existence of infinitely small quantities resemble in any way the existence proofs in non-standard analysis.

Secondly, in the calculus of differentials as it was created by Leibniz, the set of infinitesimals consisted of differentials of first, second, third, etc., order. Two differentials of the same order had a finite ratio; a first-order differential to the power k was of order k. In non-standard analysis the set of infinitesimals is more extensive; for any choice of first-order infinitesimals there will be infinitesimals of non-integer order. Before Euler,[57] no mathematicians conceived of infinitesimals of intermediary orders. This implies that, for analysis in

[54] See for instance the manuscript of I. Lakatos, 'Cauchy and the continuum: the significance of non-standard analysis for the history and philosophy of mathematics' (posthumously ed. J. P. Cleave, introd. R. Hersh), *Mathematical Intelligencer*, I (1978), 148–61.

[55] J. P. Cleave, 'Cauchy, convergence and continuity', *British Journal for the Philosophy of Science*, XXII (1971), 27–37, is an example.

[56] For a more detailed discussion see H. J. M. Bos, 'Differentials, higher order differentials and the derivative in the Leibnizian calculus', *Archive for the History of Exact Sciences*, XIV (1974), 1–90 (pp. 81–6).

[57] For Euler's study on infinitesimals of non-integer order cf. Bos, 'Differentials' (see note 56), 84–6.

the first half of the century, and that of the working mathematician throughout the century, the very concept of the set of infinitesimals was quite different from non-standard analysis.

A third difference lies in the fact that non-standard analysis deals with numbers and functions, whereas, at the beginning of the eighteenth century at least, analysis dealt primarily with geometrical magnitudes and variables. I shall return later to the difference between variables and functions.

From the viewpoint of non-standard analysis it is natural to concentrate primarily on the history of the concepts of differentials, fluxions, derivatives and infinitesimals. In this respect there is little difference from Boyer's approach. But for gaining a deeper insight into the history of the foundations of analysis this approach is too restricted. Derivatives are derivatives of *functions,* differentials are differentials of *variables,* fluxions are fluxions of so-called *fluents,* which are variables conceived as varying with respect to an abstract time. A study of the foundations of eighteenth-century analysis can only be successful by studying these concepts too. Both concepts, variable and function, have formed the subject of recent research. As to the function concept, its role in eighteenth-century analysis may be summarized as follows.

The first half of the century witnessed a gradual separation of analysis from its geometrical origin and background. In the seventeenth century analysis originated as a collection of methods for solving problems about curves. The methods involved algebraical rules and formulae. Leibniz and Newton introduced symbolisms for infinitesimal operations within this body of algebraical techniques. But analysis remained a collection of methods only; the problems to which it was applied were geometrical, as well as the solutions which the methods helped to find.

In the eighteenth century interest shifted from the geometrical background, notably the curve and the geometrical variables related to the curve, to the analytical expressions and the formal algebraical operations with such expressions. In this process of the 'degeometrization' of analysis it was natural that one concept emerged as central, namely the analytical expression involving one variable quantity and for the rest only numbers and constant quantities. This concept received the technical term 'function' and it acquired a central place in analysis when Euler used it as the fundamental concept in his *Introduction to the Analysis of Infinites* (1748).[58] The concept is usually called the Eulerian concept of function.

[58] See note 10. The concept was soon extended to include expressions involving more than one variable quantity.

The application of this concept to the study of geometrical curves and especially to physical problems soon showed that it was too restricted. In particular the 'arbitrary functions' occurring in the solutions of partial differential equations led, in the second half of the eighteenth century, to debates on these restrictions and on the properties which one might or might not presuppose for the functions in analysis.

These discussions have received quite a bit of attention from historians of mathematics. Truesdell, Ravetz and Grattan-Guinness have written about them.[59] Youschkevitch has recently dealt with the general question of the emergence of the function concept in an article on the history of that concept up to the middle of the nineteenth century.[60] The result of that study is a change to the usual picture of the history of the function concept. Youschkevitch shows that the general concept of function, usually attributed to Dirichlet and Lobachewsky in the 1830s, is in fact a creation of the eighteenth century: Euler already in 1755 gave a definition as general as those of Dirichlet and Lobachewsky.

At this point we should consider a topic which has been repeatedly touched upon, namely the influence of modern scientific viewpoints on the understanding of mathematics and science in history. In the cases of both rational mechanics and non-standard analysis we saw that recent developments (the axiomatization of rational mechanics, the creation of non-standard analysis) served as inspiration for new approaches to the history of mechanics and analysis respectively. We also saw that the dangers inherent in these new approaches are connected precisely with their modern scientific starting point; that starting point may cause too strong a concentration on certain concepts and developments and thus produce a unbalanced picture.

In Boyer's picture of the history of the calculus also the contemporary scientific point of view caused a selective approach to the subject: his primary interest in the foundations of the calculus and his concentration on the concepts of limit and derivative are in agreement with the view, dominant in the 1930s, of what the calculus was really about.

These three cases illustrate the point that inspiration from modern developments in science may be an innovative force behind new

[59] Truesdell, *1960* (see note 21), 243–7; J. R. Ravetz, 'Vibrating strings and arbitrary functions', in *Logic of Personal Knowledge: Essays presented to M. Polanyi on his 70th birthday* (1961), 71–88; I. Grattan-Guinness, *The Development of the Foundations of Mathematical Analysis from Euler to Riemann* (1970), 6–12.

[60] A. P. Youschkevitch, 'The concept of function up to the middle of the 19th century', *Archive for the History of Exact Sciences*, XVI (1976), 37–85.

trends in historiography of science yet simultaneously a cause of danger and a ground for critique against these trends. This tension cannot be avoided. After all, the relation to the present is the ultimate justification of historical research, and in the case of the history of science the relation to the present centres on aspects of modern science. Hence there will always be the danger of considering developments and concepts in earlier science only in so far as they embody or foreshadow concepts and aspects now important and familiar.

Also, in the study of the history of mathematical science a great part of the language in which the results of that study are reported, is mathematical. But the presentation must be comprehensible to readers with a recent mathematical training, and therefore a tension between the modern mathematical language of the presentation and the older mathematical arguments which are reported, is unavoidable.

Thus tension will always exist between the attempt to understand older mathematics on the one hand, and the interests and language of modern mathematics on the other. But that does not mean that arguments about the balance between these two aspects of historical work are unimportant. In the case of trends inspired by modern rational mechanics and non-standard analysis the balance dips decidedly to the side of modern interest and language. Here the concepts of eighteenth-century mechanics and analysis are taken seriously only in so far as they involve or lead up to modern concepts. The studies mentioned above on the concept of function take a stronger interest in the mathematical concepts as they were in the past, although Youschkevitch's central question – the emergence of the modern function concept – is still completely determined by the modern point of view.

I am myself involved in an approach to seventeenth- and eighteenth-century analysis which may be characterized by its making a stronger attempt to take the older concepts seriously as they were, and I want to conclude this chapter with some remarks on the results and expectations of that approach.[61]

One of the concepts to be taken seriously is the predecessor of the function concept. For if the concept of function was formulated explicitly in the eighteenth century, does that mean it had already been there before, implicitly, but not yet formally defined, or did it replace another concept? I think that there was indeed another concept some of whose roles the function concept came to assume. This was the concept of a variable quantity, or 'variable' for short. Variables are, for instance, the abscissa, ordinate, arc-length and subtan-

[61] Cf. Bos, 'Differentials' (see note 56).

gent with respect to a curve. Or, in a physical problem, the time, the velocity and the space traversed.

Variables are not functions. The concept of function implies a unidirectional relation between an 'independent' and a 'dependent' variable. But in the case of variables as they occur in mathematical or physical problems, there need not be such a division of roles. And as long as no special independent role is given to one of the variables involved, the variables are not functions but simply variables.

Variables do not occur alone: they occur in problems. These were, at the beginning of the eighteenth century, usually geometrical or mechanical. Hence the changeover from variables to functions during the century is directly connected with the process of degeometrization of analysis, the separation of analysis from its base in geometry.

I have found that an insight into the special role of variables (and their difference from functions) is very fruitful for the understanding of the early infinitesimal analysis as it was practised by Leibniz and his followers from the 1690s till well into the eighteenth century. Some details may illustrate this.

In the Leibnizian calculus the second-order differential of a variable is defined as the difference between two successive first-order variables:

$$\mathrm{dd}x = \mathrm{d}x^{\mathrm{I}} - \mathrm{d}x, \quad \mathrm{dd}y = \mathrm{d}y^{\mathrm{I}} - \mathrm{d}y, \quad \mathrm{dd}s = \mathrm{d}s^{\mathrm{I}}, \text{ etc. (see Fig. 1).}$$

At each point of the curve the first-order differentials are determined by the nature of the curve, in the sense that their ratios ($\mathrm{d}x : \mathrm{d}y : \mathrm{d}s$) are given by the slope of the curve. But the second- and higher-order differentials are not so far determined. It is possible, irrespective of the nature of the curve, to suppose $\mathrm{d}x = \mathrm{d}x^{\mathrm{I}}$, or $\mathrm{d}y = \mathrm{d}y^{\mathrm{I}}$, or $\mathrm{d}s = \mathrm{d}s^{\mathrm{I}}$, and hence $\mathrm{dd}x = 0$, or $\mathrm{dd}y = 0$, or $\mathrm{dd}s = 0$.

Fig. 1

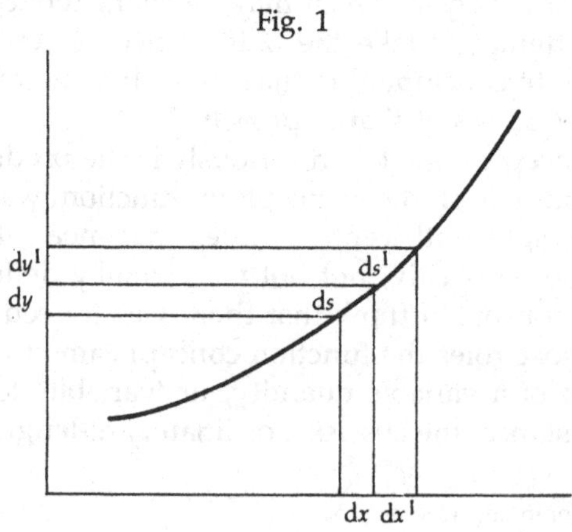

If problems about curves are analysed in terms of functions, this indeterminacy of higher-order differentials is removed by supposing the first-order differentials of the independent variable equal to each other. Thus if x is the independent variable, $\mathrm{d}x$ is considered constant, $\mathrm{dd}x = 0$, the other variables are functions of x, and their first- and higher-order differentials are fully determined by the nature of the curve and the choice of $\mathrm{d}x$ as a constant.

But if the analysis of such problems is performed in terms of variables there is no canonical preliminary choice of an independent variable and therefore the indeterminacy of higher-order differentials remains throughout the calculations.

In exploring these aspects further I have found that the indeterminacy of higher-order differentials determines many interesting aspects of the techniques of the early calculus, techniques that have since vanished. I have also shown that the indeterminacy of higher-order differentials was later experienced as a disadvantage. Euler in fact worked out a method to delete higher-order differentials entirely from the calculus. For this he had to use differential quotients, and so the indeterminacy of higher-order differentials was one of the factors in the emergence of the concepts of differential quotients and derivatives.

In this connection I may mention another point. The first half of the eighteenth century witnessed the emergence of the theory of partial differentiation and partial differential equations. The infinitesimal calculus, which was set up primarily for one-dimensional problems (especially curves) was now extended to cover two- (or more-) dimensional problems as well. This transition implied considerable conceptual difficulties. One of these was that the old concept of the differential as the difference between two successive infinitely close values of the variable is no longer tenable if the variable ranges over a two-dimensional domain. This transition is very little studied, but recent research[62] suggests that a number of aspects of the early history of partial differentiation can be put into perspective by considering the changing roles of the concepts of variable and function in the process.

In particular the importance of problems concerning families of curves and their trajectories now becomes clear. These problems were extensively studied in the early eighteenth century. They formed an intermediate stage between two types of problem: namely, on the one hand those about curves, which were the paradigm problems for the earliest form of the infinitesimal calculus, and on the other hand those

[62] By Mr S. Engelsman of Utrecht.

about two- (or more-) dimensional problems (such as surfaces), which were later studied with the help of partial differentiation techniques. The basic rules of differentiation of functions of several variables were developed in connection with trajectory problems. There is also a strong link between trajectory problems and the articles by Euler from 1740[63] in which partial differential equations occur for the first time. Until recently, it has remained quite unclear which problem Euler wanted to solve in these articles. However, it now appears that Euler's study of partial differential equations can be explained as a logical outcome of his (and others') studies on trajectories and analytical representation of families of curves.

Consideration of the concepts of eighteenth-century analysis as they were, and hence attention to the roles of variables and functions, to degeometrization and to types of problem, may bring some further structure to the picture of analysis and may suggest new problems for research.

With respect to the organization of the picture, three phases are discernible in the development of analysis. In the first phase, discussed above, which covers most of the first half of the eighteenth century, the central problem was that of adjusting the new methods to more-dimensional problems. Simultaneously, analysis went through a process of degeometrization. In the second phase, around the middle of the century, the concept of function as an analytical expression gained its central position. Analysis was consolidated around this concept as a theory concerning formulae and rules of calculating, remote from any geometrical background. In the third phase the practitioners of analysis were confronted with the contradictions of the exclusively analytical approach. This occurred in the case of the arbitrary functions in the solutions of partial differential equations. It also occurred with respect to the foundations of the calculus; it proved impossible to solve this question without having recourse to geometrical arguments about quantity in general.

This, tentative, distinction of phases suggests problems for further historical research. For the first phase I have already mentioned the theme of trajectories. For the second phase it may be fruitful to pursue the implications of the exclusively analytical approach. For instance, Euler's dealings with infinite series and Lagrange's hope for a purely analytical foundation of the calculus by means of power series

[63] L. Euler, 'De infinitis curvis eiusdem generis seu methodus inveniendi aequationes pro infinitis curvis eiusdem generis' and 'Additamentum ad dissertationem de infinitis curvis eiusdem generis', *Commentarii Academiae Scientarum Imperialis Petropolitanae,* VII (1734/5; publ. 1740), 174–200, *Opera omnia,* ser. 1, XXII, 36–75.

may well become more transparent if it is recognized that questions of convergence and existence have a quite different meaning in a formal analysis which primarily considers formulae and rules of calculations.

As to the contradictions emerging in the third phase, the debates on arbitrary functions have already attracted much attention from historians, as has the question of the foundation of the calculus. But the same tensions between formalism and existence-questions occur in the case of implicit functions. In all parts of analysis, methods were used which presupposed the existence of implicit functions. A study of the role of this assumption may reveal more about the contradictions with which analysis was confronted in the second half of the eighteenth century and which formed the motives for making analysis more rigorous in the nineteenth.

V. The context of the mathematical sciences

In the preceding section I have characterized one approach to the history of mathematics as 'taking the concepts seriously'. But this approach, centring on concepts, has in common with the earlier-mentioned approaches and trends a concentration on the internal developments of the mathematical sciences. To conclude this chapter I want to mention another trend in the historiography of the mathematical sciences which, in contrast, may be characterized as 'taking the context seriously'.

The institutional, economic and political context of past science is receiving an increasing amount of attention from historians of science. This has to do, among other things, with the fact that questions about the context of science have become increasingly crucial for the scientific enterprise now. Present-day science is challenged: it is no longer the austere activity of detached researchers, beyond the need of justification. Questions of relevance are asked, not only by disillusioned students but also in foundations and agencies that award grants. And, in addition, there are the questions about the social conditions and social conditioning of science, of its role as a factor in the economy, and of the social responsibility of science and the scientists.

In these questions about the function of science, as well as in the related philosophical disputes about the nature of science, revived by the arguments of Popper, Kuhn, Lakatos and others, history is much used as evidence. This has, as it were, created a new market for the history of science – the heightened interest in the subject among university students is one example of this. And thus, partly under the influence of these questions, historians of science have turned to the social function of science in earlier ages, to its role in the economy and in the industrial revolutions, to its social base in institutions and to the social mechanisms within the community of scientists.

The debates I have mentioned have not yet had much resonance among historians of mathematics. But this is changing. Struik has always had a strong interest in the socio-economic aspects of mathematics in history, and he has made this quite clear in his *Concise History of Mathematics*.[64] Mehrtens and I have prepared an exploratory review of the questions and sources relative to the relations of mathematics and society in history.[65] As to eighteenth-century mathematics, there is Schneider's recent summary of the influences of practice in navigation, shipbuilding, geodesy and military technology on the development of mathematics in the sixteenth to eighteenth centuries.[66] A number of recent studies on the history of probability theory devote considerable attention to the context (in games of chance, but also in demography, insurance, and the beginnings of a mathematical theory of decision processes) of the development of that subject.[67] Some studies of the social groups of mathematical practitioners and the so-called 'philomaths' may also be mentioned here.[68]

[64] D. J. Struik, *A Concise History of Mathematics* (New York, 1948).

[65] H. J. M. Bos and H. Mehrtens, 'The interactions of mathematics and society in history, some exploratory remarks', *Historia Mathematica*, IV (1977), 7–30.

[66] I. Schneider, 'Der Einfluss der Praxis auf die Entwicklung der Mathematik vom 17. bis zum 19. Jahrhundert', *Zentralblatt für Didaktik der Mathematik*, IX (1977), 195–205.

[67] For instance L. E. Maistrov, *Probability Theory, A Historical Sketch* (New York, 1974; orig. Russian edn 1964); R. Rashed, *Condorcet, mathématique et société* (Paris 1974); and the series of articles by O. B. Sheynin in the *Archive for the History of Exact Sciences*, starting in VII (1971). I. Hacking, *The Emergence of Probability* (Cambridge, 1975), centres on the concepts of probability, their relation to philosophy and especially the role of those concepts of probability that eventually were not incorporated in the standard mathematical theory of probability. J. van Brakel, 'Some remarks on the prehistory of the concept of statistical probability', *Archive for the History of Exact Sciences*, XVI (1976/7), 119–36, is also useful with regard to these questions. In fact these recent studies well illustrate the fertility of an approach which takes both the concepts and the context seriously.

[68] O. Pedersen, 'The philomaths of 18th century England', *Centaurus*, VIII (1963), 238–62; P. J. Wallis, 'British philomaths – mid-eighteenth century and earlier', *Centaurus*, XVII (1972), 301–14; I. Schneider, 'Die mathematischen Praktiker im See-, Vermessungs- und Wehrwesen vom 15. bis 19. Jahrhundert', *Technikgeschichte*, XXXVII (1970), 210–42. Cf. also the important surveys E. G. R. Taylor, *The Mathematical Practitioners of Tudor and Stuart England 1485–1714* (Cambridge, 1954), and *The Mathematical Practitioners of Hanoverian England 1714–1840* (Cambridge, 1966).

So there is, if not a new trend, at least a change of interest among historians of mathematics. Has this new interest changed the picture of eighteenth-century mathematics, or can such a change be expected from it? To discuss this let me first sketch what the now-common picture of mathematical science in the eighteenth century implies about the context. In the eighteenth century mathematics was a very broad field. Outside pure mathematics (algebra, analysis, geometry) there was mixed mathematics or the mathematical arts and sciences, including optics, probability, practical and theoretical mechanics, astronomy, navigation, shipbuilding and the military sciences such as fortification and artillery.

Mathematics at that time had a more direct and intense interaction with applications than in any other period. In earlier ages the extent of the mathematical arts and sciences was much smaller, and in the nineteenth century pure mathematics became an independent discipline, creating a large distance between itself and mathematical applications, a distance which is still felt as very wide.

In the eighteenth century most of the work in pure mathematics and the highly mathematized sciences such as mechanics and astronomy took place in the academies. These characteristic eighteenth-century institutions of science were institutions of the state. They were furthered by enlightened monarchs, because a flourishing academy, with internationally famous scientists, lent lustre to their reign and served a useful function in the state. The state could profit from expert scientific advice on military, naval and economic matters as well as questions of infrastructure and education.

Nearer to mathematical practice, other institutions played an important role, the various military and civil schools educating engineers and other professionals who needed mathematics. In France, especially, many such schools flourished, and were the precursors of the Ecole Polytechnique established in 1795. There were also the mathematical societies, formed by mathematical practitioners, instrument-makers, and teachers of course in mathematics.

Within this wide field of mathematics – from pure analysis to practical calculations, from the academies and the schools to the workshops of the mathematical practitioners and instrument-makers, and to the magazines for gentlemen and ladies interested in matters mathematical – were interactions of all sorts. A number of famous problems whose solution was felt to be desirable by economic or political powers and which inspired mathematical scientists to theoretical studies were major themes in the pattern of such interactions. These were the problem of longitude (that is the determination of a ship's position at sea), the question of the form of the earth, the paths of projectiles under resistance of the air, and the design of ships. The actual influence of theoretical results in the case of these problems is very difficult to assess. As regards the longitude problem the influence was certainly there; in the case of the other problems it is doubtful whether insights gained through mathematical theory effectively influenced practice before the nineteenth century.

So far the picture; what changes in it may we expect from a new interest in the social context of mathematics and the relations of mathematical theory to practice? Several parts of the picture, especially those concerning the use of mathematical knowledge in practice, are still very vague. Let me mention two themes which from the new point of view take on interest and whose study could provide more perspective on the empty parts of the picture.

One theme is mathematical teaching. Not very much is known about the contents and function of mathematical teaching at the engineering schools for instance.[69] Yet towards the end of the century the requirements of didactic presentation in teaching began to influence the theories themselves. It would be helpful to know more about the earlier phases of that process.

Another theme is that of the professional group of the mathematical practitioners. Their role in the transmission and use of mathematical knowledge and in determining the status of mathematical knowledge has certainly been great. At the end of the eighteenth century, in the period of the Industrial Revolution, the profession gradually disappeared. The study of this group, their work and their organizations, will lead to a better understanding of the ways mathematical knowledge functioned in practice.

What makes the study of these kinds of questions difficult is that they require quite different methods from those commonly used in the study of the history of mathematics. Concentration on the mathematics itself, on concepts, problems, theories, does not yield answers to these questions. Other methods are needed, for instance sociological, or collective biographical methods. For these the historians of mathematics can turn to their colleagues in history of science who already have a somewhat longer experience in dealing with these questions.

[69] R. Taton (ed.) *Enseignement et diffusion des sciences en France au XVIIIe siècle* (Paris, 1964) is a most valuable source but it hardly permits a view on the actual content of mathematical teaching.

VI. Conclusion

In this survey of recent trends in historiography of eighteenth-century mathematical science I have emphasized the interactions between historical research and the present-day interests of the community of mathematical scientists. In the case of mathematics this interaction has always been strong, as is evidenced by the longstanding interest in the auxiliary role of history of mathematics in mathematical teaching. The interaction is valuable and fruitful, suggesting new questions and providing historians of mathematics with a demanding audience.

On the other hand, one of the effects of this situation is the often rather esoteric character of research in the history of mathematics, which makes it inaccessible to those without a special mathematical training. Through this, history of mathematics stands somewhat apart within the field of history of science in general. This is a regrettable situation. But in the trends I have mentioned there are some elements that may help to bridge the gap between history of mathematics and history of science and technology in general. These are: interest in the interaction between mathematics and the problems of natural science, interest in concept formation, and interest in the social context of the mathematical sciences.

9

Experimental natural philosophy

J. L. HEILBRON

Contents

I. Historiographical considerations 358
II. The Case of electricity 366
III. Lessons and extrapolations 375
IV. Desiderata 383

During the eighteenth century natural philosophy – or *fisica, physique, physica, Naturlehre* – broke loose from the place in the organization of knowledge that it had occupied since antiquity. The scientific revolutionaries of the seventeenth century, however much they altered the principles and doctrines of physics, had left it with the purpose, method and coverage assigned to it by Aristotle. Eighteenth-century ferments decomposed and recombined it, drove off old parts, fixed new ones, and restructured its bonds with the body of knowledge.

We do not find an answering ferment among historians of eighteenth-century physics. That is not because they are few or idle. Perhaps as many as one historian of science in ten works on eighteenth-century physics.[1] Almost 10 per cent of the articles published during the past twenty years in the leading general journals – *Isis* and *Revue d'histoire des sciences* – concern the subject. About one half of these articles relate to experimental physics and the making of instruments. These figures are representative.[2] Over the past fifty years the moderate but sustained investigation of eighteenth-century experimental physics has produced over 300 books and papers that are noticed in the *Isis* lists or in the bibliographies in the *Dictionary of Scientific Biography*.

This literature contains much information of great value. But it offers little in the way of helpful generalization or periodization. In

[1] One hundred of the one thousand members of the History of Science Society who reported their interests in its *Directory of Members*, comp. by R. Regan (Washington, 1977), mentioned eighteenth-century physics.

[2] Between 6 and 10 per cent of the papers published in *Annals of Science* 1936–69, *Archives internationales d'histoire des sciences* 1962–76, and *Notes and Records of the Royal Society of London* 1938–65, pertain to eighteenth-century physics, about 4 per cent to experimental physics.

only one case – electricity – has the development of eighteenth-century experimental physics been studied in detail; over 40 per cent of the literature has been devoted to it, about as much as to meteorology, optics, thermodynamics, pneumatics, and magnetism taken together. This imbalance and the dearth of useful general analysis have dictated the form of this essay. First, in section I, I discuss how the eighteenth century might be rated a distinctive period in the history of experimental physics. Next, in section II, comes an outline of the history of electricity, from which, in section III, a few generalizations about eighteenth-century physics are drawn. The chapter concludes with suggestions for further work.

I. Historiographical considerations

The practices of the last general historians of physics may help us approach the problem of periodization. All began their accounts of the modern era with Galileo, the 'Columbus of science', who brought physicists across the gulf of scholasticism to the shore of the true philosophy.[3] From his method of confirmed experiment and mathematical analysis[4] the tree of physics sprang, and waxed exceedingly, 'preserv[ing] its old wood and build[ing] upon it through the cambian layer of new facts and better fitting theories'.[5] How did the tree grow? What were its periods, epochs and eras? What distinguishes the eighteenth century?

Earlier proposals[6]

One answer is that no general principles, no systematic chronology, moderated the growth of physics. Its branches shoot independently; the true history of physics is the sum of the histories of its subfields.[7] Nor have the several histories any natural punctuation. 'Event follows event in unbroken series, never offering a resting place for convenient division.'[8] The only natural way to describe eighteenth-

[3] Ferdinand Rosenberger, *Geschichte der Physik* (3 vols., Braunschweig, 1882–90), II, 5.

[4] J. C. Poggendorff, *Geschichte der Physik* (Leipzig, 1879), 6; E. Gerland, *Geschichte der Physik* (Munich and Berlin, 1913), 5.

[5] H. Crew, *The Rise of Modern Physics,* 2nd edn (Baltimore, 1935), ix.

[6] The weakness of the literature may be gauged from E. Olszewsky, 'Les problèmes de périodisation dans l'histoire de la science et de la technique', *Archives internationales d'histoire des sciences,* XIII (1960), 110–14.

[7] E. Hoppe, *Geschichte der Physik* (Braunschweig, 1926), iii–iv.

[8] E. Gerland and F. Traumüller, *Geschichte der physikalischen Experimentierkunst* (Leipzig, 1899), 5.

century physics on this principle is to follow the history of each branch in turn, from 1700 to 1800, without pause or emphasis. Abraham Wolf adopted this arrangement for his *History of Science, Technology, and Philosophy in the 18th Century,* which contains the best recent general account of eighteenth-century physics with which I am acquainted.[9]

A second answer admits that the physics of the modern era has a few distinctive periods, for example from Galileo to Newton, from Laplace to the conservation of energy, from conservation to the end of the nineteenth century. As for the eighteenth century, it obtains a character by subtraction. It has no grand syntheses, no revolutions; its unifying element is mediocrity, the want of a worthy successor to Newton. This widespread view has not yielded a fruitful historiography.[10]

A third answer accepts division into periods but denies that the eighteenth century qualifies as one. The immortal Poggendorff, without whose *Handwörterbuch* we should scarcely know where to begin, found greater unity in the period from 1660 to 1750 than in that from 1700 to 1800. He observed that leadership in physics passed from Italy to England and France about 1660. The next hundred years saw the rise, dominance and fall of Newton, '[who] in a certain sense may be regarded as the keystone of the old physics'. Newton completed the sciences begun by the ancients: mechanics, astronomy and optics. Electricity, magnetism and chemistry are modern subjects whose coming of age after 1750 opened a new era in the history of physics.[11] Ferdinand Rosenberger saw the dynamics of advance in modern physics as an alternating imbalance in the partnership of experiment and mathematics. Between 1650 and 1690, and again between 1750

[9] Wolf's book was published in 1938, re-edited in 1953, and reprinted in 1961. The histories of Crew (see note 5), Hoppe (see note 7), F. Cajori (*A History of Physics* (New York, 1929)), and E. Gerland (*Geschichte der Physik* (Munich and Berlin, 1913)) are also arranged by branches.

[10] E.g., Cajori *History of Physics* (see note 9), 106; J. D. Bernal, *Science in History* (1965), 358; D. H. Hall, *History of the Earth Sciences during the Scientific and Industrial Revolutions* (Amsterdam and New York, 1976), 157.

[11] Poggendorff, *Geschichte der Physik* (see note 4), 6–7. Cf. T. S. Kuhn, 'Mathematical vs. experimental traditions in the development of physical science', *Journal of Interdisciplinary History,* VII (1976), 1–31. A similar periodization (1660–1750) is defended on sociological grounds by G. Böhme and W. Van den Daele in G. Böhme *et al.* (eds.), *Experimentelle Philosophie: Ursprünge autonomer Wissenschaftsentwicklung* (Frankfurt am Main, 1977), 152–9, 190–219; and a strong inflection occurs *c.* 1770 in Rainoff's curves (T. J. Rainoff, 'Wavelike fluctuations of creative productivity in the development of West-European physics in the 18th and 19th centuries', *Isis,* XII (1929), 287–319 (pp. 292–3, 297).

and 1780, experiment dominated; in the interim it was mathematics, or the elaboration of mechanics and the gravitational theory.[12]

The fourth answer is that eighteenth-century physics had unity and importance as the working out, and ultimate triumph, of 'Newtonianism'. This proposition now commands wide assent; one inclines to characterize the history of Enlightenment physics, as one does the progress of its politics, economics and technology, as a conquest of Europe by England. Newtonianism victorious over darkness and Descartes is the keynote of the latest general account of eighteenth-century science.[13] What choice was there? 'The physical science of the eighteenth century could only be Newtonian, of course.'[14]

Three objections can be brought against characterizing eighteenth-century physics as Newtonian. Firstly, to make the tag fit even approximately, 'Newtonian' must be taken so broadly as to be useless in historical analysis.[15] There were, to begin with, at least two distinguishable public Newtons, who inspired quite different groups.[16] Disciples squabbled over fundamental points in his theory of matter; if one group was orthodox, its opponents could not have been Newtonian.[17] Secondly, had all the disciples been sectaries of a common religion, or physicists of the same paradigm, they could not have

[12] Rosenberger, *Geschichte der Physik* (see note 3), II, 1–5, 362–5. Charles Fabry, 'Histoire de la physique', in G. Hanotaux (ed.), *Histoire de la nation française* (15 vols., Paris, 1924), XIV, 167–418 (pp. 209, 245–6), also breaks the eighteenth century at 1780, after which (he says) precise measurement and quantitative theory came in.

[13] R. Taton (ed.), *Histoire générale des sciences* (4 vols., Paris, 1957–64), II, 422–3. Cf. Fabry, 'Histoire de la physique' (see note 12), 209, and G. Gusdorf, *Les sciences humaines et la pensée occidentale*, IV, *Les principes de la pensée au siècle des lumières* (Paris, 1971), 151–79.

[14] C. C. Gillispie, *The Edge of Objectivity* (Princeton, N.J., 1960), 152. Cf. H. Dingle, 'Physics in the 18th century', in A. Ferguson (ed.), *Natural Philosophy Through the Eighteenth Century and Allied Topics* (1948), 28–46 (p. 29): 'The dominant fact that directed the whole of 18th century science was that Newton had succeeded in formulating laws of motion'.

[15] Cf. R. Hahn, *Laplace as a Newtonian Scientist* (Los Angeles, 1967), and R. Home, '"Newtonianism" and magnetism', *HS*, XV (1977), 252–66. An example of the difficulty: Gusdorf, *Les sciences humaines* (see note 13), 168–9, 176, makes experimental physics flow from Newton and cites J. A. Nollet as its finest exponent; I. B. Cohen, *Franklin and Newton* (Philadelphia, 1956), 388 *et passim*, takes Nollet, the opponent of Newton's follower Franklin, as arch anti-Newtonian.

[16] This distinction is developed by Cohen, *Franklin and Newton* (see note 15).

[17] Cf. Y. Elkana, 'Newtonianism in the 18th century', *British Journal for the Philosophy of Science*, XXII (1971), 297–306; P. M. Heimann, 'Newtonian natural philosophy and the scientific revolution', *HS*, XI (1973), 1–7.

made their century Newtonian. The continent had, and retained, its own traditions. Many streams besides Newton's great river sprang from the philosophical watershed of the seventeenth century. They ran parallel for a time. Channels then opened between them; courses shifted, the waters mingled and the torrent of classical physics poured forth.

Thirdly, with reference particularly to experimental physics, its qualification as Newtonian repeats a pervasive fault in our historiography: the tendency always to make general theory, or world-view, or deep methodological principle the driving force in the growth of science. The metaphysics or foundations of the paradigms, or of the research programmes, or of the personal 'dictionaries'[18] supposed to guide scientists may be too remote from experimental work to order it in useful ways. In the case of electricity, the experiments done early in the eighteenth century cannot be distinguished as 'Newtonian' or 'Cartesian', nor can the instruments that provided the unexpected discoveries that forced advance. And the same is true, although in a different sense, of the hypotheses and models devised to represent the phenomena. For although one recognizes that models incorporating vortices derived ultimately from Descartes, while those invoking springy spirits probably owed something to Newton, yet in practice all such qualitative models came to much the same thing, aether being to the one side what subtle matter was to the other.

Interpreting Enlightenment experimental physics as the working out of a grand programme not only does not fit what is known, but may conceal points of importance. As Ludovico Geymonat observed, the fact that experimental physics coupled so weakly with fundamental principles in the eighteenth century is itself important evidence about the 'painful construction of modern science'.[19] The means of production of physical knowledge and the professional circumstances of the producers then decisively altered the direction of physics.[20]

An overlooked possibility

The difficulty in characterizing eighteenth-century physics has its root in the thing itself. It is difficult to grasp because it constantly changed. In 1700 traditional physics, the physics of the schools, was

[18] For the notion of the dictionary, a shifting, one-man paradigm, see E. Bellone, *Il mondo di carta* (Milan, 1976), 17–25, 142–61.

[19] L. Geymonat, *Storia del pensiero filosofico e scientifico*, III, *Il settecento*, 2nd edn (Florence, 1973), 235.

[20] Cf. A. Kauffeldt, 'Zur Periodisierung der Geschichte der Naturwissenschaften: I', Dresden, Technische Hochschule, *Wissenschaftliche Zeitschrift*, VII (7) (1957/8), 146–55.

inclusive, qualitative and literary.[21] It covered all branches of natural science from celestial mechanics through biology to psychology. It sought the essences or principles of things, not information about their sizes, speeds positions or other 'accidents'. Those whose primary concern was measurement or calculation did mixed mathematics, not physics. No more did the traditional physics authorize prolonged and promiscuous experiment. It had no need for special apparatus for exploring nature, and even good reason to eschew it. By making nature act unnaturally, that is, uncommonly, instruments might mislead physicists intent on knowing how she acts habitually, 'or for the most part'.[22]

By 1800 physics had jettisoned the biological sciences and distanced itself from most of chemistry. Physicists contested jurisdiction over former branches of applied mathematics, such as optics and mechanics. And, above all, they insisted upon subjecting to measurement or experiment everything they could lay their hands on. No doubt much of their 'experimenting' was nothing more than parlour tricks; the *Almanac dauphin* for 1777 lists as *'physiciens'* a showman and a magician, as well as two disciples of Nollet. Still, even the quacks and jobbers serve as evidence of the transformation they exploited and caricatured: by 1800, indeed by 1750, 'true physics ha[d] become experimental physics'.[23]

The unity and distinction of eighteenth century physics must be sought in precisely what makes it most difficult to grasp: in its process of redefinition, in its changing scope and technique. The older general historians missed this point because they aimed not to find the historical forces at play, but 'to clarify the fundamental concepts [of modern science] and exhibit their origins'.[24] They looked for their material in earlier contributions to the subjects recognized in their

[21] Cf. P. Allen, 'Scientific studies in the English universities of the seventeenth century', *JHI*, x (1949), 219–53; H. F. Fletcher, *The Intellectual Development of John Milton* (2 vols., Urbana, Ill., 1956–61); E. G. Ruestow, *Physics at 17th and 18th Century Leiden* (The Hague, 1973); R. Taton (ed.), *Enseignement et diffusion des sciences en France au XVIIIe siècle* (Paris, 1964), 146.

[22] Aristotle, *Physica*, II. 5 (196^b 13–14).

[23] Memorandum of the municipality of Le Havre, 1762, quoted by A. Anthiaume, *Le collège du Havre* (2 vols., Le Havre, 1905), I, 225. Cf. H. Schimank, 'Die Wandlung des Begriffs "Physik" während der ersten Hälfte des 18 Jahrhunderts', in K. H. Manegold (ed.), *Wissenschaft, Wirtschaft und Technik* (Munich, 1969), 454–68; H. Schipperges, 'Zur Bedeutung von "physica" und zur Rolle des "physicus" in der abendländischen Wissenschafts-geschichte', *Sudhoffs Archiv*, LX (1976), 354–74.

[24] F. Dannemann, *Die Naturwissenschaften in ihrer Entwicklung und ihrem Zusammenhänge* (4 vols., Leipzig and Berlin, 1910–13), IV, 2.

time as constituting physics. They omitted much that the eighteenth century included; they picked random pieces and scattered episodes, which necessarily summed to a nondescript whole.

A more promising approach is to work from a definition of physics – or, for our purposes, of experimental physics – common in the late eighteenth century. The process by which that meaning grew out of the earlier, literary one would be the theme of our history. Coherence would come not from feigning equal conceptual advances in all branches of physics or from imposing a general theoretical framework upon them, but from exhibiting factors common to the development of subjects identified as *'eigentliche Physik'*, *'physica stricte talis'*, or 'experimental natural philosophy' in the later eighteenth century.

Finding a representative definition will not be easy, for to judge it we must know much of the history that it is to help us to write. As a first approximation I propose to take the body of knowledge implicitly defined in a university textbook that satisfies the following conditions. It must have been popular, authoritative and influential. It should draw on Newton but miss his subtlety, and utilize continental work without regard to doctrine. It should illustrate theory with experiment whenever possible and use numbers for measurement but not for calculation. As for coverage, it must omit the biological sciences and attend to, at a minimum, mechanics, optics, fluid mechanics, pneumatics, heat, meteorology, geophysics, electricity and magnetism. A justification of some of these last requirements may be in order.

Firstly, as to mathematics, almost all textbook writers from 'sGravesande to Lichtenberg insist upon its importance for physics both in the expression of laws and in the recording of data. Further mathematical elaboration was not, however, the business of physicists. Once experiment had yielded a general principle, such as universal gravitation or the fundamental theorem of hydrostatics, the theory raised on the principle became the toy of mathematicians.[25]

A minority of physicists professed to disparage mathematics. Diderot defined experimental physics as natural science that could not be quantified. He believed that most matters of interest were qualitative, and looked forward to the expiration of the race of geometers.[26] W. C. G. Karsten, professor of mathematics and physics at the Uni-

[25] See articles 'Expérimental' and 'Physico-mathématiques' in D. Diderot (ed.), *Encyclopédie, ou Dictionnaire raisonné des sciences, des arts, et des métiers* (Geneva, 1777–9).

[26] H. Dieckmann, 'The first edition of Diderot's *Pensées sur l'interpretation de la nature*', *Isis*, XLVI (1955), 251–67 (p. 252); L. G. Krakeur and R. L. Krueger, 'The mathematical writings of Diderot', *Isis*, XXXIII (1941), 219–32 (p. 229).

versity of Halle, likewise insisted that mathematics dealt with quantities, physics with qualities: 'it is absurd to treat the same subject as applied mathematics and as physics'.[27] At the Coffee House Physical Society, they banned everything that smelt of mathematics. An errant member once tried to present an 'astronomical paper'; his fellows refused to hear him. 'The constitution of the Society [they said] does not extend to the consideration of subjects that may lead to mathematical disquisition, but is confined to experimental philosophy.'[28]

The disparagers and the promoters of mathematics in physics did not always differ in practice. The level of analysis in physics was commonly very low. General textbooks seldom invoked more than the laws of proportion. Few research papers employed the calculus. Numbers were used primarily to record data,[29] seldom as material for elaborate calculations or as checks of a quantitative theory.

As to the physical subjects required in our representative textbook, they have been identified by inspecting two dozen texts published between 1735 and 1790. Perhaps only the inclusion of meteorology requires comment. Meteorology belonged to experimental physics by virtue of its instrumentation,[30] and because its main concerns – the aurora, thunder and lightning, winds, precipitation of all kinds, and the constitution of the atmosphere – related closely to the study of heat, electricity and pneumatics, to which no one denied a place in physics. The tie became closer as the century progressed, after good instruments, easily deployed, became plentiful and the advance of pneumatics had revealed the complexity of 'air'.

Pneumatics had always occupied a place in the van of physics. A learned modern historian considers that experimental physics 'in the strict sense' had its origin in the seventeenth century, in the determination of the properties of air.[31] Historians writing around 1800 emphasized pneumatics above all other branches of eighteenth-century

[27] Quoted by J. C. Fischer, *Geschichte der Physik* (8 vols., Göttingen, 1801–8), VI, 2–3; cf. W. C. G. Karsten, 'Vom eigentlichen Gebiet der Naturlehre', in his *Physisch-chemische Abhandlungen* (Halle, 1786).

[28] Minute Book, Museum for the History of Science, Oxford.

[29] Cf. L. Tilling, 'Early experimental graphs', *British Journal for the Philosophy of Science*, VIII (1975), 193–213.

[30] Cf. W. E. Knowles Middleton, *The History of the Barometer* (Baltimore, 1964); *idem*, *A History of the Thermometer and its Use in Meteorology* (Baltimore, 1966); *idem*, *Invention of the Meteorological Instruments* (Baltimore, 1969).

[31] M. Gliozzi, 'Le origini della fisica sperimentale: la determinazione del peso specifico dell'aria', *Periodico di matematico*, XI (1931), 1–10.

physics. Friedrich Murhard, for example, began and ended his truncated *Geschichte der Physik* (1798–9) with ballooning and barometry. Johann Carl Fischer and Antoine Libes divided their accounts of eighteenth-century physics into two major epochs separated by 'la naissance de la chimie pneumatique'.[32]

The discovery of the different species of air made acute the problem of the constitution and activity of the atmosphere. Principles derived from the study of heat (evaporation and condensation) and electricity (lightning) as well as from pneumatics seemed directly applicable to the problems of meteorology.[33] Quantities of data stood ready for interpretation. To borrow a conception from Gehler's *Physikalisches Wörterbuch,*[34] meteorology was 'applied physics, in the strict sense': it used principles obtained from a study of physics to explain or elucidate the properties of the atmosphere. On precisely the same ground geophysics – physical geography, terrestrial magnetism, the origins of springs and fountains, etc. – qualified as physics.

A textbook satisfying our criteria is Pieter van Musschenbroek's *Introductio ad philosophiam naturalem,* published in 1762, reprinted in 1768, translated into French in 1769. The choice has the sanction of the most knowledgeable historian of eighteenth-century physics,[35] and may be further recommended by Musschenbroek's place in Enlightenment pedagogics. He was a prolific writer of authoritative texts, of which the *Introductio* was the last and longest. An earlier version, the *Essai de physique* (1739), had already won a reputation for comprehensiveness: 'despite its modest title, the book is no less than a learned course of physics, in which all the subjects of the science are treated with as much depth as clarity'.[36] Musschenbroek mediated between Newtonianism, as represented by 'sGravesande, and continental traditions, which he had studied both as physicist and physician. In multiplying measurements and experiments he had no superior; his books filled with tables and drawings initiated genera-

[32] Fischer, *Geschichte der Physik* (see note 27), VIII, lviii; A. Libes, *Histoire philosophique des progrès de la physique* (4 vols., Paris, 1810–13), I, 1. Cf. C. von Klinckowstroem, 'Die Anfänge der physikhistorischen Forschung', *Archivio di storia della scienza,* IV (1923), 113–22.

[33] Libes, *Histoire philosophique* (see note 32), IV, 180, makes this point explicitly.

[34] From the article 'Physik' in VII, part 1 (2nd edn, Leipzig, 1833), 501.

[35] Fischer, *Geschichte der Physik* (see note 27), IV, 4–5.

[36] P. Brunet, *Les physiciens hollandais et la méthode expérimentale en France au XVIIIe siècle* (Paris, 1926), 122, quoting the *Journal des savants.* Cf. A. Savérien, *Histoire des philosophes modernes,* VI, *Histoire des physiciens* (Paris, 1768), 345–98, and C. de Pater, 'Petrus van Musschenbroek (1692–1761), a Dutch Newtonian', *Janus,* LXIV (1977), 77–87.

Table 1. *Coverage of eighteenth-century experimental physics*

	Topics	Per cent of space in representative text[a]		Per cent of historical literature[b] 1927–1965	1966–1977
1.	Mechanics Statics, dynamics, percussion; machines, friction; compound motion, pendula, projectiles; collisions; strength of materials	24[c]		12	13
2.	Fluid mechanics Surface tension, capillarity; hydrostatics, specific gravity; hygrometry; fluid flow	8	} 13 (2 and 2a)	2	1
2a.	Water Physical properties, forms; specific gravity; freezing, boiling, evaporation	5			
3.	Light Geometrical optics; colours, Newton's rings; the eye, optical instruments; luminescence	13		7	12
4.	Air Specific gravity, density; pneumatic instruments; relation between volume and pressure; sound	10		12	13
5.	Fire Pyrometer; burning mirrors and lenses; thermometers; melting points; rate of absorption of heat, calorimetry; nature of flame, heat and fire	6		15	9

(*continued*)

	Topics	Per cent of space in representative text[a]	Per cent of historical literature[b]	
			1927–1965	1966–1977
6.	Electricity	6	43	38
7.	Magnetism	3	2	3
8.	Meteorology	16	7	11
	Cloud formation, rain, hail, snow, dew; rainbows, coronas, halos, parhelia; auroras, zodiacal light, lambent fire, lightning; winds			

[a] *Introductio ad philosophiam naturalem*, 2nd edn (Padua, 1768).
[b] From *Isis* critical bibliographies, 1966–77, and *Cumulative Bibliography*, 1927–65; and articles on experimental physicists in *Dictionary of Scientific Biography*.
[c] Omits section on theory of matter and definitions (9 per cent).

tions into the study of experimental physics, 'the furthest limit, the highest peak, of science'.[37]

The coverage of the *Introductio* is indicated in Table 1, which suggests Musschenbroek's emphases by reporting the fraction of space he allotted to each major topic. The table also suggests the division of labour among modern students of eighteenth-century experimental physics: the numbers in the two right-hand columns are the percentages of books and articles they have published in the past fifty years on the topics considered in the *Introductio*. It appears that historians have given much less attention to the experimental side of mechanics and fluid mechanics, somewhat less to meteorology, and much more to electricity than did our representative text.

II. The case of electricity

Chief among the reasons that historians have attended more to electricity than to other branches of eighteenth-century physics is the

[37] 'Oratio de methodo instituendi experimenta physica', in *Tentamina experimentorum naturalium captorum in Academia del Cimento* (Leyden, 1731), x, quoted by Ruestow, *Physics at Leiden* (see note 21), 132.

magnitude of its advance. Little was inherited from earlier times: a misleading definition of electricity; a few facts about electrical attraction; an inventory of bodies able to perform it; and two rival theories about its true cause. By 1800 electricians had abandoned the search for true causes, worked out the principles of electrostatics, established the basis for a mathematical theory, and opened the vast new domain of galvanism.

Secondly, much apparatus has survived, some of it still handsome and serviceable; collectors of antique instruments, historians of technology, curators of museums, have devoted themselves to its study.[38] Thirdly, the Enlightenment liked to play with electricity. Philosophers and duchesses, dragoons and barmaids, patronized makers of electrical instruments, public lecturers, showmen, vagabonds, jugglers, anyone who could perform amusing or instructive electrical games. Social historians have been attracted by this peculiar phenomenon.[39] Fourthly, electricity had its uses, even in the eighteenth century, and not only (as Franklin said) to make theoreticians humble. Electrical shocks seemed to help the palsied and paralytic.[40] Electrical conductors, properly deployed, protected against damage by lightning.[41]

Finally, the path of the historian of electricity has been smoothed by scores of collectors, librarians, and other historians. One can write an exhaustive history of electrical theory from the materials assembled by Ronalds, or by Clark, or by Dibner.[42] Excellent bibliographies, of which the latest assesses the libraries of Philadelphia,[43] list as much or more than one requires. As for histories, we have them from

[38] E.g., W. C. Walker, 'The detection and estimation of electric charges in the 18th century', *AS*, I (1926), 66–99; W. D. Hackmann, *John and Jonathan Cuthbertson: The Invention and Development of the 18th-Century Plate Electrical Machine* (Leyden, 1973).

[39] E.g., D. Mornet, *Les origines intellectuelles de la révolution française* (Paris, 1947); A. E. Musson and E. Robinson, *Science and Technology in the Industrial Revolution* (Toronto, 1969).

[40] A. Schmid, *Bibliographisches zur Frühgeschichte der Elektrizität und ihrer medizinischen Anwendung* (Bern, 1931); P. G. Ritterbush, *Overtures to Biology* (New Haven, 1964).

[41] P. Brunet, 'Les origines du paratonerre', *RHS*, I (1948), 213–53; I. B. Cohen, 'Prejudice against the introduction of lightning rods', *Journal of the Franklin Institute*, CCLIII (1952), 393–440.

[42] Available at the Institute of Electrical Engineers (London), the American Institute of Electrical Engineers (New York), and the Smithsonian Institution (Washington), respectively.

[43] E. G. Gartrell (comp.), *Electricity, Magnetism and Animal Magnetism: A Checklist of Printed Sources, 1600–1850* (Wilmington, Del., 1975).

the middle of the eighteenth century, from the moment that electricians began to suspect that they were participating in great events.[44]

These events may be divided into three periods, each characterized by a particular set of values of three variables: content, method and support.[45] During the first period, 1700–1740, electricity rose from an undistinguished variety to a sub-species in the genus 'experimental physics'. Seventeenth-century writers had treated it under 'attractions', along with magnetism and the glance of the basilisk, or under whatever rubric they put amber, their prime electrical generator. The chance discovery of electro-luminescence early in the eighteenth century suggested a connection between electricity and fire, under which it received notice in 'sGravesande's influential textbook. Discovery about 1730 of the basic regularities of electrostatics gave electricity a prominent place in the memoirs of the leading scientific academies. Most of the work that brought about this promotion was done by men closely associated with these institutions.

In the second period, 1740–1760, qualitative information increased rapidly. Electricity commanded its own monographs and independent sections in textbooks. Pleasant theories, likely stories with little or no mathematics, were devised that almost fit the facts. Amusing demonstrations, displays, audience-participation games, spread interest in electricity throughout the polite world. The interest was not merely frivolous. Electricity appeared to cure paralysis, cause earthquakes, and give the force of thunderbolts. Everyone contributed: academicians, university professors, the common man.

In the third period, 1760–1790, the qualitative theories and explorations gave way to phenomenological or instrumentalist description, to mathematical formulation, and to precise measurement. The rate of invention of instruments increased. Electricity had textbooks of its own, and book-length bibliographies. The common man retired, as did the casual academician; a few professors and salaried academicians dominated the subject.

Let us put some flesh on these bony generalizations.

[44] To mention only the most important: D. Gralath, 'Geschichte der Elektrizität', *Versuchen und Abhandlungen der Naturforschenden Gesellschaft, Danzig*, I (1747), 175–304, II (1754), 355–460, III (1757), 492–556; J. A. Sigaud de la Fond, *Précis historique et expérimental des phénomènes électriques* (Paris, 1781); J. Priestley, *The History and Present State of Electricity* (2 vols., 1767; 3rd edn, 1775; repr., 1966).

[45] The following account is drawn from my book, *Electricity in the 17th and 18th Centuries: A Study of Early Modern Physics* (Berkeley, 1979), in which full documentation, and acknowledgement of debts to other historians, will be found.

Period 1: 1700–1740

The study of electricity was revived early in the eighteenth century by Francis Hauksbee, curator or demonstrator to the Royal Society of London. The subject had not invited his attention: the known effects of electricity were too few and too weak to provide material for the 'experiments' with which he had to entertain the Society each week.[46] He came to electricity via ingenious and imposing demonstrations of luminosity produced by attrition. The need for big effects visible at a distance inspired him to replace the usual electrical generators and detectors – pieces of amber and bits of paper – with glass tubes and lampblack or leafbrass. The new apparatus served until the introduction of the electrical machine about 1740.

With his relatively powerful and sensitive equipment Hauksbee almost changed the definition of electricity, 'the property of attracting when rubbed'. He noticed that the tube sometimes threw off the lampblack it had drawn. Soon systematic writers, such as 'sGravesande, were crediting electricity with both attraction and repulsion. Another important consequence of Hauksbee's improvements was the recruitment of Stephen Gray, a dyer by trade, into the small band of electricians. Eventually Gray, armed with a glass tube, discovered that electricity could be communicated from one body to another, and even transmitted over considerable distances. How far? In attempting to find out, Gray discovered the distinction between conductors and insulators.

The miscellaneous results of Hauksbee and Gray seemed capricious and contradictory. They engaged the interest of a man who detested mysteries, Charles François de Cisternay Dufay, a pensionary of the Paris Academy of Sciences. Dufay worked out the basic regularities of electrostatics: that all bodies save flame can be electrified; that repulsion occurs after attraction and the communication of electricity; that there exist two sorts of electricity; that bodies with the same sort repel one another and with dissimilar sorts attract; and that moisture promotes conduction and weakens insulation. Dufay raised knowledge of electricity to the point where it could invite and reward prolonged scrutiny by professional savants. It was not long before the subject fell into the hands of German professors.

The earliest of these pedagogues, Georg Matthias Bose, mechanized the generation of electricity. Irked by the labour required to repeat Dufay's experiments with the tube, he fashioned a piece of

[46] Cf. H. Guerlac, 'Sir Isaac and the ingenious Mr Hauksbee', *Mélanges Alexandre Koyré* (2 vols., Paris, 1964), I, 228–53.

an alembic into a globe to be spun against the hand. Like Hauksbee, Bose was concerned to create demonstrations that would impress his viewers and promote himself: success materialized in fees, and fees made the difference between penury and comfort for a German professor. Typical of Bose's playfulness was the *Venus electrificata,* an insulated electrified young lady whose kisses would be felt long after they were enjoyed.

It was in the 1740s, under the inspiration of Bose, that 'electricity took the place of quadrille'. There was no end to the inventiveness of the German electricians. One, C. F. Ludolff, M.D., a member of the Berlin academy, ignited spirits with an electric spark. Everybody rushed to see physicists throw thunderbolts from their fingers. Bose proposed that they electrify water, and draw sparks from it: it would be odd and amusing to get fire from water, one element from another. Several persons tried; two unpracticed amateurs succeeded beyond all expectation. Their diversion, which became known as the Leyden experiment, nonplussed established electricians.

Period 2: 1740–1760

The theory universally held in 1745 attributed electrical effects to a material emanation exhaled by electric bodies when rubbed. One likened these emanations to sticky elastic strings, to stiff threads, to vortices, atmospheres and halos; their parts, the 'electrical effluvia', were fine atoms, mini-vortices, or particles repelling one another over microscopic distances. The dominant model in 1745 was the work of Dufay's protégé Nollet, an academician and a public showman. According to Nollet, effluvia projected from an electrified body effect repulsion and stimulate an influx of effluvia from surrounding objects, which causes attraction.

The theory of effluvia had much to recommend it. The electrical vapour directly affected the senses: one saw and heard it in sparks, felt it in shocks, smelled it around working electrical machines. And it revealed itself to the reason: how could one understand an interaction between two separated objects unless a material bond stretched between them? The counter-example of the gravitational theory did not seem pertinent: firstly, because of the evident existence of effluvia; secondly, because electrical theory had not been quantified, the only rationale, it was said, for Newtonian mathematical instrumentalism;[47] thirdly, because electricity, which exhibits itself fitfully, briefly, and

[47] J. Le Rond d'Alembert, 'Essai sur les élémens de philosophie', (1759), in *Oeuvres*, ed. J. F. Bastien (18 vols., Paris, 1805), II, 421–6.

only after preliminary treatment, is not analogous to immediate, universal graviation.

The theory implied that effluvia pass readily through glass, for threads within a glass bottle will follow the motion of a tube outside. *Sed contra,* glass cannot be permeable, for it insulates charged gun barrels or current-carrying strings. This apparent conflict produced the awkward doctrine that the electrical transparency of glass depends upon its thickness. To try to accomplish Bose's trick of electrifying water, the informed electrician would use a thick glass vessel, or a thin one placed upon an insulating stand.

The Leyden experiment, the invention of neophytes, worked only with water in thin glass vessels electrified while grounded. The thinner the glass and the better the grounding, the greater the electrification; the experiment worked best under precisely those circumstances in which theory foresaw failure. None of the established electricians could offer a plausible explanation of the behaviour of the Leyden jar. That did not prevent them from adding it to their collection of practical jokes.

Benjamin Franklin took up the study of electricity probably in the winter of 1745/6, without knowledge of the Leyden jar and its problems. He was guided by an authoritative account of the discoveries, inventions and amusements of the German professoriate, drawn up by Albrecht von Haller and published in French in a review conducted by Dutch academics. Franklin saw a translation of it in the issue of the *Gentleman's Magazine* for April 1745. He hit on the idea of electricity's existing in two states, positive and negative, in trying to interpret the results of an experiment he took from Haller. He did not at first try to devise a mechanics of the electrical matter in European style.

The system of plus and minus electricity had a logic of its own. In contrast to all other systems, Newtonian or Cartesian or mongrel, it made provision for two qualitatively different electrical states. When Franklin turned to analyze the Leyden jar he found his system perfectly adapted: while the Europeans could assign only differences in quantity or power to the electricities on either side of the jar, Franklin made them equal but opposite in kind. Hence he could explain why an exploded jar showed no electricity. To account for the peaceful coexistence of the opposite charges on an insulated jar, he had to assume them to be rigidly separated: glass must be absolutely impermeable to electrical matter. How then can one understand the motion of threads in a bottle? Franklin distinguished the 'repellency' of the electrical matter from its presence. He thereby introduced into electrical theory the notion of action across sensible distances, or at least across gaps equal to the thickness of glass vessels.

Franklin himself did not push the analogy between electricity and gravity: he retained most of the old theory and understood effluvia to stand around positively charged bodies in conformal atmospheres. The reworking of electrostatics in the manner of the gravitational theory was accomplished not by the English or their colonials, but by continental professors and academicians.

Franklin's ideas about electricity, together with his proposal to demonstrate the identity of lightning and electricity, were published in London in 1751. The following spring the demonstration succeeded near Paris. News of it inspired a professor in Turin, Giambattista Beccaria, who became chief of the first generation of European Franklinists. His books, which Franklin esteemed as models of method, systematically re-expressed the rules of plus and minus electricity and illustrated them with new experiments. No more than Franklin, however, could Beccaria free himself from the concept of electrical atmospheres.

Period 3: 1760–1790

The replacement of the mechanistic pictures by formulations akin to the theory of gravitation may be followed on two levels. Underlying investigations peculiar to electricity was a general relaxation of scruples against multiplying actions at a distance. A new generation of physicists did not care about the nice distinctions that Newton had tried to draw between theory and model, or about the issues disputed by Cartesians and Newtonians. This relaxation was partly brought about by mathematicians eager to calculate, and used to admitting what fictions they required to do so. It also derived partly from the discovery that electricians could not determine whether electricity came in one fluid, or two, or none. But also, one just grew used to such fictions as imponderable fluids and distance forces. 'Experience domesticates them.'[48]

The special investigations that forced the Newtonianizing of electricity arose from efforts to understand the operation of two instruments developed from hints given by Franklin: detectors of atmospheric electricity and the dissectible condenser, or electrophorus. Contrary to Franklin's prediction, an insulated detector often charged negatively in a storm, and might electrify with either sign in serene weather. John Canton, F.R.S., a London schoolmaster, understanding that induction as well as conduction was at work, invented elegant laboratory simulations of the behaviour of detectors exposed to charged clouds. These experiments, as refined by Franklin, inspired F. U. T. Aepinus, then of the Berlin Academy of Sciences, and his

[48] A. Volta, *Epistolario* (5 vols., Bologna, 1949–55), II, 510–11.

student J. C. Wilcke. Aepinus perceived an analogy between the air gap from 'cloud' to 'detector' in Canton's set-up and the glass in a parallel-plate condenser. He and Wilcke built a condenser with air as dielectric. It made atmospheres or effluvia incompatible with Franklin's beautiful analysis of the Leyden jar: the supposed atmosphere of the positive plate would extend through the air to the negative, and short the condenser internally. Aepinus developed an instrumentalist description of electricity in analogy to the gravitational theory, representing by an unknown function of the distance the postulated law of force between elements of the electrical fluid.

Aepinus' work, published in 1759, was hard reading for the few electricians able to procure it. His message came forth again in 1775, wrapped up in a device they could understand: Volta's electrophorus, the last of the line of separable condensers that derived from Franklin's attempts to locate the charge of a Leyden jar. One easily assimilated the air gap between the electrophorus's cake and separated shield to the dielectric of an intact condenser. The advantage of analyzing the resulting induction phenomena in terms of distance forces could be made tangible in useful machines. Volta rearranged the electrophorus as a condensing electroscope; others, following his lead, invented 'doublers' of electricity that anticipated the influence machines of the nineteenth century.

Once distance forces were accepted, the 'law of electrical force' – the diminution of the attraction f with distance r from a charged body – became a desideratum. Such a law had been sought as early as 1746. C. G. Kratzenstein, who became professor of physics at the University of Copenhagen, then obtained $f = 1/r$, within an accuracy of 250 per cent, which he deemed satisfactory. But neither Kratzenstein, nor any of the early measurers, sought the quantity Aepinus had in mind, namely the elementary, unobservable force between pairs of particles of a hypothetical electrical matter. The Kratzensteins offered the macroscopic, measurable interaction between ponderable charged bodies as the law of electric force. Such a 'law' is peculiar to the geometry of the experiment. To go from it to the hypothetical, but universally applicable, elementary law requires a clear conception of the relation between the theoretical entities and the measurable quantities, and an experimental situation that permits inferences from one to the other. The first men to succeed in the case of electricity – John Robison, Henry Cavendish, and C. A. Coulomb – resorted to the same mathematical instrumentalism Aepinus used.

The fixing of the elementary law of electrical force and the quantification of the relationship implicit in the qualitative analysis of the electrophorus (charge = capacity × tension) raised electrostatics to a height from which Poisson and Green could send it into the heaven of applied mathematics. As this was about to happen, Volta, following up Galvani's grisly experiments on decapitated frogs and applying to them the instruments and concepts of electrostatics, opened a new division of electrical studies more accessible to physicists than the science of Aepinus, Cavendish and Coulomb.

III. Lessons and extrapolations

The role of professors

Most of the electricians mentioned in the preceding sketch were affiliated to either a national academy of science or a university. The prominence of academicians in the advancement of eighteenth-century physics will occasion no surprise: academies were established to encourage original work. Although as yet there exists no study of physics as cultivated by or at these institutions, recent literature gives much pertinent information about organization, procedures and goals.[49] What may surprise is the importance of universities in the history of electricity; between one third and one half of the electricians of the eighteenth century were professors. This finding conflicts with the common assessment that the universities did not take a prominent part in the development of experimental physics before the nineteenth century. There is little literature dealing directly with physics in eighteenth-century universities.[50] Much information can be found, however, in biographies, institutional histories, government documents, travel books, correspondence, and Enlightenment pedagogics. A study of this material, now underway at Berkeley, has yielded the following preliminary results.

[49] E.g., P. Barrière, *L'académie de Bordeaux . . . au XVIIIe siècle* (Bordeaux, 1951); J. A. Bierens de Haan, *De Hollandsche Maatschappij der Wetenschappen, 1752–1792* (Haarlem, 1952); V. Boss, *Newton and Russia: The Early Influence, 1698–1796* (Cambridge, Mass., 1972); R. Hahn, *The Anatomy of a Scientific Institution: The Paris Academy of Sciences, 1666–1803* (Berkeley, 1971); S. Lindroth, *Kungl. Svenska Vetenskaps-akademiens historia, 1739–1818* (2 vols., Stockholm, 1967); A. S. Vucinich, *Science in Russian Culture: A History to 1860* (Stanford, Cal., 1963). Cf. R. E. Schofield, 'History of scientific societies: needs and opportunities for research', *HS*, II (1963), 70–83.

[50] What exists celebrates physics at a single institution, e.g. [H. Kangro], 'Zur Geschichte der Physik an der Universität Freiburg i. Br.', *Beiträge zur Freiburger Wissenschafts und Universitätsgeschichte*, XVIII (1957), 9–22; A. Leide, *Fysiska institutionen vid Lunds universitet* (Lund, 1968); H. Schimank, 'Zur Geschichte der Physik an der Universität Göttingen vor Wilhelm Weber', *Rete*, II (1974), 207–52; C. Schmidt-Schönbeck, *300 Jahr Physik und Astronomie an der Kieler Universität* (Kiel, 1965).

cists changed dramatically during the course of the eighteenth century. At the beginning, they lectured without demonstrations or mathematics, and had little allegiance to their discipline. Where regenting was in force, as in the Scottish universities and the Jesuit colleges, a professor might teach physics one year in three or four. In the German Protestant universities many 'physicists' were clerics or physicians awaiting preferment in their professions. In Italy most belonged to religious orders, which divided their loyalties and obligations.

Regenting had disappeared from the Scottish universities by mid-century, and from the larger Jesuit colleges in France by the time of the expulsion of the Order. Meanwhile in the German Protestant universities the physics chair became a desideratum in its own right, not merely (as it was in 1700) a station on the road to a medical or pastoral career. Two thirds of the chairholders during the years 1700 ± 25 were physicians, and of these one half left for careers in medicine. By 1800 ± 25 the fraction of chairholders with medical training had fallen to one quarter, and of these only one individual (of thirteen) went into doctoring; the rest left office by death or retirement. As physicians relinquished the physics chair M.A.s in philosophy, men who usually knew some mathematics, took their place. In the twenty-five years around 1700 (1700 ± 25) these M.A.s made up 20 per cent of the professoriate; in 1800 ± 25, 75 per cent.

The chief proximate cause of the increasing identification of physics professors with the discipline of physics was that powerful agent, the demonstration experiment. As we know, its use began to spread in the 1720s, encouraged and directed by the books of 'sGravesande and Musschenbroek. It was a commonplace by 1760, even in the universities of Italy and Austria, then recently pushed into the eighteenth century by the reforms of Maria Theresa.

The acquisition, maintenance and operation of the apparatus needed both skill and money. At most Protestant universities the professor furnished his own equipment, and maintained it as he could. Capital was required; if the experience of Bose, who had to 'pay ready money for everything, from my air pump to my funnel', was typical, professors of physics were not regarded as good credit risks.[51] Those who could make the initial investment tried to recover it by lecture fees, some of which would be ploughed back into instruments. Financial considerations pressed the professor to improve and expand his performances. An energetic lecturer, who did not

[51] Bose to Samuel Formey, 26 July 1750, in Formey Papers, Deutsche Staatsbibliothek, Berlin.

disdain playing the showman, could do well. Lichtenberg increased his audience threefold between 1777/8 and 1784, while total enrolments at Göttingen stagnated. Eventually he took as much in lecture fees as in salary.[52]

The German Protestant universities as a rule did not begin to acquire substantial collections of instruments until their professors, who had made the initial investments and had suffered the depreciation, died off. Profiting from bargain prices, Leyden bought Musschenbroek's apparatus, Altdorf Sturm's, Göttingen Lichtenberg's, and so on;[53] responsibility for further purchases was usually left to the new professor, whose apparatus might in turn be acquired, always below cost, by his university. These bargains brought with them the obligation and expense of maintenance, storage, and, eventually, modernization. By 1790 many Protestant universities had a *cabinet de physique,* a regular small budget for additions, and a mechanic to keep the equipment in order and to set it out before lectures.

Catholic universities could not usually follow the Protestant pattern of acquisition, for their clerical professors had neither adequate salary nor, ordinarily, the right to charge fees. The institutions themselves or extramural benefactors furnished the apparatus. Since Catholic schools introduced experimental demonstrations later than Protestant ones, both acquired *cabinets* at about the same time, allowance being made for the private collecting of the first generation of Protestant professors.

Although most of the equipment of the professor of experimental physics was made only for demonstration, a few expensive pieces, such as the electrical machine and the air pump, could also serve for research. An ambitious man might transform these pieces into experimental apparatus, his storage and maintenance facilities into a laboratory, and his mechanic into a laboratory assistant. The stage was set for the demonstrator of physical experiments to become an experimental physicist.

An example of what might be done in this way is Volta's arrangement at the University of Pavia, to which he went as professor of experimental physics in 1777. He won the chair partly on the strength of his reputation as a discoverer and inventor. To encourage him to

[52] G. C. Lichtenberg, *Briefe,* ed. A. Leitzmann and C. Schüddekopf (3 vols., Leipzig, 1901–4), II, 127, 228, 335. Cf. D. B. Herrmann, 'Georg Christoph Lichtenberg als Herausgeber von Erxlebens Werk *Anfangsgründe der Naturlehre*', *NTM,* VI:1 (1969), 68–81, and VI:2 (1969), 1–12.

[53] Documentation for these and other assertions about building up university *cabinets* is given in my book *Electricity* (see note 45).

further achievement and also to broaden his education, the Austrian governors of Lombardy and its university sent him on visits to centers of learning beyond the Alps. He met foreign savants and patronized their instrument-makers, all at government expense. In Pavia he enjoyed the services of a mechanic and a small budget for the purchase or repair of apparatus.

In 1785 Volta negotiated successfully for still more perquisites. He wanted a lecture hall, new storage facilities, preparation rooms, in short, an institute of physics; and he wanted more money. His justification of these requests shows how matters stood. More room, he said, would allow '[me] to busy myself in research and to give private courses on it to capable students'; more money would bring 'all my talents [to bear] on advancing the science I profess, and the instruction of students of it'.[54] Note the order in which he put his obligations: first, research; second, training of advanced students; third, promoting physics; fourth, what we might call undergraduate instruction.

No doubt Volta's conception of the professorial role went against conventional opinion. Research was the business of academicians, members of learned societies, not of professors. As M. C. Förster, the author of the centennial history of the University of Halle, put the point in 1799: 'A professor by no means needs to discover new truths or to advance his science. Should he do so, he is in fact more than a professor, he has done opera supererogationis', he has worked beyond the call of duty.[55]

Yet there is no doubt that, by 1790, at leading universities like Pavia, Halle and Göttingen, calls and promotions came most easily to those who had done their bit for science. As early as 1750 Tobias Mayer's appointment to Göttingen specified that he do research as well as teach, and later, to counter an offer from the Berlin Academy, the university met his demands for research facilities.[56] Göttingen professors were expected to write, and did so, 'as if the entire empire of letters acknowledged their academic sceptre'.[57] Those who did not measure up, who neither wrote nor researched, were as out of place

[54] Volta, *Epistolario* (see note 48), III, 283–330.

[55] Förster, *Übersicht der Geschichte der Universität zu Halle* (Halle, 1799), 2–3.

[56] Quoted in E. G. Forbes, 'Tobias Mayer, zur Wissenschaftsgeschichte des 18. Jahrhunderts', *Jahrbuch für Geschichte der oberdeutschen Reichsstädte*, XVI (1970), 132–67 (pp. 149, 155).

[57] Mayer to Euler, 6 October 1754, in I. K. Kopelevich, 'Perepiska Leonarda Eilera i Tobiasa Maiera', *Istoriko-astronomicheskie issledovaniia*, V (1959), 271–444 (p. 414).

among Göttingen professors as (to quote one of them) 'mouse turds among peppercorns'.[58]

The initiative of instruments

The key advances in electrical theory started from improvements in instrumentation made partly by chance and partly by design. The tube and globe were introduced to enhance effects for public demonstrations. The Leyden jar came when neophytes, attempting to create a display, arranged standard pieces of apparatus in a novel way counterindicated by contemporary theory. The properties of the condenser most puzzling to the older electricians became the chief support of Franklin's system. The residue of the theories of atmospheres and effluvia, from which Franklin and other non-mathematical electricians could not free themselves, were excised by the electrophorus. The century ended with the invention of the pile.

The history of the electrometer provides another example of the initiative of inanimate objects. At first electricians took the spreading of threads hung from charged bodies as an index of electrification. The index was easily improved: the threads gave way to a rule and wire, and to straws or gold leaves diverging before calibrated scales. These devices required explanation. What did they measure? Amount of electrical matter, or its 'force', 'compression', 'head', 'degree', or 'tension'? How did the effort to charge an object, as measured by the number of turns of the electrical machine, relate to the angle indicated by the electrometer? What bearing had either upon the size of the shock received, or the length of the spark produced, or the amount of wire melted, by the object's discharge? These questions could not be neglected. The physicist wanted to relate measured quantities to theoretical entities. The up-to-date physician needed to meter the shocks of his patients. The manufacturers, not to mention the purchasers, of apparatus required standard ratings for the comparison of electrical machines. Efforts to understand what electrometers measured brought electricians to clear concepts of quantity of charge, capacity, and electrostatic tension or potential.

The capability and cost of instruments also affected the pace and nature of conceptual innovation. As for capability, something is known about the power of electric generators and of the effects at-

[58] A. G. Kästner, *Briefe aus sechs Jahrzehnten, 1745–1800* (Berlin, 1912), 215–18; cf. C. H. Müller, 'Studien zur Geschichte der Mathematik an der Universität Göttingen', *Abhandlungen zur Geschichte der mathematischen Wissenschaften*, XVIII (1904), 51–143 (pp. 135–6).

tainable with them.[59] We do not know enough about ancillary apparatus, such as the air pumps used to create 'vacuum' for electrical experiments. The effects obtained depend sensitively on the degree of exhaustion. Historians have doubtless credited eighteenth-century electricians with observations they could not have made, and blamed them for missing what they could not have seen.[60] A case in point is whether vacuum insulates or conducts.

Rising costs helped to redefine the corps of electricians after 1750. Hauksbee's tubes and lampblack had cost nothing. The best generator commercially available in the 1780s, Cuthbertson's three-foot plate machine, cost about the annual salary of a well-paid professor of physics.[61] The casual and occasional observer, who had played an important role in the study of electricity early in the century, found the ante too high at its end.

Learning by doing

Worrying whether space is full or empty, whether a body acts by pushes or powers, whether the principles are one, two or many, did not lead electricians to the tube, the condenser, the electrophorus, or the pile. No more did discussion of the fundamental nature of force, or of the activity or passivity of matter, bring them to understand, accept and exploit Newton's mathematics of forces.[62] Their mathematical instrumentalism was a *pis aller* resorted to after other approaches had failed. They did not so much adapt the method of the *Principia* to the problems of terrestrial physics as rediscover it. This proposition may be illustrated by efforts to measure magnetic 'force'.

According to the usual Newtonian apologetics, as presented by a 'sGravesande or a Desaguliers, 'force' signifies effect, not cause: gravitational 'force' is the centripetal acceleration of planets, moons and stones, not the power that moves them towards one another. The 'law of graviation' meant the distance dependence of the acceleration of bodies approaching one another by virtue of the unknown cause of gravity.

[59] E.g., B. Finn, 'Output of 18th-century electrical machines', *BJHS*, v (1971), 289–91; H. Prinz, 'Dalla "commozione elettrica" alle affascinanti prospettive dell'elettricità immagazzinata nel condensatore statico', *Museoscienza*, xi:5 (1971), 3–27.

[60] E.g., G. F. J. Tyne, *Saga of the Vacuum Tube* (Indianapolis, 1977), 22–23.

[61] Hackmann, *John and Jonathan Cuthbertson* (see note 38), 52.

[62] Hence the general irrelevance to our inquiry of work like P. M. Heimann and J. E. McGuire, 'Newtonian forces and Lockean powers: concepts of matter in 18th century thought', *Historical Studies in the Physical Sciences*, iii (1971), 233–306, and R. E. Schofield, *Mechanism and Materialism: British Natural Philosophy in an Age of Reason* (Princeton, N.J., 1970).

Taking this conception over to magnetism, the first force-measurers sought a law that would give the dimunition with distance of the macroscopic, measurable acceleration of a small magnet or needle under the influence of a lodestone. They measured the magnet's acceleration by the gravitational 'force' (acceleration!) needed to balance it. Proceeding in this way they obtained nothing of value: the law did not appear to depend on any simple function of distance. Moreover, it varied with the geometry of the experiment. Musschenbroek, the most tireless of these measurers, had no idea what to do with his results. 'In such darkness', he said, 'it is best to suspend judgement, and to relate our observations [for the use of] a wiser and more serious age.'[63]

Musschenbroek guessed that the key to the problem was buried somewhere in the *Principia*. He could not find it. His wiser and more serious successors perceived that he had mixed up the measurable, macroscopic, particular force-effect of Newtonian apologetics with the hypothetical, microscopic, universal force-cause of Newton's gravity. Confusion came the easier because the bodies met with in astronomy can be considered to be spheres; in this approximation, as Newton had proved in a celebrated theorem, the distance dependence of force-effect and force-cause is the same. In the case of the irregular shapes and short distances used in the measurement of magnetic 'force', no inference to a general law was possible.

It took many years before physicists understood that to parallel the procedure of the gravitational theory they had to postulate an unobservable, universal force between magnetic elements, measure the macroscopic 'force', and infer or 'transdict' the universal law.[64] Coulomb's success depended on a clear conception of transdiction and on a supply of long magnetized needles with well-defined poles able to serve the process. When physicists finally took possession of this old-new method, they naturally scoffed at those who had failed to grasp it. What more useless, asks Lambert, who himself had not yet mastered it, than the observations of magnetic force made by 'a celebrated Leyden professor'?[65]

Although the *Principia* always stood ready as guide, physicists did not see how to use its methods to measure magnetic or electric force

[63] Musschenbroek, *Dissertatio physica experimentalis de magnete*, 2nd edn (Vienna, 1754), 17, 26–7, 37–8.

[64] For transdiction see M. Mandelbaum, *Philosophy, Science and Sense Perception: Historical and Critical Studies* (Baltimore, 1964), 61–2.

[65] Lambert to Euler, 4 April 1760, in K. Bopp, 'Leonhard Euler's und Johann Heinrich Lambert's Briefwechsel', *Abhandlungen der Akademie der Wissenschaften, Berlin, Phys.-Math. Kl.*, II (1924), 13.

until after 1750. What then opened their eyes? For one, mathematicians without the usual scruples of physicists regarding truth, plausibility and *Anschaulichkeit* turned their attention from the newly completed astronomy and mechanics to the backward subjects of experimental physics. Second, the accumulation of data, especially in the case of electricity, made it increasingly difficult to work with the old models. Third, improved apparatus could be made to deliver the precise results needed to confront theory with experiment. Fourth, a new generation came of age with a new style, which can be found not only in physics but also in Kant's epistemology, in artificial classification in botany, and doubtless elsewhere. As Nollet wrote in rejecting a piece of physics in the old style, 'the Academy is getting more and more difficult about this way of philosophizing'.[66] Why? The present state of the ferment of knowledge does not allow a definitive answer.

Context

I have mentioned that during the early eighteenth century the association of glows and sparks with electricity caused it to be reclassified from mineralogy (as a property of amber) or occult quality (as an example of attraction) to a sort of 'fire'.[67] Further knowledge made it possible to distinguish it from its erstwhile fellows, of which one, mineralogy, soon dropped from physics. After 1750 electricity moved towards mathematics, at first in the sense of orderliness (as in Franklin's system of electrostatics), then as quantification. These last steps deserve further notice.

About 1750 d'Alembert, despairing of yoking electricity to his favourite subject, left it to experimental physics. 'That is the method that must be followed with phenomena the cause of which reason cannot help us [find].' In the 1770s Lichtenberg declared that the non-mathematical electrician had played his part: '[electricity] has more to expect from mathematicians than from apothecaries'. Ten years later Karsten claimed electricity for mathematical physics.[68] The same progression can be followed in the classification of electricity in the *Commentarii* (later *Acta*) of the Petersburg Academy. From 1726 to 1746 the journal had two classes, mathematics and physics; electricity naturally fell to the latter. In 1747 a new class was added, 'physico-

[66] Nollet to E. F. Dutour, 13 March 1769 (Burndy Library).

[67] G. E. Hamberger's important text, *Elementa physices* (1727, 1735), combines the schemes: electricity comes under 'fire - air - earth' as a property of complex minerals. On the same principle, optics falls to 'fire–earth' (as a property of glass), meteorology to 'fire–air–water', etc.

[68] D'Alembert, in *Encyclopédie* (see note 25), art. 'Physico-mathématiques'; P. Hahn, *Georg Christoph Lichtenberg und die exakten Wissenschaften* (Göttingen, 1927), 41; Karsten, 'Vom eigentlichen Gebiet der Naturlehre, (see note 27), 151.

mathematics', which took optics, hydraulics, heat, magnetism and, despite d'Alembert, electricity. The arrangement persisted until 1790, when 'mathematics' and 'physico-mathematics' united. A similar change in classification had occurred in the reorganization of the Göttingen Society of Science in 1777–8.

These facts evidently apply to the emergence of electricity as an independent sub-discipline. They also have a larger significance. They indicate that the contexts in which classifiers, organizers and textbook writers placed the various topics of physics are important evidence for the shifts in meaning and scope that physics underwent during the eighteenth century.

IV. Desiderata

The first order of business is to discover how far the generalizations and periodization derived from the history of electricity apply to other branches of eighteenth-century physics. The most promising appear to be optics and the subjects discussed under 'fire' and 'air'. Optics should not be difficult to follow. Under 'fire' and 'air', however, were found not only subjects now classified as physics, such as thermodynamics and acoustics, but also subjects now assigned to chemistry and meteorology. Coming to grips with this material will be difficult, as familiar distinctions will have to be obliterated.

Despite its relatively easy access, the historiography of eighteenth-century optics has not advanced far from where Wolf left it. Ronchi gives three times as much space to Descartes and Grimaldi as he does to the entire eighteenth century.[69] The latest study, by H. J. Steffens, seldom descends from speculations about the nature of light to the experiments and mathematical descriptions that were the mark of Enlightenment optics.[70] And since, in the narrowest tradition of Anglo-American scholarship, he restricts himself to England, his book is not even useful as an inventory of speculations. The coverage afforded by Steffens and Ronchi may be gauged from their omission of Lambert's *Photometria*.

Although some modern scholars have not been able to discover much of interest in eighteenth-century optics, Priestley and Fischer managed to write many hundreds of pages about its development.[71]

[69] V. Ronchi, *Histoire de la lumière* (Paris, 1956).

[70] H. J. Steffens, *The Development of Newtonian Optics in England* (New York, 1977).

[71] P. A. Pav, 'Eighteenth-century optics: the age of unenlightenment' (Unpublished Ph.D. thesis, Indiana University, 1964); J. Priestley, *The History and Present State of Discoveries Relating to Vision, Light and Colours* (1772); Fischer, *Geschichte der Physik* (see note 27). Cf. the fine review by G. N. Cantor, 'Georgian optics', *HS*, xvi (1978), 1–21.

Why have the moderns added so little? One answer is that, when judged against the accomplishments of the next heroic age, when the wave theory was established, the development of optical *theory* in the eighteenth century appears meagre and uninteresting. But if one attended to the shift of optics in the body of knowledge; to the changing loci and groupings of its students; to the opening of new fields, such as photometry, or the elaboration of once borderline ones, like luminescence; to the invention or improvement of instruments; to the applications of lenses and mirrors in other branches of physics; one might find eighteenth-century optics an intriguing and important study.

With respect to 'fire', although no good general history exists, some excellent monographs point the way. The most important qualitative theory, Boerhaave's, has been carefully studied.[72] There is an exemplary account of the invention of the concepts of specific and latent heat, and of the experiments made to measure them.[73] Thermometry continues to fascinate.[74] Some literature on the history of chemistry is pertinent.[75] And recent work on the caloric theory, although centred on the ninteenth century, gives useful orientation for one interested in Enlightenment ideas about heat.[76] There appear to be close analogies between the histories of electricity and of 'fire'.

For 'air' and its species the history of chemistry must again be consulted. On its physical properties, its dilation under heat, its effect on the barometer, its purity, its equation of state, etc., there is also useful literature.[77] The historians of acoustics have scarcely been heard from. As for instrumentation, the barometer has been extensively studied, the air pump not. Much remains to be done on the wider subjects of 'air' – evaporation, the physical state of the atmosphere, the cause of winds, etc. One will need to consult the history of meteorology, which is not yet written.

Studies of the interaction of physicists and their institutions would be welcome. For example, one might examine the range of physics as practised by the regular attendees at the meetings of the royal

[72] H. Metzger, 'La théorie de feu d'après Boerhaave', *Revue philosophique*, CIX (1930), 253–85.

[73] D. McKie and N. H. de V. Heathcote, *The Discovery of Specific and Latent Heats* (1935).

[74] E.g. Middleton, *A History of the Thermometer* (see note 30).

[75] See M. Crosland, this volume, pp. 389–416.

[76] E.g., D. S. L. Cardwell, *From Watt to Clausius* (Ithaca, N.Y., 1971).

[77] E.g., R. Waterman, 'Eudiometrie', *Technikgeschichte*, XXXV (1968), 293–319; F. Grassi, 'I lavori del Volta e del Gay-Lussac per l'azione del calore', *Rendiconti, Istituto lombardo di scienze e lettere*, LX (1927), 505–34, and 'I lavori del Volta e del Dalton su le tensione dei vapori', *ibid.*, 535–66.

societies of London and Stockholm, or of the pensionaries of the Berlin, Paris and Petersburg academies. Studies along these lines are underway at Berkeley. A very useful preliminary work would be the publication of the topics discussed at the London Society and the Paris Academy throughout the *ancien régime*. Their potential value to scholars may be judged from the published *Procès verbaux* of the Petersburg, the *Registres* of the Berlin, and the truncated *History* of the London societies.[78]

Another deserving subject is the academic prize-competition. As preliminaries one needs a census of prizes offered throughout Europe, their values, identifications of the winners, and rough descriptions of the winning entries.[79] From such data one can develop a prosopography of successful competitors and distinguish trends in the nature of the questions and the quality of the answers. Some of these competitions affected the substantive development of physics. For example, the two big mid-century prizes in electricity – the Berlin prize of 1745 and the Petersburg one of 1755 – misdirected work in the field by encouraging the construction of elaborate theories of the mechanics of effluvia.

There is scattered literature about ancillary topics that needs to be sifted and summed. An example is the role of the instrument-maker and the cost and supply of his apparatus.[80] Data from the many biographical sketches should be tabulated and compared; the recent welcome study of Benjamin Martin, for example, would have been still more welcome had its author systematically exploited the secondary literature.[81] This is not merely to recommend group biography. From the existing sketches, catalogues of surviving apparatus, and published auction records and price lists, one can estimate the growth of the instrument trade and the multiplication of instruments. The trade expanded sharply at just the time – the 1770s and 1780s – when electrical machines grew prodigiously in power and when electrometers, thermometers and barometers became accurate and intercomparable. With whom, or with what, did the initiative lie?

[78] Akademiia nauk, SSSR, *Procès verbaux des séances de l'Académie impériale depuis sa fondation jusqu'à 1803* (4 vols., St Petersburg, 1897–1900); E. Winter (ed.), *Die Registers der Berliner Akademie der Wissenschaften, 1746–1766* (Berlin, 1957); T. Birch, *History of the Royal Society of London* (4 vols., 1756–7).

[79] Barrière, *L'académie de Bordeaux* (see note 49), has made some forays beyond A. F. Delandine, *Couronnes académiques* (2 vols., Paris, 1787).

[80] An excellent starting point is M. Daumas, *Les instruments scientifiques aux XVIIe et XVIIIe siècles* (Paris, 1953).

[81] J. R. Millburn, *Benjamin Martin: Author, Instrument-Maker, and 'Country Showman'* (Leyden, 1976).

Another sort of integration should be practised on the published correspondence of eighteenth-century physicists. A catalogue of this material is a prime desideratum; it will assist – and embarrass – us into using what lies under our noses. Even so great a treasure as Volta's correspondence, which is filled with information about physics and its institutions in northern Europe as well as in Italy, has not been exploited outside the small circle of Volta scholars. Nor have historians of physics dredged the latest edition of Voltaire's letters. If we do not change our ways the correspondence of Euler, now being edited, will also be under-utilized. One need not be working on or over Volta or Voltaire or Euler or anyone with whom they were in direct communication to learn something pertinent from the letters: a feel for the scientific culture of the time that can scarcely be obtained otherwise.

Two other desiderata concern topics with which we began: relations between natural philosophy and mathematics, and between the eighteenth century and its neighbours. As to the first, classification of textbooks and monographs by mathematical level would be very useful. The connection between precise measurement and quantification of theory also needs further study. It would be pleasant to know something about the uses of numbers in tables and graphs, about the reliability accorded to measurements, and about estimates of experimental error.

The join between eighteenth-century natural philosophy and that of the seventeenth must be made with constant attention to continental traditions independent of Newton's work and British institutions. The join with the nineteenth requires equal vigilance. The common notion that physical science started anew with the Ecole Polytechnique or the University of Berlin or the Royal Institution must be exploded.[82] In history of science, as in political history, the similarities between the Restoration and the *ancien régime* may be more significant than the differences. Preliminary study suggests that the rapid growth of experimental physics in the nineteenth century began about 1830, when its institutions had recovered to the level reached at the leading continental universities on the eve of the French Revolution.

In conclusion the available general histories of experimental physics in the eighteenth century derive from the work of physicist-historians who were more concerned with progress than with process. They did not worry much about periodization and took for granted that 'physics' referred to the same range of subjects in the eighteenth

[82] A recent exaggerated example of this thesis is M. Berman, *Social Change and Scientific Organization: The Royal Institution, 1799–1844* (Ithaca, 1978).

century as it did in their own. Recent study of the history of electricity has established the basis of a periodization of eighteenth-century experimental physics. Other branches now need attention, with appropriate cross-national comparisons and due regard for the roles of institutions and instrumentation. At the current rate of work, these branch histories can be completed in a decade or two. Then we can attempt to improve upon Wolf.

10

Chemistry and the chemical revolution

MAURICE CROSLAND

Contents

I.	Introduction	389
II.	Boyle or Lavoisier?	392
III.	Chemistry as a branch of natural history?	395
IV.	Affinity	396
V.	Gases	398
VI.	The chemical revolution	402
VII.	Phlogiston and oxygen	405
VIII.	The new theory and the new nomenclature	407
IX.	Some applications of chemistry	412
X.	After the chemical revolution	414

I. Introduction

The eighteenth century was a period of particular importance for chemistry since it witnessed a transformation usually described as 'the chemical revolution'. At the end of the seventeenth century writers of chemistry books continued to think it necessary to apologize for their study, since it was still confused with alchemy, and to explain that: 'Chymistry is a true and real Art, and (when handled by prudent Artists) produceth true and real effects'.[1] Yet by the beginning of the nineteenth century chemistry was seen by many people as the outstanding example of a successful science, and one which was recruiting the best brains. As the astronomer Delambre remarked in 1808: 'the revolution recently brought about in chemistry could not happen without turning many experimentalists a little out of their ordinary course, when they saw in a neighbouring science a road opened that promised more numerous discoveries'.[2] As the 'chemical revolution' occurred in the second half of the eighteenth century, it is appropriate for this chapter to focus on the period after 1750, while giving some consideration to the wider period beginning

[1] *The Works of Geber Englished by Richard Russell*, 1678, new edn (1928), The Translator to the Reader, xxxvii.

[2] Delambre, *Rapport historique sur les progrès des sciences mathématiques depuis 1789 et sur leur état actuel* (Paris, 1810), 31–2.

with the influence of Issaac Newton (d. 1727) and culminating in the reception and development of the work of Antoine Laurent Lavoisier (1743–1794).

Since the late eighteenth century chemistry has been recognized as one of the basic sciences and, as science came increasingly to be taught in universities, colleges and schools, chemistry nearly always constituted a major subject. Histories of chemistry were written for those who were studying or had studied that particular discipline. It will not be surprising, therefore, that the history of chemistry has often been studied in isolation from the history of other sciences,[3] and it has nearly always been studied without relation to the history of ideas or to general history. This, however, is less true for biographical studies of chemists and one of the ways in which these have been especially valuable in the history of chemistry is that they have encouraged science historians to look at chemistry in its intellectual and social context. This is particularly the case for many studies of Joseph Priestley and Lavoisier. Indeed in the case of Priestley there are aspects of interest to historians of philosophy, and a series of recent articles[4] has raised questions about the epistemological, metaphysical and methodological parameters that regulated his science. Priestley believed that chemistry could throw light on the relationship between matter and spirit and this obviously had important theological implications. It may also be time to look at the metaphysical foundations of Lavoisier's thought, where the sensationalist philosophy of Condillac provides a promising starting point.[5] Whether Priestley's chemistry and natural philosophy can be better understood by reference to his Unitarian ministry and his radical politics is a question still under discussion,[6] but the continued concern of Lavoisier with practical problems and his death under the guillotine has forced his successive

[3] The best example is the standard reference work, J. R. Partington, *History of Chemistry* (4 vols., 1961–70). See vol. III for the eighteenth century.

[4] J. G. McEvoy, 'Joseph Priestley, "aerial philosopher": metaphysics and methodology in Priestley's chemical thought from 1762 to 1781'. I. *Ambix*, XXV (1978) 1–55; II. *ibid*, 93–116; III. *ibid.*, 153–75; IV. *ibid.*, XXVI (1979), 16–38.

[5] A preliminary discussion of the wide-ranging influence of Condillac on Lavoisier is given in M. P. Crosland, 'The development of chemistry in the eighteenth century', *Studies on Voltaire and the Eighteenth Century*, XXIV (1963), 369–441 (pp. 416–21). It is high time that someone undertook a more thorough treatment of the whole question of chemistry, chemists and the Enlightenment that would supersede this monograph, which tentatively explored the relationship of chemistry to various figures in the Enlightenment.

[6] R. E. Schofield, 'Joseph Priestley, natural philosopher', *Ambix*, XIV (1967), 1–15, and a comment by J. G. McEvoy, *Ambix*, XIV (1968), 115–23. See also A. D. Orange, 'Oxygen and one God: Joseph Priestley in 1774', *History Today*, XXIV (1974), 773–81.

biographers to relate his life and work in some way to the history of France.[7] The publication of the correspondence of chemists is another factor encouraging science-historians to consider their subject in a wider context.[8]

Even within science, historians of chemistry have sometimes taken too narrow a view of what constitutes chemistry and ignored the implications of natural philosophy (or 'physics') on the one hand and of the biological sciences[9] on the other. These 'neighbouring' sciences may be profitably considered not simply as background but because they have immediate relevance to what was called chemistry in the eighteenth century. It is too easy to take nineteenth-century boundaries for granted, but one of Lavoisier's great achievements was to question the accepted methodology of the chemists of his time and to apply the methods of physics, particularly the use of the balance, to chemistry. Indeed for many of Lavoisier's contemporaries he was more clearly *physicien* than *chimiste* in the traditional sense. Pneumatics, which Stephen Hales (1677–1761) might have regarded as a part of natural philosophy, had clearly become a part of chemistry by the time of Priestley (1733–1804). Finally the subject of heat, which is now thought of as physics, was more likely in the eighteenth century to be regarded as a part of chemistry. It was an important part of the lectures of Joseph Black (1728–1799), who is characterized as a 'heat theorist' in a recent study.[10] Studies of heat or 'caloric' were also prominent in the work of Lavoisier.[11]

[7] E. Grimaux, *Lavoisier, 1743–1794, d'après sa correspondance, ses manuscrits, ses papiers de famille et d'autres documents inédits* (Paris, 1888). D. McKie, *Antoine Lavoisier, Scientist, Economist, Social Reformer* (1952). L. Scheler, *Lavoisier et la révolution française*, II (Paris, 1956; revised 1957); II (with W. A. Smeaton, Paris, 1960). L. Scheler, *Lavoisier et le principe chimique* (Paris, 1964). H. Guerlac, *Antoine-Laurent Lavoisier, Chemist and Revolutionary* (New York, 1975), an illustrated version of his article on Lavoisier in C. C. Gillispie (ed.), *Dictionary of Scientific Biography* (16 vols., New York, 1970–80), VIII, 66–91.

[8] An excellent example is the series of letters of Priestley edited with a linking commentary by R. E. Schofield: *Scientific Autobiography of Joseph Priestley*, (Cambridge, Mass., 1966). See also a collection of letters received from foreign scientists by Torbern Bergman: G. Carlid and J. Nordiström, *Torbern Bergman's Foreign Correspondence* (vol. I only, Stockholm, 1965).

[9] For a view of eighteenth-century chemistry as a branch of natural history see below.

[10] A. L. Donovan, *Philosophical Chemistry in the Scottish Enlightenment: The Doctrines and Discoveries of William Cullen and Joseph Black* (Edinburgh, 1975), 221.

[11] Time precludes any proper study of caloric here. See, however, R. Fox, *The Caloric Theory of Gases from Lavoisier to Regnault* (Oxford, 1971); R. G. Morris, 'Lavoisier and the caloric theory', *BJHS*, VI (1972), 1–38. See also H. Guerlac, 'Chemistry as a branch of physics: Laplace's collaboration with Lavoisier', *Historical Studies in the Physical Sciences*, VII (1976), 193–276.

II. Boyle or Lavoisier?

Some historians of chemistry have suggested that modern chemistry started in the late seventeenth century with Robert Boyle; but although Boyle's 'chymistry' marks the effective end of alchemy, and although in his *Sceptical Chymist* (1661) he launched some effective criticisms of traditional concepts, notably the Aristotelian theory of four elements and the Paracelsian theory of three principles, he was not able to replace these theories with any more useful theory. His idea of corpuscles was a reflection of the mechanical philosophy of the seventeenth century which had little useful explanatory power in chemistry and was soon rejected in the eighteenth century.

One great problem for historians of science is the relation of the 'chemical revolution' of the eighteenth century to what is often called '*the* scientific revolution' of the seventeenth century. Herbert Butterfield refers to 'the postponed scientific revolution in chemistry'.[12] Perhaps the worst aspect of this is the assumption that a particular approach which was successful in the seventeenth century for astronomy and physics was equally applicable to chemistry. Butterfield seems to think that a new chemistry could have arisen in the late seventeenth century with Robert Boyle. He says that 'a number of Boyle's important results and conclusions dropped out of sight and had virtually to be rediscovered a hundred years or so later'. It is true that Lavoisier repeated one or two of Boyle's experiments on the calcination of metals more carefully, but experiments by themselves prove nothing. The key to the 'chemical revolution', oxygen, depended on the appreciation of the gaseous state, which was one of the great achievements of the eighteenth century and something that could not have been easily assimilated a hundred years earlier. Butterfield goes further and writes that 'the work of the *Sceptical Chymist* had to be done over again', but he agrees that Boyle's most important influence was negative. Although Boyle is often credited with introducing the modern concept of an element, his definition of elements was not original and he spoke of elements only to reject them as the foundations of his chemistry.[13] Although the Victorian judgement of Boyle as 'the father of modern chemistry' is no longer accepted,[14] he

[12] *The Origins of Modern Science, 1300–1800* (1949), ch. 2. Despite the implied criticism above, Butterfield deserves great credit for bringing more closely together general history and the history of science.

[13] E.g. T. S. Kuhn, 'Robert Boyle and structural chemistry in the seventeenth century', *Isis*, XLIII (1952), 12–36.

[14] M. Boas [Hall], *Robert Boyle and Seventeenth-century Chemistry* (Cambridge, 1958), 3.

is sometimes seen as someone who 'prepared the way for Lavoisier'.[15] Thus if Boyle was not the Newton of the science he was at least its Galileo.[16] If one is interested in the way in which the chemistry of the early eighteenth century was indebted to seventeenth-century work, a very wide-reaching study is that of Hélène Metzger, a pioneer French scholar, to whom many later historians of chemistry owe a great debt.[17]

Sometimes British and American authors seem to have presented a distinctly 'Anglo-Saxon' view of the history of chemistry as the work of great men, led by Boyle, Newton and Dalton. Partington refers to the 'jungle of the Theory of Phlogiston' which he contrasts with 'the path of true discovery opened out by Boyle, Hooke and Mayow'.[18] In an eighteenth-century study we should add the names of the leading 'pneumatic' chemists: Hales, Black, Cavendish and Priestley, all British. It is undisputed that without their study of gases Lavoisier's chemical revolution would not have been possible.

The approach may depend as much on the subject of study as on the historians. There has been some consensus among scholars who have focused on individual British chemists or who have explicitly favoured a biographical approach to the study of chemistry in Britain. In the first place there has been general agreement over the past ten to twenty years among students of Black, Priestley and Dalton that the traditional categories of history of chemistry are insufficient to understand the conceptual framework in which these men operated and attempts have been made to relate their work to Newtonian matter theory. Schofield, for example, himself trained in physics, feels that one should regard Priestley more as a physicist or 'natural philosopher' than a chemist.[19] Donovan, in a recent study of Black and his teacher William Cullen, tries to relate them to British Newtonianism but concludes that 'Cullen's commitment to medicine and chemistry drove him away from the Newtonianism of Desaguliers and the mechanical philosophers'.[20] He is therefore left with Cullen simply as embracing a vaguer 'methodological Newtonianism'. Black is described as a Newtonian. Donovan, however, writes: 'Although Black believed that future chemical theories would be founded upon

[15] *Ibid.*, 222.

[16] *Ibid.*, 232.

[17] Hélène Metzger, *Les doctrines chimiques en France du début du XVIIe à la fin du XVIIIe siècle* (Paris, 1923); *idem*, *Newton, Stahl, Boerhaave et la doctrine chimique* (Paris, 1930).

[18] *Short History of Chemistry*, 3rd edn (1957), 84.

[19] R. E. Schofield, 'Joseph Priestley, natural philosopher', *Ambix*, xiv (1967), 1–15.

[20] Donovan, *Philosophical Chemistry* (see note 10), 29.

Newtonian conceptions of force, he knew that chemists must continue to identify and classify the substances they study'.[21]

It has been also pointed out that Black resisted attempts of the extreme Newtonians to reduce chemistry to mechanics.[22] One of the problems in any final assessment of Black as a Newtonian is that the principal source of Black's system of chemistry, his lectures, were only published posthumously and we must beware of the bias of his editor, John Robison. It is understandable that considerable interest should have been focused particularly by McKie on different manuscript copies of Black's lectures as taken down by his students.

The British natural philosopher John Dalton, whose atomic theory (since it belongs to the opening years of the nineteenth century) falls outside the terms of reference of this chapter, has been recently studied by Thackray. Dalton's chemical atomic theory was based on Lavoisier's chemical elements, but Thackray, instead of exploring this avenue, has preferred to consider Dalton's relationship to the Newtonian tradition in Britain. In conscious reaction to such 'old-fashioned' historians of chemistry as Partington, Thackray has studied Newton's matter-theory in the widest terms of contextual history of the eighteenth century without claiming that much of this eventually contributed to the main stream of chemistry. There is, however, the implication that chemists *should have* been concerned with matter-theory as one of the fundamentals of chemistry. The fact that Lavoisier ignored such speculations as 'metaphysical'[23] and based his chemistry solidly on what he could observe, collect or measure in the laboratory, earns for the Frenchman the judgement that he was concerned only with the 'superstructure' of chemistry.[24] This is perhaps to attempt to impose a foreign and ultimately sterile framework on one of the more fruitful approaches to chemistry in the eighteenth century.

One aspect of matter-theory which had always concerned chemists was composition. Up to the seventeenth century it was widely held that the four Aristotelian elements were present in all matter but in

[21] *Ibid.*, 215.

[22] Black rejected the use of 'lever diagrams' to represent chemical reactions because of the mechanical implication of levers. M. P. Crosland, 'The use of diagrams as chemical "equations" in the lecture notes of William Cullen and Joseph Black', *AS*, xv (1959), 75–90.

[23] *Traité élémentaire de chimie* (Paris, 1789), trans. *Elements of Chemistry*, (Edinburgh, 1790; repr. New York, 1965), Introduction, xxiv.

[24] A. Thackray, *Atoms and Powers: An Essay on Newtonian Matter-Theory and the Development of Chemistry* (Cambridge, Mass., 1970). The last chapter of Thackray's book is devoted to the triumph of John Dalton.

different proportions. Thus any liquid would contain a large proportion of water. A more pragmatic view of composition related a substance to the substances required to produce it. For example, 'corrosive sublimate' (a chloride of mercury) was related to mercury and spirit of salt. The 'furniture' of chemistry was a comparatively small number of substances, later to be distinguished as elements and compounds. The chemistry best understood was the chemistry of the mineral kingdom, corresponding to inorganic chemistry. Some work was done in vegetable and animal chemistry but the complexities of this area made progress difficult. It was only in mineral chemistry that we find dramatic progress in the eighteenth century by the differentiation of known substances and the discovery of new ones; thus the alkalis were divided into 'mild' (carbonates) and 'caustic' (hydroxides), soda was distinguished from potash, and many new metals were discovered, including nickel, cobalt, manganese, tungsten and platinum. This basic understanding of chemical composition by improved methods of analysis was completed by the chemical revolution. Lavoisier adopted an analytical approach and systematically sought out the 'constituent principles' of substances.

III. Chemistry as a branch of natural history?

From the standpoint of the twentieth century or even of the nineteenth century it is only too easy to classify chemistry as a physical science. By the end of the eighteenth century it certainly had a good claim to be so considered, but not at the beginning. One of the greatest chemists of the early eighteenth century, G. E. Stahl (1660–1734), explicitly rejected the physical and mathematical approach to nature in favour of a vitalistic philosophy which links him more with the traditions of earlier centuries than with any Newtonian-inspired chemist.[25] The crude physical approach of Stephen Hales, who measured the 'air' given off from different substances without considering the *qualitative* differences between them, provides a sufficient reminder that the basic distinction for chemists is between different kinds of matter. Chemistry was therefore necessarily concerned with classification before quantification.[26] The basic categories were classes such as metals, earths, acids and alkalis. A second way in which chemistry must be viewed within the terms of natural history is in relation to

[25] See A. G. Debus, *The Chemical Philosophy: Paracelsian Science and Medicine in the Sixteenth and Seventeenth Centuries* (2 vols., New York, 1977), II, 469, 548.

[26] Some ideas on classification in chemistry are given in D. M. Knight, *Ordering the World* (1980), and in Crosland, 'The development of chemistry' (see note 5), 400–2.

the three 'kingdoms': animal, vegetable and mineral. Seventeenth-century chemistry textbooks had usually been presented in this way and the division continued in the eighteenth century. The mineral kingdom included such substances as metals, sulphur, vitriol and common salt. Vinegar and potash (obtained from wood ashes) would be classified as 'vegetable', while the animal section might contain anything derived ultimately from that kingdom, from blood to phosphorus (from urine). The study of mineral substances had an obvious overlap with mineralogy, in which the emphasis was on the collection and classification of specimens. It is of little importance to our argument that classification of minerals according to their crystalline form was often an alternative to their classification according to their chemical composition. Fourcroy published a book called *Leçons élémentaires d'histoire naturelle et de chimie* (1782) which went through several editions and in which he made use of the classification of minerals of his future colleague at the Jardin du Roi, Daubenton. Earths and stones are presented in classes, genera, species and varieties.

But it was not only the link with mineralogy that made much early eighteenth-century chemistry similar to botany, it was the growing importance of the class of salts. Salts, made by the action of an acid on an alkali or metal, consisted of two parts, and after Linnaeus had imposed his binomial nomenclature on botany, it was natural to consider these parts as the genus and the species. The Latin binomial nomenclature of Linnaeus was adopted by Bergman for naming salts and this was translated into French by Guyton de Morveau. This gave such vernacular names as 'vitriolated potash' or 'nitrated copper' and it was not long before this nomenclature developed into the modern form of e.g. copper nitrate. Thus there is an historic link between the names of salts and the natural history tradition, one which survived the new physicalist approach of Lavoisier.

IV. Affinity

It used to be common to speak of theories of phlogiston as if they were the only theories of eighteenth-century chemistry before Lavoisier. A substantial amount of research over the last twenty years has focused on a quite different tradition in eighteenth-century chemistry, one based on theories of affinity.[27] Much of this was linked with the influence of Isaac Newton. Many chemists, who gave

[27] A. M. Duncan has written several articles on chemical affinity, including 'Some theoretical aspects of eighteenth-century tables of affinity', *AS*, XVIII (1962), 177–94, 217–32.

little thought to the problems of the structure of matter, were fascinated by the idea that it might be possible to explain chemical reactions in terms of attractive forces similar in some ways to gravitational forces. Although Newtonians continued to speculate throughout the whole century on these short-range forces, and there were various attempts to quantify them,[28] the most profitable development of this theory was the construction of affinity tables. Indeed one did not have to be an avowed Newtonian in order to accept a vaguer but probably more useful concept of affinity which meant simply that if a reaction $AB + C \rightarrow AC + B$ took place, this was an indication that the affinity of A for C was greater than that for B. This was the position of Etienne François Geoffroy, whose table of 1718 in sixteen columns[29] provided a model during much of the eighteenth century. Each column showed an order of displacement of the chemical substances listed in it.

A great deal of effort over the next sixty years went into expanding, enlarging and correcting Geoffroy's original table. Finally Bergman in 1775 published separate affinity tables of fifty-nine columns for reactions in 'the dry way' (i.e. by heating in a crucible) and 'the wet way' (i.e. in solution). Although these tables were intended to provide a compact description of reactions, they were useful at the even more fundamental level of providing lists of substances, the basic 'materials' of the chemist. The increasing understanding of materials is reflected in these tables, so that what had been simply 'absorbent earth' for Geoffroy in 1718 was to be distinguished in later tables as lime, magnesia, alumina, etc. Analysis is an aspect of chemistry often taken for granted but, unless such distinctions had been made earlier in the century, it would have been impossible for Lavoisier by 1789 to draw up a list of some thirty simple substances.

Although there were chemists, like Bergman, for whom affinity was the main organizing principle of their science, affinity also featured in the work of many other chemists who are usually associated with other fields. Thus Joseph Black's dissertation on magnesia alba, which contains his discovery of 'fixed air' (carbon dioxide), could alternatively be regarded as an exercise within the Newtonian tradition of attractive forces. Part 2 of the dissertation refers repeatedly to competing attractions and the final conclusion is that a further column should be added to Geoffroy's table. Because Lavoisier was to advance chemistry in quite a different way, it is sometimes overlooked that he too worked within the affinity framework. His collab-

[28] Thackray, *Atoms* (see note 24), ch. 7.

[29] *Mémoires de l'Académie Royale des Sciences* (1718), 202–12.

oration with the applied mathematician, Laplace had given him to hope that

> Perhaps one day the precision of the data might be brought to such a perfection that the mathematician in his study would be able to calculate any phenomenon of chemical combinations in the same way, so to speak, as he calculates the movement of the heavenly bodies.[30]

Yet Lavoisier himself appreciated that affinities could be altered by heat and were not therefore the constants that had been assumed. Berthollet was to take this aspect of affinities much further and to show that they could apparently be altered by a whole range of circumstances and particularly by the quantities present.[31] Thus a large quantity could compensate for a weak affinity. Under such fundamental criticisms the ambitions of the affinity theorists wilted and the theory faded away – there were just too many variables to allow a theory that was useful let alone exact. Chemistry in fact was to make progress not by the quantification of forces but by the quantification of units of matter, in the atomic theory which Dalton published in 1808.[32]

V. Gases

The concept of the gaseous state, which it is so easy to take for granted today, was not achieved without considerable difficulty. Whereas it is not difficult to arrive at the concept of solid or liquid as a generalization from experience of several distinct species of solids or liquids, the idea that there could be several distinct species of gas was one of the great achievements of natural philosophy or chemistry in the eighteenth century. It is true that the term 'gas' was coined as early as the seventeenth century by van Helmont, but for Helmont 'gas' was a wild spirit not capable of being contained within a vessel. For Boyle and Newton anything now considered a gas was considered as 'air', a perpetuation of the Aristotelian element air. Boyle thought of air exclusively in terms of its physical properties, notably elasticity, and the question he asked of air was not of what kind it was (since this did not arise) but whether it was a 'true and permanent air' or a merely a vapour like steam, which would soon condense. For Newton the main interest lay in the fact that for air to be expelled

30 'Mémoire sur l'affinité du principe oxygine' (1783), Lavoisier, *Oeuvres* (6 vols., Paris, 1762–93), II, 550.

31 Berthollet, *Recherches sur les lois d'affinité* (Paris, 1801); *idem*, *Essai de statique chimique* (Paris, 1803; modern repr. New York, 1972).

32 Thackray, *Atoms* (see note 24).

from a solid or liquid it would have to be changed from an attractive state to a repelling state; also, that the solid or liquid could apparently contain within itself many times its own volume of air.[33]

It was this query of Newton's which stimulated Stephen Hales, man of Kent, graduate of Cambridge and vicar of Teddington, to carry out a systematic series of experiments which he published in his *Vegetable Staticks* (1727). Hales carefully and accurately measured the 'air' given off by heating various substances including nitre (potassium nitrate) and powdered coal. We know that the 'air' he collected in these experiments must have been oxygen and coal gas respectively, but Hales was so convinced that there was only one kind of air that, after measuring the volume of each, he threw it away without examining it! Yet Hales concludes with a plea that air should be taken more seriously as one of the elements:

> may we not with good reason adopt this now fixed, now volatile Proteus among the chemical principles . . . notwithstanding it has hitherto been overlooked and rejected by chemists[34]

If chemists had studied air, they could 'have found their researches rewarded with very considerable and useful discoveries'.

The air which Newton had described as fixed in bodies was also the subject of study of Joseph Black. In his doctoral thesis on magnesia alba Black studied a group of substances which we would call carbonates and concluded that, when they were either heated or treated with acids, they gave off 'air' which he described as 'fixed air'. The important thing about this conclusion was that, despite his conservative Newtonian nomenclature, he had distinguished a kind of air which was essentially *different* from common air. 'Air' now had a plural and the way was open for Henry Cavendish to prepare a further type of air: 'inflammable air' obtained by the action of a dilute acid (spirit of vitriol) on a suitable metal such as zinc. The title 'father of pneumatic chemistry' is generally reserved for Joseph Priestley, who was to prepare more gases than anyone else. They included ammonia gas, hydrochloric acid gas, nitric oxide, nitrogen dioxide, carbon monoxide and oxygen (if it is permitted to use the nomenclature of a later period than Priestley's). The importance of gases in the chemistry of the 1770s can be inferred from the fact that, whereas the first edition of Macquer's important *Dictionnaire de chymie* of 1766 contained only half a page on the subject, in the second edition (1778) this was increased to a hundred pages.

The understanding of gases was to be crucial to the chemical revo-

[33] *Opticks*, 4th edn (1730; modern repr. New York, 1952), query 31, 395–6.
[34] Stephen Hales, *Vegetable Staticks* ([1727]; modern repr. 1961), 179–80.

lution of Lavoisier, both the understanding of the relation of gases to other forms of matter and the study of one particular gas which Lavoisier was to call oxygen. As regards the general idea of a gaseous state, Lavoisier in 1777 was experimenting with various volatile fluids and showing that, by suitably raising the temperature or lowering the pressure, they could turn into vapour.[35] Turgot had already spoken of the three states of matter[36] and Lavoisier was to develop the great organizing concept that all matter could exist in either the solid or the liquid or the gaseous form according to the amount of heat present.

Lavoisier prepared oxygen (which he first called 'the salubrious part of the air') by heating red mercury calx (mercury oxide), the same method used earlier by Priestley. But in order to show that this gas combined with metals to form the corresponding calx (oxide) he devised an experimental demonstration which has become a classic.[37] He heated a small quantity of mercury in a retort with a long bent neck connected to a bell jar containing a measured volume of air. It was noticed that the volume of this air gradually diminished from 50 to 42 cubic inches at the same time as scales of a calx formed on the surface of the heated mercury. This residual air did not support combustion and Lavoisier called it 'mephitic air' or 'azote'; it was later to be called nitrogen. When the calx formed in the experiment was carefully heated, it reverted to metallic mercury and produced 8 cubic inches of 'vital air' in which a candle burned brightly.

Some writers have made much of a supposed event, the 'discovery of oxygen'. If any single experiment can be characterized as 'the discovery of oxygen' perhaps it is this one, but the experiment is mainly noteworthy as a demonstration of the oxygen theory of combustion in the context of respiration studies and the clear demonstration that the principle which combined with metals on calcination was distinct from 'fixed air' (carbon dioxide), with which Lavoisier had at first confused it.

It has only been in the 1970s that scholars have begun seriously to trace the central importance of the understanding of the gaseous state to the 'chemical revolution'. Students of Lavoisier have traditionally asked what caused him (apparently in 1772) to turn from geology to

[35] For a perceptive analysis of this memoir see H. Guerlac, 'Chemistry as a branch of physics' (see note 11), 207.

[36] *Encyclopédie*, VI (1756), 274–85, art. 'Expansibilité': see M. P. Crosland, 'The development of the concept of the gaseous state as a third state of matter', *Proceedings of the Xth International Congress of the History of Science* (Ithaca, 1962), 851–4.

[37] 'Expériences sur la respiration des animaux' (1777), in Lavoisier, *Oeuvres* (see note 30), II, 174–83.

the study of combustion.[38] Since, however, J. B. Gough drew attention to the plan of Lavoisier in 1766 to study in turn each of the four Aristotelian elements[39] attention has rightly focused on Lavoisier's studies of 'air'. Siegfried has suggested that Lavoisier found his model for the gaseous state in solutions,[40] but the main value of his paper is in focusing attention on the importance of the gaseous state to the whole of Lavoisier's chemistry.

If the most spectacular (and necessary) step forward in the eighteenth century was in the preparation not of solids or liquids but of gases, we might expect there to have been new types of apparatus introduced to collect and study gases. Thus, whereas Mayow (1674) had collected 'air' in the same vessel in which it was generated, Hales collected his airs in a separate receiver over a pneumatic trough-apparatus which was adapted and improved by Brownrigg, Cavendish and Priestley.[41] It is worth pointing out, however, that Black did not use such apparatus and his 'discovery' of 'fixed air' (carbon dioxide) depended not on collecting it but on inferring its existence from weight changes.[42] In other words the essential apparatus used by Black was the chemical balance, which became the fundamental instrument of chemical investigation in the hands of Lavoisier.[43] Of course the apparatus had to be used effectively and Lavoisier did this by assuming the principle of conservation of mass. This principle of conservation included gases as a form of gross matter but was embarrassed by the fact that the matter of heat or 'caloric' appeared not to have weight. Lavoisier was able to use his considerable wealth to have a variety of special apparatus constructed. Some of this apparatus, e.g. for the storing of gases, made a definite contribution to experimental work but other pieces were unnecessarily complex.[44]

[38] The best presentation of this is given in H. Guerlac, *Lavoisier – The Crucial Year* [i.e. 1772] (Ithaca, 1961).

[39] J. B. Gough, 'Lavoisier's early career in science: an examination of some new evidence', *BJHS*, IV (1968–9), 52–7.

[40] R. Siegfried, 'Lavoisier's view of the gaseous state and its early application to pneumatic chemistry', *Isis*, LXIII (1972), 59–79 (p. 60).

[41] J. Parascondola and A. J. Ihde, 'History of the pneumatic trough', *Isis*, LX (1969), 351–61. The importance of the gaseous content of mineral waters in Britain is examined by J. Eklund, 'Of a spirit in the water: some early ideas on the aerial dimension', *Isis*, LXVII (1976), 527–50.

[42] The gas was also passed through limewater in which, unlike ordinary air, it formed a precipitate.

[43] For a discussion of Lavoisier's apparatus see chs. 5 and 6 of M. Daumas, *Lavoisier, théoricien et experimentateur* (Paris, 1955).

[44] For a criticism of Lavoisier's apparatus see T. H. Lodwig and W. A. Smeaton, 'The ice calorimeter of Lavoisier and Laplace and some of its critics', *AS*, XXXI (1974), 1–18.

Lavoisier was one of several members of the Academy of Sciences who used a large burning glass as one method of attaining high temperatures.

VI. The chemical revolution

The traditional interpretation of eighteenth-century chemistry has been in terms of a 'chemical revolution' associated with the name of Lavoisier, a revolution which completely changed chemistry in the last quarter of the century. The chemical revolution provides so obvious an example of a complete conceptual shift that it was one of the prime examples seized on by T. S. Kuhn to argue that such revolutions play a key part in scientific change.[45] Other philosophers have used the chemical revolution to argue for their own views on philosophy of science. Thus Lavoisier's achievements have been cited to support the views of Imre Lakatos on scientific research programmes and to discredit the views of other philosophers of science.[46] For the 1980s the chemical revolution remains as a fact, although there has been considerable, if mostly indirect, discussion by historians about what exactly constituted the revolution. The most direct comment has been by Christie, who writes:

> Fifteen years ago the chemical revolution was to do with the replacement of the phlogiston theory of combustion and calcination by the oxygen theory; nowadays a more likely reply [to the question of what the chemical revolution was about] might be that it had to do with a general theory of the aeriform state and a particular theory of acidity.[47]

Various possible interpretations of the 'revolution' will be considered after we have looked at the origins of the concept of a revolution in eighteenth-century chemistry and the role of Lavoisier in it.

The term 'revolution' was first applied to the great changes taking place in chemistry by Lavoisier himself. His use in 1773 of the phrase 'une révolution en physique et en chimie' about his own work is an interesting reflection of the ambition of the young scientist and his

[45] T. S. Kuhn, *Structure of Scientific Revolutions*, 2nd edn (Chicago, 1970). The chemical revolution of the eighteenth century is discussed on 69–72.

[46] A. Musgrave, 'Why did oxygen supplant phlogiston? Research programmes in the chemical revolution', in C. Howson (ed.), *Method and Appraisal in the Physical Sciences* (Cambridge, 1976), 181–209.

[47] J. R. R. Christie (University of Leeds), 'The chemical revolution in Scotland', discussion paper for the British Society for the History of Science meeting at Leicester, April 1979.

complete lack of modesty.[48] But the importance of the changes was also recognized by his contemporaries. Macquer and Fourcroy in 1782 could see these changes as 'a great revolution' even though neither had been been won over. Of particular interest is the parallel drawn between the political revolution of 1789 and the changes which had taken place in chemistry. Unfortunately the documentation on this is slight[49] and is obscured by the fact that the phrase 'chemical revolution' was already current. Condorcet, however, was one of those who accepted the analogy between French politics and science.[50] It was really through the celebration in 1889 of the centenary of the political revolution that the phrase became permanently established. In December 1889 the chemist Marcellin Berthelot, who had only just been elected as secretary of the Académie des Sciences, decided that it would be appropriate for the Academy to celebrate the centenary by an *éloge* of one of its most distinguished members, the first *éloge* of Lavoisier to be read before the Academy. Berthelot was able to draw on the biography of Lavoisier by Grimaux, which had appeared in the previous year.[51] He supplemented this, however, with some detailed research on Lavoisier's scientific work and finally presented the result in a book published the following year with the title *La révolution chimique: Lavoisier.*

The traditional presentation of eighteenth-century chemistry has been in terms of individuals, especially the two 'giants' Priestley and Lavoisier. This approach has not only been that of the science-historian but also of philosophers of science who have found it convenient to regard these two as spokesmen for two opposing theories.[52] Priestley is a particularly attractive character to choose

[48] For the first usage of the term revolution among French chemists see H. Guerlac, 'The chemical revolution: a word from Monsieur Fourcroy', *Ambix*, XXIII (1976), 1–4.

[49] A heavily annotated copy of Chaptal's *Tableau analytique du cours de chymie* (Montpellier, 1783), sold at Sotheby's as Lot 1144 on 2 May 1979, with notes obviously written in the midst of the revolutionary turmoil, makes this comparison explicitly.

[50] *Esquisse d'un tableau historique des progès de l'esprit humain* (Ninth Stage). I. B. Cohen has pointed out that Condorcet in his *éloges* of deceased academicians made several references to 'revolutions' in science: 'The eighteenth-century origins of the concept of scientific revolution', *JHI*, XXXVII (1976), 257–88 (p. 280).

[51] Grimaux, *Lavoisier* (see note 7).

[52] Sir P. J. Hartog, 'The newer views of Priestley and Lavoisier', *AS*, V (1941–7), 1–56. S. E. Toulmin, 'Crucial experiments: Priestley and Lavoisier', *JHI*, XVIII (1957), 205–20.

because his main work is readily available in published form and in English, both in books and in the *Philosophical Transactions*. Also there is a particular irony in the fact that the king-pin of Lavoisier's theory, oxygen, had previously been prepared by Priestley, though he regarded it as 'dephlogisticated air'. But what of Black, Cavendish, Kirwan and other major chemists from the British Isles?[53] Should they not be taken into consideration, together with less well-known figures, in discussing the phlogistic reaction to Lavoisier's work? We may expect more research here as research moves towards the idea of a chemical community.[54] A study of the University of Edinburgh provides one example of such a community.[55]

There is even greater evidence for the existence of such communities in France. Paris was a focus for scientific research much more than any one city in Britain. Lavoisier was in constant touch with colleagues both at the Academy of Sciences and in his own laboratory at the Arsenal. Lavoisier himself admitted as much:

> If at any time I have adopted, without acknowledgement, the experiments or the opinions of M. Berthollet, M. Fourcroy, M. de la Place, M. Monge, or in general of any of those whose principles are the same with my own, it is owing to this circumstance, that frequent intercourse, and the habit of communicating our ideas, our observations, and our way of thinking to each other, has established between us a sort of community of opinions, in which it is often difficult for everyone to known his own.[56]

It should be noted that Lavoisier wrote this not near the beginning of his career but near the end of it, in fact seventeen years after his first experiments on which the oxygen theory was based. The first two scientists mentioned were prominent converts to the new theory, the second two were eminent physical scientists with whom he had collaborated in the 1780s. Lavoisier's collaboration with Guyton, Berthollet, and Fourcroy in the publication of the *Méthode de nomenclature chimique* had the effect of making some people think of the oxy-

[53] Richard Kirwan's *Essay on Phlogiston* (1787) provides the focus of an interesting debate within the covers of one book. It was translated into French by Madame Lavoisier (1788) with a criticism of the text by all the leading French chemists and then published as a second edition in English in 1789 with a reply by Kirwan. By 1791 Kirwan had accepted the new theory.

[54] For an interesting quantitative study by a sociologist, see H. Gilman McCann, *Chemistry Transformed: The Paradigmatic Shift from Phlogiston to Oxygen* (Norwood, N. J., 1978), e.g. ch. 3.

[55] Christie, 'Scotland' (see note 47).

[56] *Elements of Chemistry* (see note 23), xxxiii–xxxiv.

gen theory too as a collaborative enterprise, and it was sometimes called 'the French theory'. Lavoisier, however, reacted immediately to this: 'This theory is not, as I have heard it said, the theory of the French chemists, it is mine and it is a property which I claim from my contemporaries and from posterity'.[57] We must agree that the basic oxygen theory was indeed largely the work of one man and this justifies the considerable literature which has built up about him.[58] However that is not a justification for studying Lavoisier in isolation. It is to be hoped that he will be studied in relation to his scientific education,[59] his colleagues, both outside chemistry (Laplace, Monge) and within (Guyton, Berthollet, Fourcroy), his younger followers (Séguin, Adet) and his opponents (Baumé, De Lamétherie). Lavoisier needs to be studied not simply as an individual but as a member of the Academy of Sciences,[60] which was the centre of his scientific life from the time of his election in 1768 at the early age of 24 to the final years of his life when he made strenuous efforts to save the Academy from iconoclastic elements within the French Revolution. Finally it needs to be remembered that Lavoisier was not alone in his dissatisfaction with the phlogiston theory, which had many variants.[61] The situation may be clearer when further work has been done on the study of the chemical community in the eighteenth century.

VII. Phlogiston and oxygen

According to Stahl's phlogiston theory inflammable substances and metals contain phlogiston which is given off on combustion or calci-

[57] Lavoisier, *Oeuvres* (see note 30), II, 104. This passage was probably written in 1792.

[58] Shortage of space precludes any attempt at a full listing. A valuable source is W. A. Smeaton, 'New light on Lavoisier: the research of the last ten years', *HS*, II (1963), 51–69. It would be useful to have a further analysis of Lavoisier scholarship covering the period 1963–80. Meanwhile one may consult references up to 1972 in the extensive list of secondary sources in the bibliography of Guerlac, *Lavoisier* (see note 7).

[59] See H. Guerlac, 'A note on Lavoisier's scientific education', *Isis*, XLVII (1956), 211-16. For a study of Lavoisier's teacher, Rouelle, see R. Rappaport, 'G.-F. Rouelle: an eighteenth-century chemist and teacher', *Chymia*, VI (1960), 68–101.

[60] The standard source for the history of the Académie des Sciences in the eighteenth century is: R. Hahn, *The Anatomy of a Scientific Institution: The Paris Academy of Sciences, 1666–1803* (Berkeley, L. A., and London, 1971).

[61] C. Perrin, 'Early opposition to the phlogiston theory: two anonymous attacks', *BJHS*, V (1970–1), 128–44.

nation. The definitive series of articles by Partington and McKie[62] concentrated on the question of the weight of phlogiston. Once Lavoisier had shown that phosphorus, sulphur and several metals *increased* in weight on combustion or calcination how could chemists still accept the idea of phlogiston being *given off*? Partington and McKie discuss the concept of negative relative weight, put forward by Guyton de Morveau in 1772. This idea, which is supported by reference to the medium in which weighing takes place, is quite distinct from the ideas of negative absolute weight advanced, for example, by the German chemist Gren in 1789. Of course such physical considerations as weight had not been considered vital or even relevant by many earlier chemists, who had been more concerned with essences than physical attributes, but Lavoisier helped to create a new chemistry in which number, weight and measure were basic parameters. This sometimes resulted in a situation in which 'traditional' chemists and the new 'physical' chemists argued past each other with no common frame of reference, a classic case of 'incommensurability'. Another contrast could be made between 'philosophical' chemists like Priestley, who wanted a chemical theory of matter consistent with his natural philosophy and possibly linked with his theological ideas, and Lavoisier, who rejected such 'metaphysical' ideas and took a purely operational view of his work. Priestley's favourite method of arguing was to challenge his opponents to explain one of his own experiments in terms of their theory. It was particularly difficult for Lavoisier and his followers to explain how 'inflammable air' (which they presumed to be hydrogen) could be given off from charcoal and it was not until 1801 that William Cruickshank showed that this was carbon monoxide and thus fitted in perfectly with the oxygen theory. But by then the oxygen theory was widely accepted on other grounds, this particular experiment having been relegated to the status of an anomaly.

A more crucial problem had been that of the composition of water. It had been easy for supporters of the phlogiston theory to explain how 'inflammable air' could arise from the action of a metal on a dilute acid but not for Lavoisier. It was Cavendish's experiments on the combustion of inflammable air which gave Lavoisier the vital clue in 1783. He was able to synthesize water from 'inflammable air' (hydrogen) and 'vital air' (oxygen) and he could now explain the inflamm-

[62] J. R. Partington and D. McKie, 'Historical studies on the phlogiston theory'. I. 'The levity of phlogiston', *AS*, II (1937), 361–404; II. 'The negative weight of phlogiston', *AS*, III (1938), 1–58; III. 'Light and heat in combustion', *AS*, III (1938), 337–71; IV. 'Last phases of the theory', *AS*, IV (1939–40), 113–49.

able air evolved in the action of dilute acids on metals as coming from the water.[63]

From this point onwards the oxygen theory provided a superior theory of composition to the phlogiston theory. In combustion alone there was not much to choose (apart from the weight criterion) between a theory that said that phlogiston was given out to one that said that oxygen was taken in and caloric was given out. But as the idea of phlogiston came under attack, different supporters of the theory responded in different ways. Often they adapted the theory to experimental evidence in the most *ad hoc* manner. As Lavoisier complained in 1785:

> Chemists have made of phlogiston a vague principle which is not at all rigorously defined and which consequently adapts itself to all explanations for which it is required. Sometimes this principle has weight, sometimes it does not; sometimes it is free fire, sometimes it is fire combined with the earthly principle; sometimes it passes through the pores of vessels, sometimes they are impenetrable to it. It explains at the same time causticity and non-causticity, transparency and opacity, colour and the absence of colour. It is a veritable Proteus which continually changes its form.[64]

In short a principle which explains everything explains nothing. It would, however, be a mistake for historians to emphasize the irrationality of phlogiston and on the whole they have not done so. It was obviously in Lavoisier's interest to exaggerate the deficiencies of the rival theory and the superiority of his own.

VIII. The new theory and the new nomenclature

Lavoisier's theory was sometimes described by his contemporaries as the 'antiphlogistic' theory, although in view of what we have said about his indebtedness to the British pneumatic chemists, the alternative contemporary label of 'la doctrine de gas'[65] is more revealing. It used to be presented by historians as essentially a new theory of combustion. Certainly Lavoisier showed that combustion should be interpreted as the taking in of oxygen rather than the giving out of

[63] Lavoisier considered the acid as an oxide, whereas what we call acid would be oxide plus water (e.g. $SO_3 + H_2O = H_2SO_4$). Therefore in our terms the hydrogen comes from the acid.

[64] Lavoisier, 'Réflexions sur le phlogistique', *Oeuvres* (see note 31), II, 623–55 (p. 640).

[65] Fourcroy, *Leçons elementaires d'histoire naturelle et de chimie* (2 vols., Paris, 1782), I, 22.

phlogiston. But Lavoisier was too impressed by the heat evolved in combustion not to include this in his theory. Thus his oxygen gas was really a compound of oxygen principle and caloric, the latter being liberated on combustion. Some historians have suggested that caloric played a central role in Lavoisier's oxygen theory.[66]

The oxygen theory could, however, be regarded equally as a theory of acidity.[67] As early as 1772 Lavoisier spoke of acids as containing 'air' and he later coined the term oxygen from two Greek words meaning 'acid-producer' because he was convinced that oxygen was the principle of acidity. This is an area where his new theory had superior explanatory power. In terms of explaining combustion the oxygen–caloric theory was merely an alternative to phlogiston and it can even be argued that when Lavoisier said caloric was given off he was merely describing phlogiston under a new name.[68] This obviously weakens the claim of Lavoisier (and most historians of chemistry) that his theory introduced a sharp discontinuity. It was a great help to Lavoisier's posthumous reputation that he should exaggerate the novelty of his contribution.

Lavoisier can, however, claim credit for the first chemical theory of acidity. Impressed by the fact that sulphur, phosphorus and carbon all formed acids which excess of oxygen, he generalized that all acids contained oxygen. Such counter-examples as 'muriatic' (hydrochloric) acid defied analysis at the time and so the balance of evidence was certainly in favour of the new theory. Yet Lavoisier was obviously imprudent in associating irrevocably his theory of acidity and his theory of combustion. The oxygen theory of acidity is sometimes treated as an embarrassing mistake,[69] but it is probably more useful to consider it as a valid theory of limited applicability; it certainly had some predictive value.[70] In so far as he was describing oxy-acids, Lavoisier had an adequate theory.

Attention to the oxygen theory, whether of combustion, caloric, acidity or the more general concept of oxidation, has diverted atten-

[66] Morris, 'Caloric theory' (see note 11).

[67] M. P. Crosland, 'Lavoisier's theory of acidity', *Isis*, LXIV (1973), 306–25. H. LeGrand, who had recently completed a thesis on Berthollet, takes a rather different approach in: 'Lavoisier's oxygen theory of acidity', *AS*, XXIX (1972), 1–18. There had previously been a discussion of theories of acidity in the chemistry section of the 13th International Congress of the History of Science in Moscow in 1971.

[68] Caloric, however, unlike phlogiston, could be measured.

[69] Partington, *History of Chemistry* (see note 3), III, 377.

[70] Lavoisier proposed a useful general method of preparing acids, based on his theory. It consisted in treating the appropriate element or compound with concentrated nitric acid as an oxygenizing agent.

tion from another equally important achievement of Lavoisier, that of reinterpreting chemical composition. It is obviously simpler to associate one man with one idea, but Lavoisier affected chemistry in so many ways that it is understandable that, in exploring any one of these, historians have sometimes lost sight of some of the others. Indeed even contemporaries of Lavoisier might have selected different aspects of his work as constituting the most revolutionary aspect.[71] It has been pointed out that Joseph Black was able to discuss the chemical revolution without mentioning the overthrow of the phlogiston theory, referring instead to 'discoveries ... relating to the constituent parts or principles of natural substances'.[72]

It was a necessary corollary of the oxygen theory that a 'calx' formed in the calcination of a metal could no longer be considered as a simple substance but as a compound. Thus instead of the reaction on combustion:

$$\underset{\text{(calx + phlogiston)}}{\text{Metal}} \rightarrow \text{Calx + Phlogiston} \uparrow$$

we have

$$\text{Metal} + \underset{\text{(oxygen principle + caloric)}}{\text{Oxygen gas}} \rightarrow \underset{\text{(metal + oxygen principle)}}{\text{Metal oxide}} + \text{Caloric} \uparrow$$

Metals, therefore, were now simple substances, as were sulphur, phosphorus and carbon. Lavoisier was later to make a list of these simple substances, or elements[73] and this list (see Fig. 1) has been called 'one of the mileposts of modern chemistry'.[74] Lavoisier was able to present chemistry in a logical order starting from the elements before considering compounds of these elements. This was all the more necessary in so far as substances which had previously been thought of as simple were now considered compound and vice versa. Lavoisier had, for example, to overcome the deep-rooted prejudice which regarded acids as necessarily simple substances. He destroyed the status of air as an element, showing that it was a mixture of two

[71] Christie, 'Scotland' (see note 47). Christie complains of a tendency of historians to tidy up the conflicting and divergent views to a greater extent than is justified by the historical situation.

[72] Joseph Black, *Lectures on the Elements of Chemistry*, ed. J. Robison (2 vols., Edinburgh, 1803), I, 488.

[73] *Traité élémentaire de chimie* (Paris, 1789), 192. For a discussion of the importance of elements from Lavoisier to Humphry Davy see R. Siegfried and B. J. Dobbs, 'Composition, a neglected aspect of the chemical revolution', *AS*, XXIV (1968), 275–93.

[74] D. I. Duveen and A. S. Klickstein, *A Bibliography of the Works of Antoine Laurent Lavoisier, 1743–1794* (1954), 156.

Fig. 1. Table of elements from Lavoisier, *Traité élémentaire de chimie* (Paris 1789; reprinted in *Oeuvres* (see note 30), I, 135).

TABLEAU DES SUBSTANCES SIMPLES.

	NOMS NOUVEAUX.	NOMS ANCIENS CORRESPONDANTS.
Substances simples qui appartiennent aux trois règnes, et qu'on peut regarder comme les éléments des corps.	Lumière	Lumière.
	Calorique	Chaleur. Principe de la chaleur. Fluide igné. Feu. Matière du feu et de la chaleur.
	Oxygène	Air déphlogistiqué. Air empiréal. Air vital. Base de l'air vital.
	Azote	Gaz phlogistiqué. Mofette. Base de la mofette.
	Hydrogène	Gaz inflammable. Base du gaz inflammable.
Substances simples, non métalliques, oxydables et acidifiables.	Soufre	Soufre.
	Phosphore	Phosphore.
	Carbone	Charbon pur.
	Radical muriatique	Inconnu.
	Radical fluorique	Inconnu.
	Radical boracique	Inconnu.
Substances simples, métalliques, oxydables et acidifiables.	Antimoine	Antimoine.
	Argent	Argent.
	Arsenic	Arsenic.
	Bismuth	Bismuth.
	Cobalt	Cobalt.
	Cuivre	Cuivre.
	Étain	Étain.
	Fer	Fer.
	Manganèse	Manganèse.
	Mercure	Mercure.
	Molybdène	Molybdène.
	Nickel	Nickel.
	Or	Or.
	Platine	Platine.
	Plomb	Plomb.
	Tungstène	Tungstène.
	Zinc	Zinc.
Substances simples, salifiables, terreuses.	Chaux	Terre calcaire, chaux.
	Magnésie	Magnésie, base de sel d'Epsom.
	Baryte	Barote, terre pesante.
	Alumine	Argile, terre de l'alun, base de l'alun.
	Silice	Terre siliceuse, terre vitrifiable.

gases, one of which took an active part in combustion. Similarly the Aristotelian element of water became in the new chemistry a compound of hydrogen and oxygen. This new view of composition is not the least of the claims of the new chemistry to represent a revolution.

The state of chemical nomenclature[75] before this time had been chaotic and Guyton de Morveau made some proposals in 1782 for rationalizing chemical names.[76] When Guyton was himself won over to the new chemistry there was an ideal opportunity for founding a new chemical language which was not only rational but organized around the new theory. Lavoisier and Guyton collaborated with two other prominent French chemists, Berthollet and Fourcroy, in composing the *Méthode de nomenclature chimique* (1787). It laid down that simple substances should be given simple names and compounds should be given more complex names relating to their constituents. Thus red mercury calx became *oxide de mercure.* A system of suffixes was introduced to indicate different proportions of oxygen in a compound. For example, the name *acide sulfureux* distinguished this acid from *acide sulfurique,* containing more oxygen and previously known as 'oil of vitriol', and the names *sulfite* and *sulfate* were given to the corresponding salts. The new nomenclature was obviously theory-laden and was thus resented all the more by supporters of the phlogiston theory. Nevertheless, as the authors were the leading French chemists and they were not content with proposing the nomenclature but actually used it in their publications, even people like Priestley, then near the end of his distinguished scientific career, had to learn the new language. The new terms were soon translated and adapted into other languages and became the basis of the modern nomenclature of inorganic chemistry. In the purely pragmatic terms of the terminology used, modern chemistry therefore starts in 1787. The interesting historiographical question here is whether the abrupt change in terminology has tended to overemphasize the novelty of Lavoisier's chemistry.

In this chapter the main emphasis has been on the two most important countries for chemistry in the eighteenth century: France and Britain. It has been shown that in the period 1760–95 French chemical authors outnumbered their British counterparts by more than 3 to 1;[77] Scandinavian and German contributors followed some way behind.[78]

[75] M. P. Crosland, *Historical Studies in the Language of Chemistry,* 2nd edn (New York, 1978).

[76] 'Mémoire sur les dénominations chimiques . . .', *Observations sur la physique,* XIX (1782), 370–82.

[77] McCann, *Chemistry Transformed* (see note 54).

[78] J. R. Partington in his *History of Chemistry,* III (1962), devotes some 250 pages to French chemistry between 1750 and 1800, over 270 pages to British

This measure of activity, encouraged by the existence of institutions such as the chemistry section of the Académie des Sciences and journals such as the *Observations sur la physique* and the *Annales de chimie,* makes it less surprising that the new theory and nomenclature originated in France, and justifies the present emphasis on that country. The reception of the new theory and nomenclature in other countries has been the subject of a considerable literature.[79]

IX. Some applications of chemistry

Even the briefest survey of eighteenth-century chemistry could not exclude a mention of the applications of chemistry. Chemistry came to be valued more not only because it came to satisfy more clearly the intellectual criterion of rationality but also because it fulfilled the socially important criterion of utility. At the beginning of the eighteenth century its utility was mainly restricted to medicine, but by the end of the century it had wide applications in industry and even its relevance to warfare was beginning to emerge.

In so far as chemistry was taught in universities in the eighteenth century it was mainly in faculties of medicine, notable examples of teachers in the first half of the century being Herman Boerhaave at Leyden and William Cullen at Glasgow. It was for the degree of M.D. that Joseph Black's important dissertation on magnesia alba was written in 1754 and this provides an example of medical motivation for research in what is now usually seen as pure chemistry. Pharmacy was even more closely linked to chemistry, particularly on the continent. Lemery's *Cours de chymie* (first edition 1675), published in numerous editions and translations up to 1756, was written within the pharmaceutical tradition, the description of chemicals often ending with their medical application and the dose. Lavoisier's teacher Rouelle and his contemporary Baumé both had a pharmaceutical background. The Swedish apothecary Scheele made important contributions to chemistry with minimal resources. Such a medical or pharmaceutical context has been largely taken for granted by chemical historians. Pharmacists in Germany, who constituted a clearly defined group with a professional interest in chemistry, provided a major part of the readership of Crell's *Annalen,* one of the first jour-

chemistry from the time of Hales, but only 80 pages to Scandinavian (Bergman and Scheele) and 50 to German chemistry of that period.

[79] Duveen and Klickstein, *Bibliography* (see note 74), and Crosland, *Language of Chemistry* (see note 75), 193–214, provide a preliminary guide. R. G. A. Dolby treats this as an example of transmission: 'The transmission of science', *HS,* xv (1977), 1–43 (p. 11).

nals specializing in chemistry.[80] The *Annales de chimie,*[81] the journal of the new Lavoisierian chemistry, also has a pharmacy dimension, although it was equally concerned to advocate the utility of the new chemistry to industry.

Some chemical arts and primitive chemical industries had been practised since very early times. Methods were empirical with little chemical theory to guide the artisan. The extraction of alkalis (to make glass) from marine plants was one process which underwent transformation in the eighteenth century. New sources were sought and chemistry was applied, for example, in finding a method of making 'artificial' soda from sea salt (Leblanc process, 1789). 'Saltpetre' or 'nitre', the basic ingredient of gunpowder, was known to be a compound of potash and nitric acid, and Lavoisier took this analysis further by analysing nitric acid. The burning of sulphur (to which some nitre was later added) to make 'spirit of vitriol' (sulphuric acid) was practised a few generations before Lavoisier showed that this was essentially a process of oxidation. Spirit of vitriol had been used to counteract the alkaline lye used in the tedious process of bleaching by repeated exposure to the atmosphere, which before the 1780s used to take weeks or even months. Berthollet's discovery of the bleaching action of 'oxygenated muriatic acid' (chlorine) revolutionized bleaching and raised further problems for chemists of how the gas should be applied and how the strength should be measured. Although some industrial processes continued without reference to theory, future assessments of the chemical revolution will no doubt make more serious attempts to relate the advance of chemical theory and industrial practice.[82]

If the application of the new chemistry to industry in the 1780s had not made people more aware of the utility of science, the circum-

[80] K. Hufbauer, 'The formation of the German chemical community, (1700–1795)' (unpublished Ph.D. thesis, University of California at Berkeley, 1970).

[81] The present writer is planning a study of the *Annales de chimie* (1789–1815) and its successor, the *Annales de chimie et de physique* (1816–). For a study of the early *Annales de chimie* based on published sources see S. Court, 'The *Annales de chimie* 1789–1815', *Ambix,* XIX (1972), 113–28. For the pharmacy element see S. Court and W. A. Smeaton, 'Fourcroy and the *Journal de la société des pharmaciens de Paris*', *Ambix,* XXVI (1979), 39–55.

[82] There are several chapters on industrial chemistry in Britain in the eighteenth century in A. E. Musson and Eric Robinson, *Science and Technology in the Industrial Revolution* (Manchester, 1969). Henry Guerlac gives a useful account of technical chemistry in France in 'Some French antecedents of the chemical revolution', *Chymia,* v (1959), 73–112. For a well-researched account of French chemical industry in the late eighteenth century see J. Graham Smith, *The Origins and Early Development of the Heavy Chemical Industry in France* (Oxford, 1980).

stances of the French Revolutionary wars might have done so, at least in France. That country, deprived after 1793 of her traditional imported raw materials, had to fall back on her own resources, taking advice from the most eminent chemists. Guyton, Berthollet, Fourcroy and Chaptal were all engaged in the extraction of saltpetre for gunpowder, and a process devised by Fourcroy was used to increase the proportion of copper in church bells so that the resulting bronze could be used for cannon.[83] Thus when the emergency was over chemists, and through them chemistry and physical science generally, had acquired a new prestige in France.

X. After the chemical revolution

Lavoisier's theory gave chemistry a great impetus. His *Traité élémentaire* provided a useful introduction to any young man who was interested in the subject, and both Ampère in France and Humphry Davy in England learned the new chemistry in this way. The table of elements provided a basis for the understanding of the mineral world and stimulated the search for new elements. Lavoisier had defined elements as bodies which chemists of the time had not found means of decomposing and this provisional definition was a particular encouragement to chemists to add to the list. The electric battery of Volta (1800) provided a new tool with which Davy was able to isolate potassium, sodium and other metals from their respective compounds.

Another legacy of Lavoisier came through his treatment of vegetable and mineral substances. Although he thought that he was dealing with only a small number of substances, he did appreciate that what these substances have in common is the elements carbon, hydrogen and oxygen and that the fundamental problem was to ascertain the proportion of these elements. Although this is by no means the end of the problem of differentiation, it was the start that organic chemistry needed. Lavoisier's method of analysis by oxidation was extended and perfected in the first few decades of the nineteenth century by Gay-Lussac, Thenard, Berzelius and Liebig. Lavoisier's concept of a radical was to be of importance in the establishment of organic chemistry. The collaboration of Lavoisier and Laplace and their 'Mémoire sur la chaleur' of 1783[84] effectively established the foundations of thermochemistry, which in the nineteenth century was to provide basic information about chemical reactions and the stability of compounds.

[83] W. A. Smeaton, *Fourcroy: Chemist and Revolutionary* (1962), 120–1.
[84] Guerlac, 'Chemistry as a branch of physics' (see note 11).

But Lavoisier's heritage was not entirely a success story. The material nature of caloric was soon questioned and the entry 'caloric' was quietly dropped from the list of elements together with light. The other great 'mistake' of Lavoisier was to consider oxygen as the principle of acidity. During Lavoisier's lifetime Berthollet had pointed to the existence of certain compounds with acidic properties, such as hydrogen sulphide, which did not seem to contain any oxygen. But it was left to Gay-Lussac on the basis of prussic acid (1815) to introduce the class of *hydracids,* which represented an acknowledgement that oxygen was by no means always present in acids. Nevertheless this was presented not as a revolution but rather as a small adjustment within Lavoisier's new system, and Gay-Lussac and his collaborator Thenard may be considered as representing the main stream of Lavoisierian chemistry in France in the early nineteenth century.[85]

They were succeeded by J. B. Dumas, who gave prominence to the work of Lavoisier in the lectures he gave at the Collège de France in 1836 and who took responsibility for editing the complete works of Lavoisier.[86] In 1868 the Alsatian chemist Wurtz raised a storm by using Lavoisier's achievement as the basis of a claim that chemistry was essentially a French science.[87] This was intended to belittle the very real achievements of chemists in Germany in the nineteenth century and was particularly resented in the German states which were then moving towards unification under Prussia to become the predominant military power in Europe. Fortunately twentieth-century studies of Lavoisier have largely shaken off this nationalistic image and French scholars carrying out research on Lavoisier[88] and who might have been suspected of nationalistic sentiments have been out-numbered by American and British authors.[89] There is clearly no parallel to the cult of the eighteenth-century Russian chemist

[85] M. P. Crosland, *Gay-Lussac: Scientist and Bourgeois* (Cambridge, 1978); *idem, Les héritiers de Lavoisier* (Paris, 1968).

[86] Dumas edited only vols. I–IV. Vols. V and VI are edited by E. Grimaux.

[87] 'La chimie est une science française. Elle fut constituée par Lavoisier d'immortelle mémoire': 'Histoire des doctines chimiques', *Dictionnaire de chimie pure et appliquée* (3 vols., Paris, 1868–73), I, i.

[88] The major problem of editing Lavoisier's correspondence was entrusted to René Fric, an amateur whose willingness to undertake the task hardly compensated for his poor achievement. At the time of his death Fric had produced only three fascicules out of the projected eight: see R. Fric (ed.), *Oeuvres de Lavoisier: Correspondance* (3 fascicules, Paris, 1955–64). There is little evidence of Lavoisier being used in the twentieth century as a vehicle for nationalistic purposes by the French government of any of its agencies.

[89] This is obvious from the names of recent authors cited by Smeaton, 'Lavoisier' (see note 59), and Guerlac, *Lavoisier* (see note 7).

Lomonosov, who has inspired a large literature mainly originating from the Soviet Union.[90]

But Lavoisier had neglected to consider the units of matter, and the atomic theory which John Dalton published in 1808 has some claim to have completed the eighteenth-century chemical revolution. Dalton took an important step in assigning weights to the atoms of each chemical element. Thus the basic chemistry of the nineteenth century was not simply to be based on elements and compounds but also on atoms and compound atoms or molecules.

There is something of a hiatus in studies of the history of chemistry immediately after Lavoisier. Most historians of chemistry who have examined the 'chemical revolution' have been concerned with its origins rather than its influence. This is perfectly proper and is certainly a logical way to begin. However historians of eighteenth-century chemistry should be encouraged to extend their studies forward in time. Also historians of nineteenth-century chemistry might consider an earlier perspective; there could then be a fruitful interaction between the two groups.

Acknowledgements

I should like to thank Dr W. A. Smeaton, who read and criticized a first draft of this chapter; also Mr R. G. A. Dolby, who made various useful suggestions.

[90] There is, however, an English translation of one study by a leading Lomonosov scholar: B. N. Menshutkin, *Russia's Lomonosov: Chemist, Courtier, Physicist, Poet* (Princeton, N.J., 1952).

11

Mathematical cosmography

ERIC G. FORBES

Contents

I. Introduction 417
II. Cartographic projections and the earth's shape 421
III. Errors in mapping and in the reduction of data 424
IV. The Cosmographical Society of Nuremberg 427
V. Cosmographical schemes 429
VI. The Cosmographical Institute 435
VII. The ferment in astronomy and navigation 437
VIII. Magnetic forces 444
IX. Reflections 446

I. Introduction

Cosmography, in the context of this portrayal of a historically neglected area of eighteenth-century science, is another name for the 'Idée Générale de la Géographie' drawn up in 1752 by Philippe Buache, son-in-law of the famous French cartographer Guillaume De L'Isle and his successor as first geographer to Louis XV. This scheme was then presented to the Paris Academy of Sciences in the form of three charts containing detailed classifications of historical, physical, and mathematical geography respectively; these were duly approved nine years later, and published in 1762. Significantly, the third chart 'Géographie Mathématique ou Astronomique' is identical, *down to the last detail,* with another entitled 'Mathematische Astronomie' in Johann Gabriel Doppelmaier's *Atlas coelestis* (Nuremberg, 1752). The two titles reflect what may at first appear to be ambiguity in the conceptions of geography and astronomy, but arise from the fact that 'astronomy' was then regarded as possessing a terrestrial as well as a celestial component. The eighteenth-century meaning may, however, be captured by substituting the word 'cosmography' to encompass both components of that broader definition. The major sub-divisions of this classificatory system may therefore be represented as shown in Fig. 1.

According to Doppelmaier, Buache, and the Paris academicians, the category 'earth' includes: the knowledge of the shape of our planet; its climate in the torrid, temperate, and polar regions; the specification of great circles such as the equator, horizon, and meridian, and small circles such as the tropics, arctic circles, and latitude circles generally. 'Heavens', on the other hand, incorporates apparent motions of stars like those reflected in the earth's diurnal and annual rotations, or observed as proper motions; celestial appearances generally; eclipses of the sun, moon, and Jupiter's satellites; and ideas concerning the system of the universe (i.e. cosmology). 'Theoretical geometry' comprises latitude and longitude observations; trigonometrical operations; scales of measurement; and descriptions of itineraries, journals, and voyages. 'Practical geometry' covers the construction of armillary spheres, celestial and terrestrial globes, land-maps, and sea-charts. Thus the link between astronomy and geometry (as here defined) introduces certain aspects of the science of navigation into the realm of mathematical cosmography.

By the end of the eighteenth century, thanks principally to the successful application of John Harrison's chronometer and the use of accurate lunar tables based on the reliable analytic theories pioneered by Leonhard Euler and the French mathematicians Clairaut, d'Alembert, Lagrange and Laplace, navigation had become a science in its own right. Once it became evident that the physical world could be interpreted in terms of the geometrical axioms of Newtonian mechanics, the distinction between physical and mathematical cosmography and their sub-divisions became blurred, being replaced by the more obvious distinction between the science of the heavens and that of the earth – astronomy and geography as we now understand

Fig. 1. The meaning and scope of cosmography.

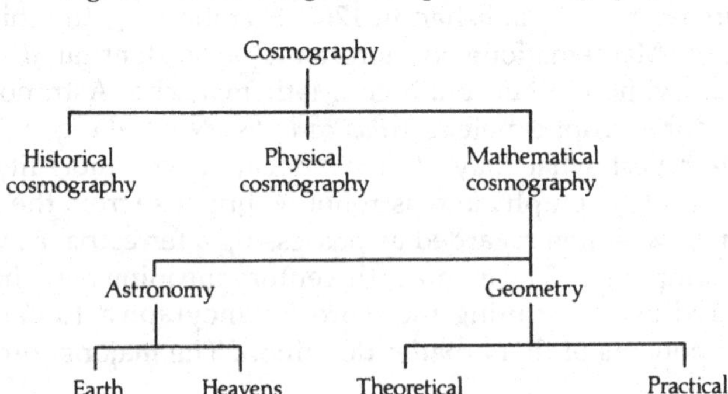

them. Each of these possessed a practical as well as a theoretical component, and a rapidly developing instrumentation appropriate to its purposes. The accompanying differentiation in methods and techniques revealed that the all-embracing vision of nature which found its substantial form in cosmography, was as unreal as the Aristotelian–Ptolemaic universe. Nevertheless, it proved to be an inspiration to a few German cartographers calling themselves the Cosmographical Society of Nuremberg, with whose activities this chapter is concerned.

The aims and achievements of this small group, whose principal members were Johann Michael Franz, Georg Moritz Lowitz and (Johann) Tobias Mayer, have hitherto remained obscure, partly because many of the facts are not to be found where one might at first expect them (viz. in Nuremberg itself) but in Göttingen, and partly because most of the primary sources are unpublished manuscripts written in at times undecipherable Gothic German script. To make matters worse, archival records in Nuremberg, consulted by the late-nineteenth-century cartographer Christian Sandler when preparing an article (cited in note 91) on these Homann *Erben*, as they were called, have since gone astray.

The discovery which made this study feasible, since it provided much of the factual material required for a reconstruction of the contributions of Franz and Lowitz to the geographical and geometrical aspects of the subject-matter, was the existence in the Göttingen University library of the *Acta* of the Cosmographical Society,[1] apparently compiled shortly before Lowitz's departure from Nuremberg for Göttingen and containing fifteen pamphlets written mainly by Franz between 1746 and 1754. Without that convenient collation it would have proved very difficult to obtain an understanding of the role of that Society as a catalyst in this scientific development. With it, many of the documents lying in the *Personalakte* of these two Göttingen professors become much more meaningful.

It is regrettable that comparatively little has as yet been done to investigate the rich archival holdings of the Göttingen repositories, and that no one has written an institutional history of the Göttingen Academy of Sciences that would throw light upon the varying nature of its activities since its origins under the presidency of Albrecht von Haller until the present time. Its membership, organizational struc-

[1] Acta Societatis Cosmographiae, seu Minorum Scriptorum a Membris Societatis annunciandi Instituti Cosmographici causa editorum in unum Volumen Collectio facta Anno MDCCLIV quo expiravit hujus Societatis Directorium Norimbergense id que ad Musas Goettingenses felicissimis Auspiciis emigravit.

ture, academic contacts with the local university, the Berlin Academy and the European cultural scene generally, are matters which are still ripe for historical exploration. A close study of the diplomatic correspondence between Göttingen and Hanover would certainly reveal much more than is currently known regarding the political links with England (and, through England, with America). For example, preserved in the University Library are over 1,500 letters from the Hanoverian Secretary for Domestic Affairs Georg Friedrich Brandes (1719–1791) to the Secretary of the Göttingen Scientific Society Christian Gottlob Heyne (1729–1812), who was simultaneously professor of rhetoric and classical philology and University librarian. These provide ample testimony to the manner in which new professorial appointments and the purchases of important new books were arranged, and insights into the actions taken to combat the sordid intrigues which Göttingen University (in common with other universities before and since) seemed to spawn. The extensive manuscript correspondence of Johann David Michaelis (1717–1791) reveals that similar functions had previously been performed by himself in Göttingen and by Gerlach Adolph von Münchhausen in Hanover. Although, after von Münchhausen's death in 1770, two elderly Privy Councillors were nominally appointed as curators of the University's affairs, they were content to let younger and more active manipulators like Brandes and Heyne take control. As will be shown in this chapter, important astronomical and navigational discoveries were initiated in Nuremberg under the stimulus of the Cosmographical Society's ideals and cartographic enterprises, before being generated in Göttingen by a chain reaction of researches subsumed within a programme of investigations defined by that Society at its inception about 1747.

The pioneering efforts of Abraham Ortelius and Daniel Cellarius in collating maps from various countries, of Gerhardt Mercator, Wilhelm and Johann Blaeu, Johann Jansson and others in systematizing and correcting these, did much to stimulate improvements in cartography. The political and historical connections were stressed by Eberhard David Hauber in 1727 when he made a special plea for rectifying the weaknesses of map-making within Germany,[2] the urgency of which had previously been made clear to him in an earlier study of ancient writers' reports on terrestrial globes. The strong tradition of globe-manufacture stemming from Martin Beheim in Nuremberg and Hieronymus Fracastorius of Verona had been con-

[2] E. D. Hauber, *Discours von dem gegenwärtigen Zustand der Geographie* (Ulm, 1727).

tinued by Guillaume de L'Isle in France, Hermann Moll in England, and Erhard Weigel in Germany; while in Nuremberg itself, Johann Georg Puschner made globes for Doppelmaier. The largest celestial globes then in existence were those of Johann Blaeu in Amsterdam and Vincenzo Coronelli in Paris, but the star-positions depicted on them were based upon the catalogues of Tycho Brahe and Johann Hevelius, which had been superseded by the *Historia coelestis Britannicae* (1725) of the first Astronomer Royal John Flamsteed. De L'Isle was among the first to base maps more firmly upon astronomical observations, although not all were good and his historical knowledge of the lands that he skilfully drew was often scanty.

II. Cartographic projections and the earth's shape

In Germany, De L'Isle's example was followed by Johann Matthias Haase, for many years Doppelmaier's right-hand man in Nuremberg, who explained the advantages (but not the theory) of a new type of geometrical projection - the horizontal stereographic projection - in a Latin pamphlet *Sciagraphia integra tractatus de constructione mapparum omnis generis* . . . (Leipzig, 1717). Subsequently Haase used it in all his maps. Its major advantage was that, since the earth's shape approximates to a sphere, it led to no appreciable distortion of the areas of lands or the distances between places, as was the case with the Mercator projection. Doppelmaier himself was professor of mathematics in the Aegydien-Gymnasium (now the Melanchthon Gymnasium) in Nuremberg and was well known in his time as a maker of sun-dials. As a close personal friend of Johann Baptist Homann, the founder of the Homann Cartographic Bureau in Nuremberg, he was influential in persuading the latter to involve that firm in the construction of small armillary spheres, pocket-globes, and other works of an astronomical as well as a geographical character.

Christian Conrad Nopitsch's allusion in 1801 to 'The Doppelmayer, or rather the Eimmart Observatory'[3] reflects the closeness of Doppelmaier's association with an observatory established in 1678 by the artist Georg Christoph Eimmart (1638–1706) in the grounds of the castle in Nuremberg, of which there is a drawing at the bottom right-hand corner of the map entitled 'Hemisphaerium coeli boreale' in Doppelmaier's *Atlas coelestis*. An impressive illustration of the instrumentation with which Eimmart equipped his observatory is con-

[3] Quoted by F. A. Nagel, 'Das Georg Christoph Eimmartische Observatorium auf der Vestnertorbastei nach einem Kupferstich vom Jahre 1716', *Fränkischer Kurier*, 29 December 1929, no. 360, 8.

tained in Christoph Jacob Glaser's *Epistola eucharistica* (Nuremberg, 1691). The most regular and conscientious users of this observatory were Eimmart's married daughter Maria Clara Müller, and a local businessman called Johann Philipp Wurzelbau, whose dedication to observation was not inconsiderable. Between them they made numerous observations of sun-spots and solar altitude measurements between 1682 and 1708.[4] Two copper-engravings by Johann Adam Delsenbach, dated 1716 and 1748, plus a manuscript index of the instruments drawn up in 1751 (shortly after Doppelmaier's death) clearly illustrate, however, that no significant changes were made to the equipment throughout the intervening sixty-year period.[5] One might therefore be justified in inferring that the Eimmart observatory can scarcely be regarded as symbolizing a development of astronomical science in Germany at this time, although the Homann Bureau kept the astronomical-geographical tradition alive throughout the early part of the eighteenth century.[6]

A final assessment of the role of that observatory and its successive directors in the European cultural scene ought, however, to be reserved until such time as the fifty-seven volumes of Eimmart's manuscripts, containing several substantial astronomical treatises and extensive correspondence, have received the scholarly attention that they undoubtedly deserve. These papers were indexed in 1781 by Christopher Theophilus von Murr,[7] before being taken from Nuremberg by a monk who deposited them in a Lithuanian monastery; from there they were transferred in 1831 to what is now the Saltykov-Schedrin Public Library in Leningrad. They are still preserved there under Signatur: *Archiv Eimarta,* but to date only three of the ten volumes of correspondence in this collection (viz. Tom. 319/1–3) have been microfilmed and studied.[8] However, the scholar concerned was interested primarily in the contemporary political scene, rather than in the scientific issues involved, and an enormous amount of work still remains to be done on this valuable astronomical manuscript collection.

[4] J. P. von Wurzelbau, *Opera astronomico-geographica* (Nuremberg, 1728).

[5] E. G. Forbes, 'Das Eimmartische Observatorium zu Nürnberg', *Sterne und Weltraum,* IX (1970), 311–15.

[6] E. G. Forbes, 'Nuremberg's astronomical heritage', *Journal of the British Astronomical Association,* LXXXI (1971), 391–3.

[7] C. T. von Murr, *Catalogus omnium operum manuscriptorum . . . dom. Georgii Christopheri Eimmart* (Nuremberg, 1781).

[8] O. Feyl, 'Ein unbekanntes Erbe der Weigel-Zeit der Universität Jena in der Sowjetunion. Der Leningrader Eimmart-Nachlass vom Ende des 17. Jhs. und seine wissenschaftsgeschichtliche Bedeutung', *Wissenschaftliche Zeitschrift der Friedrich-Schiller-Universität Jena/Thüringen,* VIII (1958/9), 41–47. A second article was here promised by the author, but failed to appear.

The leading European observatories in the early eighteenth century were, of course, those in Paris and Greenwich, committed respectively to establishing the necessary astronomical basis for cartography and for navigation. Prior to 1740 the French astronomers' refusal to accept Newton's demonstration[9] that the earth was slightly flattened at its poles owing to a combination of the centrifugal force associated with its diurnal rotation and the inverse-square law of gravitational attraction, prevented cartographers from taking account of this effect in the preparation of maps or planispheres for globes. However, after the Paris Academy's expeditions by Pierre Bouguer and La Condamine to Peru, and by Pierre Maupertuis to Tornea in Lapland, had vindicated this prediction,[10] and Alexis Clairaut's assumption of an inhomogeneity in the earth's density satisfactorily removed the discrepancy between the measured and theoretical values of the oblateness,[11] a more elaborate theory of projection had to be developed to quantify it.

The first person who sought to do this was Georg Moritz Lowitz, whom the director of the Homann Bureau, Johann Michael Franz, initially engaged for this very purpose on account of his skill in analysis. Manuscript correspondence preserved in the archives of the U.S.S.R. Academy of Sciences in Leningrad[12] reveals that Lowitz initiated a correspondence with Leonhard Euler on 26 October 1745 to request advice on the mathematical problems associated with the construction of meridians and parallels of latitude on the horizontal stereographic projection of a world-map or *Planiglobium* (1746).

Lowitz's prime concern, however, was to apply his rather elaborate new theory in drawing planispheric projections of twelve 30°-wide segments of 3-foot diameter terrestrial and celestial globes, as advertised in Franz's announcement of 15 July 1746 declaring his firm's intention of instituting a globe-factory in Nuremberg.[13] The first 100 persons who would notify the Homann Bureau of their wish to purchase such a pair of globes were to be given a substantial discount on

[9] I. Newton, *Philosophiae naturalis principia mathematica* (1687), book 3, prop. 18, theorem 16.

[10] S. Chapin, 'Expeditions of the French Academy of Sciences, 1735', *Navigation*, III (1951), 120–2.

[11] A. C. Clairaut, *Théorie de la figure de la terre* (Paris, 1743).

[12] The letters in folio 136, op. 2, no. 2 are dated 19 November 1745 (pp. 74–5), 25 December 1745 (p. 83), 8 October 1746 (pp. 164, 331–3), 30 December 1747 (pp. 374–5); that in folio 136, op. 2, no. 3 is dated 10 December 1749 (pp. 79–80); and that in folio 136, op. 2, no. 4 is dated 23 September 1750 (pp. 168–9).

[13] J. M. Franz, *Homännischer Bericht von Verfertigung grosser Welt-Kugeln* (Nuremberg, 1746); also published in French as *Avertissement des héritiers de Homann, sur la construction de grands globes* (Nuremberg, 1746).

the price that others would be required to pay. As soon as the target figure of 100 subscribers should be reached, a second advertisement specifying the precise price and delivery date would be circulated. An advance of one-quarter of the set price was expected, the remaining three-quarters to be paid on delivery.

Meanwhile, having with Euler's help developed his rigorous theory, Lowitz had discovered that the effect of the earth's oblateness was equivalent to only one-sixth of an inch on the planispheres of a 3-foot diameter globe, and so could be ignored altogether. Nevertheless, in the course of their correspondence he had obtained useful advice from Euler on how to represent asterisms on the celestial globes. The difficulty was that the projection of these figures from maps on to the segments of a sphere caused right and left to be transposed. John Flamsteed, Giovanni Cassini and Georg Christoph Eimmart had all regarded the asterisms as real bodies rather than transparent ones, but had distorted their shapes in order to retain the names given to the stars by Claudius Ptolemy (2nd century A.D.). Lowitz, guided by Euler, decided to remove these distortions without departing from tradition, even though the rational geometrical solution to this problem would have been to represent the asterisms as laterally inverted.

A pamphlet entitled 'Kurze Erklärung über zwey Astronomische Karten von der Sonnen- und Erd-Finsternis den 25. Julius 1748...', translated into French by Joseph De L'Isle, Guillaume's younger brother, and read to the Paris Academy of Sciences on 23 and 27 March 1748, contains evidence that Lowitz was applying both stereographic and orthographic projections to the earth, and ascribing small inconsistencies between the two sets of calculations to the earth's non-spherical shape, in accordance with the results of the French expeditions referred to above. This is why De L'Isle was so interested in explaining this piece of research to the Paris Academy. Another interesting fact announced in the title of the first of the two eclipse maps was that the calculation of the instant of the new moon had been made with the help of Euler's new solar and lunar tables published in *Opuscula varii argumenti* I (Berlin, 1746), from which the corresponding ephemerides in the Berlin *Calendar* for 1746 had been prepared.

III. Errors in mapping and in the reduction of data

Meanwhile, Lowitz had been joined by another colleague with similar talents to himself, attracted by Franz's advertisement for the globes and taken into his employment. This was Tobias Mayer from

Esslingen, who had previously spent two years working for the Augsburg cartographic firm of Matthias Seutter with which the Homann Bureau had close personal and business contacts.[14] He too was engaged at much the same time on the closely analogous problem of using the conventional orthographic projection to depict the moon as seen from the earth, and the stereographic projection for depicting the earth as it would be seen from the moon, in the construction of a map of a lunar eclipse predicted for 8/9 August 1748 (reproduced as table XXI in Doppelmaier's *Atlas coelestis*).[15] The inconsistency found between the predicted times of this eclipse at Berlin, derived from these alternative constructions with the aid of Euler's data, implied that it was not permissible to employ the orthographic projection as had been done hitherto; for, by so doing, one would be effectively neglecting the lunar parallax and thereby placing an unnecessary limitation on the accuracy attainable using the lunar eclipse method of longitude determination. Mayer was also dismayed to discover, when preparing this representation of a partial lunar eclipse, that even the best lunar maps then in existence used different nomenclatures and contained substantial errors. He found Francesco Grimaldi's lunar diagram in Riccioli's *Almagestum novum* (Bologna, 1651) to be inferior to that in Johann Hevelius's *Selenographia* (Danzig, 1662), both from the point of view of the accuracy in depicting the moon's topographical features and in the quality of the copper-engraving. The fact that the markings on the moon's surface appear at some times nearer to, and at others farther from, its limb – a phenomenon known as lunar libration – constituted an additional source of inaccuracy for which no reliable quantitative correction had previously been made. Thus he decided to embark on a thorough observational investigation of the moon's libration using a 9-foot-long telescope with a glass-micrometer of his own design situated in the focal plane; the latter instrument is described in detail in the *Kosmographische Nachrichten und Sammlungen auf d. J. 1748* (Nuremberg, 1750).[16]

In the seventh edition of the *Encyclopaedia Britannica*, XIV (Edinburgh, 1842) James Browne, writing with hindsight gained through Carl Friedrich Gauss's mathematical discoveries and in the light of Joseph Delambre's technical appraisal of Mayer's contributions to as-

[14] C. Sandler, *Johann Baptista Homann, Matthäus Seutter und ihre Landkarten: ein Beitrag zur Geschichte der Kartographie* (Amsterdam, [1963]).

[15] T. Mayer, 'Vorstellung der in der Nacht zwischen dem 8. und 9. Aug. 1748 vorfallenden partialen Mondfinsternis...' (1748).

[16] 'Abhandlung über die Umwälzung des Mondes um seine Axe, und die scheinbare Bewegung der Mondsbeschreibung aus neuen Beobachtungen gelegt wird', *op. cit.*, 52–183.

tronomy in his *Histoire de l'astronomie au dix-huitième siècle* (Paris, 1827), affirms Mayer's priority in the use of this method of analysis, and remarks that the treatise on the lunar libration was:

> The first wherein, for a problem which appeared not to require or even to admit of more than three observations, there was imagined the method of *conditional equations,* which, instead of three observations strictly necessary, permits the employment of thousands, and conducts at once to the most certain or the most probable conclusions resulting from the totality of the observations; in fact, the errors which cannot be avoided, yet follow no certain law, are found to act each time in a different manner, and to correct themselves by mutual compensation. It is to this method that we are in a great measure indebted for the precision of the most modern astronomical tables; although it did not for some time attract the attention of astronomers, it is now generally employed; and it is thus that have been constructed from hundreds and even thousands of observations, the tables adopted by Lalande in the third edition of his *Astronomie.*[17]

This was, of course, a fortunate and totally unpremeditated development in the reduction of observational data which owed its success entirely to the fact that Gauss's analysis[18] is based upon the postulate that the algebraic sum of all random (or non-systematic) errors incurred in a series of independent measurements is zero. This postulate was taken for granted by Mayer, and was implicit in his handling of the observational data.

The problem of how to quantify the *systematic* errors incurred in geographical, astronomical, or indeed in any physical measurements presented another difficult challenge to eighteenth-century scientists. One necessary criterion for reliability was that full information should be supplied by those who made such measurements to those who would apply them. The limitations in the accuracy of maps were, to a large extent, due to the failure of astronomers to apply their own observations (the accuracy of which they themselves were in the best position to assess) to effect the much-needed improvements. First-hand knowledge was infinitely preferable to the uncritical acceptance of 'authorities'; indeed, any critical examination of published values by different astronomers of polar altitudes and hour angles would

[17] *Encyclopaedia Britannica,* 7th edn, XIV, 335.

[18] First published at the end of C. F. Gauss, *Theoria motus corporum coelestium in sectionibus conicis solem ambientium* (Hamburg, 1809), though discovered about fifteen years earlier.

reveal the substantial inconsistencies that affected the values of the latitude and longitude differences deduced from these data. One can, of course, scan respectable periodicals such as the Berlin *Calendar*[19] and the *Connaissance des temps* to obtain such information; but, as often as not, no details were given of the observers concerned, their instruments, nor the basis of the calculations leading to their results. There were numerous places for which no observations had been made at all, yet for the purposes of accurate mapping they were just as important as those for which the geographical locations had been tabulated.

IV. The Cosmographical Society of Nuremberg

Franz and his colleagues in Nuremberg were fully aware of such deficiencies, and of the urgent need to establish cartography upon a fresh and sounder foundation. A *Sammlung*, or collection of data, containing a short history of each observer, his instruments, his mode of observation, etc., was one way of effecting an improvement. Geographers needed to collaborate more with historians and physicists when establishing changes in state-boundaries, writing or rewriting descriptions of lands and places, including the climatic and other physical conditions, and assessing the significance of new sources of data. Ideally, every German state ought, in Franz's opinion,[20] to employ at least one cosmographer charged with such responsibilities and with the task of formulating a policy for the classification and selection of maps. The headquarters to which each such *'Staatsgeographus'* would report was, of course, to be the Homann Bureau. Those employees wishing to associate themselves with the promotion of this noble aim constituted the Cosmographical Society of Nuremberg, whose aims and rules are defined in the preface to the *Homannisch-Hasische Gesellschafts Atlas* (Nuremberg, 1747), which was issued as a separate pamphlet that same year.[21]

The two principal categories of membership were the Mathematical Class and Historical Class; a third Corresponding Class was intended

[19] The actual title of this annual periodical, which was being edited at this time by J. Kiess (1713–81), is 'Historisch und geographischer Calendar . . . unter Genehmhaltung der von Sr. Konigl. Maj. in Preussen in Dero Residenz Berlin gestifteten Akademie der Wissenschaften'.

[20] J. M. Franz, *Gedanken von einem Reise-Atlas und von der Notwendigkeit eines Staats-Geographus . . . (Nuremberg, 1751).*

[21] This bore the title: 'Homännische Vorschläge von den nöthigen Verbesserungen der Weltbeschreibungswissenschaft und einer disfals bey der Homannischen Handlung zu errichtenden neuen Academie'.

for persons not employed by the Homann Bureau, or interested laymen, who were expected to report on natural, social, secular and religious features characteristic of their own geographical surroundings. As members of the Mathematical Class, Lowitz and Mayer pledged:

1. To search for new and more general principles in geodetic measurement and methods of geometrical projection.
2. To demonstrate the useful application of such projective methods to celestial phenomena such as eclipses and occultations.
3. To pursue practical astronomical science in collecting old and new observations by others *and in making their own accurate observations,* for the purpose of preparing a reliable list of the geographical coordinates of places.
4. To invent, design, and construct celestial atlases, globes, orreries, and other scientific instruments and equipment; as well as to prepare written descriptions of them.
5. To make and collate accurate measurements of the variation and dip of the magnetic needle for as many places as possible, with a view to discovering the laws governing these phenomena and how they might be applied in cosmography.

Here, indeed, is a programme of mathematical, geographical, astronomical and physical research feeding on the interplay of the one subject on the other, involving the simultaneous search for new principles and for new empirical data with which both to discover and test them, and encouraging the construction of better instrumentation with which to put theory into practice. What label may we attach to such a comprehensive scheme? Francis Bacon, had he been alive, would have given it his blessing since it incorporates the break from tradition and authority and the institutional focus which was fundamental to his own philosophy of nature. It is built up too on empirical knowledge, albeit not on the facts of natural history. Franz, himself no more of a scientist than Bacon was, differs from him, however, in one significant respect: namely, his advocacy of a single guiding spirit in the form of a person with a wide breadth of knowledge and vision to direct the individual labours of all concerned.

The 'model' whom Franz wanted to emulate was, of course, Cassini de Thury, whose geodetic survey of France had cost over 300,000 livres plus his annual salary of 5,000 livres. A similar map of Germany would have cost a comparable sum – well beyond the means of the Homann Bureau – which is why Franz saw the need to cultivate the interest of royalty, or dignitaries of high social standing, in his ambitious national plan.

V. Cosmographical schemes

Franz had three means of attaining this end. One was the publication of critical annual reports on the state of cosmography in and around Germany, including the results of his colleagues' researches, of which the *Kosmographische Nachrichten und Sammlungen auf d. J. 1748* could be regarded as the first example. Another was the Cosmographical Society itself; while the third was Lowitz's globe-factory.

The subscriptions for the large globes were potentially the main source of income, since the planned atlas of Germany, though initiated by Haase over forty years previously, was still far from completion at the end of 1749 when the promised *'second avertissement'* relating to Lowitz's work on the globes, was issued.[22] It is here that one finds the rather involved equation for the projection of a spheroidal earth that was developed by Lowitz with Euler's help, though the proof itself is not given. In the second (mechanical) part of this discourse the sources being used in drawing the planispheres for the *terrestrial* globe are cited; but little progress had been made with the *celestial* globes. Part of the reason for this delay was the lack of capital required for such an expensive project, and the poor response to the first advertisement which had brought a very limited sum into the firm for this purpose; moreover, James Bradley's important discoveries of stellar aberration (1728)[23] and nutation (1748)[24] had necessitated a revision of Halley's tables,[25] a task with which Lalande was busily engaged.[26] The desirability of constructing a machine inside each celestial globe in order to reproduce the slow precessional rotation of the axis joining its poles around that through the poles of the ecliptic was also introduced as an excuse. In an effort to raise more money Franz had been involved during 1749 in an abortive attempt to interest the Austrian Kaiser in supporting his cause, the reason for the failure apparently being the latter's insistence that the members of the Cosmographical Society all become converts to the Catholic reli-

[22] J. M. Franz, *Description complète ou second avertissement sur les grands globes terrestres et celestes auxquels la Societé Cosmographique établie à Nurenberg fait travailler actuellement par George-Maurice Lovviz, de la Societé Cosmographique & desinateur des susdits globes* ([Nuremberg], 1749).

[23] J. Bradley, *PT*, xxxv (1728), 637–61.

[24] J. Bradley, *PT*, xlv (1748), 1–43.

[25] J. Bevis (ed.), *Astronomical Tables, with Precepts both in English and Latin for Computing the Places of the Sun, Moon, etc.* (1752).

[26] J. J. L. de Lalande, *Tables astronomiques . . . pour les planètes et les comètes . . . augmentées de plusieurs tables nouvelles . . . et l'histoire de la comète de 1759* (Paris, 1759).

gion.[27] This might, of course, be interpreted as a diplomatic way of saying 'no' to Franz after he had spent eighteen weeks in Vienna lobbying various high-ranking Austrian state officials and government ministers. More light might be thrown upon the actual circumstances surrounding these negotiations by an investigation of the appropriate Austrian archives.

Another person with whom he was negotiating at around this time was the Prussian Field-Marshal von Schmettau, who was hoping to persuade the Cosmographical Society to leave Nuremberg and cooperate with him in Berlin on the preparation of a large German atlas. His intention was to prepare one general map and around 380 detailed maps based on information collated during the course of the previous few years. The French triangulations had ensured the accuracy of Germany's western boundary, while Mykovini's maps of Hungary had established the eastern one. The greatest uncertainties were in the positions of towns in Westphalia, Mecklenburg, Lusace and Saxony. The plan was to begin new triangulations from a mountain called Petersberg between Halle, Eulenburg and Mansfeld, and draw a meridian line through it across Germany, the surrounding land being flat enough to permit a base-line several miles long to be established. The northern part of the meridian would reach the Baltic Sea between Wismar and Rostock, while the southern extension would pass close to Regensburg on the Danube and Wasserburg on the river Inn.[28] Progress with this project, which was sponsored by the Berlin Academy of Sciences, proved to be disappointingly slow, and fraught with difficulties. When von Schmettau himself travelled south to Kassel in 1750 to begin triangulation measurements there, the Hanoverian monarch George III ordered him to abandon his well-laid scheme and return to Berlin. This must have been a totally demoralizing blow to him, and may have hastened his death less than a year later. Further research is required to define the precise nature, purpose and political implications of this cartographic scheme, encouraged by the Berlin Academy of Sciences during the period while Frederick the Great was in effective control of its affairs.

Although Franz had turned down von Schmettau's proposal, ostensibly on the ground that it was impracticable to uproot more than fifty persons – his employees and their wives and families – from their

[27] This fact is explicitly stated in an unpublished document entitled 'Erläuterung wegen des Anlesens' (June 1755) preserved under Ph. 1/17, 4/Vb 27, item 12a: 'Betr. die Bestellung des Profess: Philos: Georg Moritz Lowitz 1755–68' in the University archive, Göttingen.

[28] This information has been drawn from an unpublished letter from von Schmettau to the Hanoverian Prime Minister, G. A. von Münchhausen, preserved among *Cod.Ms.Tob.Mayer* 15_2, p. 6, in the Göttingen University library.

accustomed homes and environment, he confides not long afterwards to the Hanoverian court-adviser C. L. Scheidt that his personnel would be much less disinclined to make such a transfer to Göttingen, where they would be under the protection of King George III.[29] Thus the call from the Hanoverian Prime Minister Gerlach Adolph von Münchhausen to Tobias Mayer on 26 November 1750 to occupy the post of professor of economy which had recently become vacant through the death of Johann Friedrich Penther, may be regarded as the first diplomatic step in a plan to realize this aim[30] – a conjecture that would appear to be borne out by similar invitations issued to Lowitz and Franz a few years later.

In 1751 Lowitz fell heir to the directorship of the Eimmart observatory, which became vacant as a result of Doppelmaier's death; he likewise assumed the latter's role as professor of mathematical and astronomical sciences at the Aegydien Academy. His inaugural lecture in that post dealt with the theme of the usefulness of higher mathematics for mankind.[31] Franz also decided to mark this auspicious occasion with an essay on the usefulness of establishing a lectureship, or teaching post, in mathematical geography at the Homann Bureau,[32] in which he stresses the need for more school-teaching in this subject and distinguishes between astronomical geography, geographical land-measurement, and geometrical projection, as its three main parts. Although Haase had pioneered the method of horizontal stereographic projection in map-making, few books provided an explanation of its rules; and those which did, tended to discuss only special cases of the general theory. Among the best books on both the practical and theoretical aspects of map- and chart-making at that time was Peter Smith's *Cosmographie* (Amsterdam, 1726), but Lowitz felt there was still a need for a textbook on mathematical geography and its related instruments that would explain the correct way to make land-measurements. Cassini de Thury's *La méridienne de l'Observatoire de Paris* (Paris, 1744) had already served as a warning to those who had previously entertained the erroneous impression that work of this nature was straightforward.

[29] Memo by C. L. Scheidt, dated Hanover 4 March 1750, in 'Betr. Sternwarte zu Göttingen', Ms. 4/Vf/1, item 15, in the University archive, Göttingen.

[30] E. G. Forbes, 'The foundation of the first Göttingen observatory: a study in politics and personalities', *Journal for the History of Astronomy*, v (1974), 22–9.

[31] An announcement to this effect is made in G. M. Lowitz, *Beschreibung eines Quadrantens der zur Sternkunde und zu den Erdmessung brauchbar ist . . .* (Nuremberg, 1751).

[32] J. M. Franz, *Als S. T. Herr Georg Moritz Lowitz öffentlicher Lehrer der Mathematik in Nürnberg den 27 December 1751 seine feyerliche Antritsrede daselbst hielte/wollte dabey die Notwendigkeit eines zu errichtenden Lehrbegriffs der mathematischen Geographie bey der kosmographischen Gesellschaft darthun u.s.w.* (Nuremberg, 1751).

The following year marks the turning point in the fortunes of the Homann Bureau and a deterioration in the personal relationships among the triumvirate of Franz, Lowitz and Mayer who, through marriage, had become brothers-in-law. Lowitz's unjust verbal accusations of plagiarism against Mayer, by putting an end to their correspondence, temporarily froze contact between them.[33] Franz's increasing concern at the delays in keeping to the advertised schedule for delivery of the large globes caused him to coax Lowitz into a legal contract wherein he agreed to accept an equal share of responsibility for all losses or profits made from maps, charts and globes produced by Franz's part of the Homann Bureau.[34] The other aspect of the agreement concerned the sharing of duties between them: Franz was to provide the financial backing and publicity, administrative arrangements for delivery, and all associated secretarial duties, while Lowitz had to produce the drawings, segments, circles, etc. for the globes and assist others (such as the copper-engravers) who were employed on this project. It may have seemed that Franz was doing Lowitz a favour but, as subsequent events were to prove, he was really protecting his own interests, by controlling the incoming funds while reducing his personal liability in case of loss. Franz's *'Troisième avertissement sur les grands globes'* (1753) reveals that since by mid-1752 only nineteen subscribers had actually paid their deposits, the income had not been sufficient to finance the capital cost of the cutting machines and materials. This poor response meant that the only way left open, if this project were to be kept alive, was to launch a wider public appeal and offer new incentives (such as prestige and financial gain) to those whose support was to be won for this cause.

These considerations caused Franz to return to the theme of state-support for scientific work.[35] He was well aware that precedents existed in the large-scale trigonometrical surveys supported by enlightened monarchs in France, Russia and Sweden, and in the Pope's authorization of a similar survey in Italy of the Ecclesiastical State that was successfully undertaken by Christoph Maire and Roger Bos-

[33] The tense situation is revealed in an undated draft of an unpublished letter from Mayer to Franz, written between 5 May 1752 and 26 April 1753, preserved under *Cod.Ms.philos.159* 'Briefe von und an J. Tobias Mayer', p. 42, in the Göttingen University library.

[34] A copy of this document, dated 2 May 1753 and bearing the signatures of Franz, Lowitz, Frau Franz, Frau Lowitz and Dr Christoph Adam Rinder (the Nuremberg lawyer who witnessed these signatures) is preserved in Lowitz's *Personalakte* (see note 27), item 22.

[35] J. M. Franz, *Der deutsche Staatsgeographus mit allen seinen Berichtungen Höchsten und Hohen Herren Fürsten und Ständen im deutschen Reiche nach den Gründsätzen der kosmographischen Gesellschaft vorgeschlagen von den dirigirenden Mitgliedern der kosmographischen Gesellschaft* (Frankfurt and Leipzig, 1753).

covich. The political usefulness and importance of enterprises of this nature had already been spelt out by Veit Ludwig von Sechendorf in his *Deutschen Fürstenstaat.* Although Franz still saw his *'Staatsgeographus'* as the person in overall control of the planning, he recognized that this ideal was most unlikely to be attainable in practice; in which case he advocated the establishment of an Ordnance Survey Office in each ducal principality, with a staff devoted to the measurement and geographical description of that area. Information regarding the techniques and instruments required for this purpose, and the precautions to be taken, could be drawn from Cassini de Thury's book (1744) or - preferably - from a short but thorough paper composed by Lowitz in response to the Austrian Kaiser's proposal to make a large-scale survey of his Empire.[36]

This 'Mathematical Specification', as it was called, began with the three-fold distinction in the realm of application of geographical science adopted by Franz: namely, the parts concerned with the rules for finding the relative positions of places (geodetic measurement); the comparison of such positions with an absolute position in the heavens (astronomical geography); and the representation of this absolute position on a plane surface (geometrical projection). Mayer too made use of a similar classification in an incomplete tract contained in his manuscript *Collectanea geographica et mathematica* (1747),[37] but failed to develop this line of thought beyond a brief description of geometrical aids before being confronted with the inconsistencies in the standards of length adopted by different writers from different nations. Lowitz now considers at length the complementary aspect of that same problem - the variations in the length of a single standard due to the effect of temperature changes - also remarking upon the effect of the stretching or contracting of paper on the scale drawn on a map. The alternative approach to improvements in cartography which did not suffer from those uncertainties was the astronomical one, preferable where large territories had to be surveyed and practicable now that the size and shape of the earth had been more accurately established. Both Lowitz[38] and Mayer[39] continued to be concerned with the need to improve both astronomical and geodetic

[36] G. M. Lowitz, *Mathematische Vorschrift, von der rechtmässigen Verfahrungsart die Länder zu messen und zu mappiren* (Nuremberg, 1749).

[37] T. Mayer, 'Von der Construction der Land-Karten mit dem Exempel einer Karte von Ober-Teutschland erkäret', now printed in E. G. Forbes (ed.), *The Unpublished Writings of Tobias Mayer* (3 vols., Göttingen, 1972), I, 45–50.

[38] Lowitz, *Beschreibung eines Quadrantens* (see note 31).

[39] T. Mayer, 'Nova methodus perficiendi instrumenta geometrica. Et novum instrumentum goniometricum', *Commentarii Societatis Regiae Scientiarum Gottingensis,* II (1753), 325–36.

instruments. It was Mayer's application of the repeating principle of angle-measurement to a circular instrument with a graduated limb, that gave rise to the Repeating Circle.[40] This invention enjoyed a brief period of popularity in France before the end of the century, first as an alternative to the Hadley quadrant and subsequently in the geodetic survey work which led to the establishment of the standard metre.[41]

Franz's experience in trying to elicit financial support from wealthy princes, noblemen and heads of state, on the other hand, had met with little success; and, at this stage, many in his position would have been content to have let matters rest. Franz, however, saw another solution in the radical scheme of conducting what he termed a 'cosmographical lottery'.[42] The prize was to be a complete set of all the Homann publications, which comprised thirteen items including: Doppelmaier's *Atlas* of 1752, the *Atlas geographicus maior* I, II and III plus a supplementary volume to it, the *Homannisch-Hasische Gesellschaftsatlas* (1747), an *Atlas historicus*, the *Atlas Silesiae* (1749), wall-maps, a school-atlas, and other maps by foreign map-makers or of foreign countries. He believed that if a large enough number of individuals, universities, libraries, museums, and similar institutions were to be enticed by such an offer, such a lottery could provide the Homann Bureau with a substantial capital investment. Yet another fund-raising venture, described in another pamphlet dated 1 January 1754,[43] was to sell 750 copies of the new 152 folio-page *Atlas von Deutschland* (1753) and an additional 100 copies of a more expensive edition of that same work. Neither scheme proved to be a success, although there were many subscribers to the former from Nuremberg itself until Lowitz left with his entire globe-factory for Göttingen and rumour spread of Franz's own impending transfer; at this stage, not surprisingly, those local subscribers became suspicious of the latter's intentions and demanded their money back. His partner, Johann Georg Ebersperger, even took up a lawsuit against him, for running the firm into debt.[44]

[40] E. G. Forbes, 'Tobias Mayer's new astrolabe (1759): its principles and construction', *AS*, XXVII (1971), 109–16.

[41] M. P. Crosland (ed.), *Science in France in the Revolutionary Era* (Cambridge, Mass., and London, 1969), 201.

[42] J. M. Franz, *Die kosmographische Lotterie was diese seye und was die Deutsche Nation für Bewegungsgründe habe, derselben forderlich zu seyn* (Nuremberg, 1753).

[43] J. M. Franz, *Avertissement touchant la publication d'un Grand Atlas de Cartes Géographiques de toutes L'Allemagne* (Nuremberg 1754).

[44] The details regarding this are to be found in: 'Acta Betr. den Abzug des Raths Frantz von Nürnberg und desfalls an ihm gemachte Forderungen des Kup-

VI. The Cosmographical Institute

Ever since Mayer's departure in 1751, Franz had been entertaining the hope that he would be able to transfer the Cosmographical Society from Nuremberg to Göttingen. Free from its association with the Homann Bureau, it would become known under the name of the Cosmographical Institute (*Institutum Cosmographicum*). The globe-factory, as an integral part of this institute, would then be re-christened the Cosmographical Mechanical Laboratories (*Laboratorii Cosmographico-Mechanici*). The reason for the use of the plural form 'laboratories' was that, in addition to Lowitz's factory for 3-foot diameter terrestrial and celestial globes, a project had been initiated in 1750 for the construction of 15-inch diameter lunar globes which was abandoned when Mayer left for Göttingen. Franz, probably with help from Mayer, had then published a report[45] in which reference was made to a dozen problems whose solutions would be facilitated by the use of such globes. In it, he even announced his intention of issuing another pamphlet with the globes, in which further uses would be explained; and was rash enough to promise delivery by Easter 1752 to those who would send immediately an advance subscription equal to half of the total anticipated cost. Only three months after his own arrival in Göttingen at the beginning of August 1755, Franz relinquished this particular responsibility to Mayer, who gladly consented after suitable conditions had been agreed between them. Franz was still hopeful that his Institute would persevere in the aim of improving maps, charts and globes using Haase's distinctive method of cartographic projection as developed by Lowitz, his own first project in Göttingen being to prepare an historical atlas while his new colleague Professor Anton Friedrich Büsching prepared an atlas for schoolchildren. They concerned themselves with political and historical geography, leaving Lowitz and Mayer to cope with the actual manufacture of the celestial, terrestrial and lunar globes.

By now, however, Franz found himself in very serious financial difficulties. The fact that several wealthy influential gentlemen had subscribed to Lowitz's globes meant that the integrity of the Hanoverian government, as the new sponsors of the Cosmographical Insti-

ferstecher Ebersperger, 1755', which constitutes a sub-section of nine items within Franz's *Personalakte* Ph. 1/18 4/Vb Nr. 23 in the University archive, Göttingen.

[45] J. M. Franz, *Bericht von den Mondskugeln, welche bey der kosmographischen Gesellschaft in Nürnberg, aus neuen Beobachtungen verfertiget werden durch Tobias Mayern. Mitgliede derselben Gesellschaft. Zu finden in der Homannischen Officin* (Nuremberg, 1750).

tute, was also at stake. Von Münchhausen, tired of Franz's unfulfilled promises and repeated requests for funds, instigated an official inquiry and summoned the two Göttingen professors to Hanover to answer questions about the state of their work. A marathon interview was held on 10–12 November 1755 at which the two assessors Balck and Scheidt were also present; their considered opinions on the matters raised and answered were set out immediately afterwards in a neatly written letter to the Hanoverian Privy Councillors.[46] Franz was judged to be at fault for his lack of business acumen, while Lowitz was less to blame since he had not received adequate funds to enable him to employ more artists and thereby reduce the time required to construct the globes. Franz owed him money, which he was ordered to pay over as soon as possible. The root causes of the troubles which the two cosmographers faced were recognized as being Franz's inability to stop planning expensive projects, and the various problems that arose as a direct result of the transfer from Nuremberg to Göttingen.

One such problem was the lack of skilled artists capable of preparing accurate reproductions of the Homann maps or making copper-engravings of these and the planispheric segments of the globes. This particular deficiency may have led Mayer to develop a theory of colour-mixing in order to ensure a greater degree of consistency between maps reproduced by different artists.[47] It may reflect the influence of Robert Boyle's *Experiments and Considerations touching Colours* (1664) or some derivative work, inasmuch as it was founded on the concept that all colours can be generated from mixtures of the three primary colours red, yellow and blue. These were visualized by Mayer in 1758 as forming the vertices of a colour-triangle, with sides containing mixtures of a pair of these colours and the interior composed of mixtures of all three.[48] A skilled artist would be able to judge visually the connection between the primary colours and a scale of colour obtained by filling in each small area of the triangle with the requisite proportions by weight of ground dry pigments. However, this introduced the subjective and rather questionable relationship between colours in pigments and colour as a physiological phenomenon. Whereas Mayer himself would have been inclined to dismiss such a distinction as an intangible philosophical problem, and to

[46] This nineteen-page report, dated 13 November 1755, is preserved among Lowitz's *Personalakte* (see note 27), item 34.

[47] E. G. Forbes, 'Tobias Mayer's theory of colour-mixing and its application to artistic reproductions', *AS*, xxvi (1970), 95–114.

[48] T. Mayer, 'De affinitate colorum'; translated into English and published in E. G. Forbes (ed.), *Tobias Mayer's 'Opera inedita'* (1971), 81–91.

argue that the question of whether or not to call 'white' a colour was a mere semantic quibble, later generations of physical chemists and psychologists were to take up such problems very seriously, with interesting and important results.[49]

Despite the errors of judgement made by Franz, and the strained relations that persisted between him and Lowitz after the Hanoverian confrontation, the terrestrial and celestial globes might still have been completed before the end of the decade had not a situation arisen over which neither had any control: namely, the advent of the Seven Years' War, whose effects began to be felt in Göttingen after its capitulation to the French in July 1757.[50] Many of the wooden stands and accessories for the globes were broken by the occupying troops. The globe-laboratory in Lowitz's home became a stable, and every available room other than the most essential living-quarters was occupied by the French soldiers.[51] To make matters worse his wife died, leaving him to look after an infant son, and he had to pay out substantial sums of money for repairs to his property. Franz also died, from a fever contracted from one of the officers billeted in his home. These were the circumstances that set the seal on the Cosmographical Institute and its globe-laboratory, and induced Lowitz to accept an offer of a professorship at the St Petersburg Academy of Sciences and begin a new career in Russia. He died with his boots on, as it were; he was cruelly murdered by a Cossack revolutionary while conducting a geodetic survey through the Ukraine in 1773.

VII. The ferment in astronomy and navigation

There was, nevertheless, an important and unforeseen ferment from Mayer's continued devotion to the aims of the Mathematical Class of the Cosmographical Society as set out in the Homann atlas of 1747. We have already seen how his projection of a lunar eclipse led him to perceive the need for a new and better lunar map. He succeeded in

[49] Johann Heinrich Lambert (1728–1777), Philip Otto Runge (1777–1810) and Johann Wolfgang Goethe (1749–1832) feature amongst the most eminent scientific, artistic and literary figures who were to occupy themselves with this theme. The chief proponent, however, was undoubtedly the Latvian-born Wilhelm Ostwald (1853–1932), whose *Farbenfibel* (Leipzig, 1916), or 'Colour-Primer', has run into no fewer than fifteen editions.

[50] See, for example, A. Schöne, *Prof. Hollmann's Chronik; daraus: Die Universität im 7j. Kriege* (Leipzig, 1887); also F. W. Unger, *Göttingen und Die Georgia Augusta* (Göttingen, 1861), 83–9.

[51] These tragic circumstances are related in a 'pro Memoria' by Geheim-Rath G. L. Böhmer, dated Göttingen 24 December 1761, in Lowitz's *Personalakte* (see note 27), item 59.

completing this before his death.[52] A photolithographic reproduction of it, and of the forty detailed drawings of the lunar surface on which it had been based, was eventually produced in 1881 in Dresden at the instigation of Wilhelm Klinkerfues,[53] a director of the new Göttingen observatory; but a half-scale copy of 20 centimeters diameter had appeared more than a century earlier in Georg Christoph Lichtenberg's *Opera inedita Tobiae Mayeri* I (Göttingen, 1775), and been reproduced in Johann Hieronymus Schröter's *Selenotopographische fragmente* (Lilienthal, 1797).[54]

The inconsistency which Mayer found between the results for the circumstances of the lunar eclipse calculated from the stereographic and orthographic projections respectively, meant in effect that future computations ought to take account of the moon's finite distance from the earth; that is, the lunar parallax. The lack of accurate knowledge concerning the magnitude of this quantity was, at the time, an obstacle to the successful application of the phenomenon of lunar occultations as a means for determining geographical longitudes; thus, true to the third aim of the Mathematical Class, he decided to investigate this problem. Following Lowitz's example he decided to seek Euler's advice, and initiated a correspondence on 4 July 1751 shortly after settling down in Göttingen. Euler, through his previous exchanges of letters both with Lowitz and with Franz, and his reading of the *Kosmographische Nachrichten und Sammlungen auf d. J. 1748,* was already well aware of Mayer's pioneering researches into lunar mapping, the projection of lunar and solar eclipses, and the lunar libration, and reacted enthusiastically to his queries on the lunar parallax.[55]

The only reliable approach to this question was through the lunar theory itself, in which the moon's distance from the earth was one of many fundamental parameters requiring evaluation. He followed a method of analysis similar to that used by Euler when investigating

[52] E. G. Forbes, 'Tobias Mayers Mondkarten', *Sterne und Weltraum,* VIII (1969), 36–9.

[53] *Tobias Mayer's grössere Mondkarte nebst (40) Detailzeichnungen* (Dresden, 1881) is still catalogued under 'Tobias Mayer's Nachlass, aufbewahrt in der K. Sternwarte' no. VIII 364, in W. Mayer (ed.), *Die Handschriften in Göttingen* (3 vols., Berlin, 1894), III, 158 – though deposited in the Göttingen University library in 1977.

[54] D. B. Herrmann, 'Georg Christoph Lichtenberg und die Mondkarte von Tobias Mayer', *Mitteilungen der Archenhold-Sternwarte Berlin-Treptow No. 72* (Berlin, 1965).

[55] An English translation of the thirty-one letters known to have passed between them during the course of the next four years may be found in E. G. Forbes (ed.), *The Euler–Mayer Correspondence (1751–1755)* (1971).

the motions of Jupiter and Saturn (1748)[56] but, as the published abstract of his lecture on the subject to the Göttingen Scientific Society on 26 April 1753[57] reveals, he made a quantitative allowance in his theory for the earth's spheroidal shape. This was a crucial refinement which Euler himself had not made in his own investigation of the lunar theory, now on the verge of publication.[58] The new solar and lunar tables that were the outcome of Mayer's researches appeared in print in the Göttingen *Commentarii*[59] at about the same time as Euler's lunar theory, and were regarded by Euler himself as a 'most admirable masterpiece in theoretical astronomy' incorporating 'everything that from a practical point of view can ever be desired'.[60]

There were two very important developments arising from the precise values for the motions and parallaxes of the sun and moon now attainable from these new solar and lunar tables. One was the 'simple and accurate method for computing solar eclipses viewed in a given place', explained by Mayer in another lecture to the Göttingen Scientific Society in 1757.[61] The simplicity lay in the fact that he had now abandoned his original intention of taking account of the lunar and solar parallaxes by adopting a stereographic instead of an orthographic projection – this would have greatly increased the complexity of the calculations required – and thus avoided the false assumption (common to both projections) that the moon's shadow on the spheroidal earth was circular. Instead, he reverted to the Ancients' use of the sun's and moon's apparent motions as the basis for determining the circumstances of the eclipse. This reduced the problem to that of finding those instants when the distance between the centres of the two luminaries was equal to the sum of their apparent semi-diameters (viz. the beginning and end of the eclipse) and when that distance was at a minimum (the middle of the eclipse). Tables were

[56] L. Euler, 'Recherches sur les irregularités du mouvement de Jupiter et de Saturne', *Recueil des pièces qui ont remporté les prix de l'Académie Royale des Sciences, depuis leur fondation,* VII (1769), 1–84. This particular prize-essay was submitted in 1748 and first published separately in 1749.

[57] *Göttingische Anzeigen von gelehrten Sachen* (1753), 51 Stück, 467.

[58] L. Euler, *Theoria motuum lunae . . .* (St Petersburg, 1753).

[59] T. Mayer, 'Novae tabulae motuum solis et lunae', *Commentarii Societatis Regiae Scientiarum Gottingensis,* II (1753), 383–430.

[60] Euler to Mayer; Berlin, 26 February 1754, in Forbes (ed.), *Euler-Mayer Correspondence* (see note 55), 79.

[61] T. Mayer, 'Methodus facilis et accurata eclipses solares computandi', of which an English translation is given in Forbes (ed.), *'Opera inedita'* (see note 48), 71–80. See also, E. G. Forbes, 'Tobias Mayer's method for calculating the circumstances of a solar eclipse', *AS,* XXVIII (1972), 177–89.

required for converting observed latitudes into geocentric ones, and for reducing the moon's semi-diameter to the equator. Errors between the predicted and observed times could be attributed to the adoption of incorrect values for the motions and parallaxes of the two bodies concerned; by adjusting these to obtain agreement, the coefficients in the equations of the lunar theory could be corrected.

The other important development referred to above lay in the consequent improvement to the predictive power of the lunar theory, itself but a means to the end of increasing the accuracy of longitude determination either on land or at sea. Mayer's comparison of the mean motions of the sun and moon obtained from the times of their conjunctions and oppositions, with those derived from observations of its easterly progress through the stars in the zodiacal region of the heavens, revealed differences that could be attributed only to the existence of errors in the star positions to which these motions were referred. Thus he corrected these positions for 100 of the brightest stars in such a way as to eliminate the inconsistency between the two independently derived values of the mean motions, and used these as a kind of grid against which the moon's previous and future motion could be mapped. In this way, the lunar theory was further improved to such a degree that he felt able to guarantee predictions based on it to within ± 1′, the same order of accuracy as that of the best lunar observations themselves.[62]

Euler now encouraged Mayer to stake a claim to the British Admiralty for one of the three bounties offered under the terms of the Act 12 Queen Anne, cap. XV (1714) 'for providing a publick reward for such person or persons as shall discover the longitude at sea'. In theory, an error of ± 1′ in the moon's position is approximately equivalent to one of ± 30 geographical miles in the place on the earth's surface from which the observation has been made, and this was sufficient to qualify for the maximum reward of £20,000. In practice, however, there were additional uncertainties in observing the moon's distance from a star with a Hadley quadrant (or Repeating Circle) which reduced the accuracy by a factor of two and thereby qualified the method only for the minimum reward of £10,000. In either case the prospects seemed good, and, though the Seven Years' War was again to intervene in causing a postponement of the sea-trial of this lunar distances method based on an improved set of Mayer's tables, the British parliament did authorize awards of £3,000 to his

[62] E. G. Forbes, 'Tobias Mayer's contributions to the development of lunar theory', *Journal for the History of Astronomy*, I (1970), 144–54.

widow and an unsolicited £300 to Euler in 1765.[63] These tables were the ones from which the lunar distances published in the first issues of the *Nautical Almanac* were calculated.[64]

As far as I am aware, no attempt has been made by modern historians of science to ascertain the extent to which this publication was actually adopted by practising oceanic navigators during the last decades of the eighteenth century as the basis for improving their knowledge of longitude at sea. A systematic survey of the numerous ships' log-books[65] preserved in the library of the National Maritime Museum at Greenwich would be a suitable starting-point for such an investigation, and throw much light upon the actual computational techniques used and the level of accuracy attained. A pilot study of this nature was, on the other hand, carried out by a former Research Assistant at that Museum, Susanna Nockholds (née Fisher), with the express purpose of establishing when ships began to use chronometers as an alternative means of ascertaining longitude.[66] Her conclusion was that such instruments were not in general use until the early nineteenth century, although they were carried prior to that time on important voyages of exploration and discovery and upon a significant proportion of East India Company ships.

Recently, Alun Davies[67] has shed further light on the situation by surveying the development of the chronometer-manufacturing sector of the British horological industry throughout its effective lifespan from the 1780s to the 1820s. This confirms that both public and private demand was reinforced by Captain James Cook's dramatic demonstration of the practical usefulness of John Harrison's invention. Among the most successful of the early chronometer-makers were John Arnold and Thomas Earnshaw who, between them, were responsible for supplying over 2,000 box and pocket-chronometers to the merchant navy, Royal Navy[68] and private individuals. Slightly

[63] E. G. Forbes, 'Tobias Mayer's claim for the longitude prize: a study in 18th century Anglo-German relations', *Journal of Navigation*, XXVIII (1975), 77–90.

[64] E. G. Forbes, 'The foundation and early development of the Nautical Almanac', *Journal of the Institute of Navigation*, XVIII (1965), 391–401.

[65] W. E. May, 'The logbooks used by ships of the East India Company', *Journal of Navigation*, LXXVII (1974), 116–18.

[66] G. W. Nockholds, 'Early timekeepers at sea: the story of the general adoption of the chronometer between 1770 and 1820', *Antiquarian Horology*, IV (1963), 110–13; *ibid.*, 148–52.

[67] A. C. Davies, 'The life and death of a scientific instrument: the marine chronometer, 1770–1920', *AS*, XXXV (1978), 509–25.

[68] J. F. Cole, 'Chronometers and the Royal Navy', *Horological Journal*, LVIII (1915), 28–35.

later, in the years between 1796 and 1820, Paul Philip Barraud[69] produced about 1,000 more instruments. Thus these three individuals undoubtedly shared the monopoly of the market until the end of the period of dynamic growth in this new industry.

Let us return, however, to the mid-eighteenth-century scene. Mayer's cartographic and astronomical investigations had made him fully aware of the significant errors still inherent in astronomical practice. The use of calculated star positions rather than directly observed ones was an unsatisfactory foundation for an astronomical theory with such an important application for oceanic navigation, and for tables which mariners required to find their longitude at sea. Thus he resolved to devote his attention to the preparation of a new catalogue of zodiacal stars, a feasible proposition now that the Göttingen observatory had finally been completed and equipped with an excellent 6-foot radius mural quadrant made by John Bird, the highly skilled mathematical-instrument-maker in London. In the course of this work he was confronted with the difficulty of establishing a quantitative relationship between the values of refraction and the density of the atmosphere. He found that the closest agreement between theory and observation was obtained by assuming a linear dependency between the density and the altitude above sea-level. How he actually derived his formula has never come to light, but Christian Bruhns, in a historical review of the various theories of astronomical refraction up to the mid-nineteenth century, has demonstrated how he might have done so.[70] The key to his rather involved mathematical proof is that it is founded upon a rule stated in Thomas Simpson's *Mathematical Dissertations* (1743), which itself rests upon the same incorrect assumption used by Mayer. There is no evidence for a link between them, although Mayer's eldest son was later to make explicit reference to Simpson's rule in *De refractionibus astronomicis* (Altdorf, 1781).

More important historically is the fact that Mayer's was the first known formula for refraction that took both thermometric and barometric variations into account. For this purpose, he made use of Edmond Halley's law that refraction is directly proportional to the height of the merucry column in a barometer.[71] A slightly amended version of Mayer's formula was later published in Nevil Maskelyne's

[69] C. Jagger, *Paul Philip Barraud: A Study of a Fine Chronometer Maker, and of his Relatives, Associates, and Successors in the Family Business, 1750–1929* (1968).

[70] C. Bruhns, *Die astronomische Strahlenbrechung in ihrer historischen Entwicklung* (Leipzig, 1861), 51–60.

[71] E. Halley, 'Some remarks on the allowances to be made in astronomical observations for the refraction of the air... With an accurate table of refractions', *PT*, xxxi (1721), 169–72.

edition of his tables (1770).[72] However, Bradley's formulae – also derivable from Simpson's rule – were more suited to logarithmic calculation and were generally adopted after being made known through Thomas Hornsby's edition of Bradley's *Astronomical Observations* (Oxford, 1798). Having quantified and eliminated the effects of instrumental errors and astronomical refractions from the zenith distance measurements, and of the clock-errors and alignment of his quadrant from the observed right-ascensions,[73] Mayer was now able to produce a reliable catalogue of the equatorial coordinates of almost 1,000 zodiacal stars.[74] Most of these positions were first reduced to ecliptic coordinates by J. A. Koch and published in the *Astronomisches Jahrbuch für das Jahr 1790* (Berlin, 1787). The first complete catalogue appeared in London by order of the Commissioners (or Board) of Longitude in 1826;[75] its contents were critically discussed by Francis Baily in a memoir of 1831.[76] Arthur Auwers's revised German edition of 1894[77] can be regarded as a fitting culmination to this particular line of development.

There was, however, another ferment created as a direct consequence of these researches, arising from Mayer's natural interest in comparing his newly acquired stellar coordinates with those contained in other star catalogues. This convinced him that some of the differences which he found to exist had to be regarded as real. His preferred choice of comparison data was observations made by the Danish astronomer Olaus Roemer on 20, 22, 23 October 1706 – exactly half of century before his own – which he was able to correct for instrumental errors thanks to the detailed description of Roemer's transit instrument (or *rota meridiana*) and observing technique provided in the second three-volume edition of Peter Horrebow's *'Basis astronomicae'*, entitled *Operum mathematico-physicorum* (Copenhagen,

72 N. Maskelyne (ed.), *Tabulae motuum solis et lunae novae et correctae; auctore Tobia Mayer: quibus accedit methodus longitudinum promota, eodem auctore* (1770), 64.

73 These aspects of his work are discussed by E. G. Forbes in Forbes (ed.), *'Opera inedita'* (see note 48), 21–4, 41–3; and by Mayer himself in *ibid.*, 62–70.

74 The English translation of Mayer's Latin lecture on this subject, entitled 'Treatise on the new catalogue of zodiacal stars', and a reprint of Lichtenberg's edition of the catalogue, is contained in *ibid.*, 92–108.

75 T. Mayer, *Astronomical Observations, made at Göttingen, from 1756 to 1761* (1826).

76 F. Baily, 'Mayer's Catalogue of Stars, corrected and enlarged; together with a comparison of the places of the greater part of them, with those given by Bradley; and a reference to every observation of every star', *Memoirs of the Astronomical Society of London*, IV (1831), 391–445.

77 *Tobias Mayer's Sternverzeichniss, nach den Beobachtungen auf der Göttinger Sternwarte in den Jahren 1756 bis 1760, neu bearbeitet von Arthur Auwers* (Leipzig, 1894).

1741). He also supplemented his own observations with those of a few non-zodiacal stars whose coordinates were given in La Caille's *Astronomiae fundamenta* (Paris, 1757).

Out of eighty stars selected from these sources, between fifteen and twenty appeared to exhibit a motion of their own; some fainter stars revealed this just as clearly as the brighter ones, whereas some very bright stars did not appear to be affected at all.[78] The solution to this behaviour was to come early in the nineteenth century when it was discovered that the intrinsic brightness of a star can vary so enormously that it might appear very faint and yet be comparatively close to us. Mayer looked for, yet failed to detect, the influence of the solar motion that William Herschel was fortunate to discover from an analysis of the proper motions of the twelve brightest stars in Mayer's list,[79] which he found conveniently tabulated in Joseph Louis de Lalande's *Astronomie*, IV (2nd edn; Paris, 1781), p. 685. A later investigation,[80] using a much greater wealth of data, proved to be less conclusive for the same reason as Mayer's pioneering study had been: namely, that the true motions were masked by the unavoidable errors still persisting in the absolute values of stellar coordinates that could not be entirely eliminated by statistical procedures.

VIII. Magnetic forces

The sole remaining aim of the Mathematical Class of the Cosmographical Society which has yet to be included in this discussion of the manifold activities of its two chief proponents Lowitz and Mayer, is that specified as the fifth and last on Franz's list: to make and collate accurate measurements of the variation and dip of the magnetic needle for as many places as possible, with a view to discovering the laws governing these phenomena and how they might be applied in cosmography.[81] The initial stimulus to Mayer to preoccupy himself with this particular problem was the occurrence of an earth-tremor on 18 February 1756, the effect of which was experienced over most of Lower Saxony. Although he quickly confirmed that this had no

[78] These conclusions are stated in Mayer's treatise on the proper motions of stars first published by Lichtenberg (1775), and in English by Forbes (ed.), *'Opera inedita'* (see note 48), 109–12.

[79] W. Herschel, 'On the proper motion of the Sun and solar system; with an account of several changes that have happened among the fixed stars since the time of Mr. Flamstead', *PT*, LXXIII (1783), 274–83.

[80] W. Herschel, 'On the direction and velocity of the motion of the Sun and solar system', *PT*, XCV (1805), 233–56; and *idem*, 'On the quantity and the velocity of the solar motion', *PT*, XCVI (1806), 205–37.

[81] See note 21.

noticeable effect upon the alignment of the mural quadrant, it encouraged him to speculate on a possible cause.[82] He knew that measurements of magnetic variation made in Augsburg during the Lisbon earthquake on 1 November 1755, and the erratic behaviour of ships' compasses about that time, might mean that terrestrial magnetism (or ultimately electricity) was responsible; but he was also aware that the force experienced by a magnetic needle varies inversely to just under the square of the distance separating the attracting and repelling poles, and that the same law applied to the oscillations of a pendulum under gravity. He concluded – logically, but erroneously (since the analogy was false) – that the observed changes in magnetic variation are controlled by the force of universal gravitation.

Using as his working hypothesis the corollary that the magnetic force varies inversely with the square of the distance between two magnetically attracting or repelling particles, he developed an impressive *Theoria magnetis* (1760), in which he establishes formulae connecting observed values of the magnetic variation and dip of a compass needle to the latitude and longitude of each place on the earth's surface at which these quantities were measured. This treatise, published for the first time in 1972[83] but known in Mayer's lifetime through an abstract on pp. 73–5 of the *Göttingen Anzeigen von gelehrten Sachen* for 21 January 1760, is of great significance in containing the first hypothetico-deductive proof and experimental verification of the validity of the inverse-square law of magnetic attraction and repulsion – antedating by a quarter of a century the well-known but much less rigorous empirical demonstration of this law by Augustin Coulomb in 1788.[84] The *Nova theoria declinationis et inclinationis acus magneticae* (1762),[85] completed by Mayer just before his death, is a follow-up of a paper by Euler in the Berlin Academy of Science's *Mémoires* for 1757,[86] which was a pioneering attempt to provide a mathematical basis for the interpretation of the lines of magnetic var-

[82] T. Mayer, 'Versuch einer Erklärung des Erdbebens', *Hannoverischen nützlichen Sammlungen* (1756), 290–6.

[83] E. G. Forbes (ed.), *The Unpublished Writings of Tobias Mayer* (3 vols., Göttingen, 1972), III, contains the Latin text of this treatise and its English translation, prefaced by a historical analysis and assessment of its contents.

[84] A. Coulomb, 'Second mémoire sur l'électricité et le magnetisme', in *Histoire de l'Académie Royale des Sciences, Année 1785* (1788); *Mémoires*, 578–611. The first and third memoirs on this subject appear in the same volume, and a fourth in that for the year 1786.

[85] Also published in Latin and English in Forbes (ed.), *Unpublished Writings* (see note 83), 95–104.

[86] L. Euler, 'Recherches sur la déclinaison de l'aiguille aimantée', *Histoire de l'Académie Royale des Sciences et Belles-Lettres, Année 1757*, XIII (1759), 175–251.

iation on Edmond Halley's *World Chart* of 1702.[87] Mayer's conclusions were, however, based upon an original analysis that took account not only of variation, but also of dip – a phenomenon which Euler had, up to this time, entirely neglected although he was to introduce it later, in 1766.[88]

The introductory remarks in the opening chapter of the *Theoria magnetis* contain the only clear expression in any of Mayer's writings of his methodological approach to scientific investigations, which might be labelled as Newtonian. He remarks, for example, that an ability to express the force and action of a magnet in mathematical terms, and to provide a geometrical definition of its observed range of operation, were important criteria of truth; and dismisses as useless and inept the vortex theory by which contemporary scientists such as Daniel and Jean Bernoulli had tried to account for magnetic action.[89] While condemning this approach as being 'altogether too Cartesian',[90] he also indicates his disapproval of the Baconian principle of inferring some general law by working backwards from the observed effects of magnetic force – apparently on the ground that a law inferred by induction from a few particular experiments could never be accorded the status of an absolute truth. A knowledge of the cause of magnetism was no more necessary than that of the cause of gravitation, elasticity, or the hardness of bodies. Only the law governing the change of force with distance from a magnet was required to explain any individual phenomenon. There can be little doubt, therefore, that Mayer departed from this world uncontaminated by the neo-scholasticism of Leibniz and Wolff, unimpressed by Baconian method or the subtle implications of Descartes' philosophy, and a stranger to the philosophical scepticism of George Berkeley and David Hume.

IX. Reflections

Franz may have been inspired by the Wolffian ideal of constructing an extensive system of knowledge out of the facts of history and the

[87] This date is inferred from the dedication to Prince George of Denmark, Queen Anne's consort, as 'Lord High Admiral and Generalissimo' – a title which he received on 11 April 1702. The isogonics are drawn for the year 1700.

[88] L. Euler, 'Corrections nécessaires pour la théorie de la déclinaison magnétique, proposée dans le XIII volume des Mémoires', *Histoire de L'Academie Royale des Sciences et Belles-Lettres, Année 1766* (1768), 213–64.

[89] D. and J. Bernoulli, 'Nouveaux principes de méchanique et de physique, tendans à expliquer la nature et les propriétés de l'aiman', *Recueil des pièces qui ont remporté les prix de L'Académie Royale des Sciences, depuis leur fondation,* v (1752), 117–44.

[90] Forbes (ed.), *Unpublished Writings* (see note 83), 65.

framework of mathematics, though laying more stress on geography as the source of data for the foundations of his intellectual edifice. Certainly, while a student at Halle he had been exposed to Wolff's influence and was intimately acquainted with Johann Christoph Homann with whom he used to have prolonged discussions on the subjects of Wolff's lectures. It would not have been at all surprising if they had made a pact to use young Homann's inheritance in an attempt to make that philosophy a reality. At any rate, this would explain why the latter was so eager to engage his student friend as his personal secretary, and to incur the wrath and indignation of his family by willing him his share of the Homann fortune.[91] Franz's obsessive pamphleteering is also consistent with such a conjecture.

As we have seen, however, this cosmographic ideal was doomed to failure, since it was not fostered by the social climate of the times. Despite Franz's enthusiastic efforts to secure an eminent patron for his schemes, Germany was too divided for any one prince or duke to contemplate sponsoring a geodetic survey extending beyond his territorial boundaries; nor could Franz's rhetoric rouse his passion for the mundane and laborious business of collecting data with which to construct better land-maps. The opportunity to possess large globes was a more fashionable bait for the nobility, but the advent of a major European war and the consequent financial strains on governmental and private purses were largely responsible for the failure of that scheme. The remaining misfortunes of the Cosmographical Society can be accredited to a combination of those factors with Franz's dictatorial attitude and his general mismanagement of funds, with Lowitz being less to blame.

The major advances in mathematical cosmography during the eighteenth century lay not in the improvement of the mathematical methods of projection, nor in globe construction, but in the advancement of knowledge of geodetic and astronomical instrumentation and measuring techniques, astronomical theories, physical phenomena such as terrestrial magnetism, and their applications to navigational theory and practice. The chief architect of these achievements was, of course, Tobias Mayer who, while indebted to Leonhard Euler for scientific advice and encouragement, and to Newton for providing the basis of his methodology, can also be counted as a disciple of Wolff inasmuch as his earliest studies were devoted to absorbing the entire content of Wolff's *Anfangs-Gründe aller mathematischen Wissenschaften* (Frankfurt and Halle, 1737–8), and popularizing it in his beautifully illustrated *Mathematischer Atlas* (Augsburg, 1745). It is

[91] C. Sandler, 'Die homännischen Erben', *Zeitschrift für Wissenschaftliche Geographie*, VII (1890), 334.

wiser, therefore, to view him like his colleague A. G. Kaestner did, as 'the measurer of land and sea and boundless space',[92] which aptly epitomizes both the nature and scope of his mature researches, and to avoid the traditional philosophical labels by classifying him instead as a pioneer of a *practical scientific* Enlightenment. By the end of the century the spirit which had inspired the rise of mathematical cosmography now revealed itself in the complementary field of physical cosmography, through dedicated universal men such as Alexander von Humboldt, while the accompanying idealism found a new expression in the romanticism of the German *Naturphilosophen.*

[92] This is translated from the opening couplet of the Latin ode prefacing Kaestner's *Elogium Tobiae Mayeri* (Göttingen, 1762).

12

Science, technology and industry

D. S. L. CARDWELL

Contents

I.	The background	449
II.	Social contexts	452
III.	The Newcomen engine	454
IV.	A classification of invention and technology	456
V.	The mechanization of the textile industries	459
VI.	Darby and the iron revolution	463
VII.	The chemical industry	465
VIII.	British technology	468
IX.	Executive engineers	475
X.	Conclusion	477

'We', wrote a late-eighteenth-century English author, 'are doing the drudgery by which the Golden Age is to profit', and, he added, 'Perhaps some other power may be discovered, as forcible and as manageable as the evaporation from boiling water – another gunpowder that may supersede the present – and other applications of mechanical powers which may make our present wonders sink into vulgar performances'.[1] These forecasts were remarkable, for the first evidently predicts, accurately enough, the wealth and good fortune of nineteenth-century England, while the second foreshadows the characteristic achievements of the twentieth century. And, moreover, they set the eighteenth century in a context; not a golden age, as the writer admitted, for he designated his own time a 'silver age', but when compared with the troubled if brilliant seventeenth century a time of consolidation and reasonable, measured progress.

I. The background

The widespread changes associated with the 'industrial revolution' in Britain in the second half of the eighteenth century became apparent

[1] William Jackson, *The Four Ages* (1798), 60, 92.

after 1815. Babbage, Dupin, Ure and others toured the industrial areas and reported on the new industries from both the social and the engineering and scientific points of view.[2] Strangely, the economists such as Ricardo and Nassau Senior seem to have been uninterested in the latter, and thus contributed to a culture gap that persisted a long time. From that period onward we can distinguish three separate academic disciplines: the history of technology, economic history and the history of science. The history of technology, particularly that of the industrial revolution in Britain, was popularized by Samuel Smiles who celebrated the triumph of the hero-engineer; the man who, by intelligence and strength of character, succeeded in the face of hostile nature in building roads, canals, railways, harbours, in laying on water supplies and in draining marshes and fen, all of which were preconditions for industrial development and therefore of social progress.[3] Smiles left no immediate successors; nevertheless interest in the history of technology has steadily increased over the last sixty years.[4] Museums of science and industry have been improved and extended and many new ones established. Successful new journals have been launched.[5] The study of large industrial relics, such as old mills, warehouses, railways, etc. has become fashionable under the rather misleading name of 'industrial archaeology'. And a number of major works dealing in whole or part with the eighteenth century have been published.[6]

Older histories of technology restricted themselves almost exclusively to the technical details of major innovations; social, economic,

[2] F. P. C. Dupin, *Voyages dans la Grande Bretagne entreprise relativement aux Services Publics de la Guerre, de la Marine et des Ponts et Chausseés en 1816, 1817, 1818 et 1819* (Paris, 1820–4); Charles Babbage, *On the Economy of Machinery and Manufactures* (1832); Andrew Ure, *The Philosophy of Manufactures* (1835). The last two are available in reprint editions.

[3] See the Introduction by Thomas P. Hughes (ed.) to *Selections from Lives of the Engineers . . . by Samuel Smiles* (Cambridge, Mass., 1966).

[4] Since, that is, the foundation of the Newcomen Society in 1919. This, as far as I know, was the first society devoted to the study of the history of technology.

[5] Such as *Technology and Culture, History of Technology.*

[6] Prominent among them are *The Chemical Revolution,* by A. & N. Clow (1952); *Histoire générale des techniques,* ed. Maurice Dumas (3 vols., Paris, 1964–9); *A History of Western Technology,* by Friedrich Klemm (1959); *Technology in Western Civilisation,* ed. Melvin Kranzberg and Carroll W. Pursell (2 vols., New York, 1967); the compendious *History of Technology,* ed. C. Singer, E. J. Holmyard, A. R. Hall and T. I. Williams (6 vols., 1954–78); *A History of Mechanical Inventions,* by A. P. Usher (Harvard, 1962), and finally the invaluable *Bibliography of the History of Technology,* by E. S. Ferguson (Cambridge, Mass., and London, 1968).

scientific and philosophical considerations were kept to a minimum or entirely excluded. That things have now changed is due, one suspects, to the influence of economic historians, not least to those who found such a rich field in the British industrial revolution. The economic historians have taught the historians of technology that invention and innovation do not take place in a social void; they are not merely the consequences of ingenuity and genius motivated by personal interest, they are dependent on the opportunities and stimuli that only society can provide.[7] An adequate explanation of any invention must include a full account of the circumstances in which it came to be made and that made it acceptable in society.

Marx and those influenced by his writings must surely have much of the credit for this. The informative Chapter 15, 'Machinery and modern industry', of *Capital* is in sharp contrast to the indifference of the classical economists.[8] This is not surprising for Marx had observed that machinery was an important factor in the creation of relative surplus value[9] and Engels had earlier pointed out that it was mechanization that had brought about the creation of the proletariat.[10] Marx himself called for a history of technology and tools, remarking that 'Technology discloses man's mode of dealing with Nature, the process of production by which he sustains his life, and thereby lays bare the mode of formation of his social relations, and the mental conceptions that flow from them'.[11]

It is, however, difficult to see how a distinctively Marxist approach could add much to our knowledge of eighteenth-century technology. It was, after all, a uniquely successful period of industrial innovation under the guidance of the ever-rising bourgeoisie. On the other hand Marxism by offering alternative, short-cut answers can be misleading. For example, the Marxist J. D. Bernal explained the apparent long delay in the development of the electric power industry in the nineteenth century by referring to distinctive features of capitalism.[12]

[7] Notable among recent economic historians who have shown great interest in the history of technology are A. E. Musson and E. Robinson, *Science and Technology in the Industrial Revolution* (Manchester, 1969) and David Landes, *The Unbound Prometheus* (Cambridge, 1969).

[8] Karl Marx, *Capital: A Critical Analysis of Capitalist Production* (Moscow, 1961). It is fair to point out that Marx was, to some extent, indebted to Babbage and Ure for his insights into the processes of mechanization.

[9] *Ibid.*, 371.

[10] Friedrich Engels, *The Condition of the Working Class in England in 1844* (1968 edn), 15, 20.

[11] Marx, *Capital* (see note 8), 372, n. 3.

[12] 'The development [of electric power technology] hung fire in a way characteristic of competitive capitalism': J. D. Bernal, *Science and Industry in the*

It is much more likely that the cause, apart from the inherent difficulties of the new technology, was the rapid development of high-pressure water systems and, later, gas engines, both of which offered flexible and economical power sources to satisfy the great majority of urban consumers.

II. Social contexts

Histories of science, like histories of technology, have in recent years tended to show a much clearer awareness of social, political and religious influences on the development of knowledge. Writers as diverse as A. N. Whitehead and R. K. Merton have contributed to this change but the most important influence was surely the dramatic rise in the public importance of science during and since World War 2. And yet the historical relationships between science on the one hand and technology and industry on the other have been largely neglected. It is difficult to escape the conclusion that the powerful influence of certain philosophers of science, in particular those brought up in that mainland European tradition that emphasized the distinction between the 'pure' science of the universities and the other sort, as practised and taught in the technical high schools, has been responsible for this breach. (It should also be remembered that the immense variety of human activities and the enormous time-span that the history of technology must cover present, in themselves, serious problems for a philosopher.) Their motive was often the laudable one of countering state encroachments on freedom of science and learning but the academic consequences have been regrettable.

On the other hand the relatively small group of Marxist and neo-Marxist writers such as B. Hessen, J. D. Bernal, S. Lilley and B. Farrington have consistently stressed the close relationship between science and industry. And, more recently, non-Marxist scholars like Thomas P. Hughes and E. Layton have made important contributions in this area.[13] However, it is still the case that the history of science

Nineteenth Century (1953), 120. In his later and more comprehensive work, *Science in History*, II, *The Scientific and Industrial Revolutions* (1969), 612–13, Bernal admits that there were 'enormous practical difficulties' to the development of electric power in the nineteenth century.

[13] Thomas P. Hughes, 'The science–technology interaction: the case of high-voltage power transmission systems', *Technology and Culture*, XVII (1976), 646–62; E. Layton, 'Millwrights and engineers: science, social roles and the evolution of the turbine in America', in E. Layton, W. Krohn and P. Weingart (eds.), *The Dynamics of Science and Technology* (Amsterdam, 1978), 61–87. These papers are good examples of recent work in this area. For a succinct discussion of the present state of the philosophy of technology see Stanley R. Carpenter,

and the history of technology are studied (and taught) independently of each other. This is surprising: surely it cannot be assumed that science, technology, industry have all developed quite separately, with little or no reference to each other? Mankind and nature are not like that. Not merely would it be inherently improbable, but historical facts and recent human experience refute it. The discussion that follows is an attempt to examine the relationships between science, technology and industry during the eighteenth century - the period when the industrial revolution began - and, as, by common consent, the most significant developments took place in Britain I shall concentrate on that country.

Few would deny that throughout the eighteenth century France was the leader in intellectual matters, as well as in fashion and taste, although many would claim that Scotland ran her a good second, at least in intellectual matters. French technology, too, was of a high standard in the age of the *Encyclopédie,* but it tended as much to academic as to practical excellence, so that a complete appraisal of French technology in the eighteenth century would require a detailed examination of many nineteenth-century achievements. However, if France was ascendant in so many respects in the eighteenth century, the fortunes of Germany and Italy, as if in compensation, were at a very low ebb. Although by the end of the next century Germany, in particular, had powerfully regained the leadership that had been hers in medieval and Renaissance times, during the eighteenth century the German states were a backwater and many of the ablest young Germans had to seek their fulfilment abroad, notably in Paris, in the exotic intellectual colony in St Petersburg and in London.

It is generally agreed that up to the beginning of the eighteenth century Britain was technologically and industrially backward. She was, however, most willing to learn from foreign teachers, and accordingly many skilled workers were encouraged to come to England from Europe in Tudor and Stuart times: miners and metallurgists, cannon founders, armourers, architects, builders, clockmakers, textile workers and millwrights came from France, Germany, Italy, the Low Countries and Switzerland; and from Holland also came the hydraulic engineers who did so much to drain the English fens and marshes.[14] As late as the beginning of the eighteenth century Daniel Defoe could admit, complacently, that the English were no good at making original inventions, although they were competent enough at

'Developments in the philosophy of technology in America', *Technology and Culture,* XIX (1978), 93–9.

[14] For an account of Dutch hydraulic engineers in England in the seventeenth century see L. E. Harris, *Vermuyden and the Fens* (1935).

copying other people's ideas and turning them to profitable account;[15] a surprising inversion of the supremacy usually ascribed to British inventiveness by British commentators today.

III. The Newcomen engine

A rather different picture emerges, however, if we take account of the invention and development of the Newcomen atmospheric pumping engine in the first decades of the century. As there are a number of competent accounts of this invention[16] a detailed description is not necessary here. Denis Papin, and before him the Abbé Hautefeuille, had conceived the idea – the brilliance of which is too easily overlooked – of harnessing the pressure of the earth's atmosphere by filling a cylinder with vapour (steam in Papin's case) and then condensing it so that atmospheric pressure drove down the piston that was fitted in the cylinder. But the business of turning this small-scale idea into an economical and powerful engine demanded at least as much genius from the English inventor Thomas Newcomen (1663–1729). The engine he devised was elaborate, self-acting and safe; it must have taken several years, at least, to develop.

Great skill, faith and courage were needed to launch the Newcomen engine. Experience gained in building and operating a replica of the original machine of 1712 at the Manchester Science Museum[17] has shown that many months' patient work must have been required before any of the early machines worked satisfactorily. A number of critical, and mutually dependent, features necessitated careful design and adjustment: the timing of the valve mechanism, the diameter of the condensing spray and the load on the pump rods to mention only three. Had we been concerned with one Newcomen engine, that put up by Newcomen himself, not far from Birmingham, then we could

[15] A. P. Wadsworth and Julia Mann, *The Cotton Trade and Industrial Lancashire, 1600–1780* (Manchester, 1951), 413.

[16] See, for example, L. T. C. Rolt, *Thomas Newcomen: the Pre-History of the Steam Engine* (1963); L. T. C. Rolt and J. S. Allen, *The Steam Engine of Thomas Newcomen* (Buxton, Derby, 1977); *idem*, 'The introduction of the Newcomen engine from 1700–1733', *Transactions of the Newcomen Society*, XIII (1969–70), 169–90; *idem*, 'Bromsgrove and the Newcomen engine', *ibid.*, XLIII (1970–1), 183–98; *idem*, 'Addendum to the introduction of the Newcomen engine', *ibid.*, 199–202; *idem*, 'The introduction of the Newcomen engine from 1710 to 1733, second addendum', *ibid.*, XLV (1972–3), 223–26; *idem*, 'The 1715 and other Newcomen Engines at Whitehaven, Cumberland', *ibid.*, 227–32.

[17] R. L. Hills, 'A one-third scale working model of the Newcomen engine of 1712', *Transactions of the Newcomen Society*, XLIV (1971–2), 63–77.

have concluded that Newcomen was a great genius (which he undoubtedly was) and the exception to prove the rule about the incompetent English. Unfortunately for this argument no sooner was the first engine at work than scores of similar ones were put up all over the country: from the lowlands of Scotland and Westmorland ('Cumbria') to London, from the coalfields of the north-east to the mines of Cornwall. Engines were also erected in Europe, in Sweden (Dannemora), Hungary (Schemnitz), France (Passy), Belgium, Austria and, before the century was out, in Russia and the Americas.[18] But nowhere were as many engines put up as in Britain and nowhere did they match the power and efficiency attained by the British machines. The unavoidable conclusion must be that this would have been quite impossible without the prior existence, from the beginning of the century, of a substantial and well-distributed population of skilled, energetic artisans – millwrights, blacksmiths, instrument-makers, etc. – who were prepared to erect new and revolutionary machines, and of men who were willing to risk their capital on such ventures.

There seems, therefore, to be a conflict of evidence about British inventiveness at the beginning of the eighteenth century, the widely reported unoriginality of the English (continental Europeans are notoriously unable to distinguish between Englishmen and Scotsmen) being inconsistent with the genius required to invent and develop the Newcomen engine and the skill needed to erect such engines and make them work satisfactorily. To resolve this conflict it is enough to point out that even acute observers often fail to read the signs of the times correctly; or, to put the point another way, there is usually a time lag before a firm trend becomes apparent, even to the most perceptive. But this does not answer the implicit – and fundamental – question of the nature and content of 'skill' in the eighteenth century. We will return to this question later; in the meantime we may reconcile the two sides by proposing that the best way in which a technologically backward nation can become progressive is by importing skilled foreigners and learning the tricks from them. In any case, importing foreign skills, as had been done so frequently in the seventeenth century, gives clear evidence of a widespread willingness to

[18] Early accounts of the Newcomen engine are given by B. F. de Belidor, *Architecture hydraulique* (4 vols., Paris, 1737–53); J. T. Desaguliers, *A Course of Experimental Philosophy* (2 vols., 1734, 1744); J. Leupold, *Theatrum machinarum generale* (Leipzig, 1724) – the Newcomen engine is discussed in II: *Theatri machinarum hydraulicarum* – and Stephen Switzer, *An Introduction to a General System of Hydrostatics and Hydraulicks* (2 vols., 1729). By far the most comprehensive and accurate account is that given by Desaguliers.

innovate, and that is at least half the struggle. The experiences of other nations tend to support this hypothesis.[19] Although a discussion of the influences that help to stimulate a willingness to innovate lies outside the scope of this essay some mention of them is necessary. On the one hand there was the strong Baconian tradition with its emphasis on knowledge and on utility. The status of Bacon's theory of science does not concern us; what does, is his clear and unambiguous assertion that new knowledge can lead to new inventions that in turn increase wealth and prosperity. The invention of the Newcomen engine was a strong confirmation of the validity of Bacon's assertion. On the other hand there were the various seventeenth-century movements that propagated the idea of improvement, notably those associated with the nonconformist sects.[20]

IV. A classification of invention and technology

A desideratum in the history of technology is, if not a theory, at least a taxonomy of invention and innovation. It is significant that the word technology was coined during the seventeenth century, during the course of the 'scientific revolution', and it is evident that over the last three hundred years the technical arts have been increasingly and most fruitfully linked with science; so much so, in fact, that in common discourse science and technology are assumed to be one and the same thing. However, the historians of technology have not, as yet, paid much attention to the nature of the inventive process, to what may be called the philosophy of the subject. For their parts the economist and the economic historian can, no doubt, explain invention and innovation as the result of demands making themselves felt and suitable skills and talent being available to satisfy those demands. Such explanations may well be adequate for the disciplines in question, but they do not help us to understand the nature of technological progress. The eighteenth century is a particularly appropriate period for such a study. The paradigmatic science of mechanics had been established in the previous century, and the political, legal and economic features of the social structure of Britain were such as to stimulate industrial innovation. But before we proceed further a short survey of the relevant aspects of contemporary science is necessary.

19 It is well known that early medieval Europe borrowed heavily from the superior Arab civilization. Up to forty years ago it was commonly said of the Japanese that they were capable only of copying European and American technology.

20 For an exhaustive and authoritative account see Charles Webster, *The Great Instauration: Science, Medicine and Reform, 1626–1660* (1975).

Eighteenth-century physical science presents a picture that is, at first sight, misleadingly simple. Although many of the branches that were to be brought to triumphant maturity by the grand mid-nineteenth-century syntheses were being actively developed, the emphases and assumed objectives were different from those accepted later on. The Newtonian synthesis was recognized as the very summit of human achievement: the riddle of the universe having been solved, the Newtonian system was reverently enshrined in the form in which the master had set it down; the young were to be initiated into Newtonian philosophy by the same route and in the same mathematical language that he himself had used. Such, at any rate, was the case in Britain. In Europe, and particularly in France, the approach was more liberal: a proper reverence for Newton did not inhibit the development of alternative, non-Newtonian concepts in mechanics, or of the familiar Leibnizian notation in calculus. In Britain the most interesting developments were in those branches of science that Newton had not, Medusa-like, petrified by his attention: that is, they were in sidereal as opposed to planetary astronomy, in chemistry, heat, electricity and magnetism and, towards the end of the century, the new science of geology.

The scope of chemistry had been defined by Hermann Boerhaave in his magisterial textbook, *Elementa chemiae*. Boerhaave, the leading figure in the Leyden medical school, together with his many pupils and disciples, ensured that chemistry should have a medical bias throughout the century. It was to a considerable extent a chemical age so far as experimental science was concerned. And Boerhaave placed heat as the central agent and subject of chemical change. It followed, inevitably, that heat, like chemistry generally, was taken out of the scope of mechanics, whether theoretical or practical. In other words the scientific study of heat was divorced from the most impressive and at the same time one of the most important developments of contemporary technology: the Newcomen engine. The common expression 'fire engine' used to describe these machines should not mislead us. Its use by writers of the period certainly does not imply that they commonly recognized *heat* as the motive agent, still less that heat was actually converted into work by the operation of the engine. The quantification of heat was not established until the mid-century at the earliest, long after the expression 'fire engine' was coined. And the notion of a heat engine did not become feasible until the foundation of modern thermodynamics in the 1840s. Instead we must interpret the expression as indicative only of one of the machine's most obvious features: the use of fire in place of wind or water.

The scientific study of electricity and magnetism, which had proved relatively intractable in terms of Newtonian mechanics,[21] was in a similar way the province of medical men, chemists and various 'experimental philosophers'. To the extent that there was a recognizable discipline called physics it differed significantly from that so confidently established during the nineteenth century. Over much of the eighteenth century 'physics' was Aristotelian in range, if not in method and metaphysic; it covered the study of motion, of development and change, and therefore included the life sciences. At the end of the century John Robison, in an authoritative article, remarked that 'Physics, then, is with us the study of the natural system, including both natural history and philosophy. The term is not indeed very familiar in our language'.[22]

In one part of what we should now recognize as physics a certain set of concepts was, to some extent, more acceptable in France than in Britain. These were the concepts associated with *vis viva*, a doctrine derived more from Huygens and Leibniz than from Descartes and Newton. They were brought to a precocious, but substantially abortive, maturity by Daniel Bernoulli with his concepts of potential and actual *vis viva* (1739).[23] Although not quite academically respectable in England the doctrine was quietly accepted by the engineers – for the concept of 'work', initially understood as the proportion of 'effort' transformed into useful 'effect' by any machine driven by wind or moving water, was plainly related to *vis viva*, or as we should say, kinetic energy. The concept was essential for the rationalization of practical mechanics, or *mécanique industrielle*. One important step on the way was James Watt's standardization of the unit of horse power in 1783.[24]

The pattern of eighteenth-century science was therefore complex: which does not simplify the study of the relations between science and industry in the first great age of industrial development. Although we must accept the dictionary definition of technology – the scientific study of industrial processes and the application of scientific

[21] I. B. Cohen, *Franklin and Newton* (Philadelphia, 1956) discusses the influence of Newtonian speculative philosophy on Franklin's researches in electricity. For an account of the failure of Newtonian mechanical philosophy applied to eighteenth-century chemistry see A. Thackray, 'Quantified chemistry – the Newtonian dream', in D. S. L. Cardwell (ed.), *John Dalton and the Progress of Science* (Manchester, 1968), 92–108.

[22] *Encyclopaedia Britannica*, 3rd edn, XIV (1797), 643. The article 'Physics' in A. Rees, *Cyclopaedia*, XXVII (1819), lists Linnaeus among the leading contributors to physics.

[23] A. J. Pacey and S. J. Fisher, 'Daniel Bernoulli and the *Vis-Viva of compressed air*', *BJHS*, III (1967), 388–92.

[24] H. W. Dickinson, *James Watt: Craftsman and Engineer* (Cambridge, 1935), 145.

ideas in industry – as authoritative, the problem remains that the relations between the natural sciences on the one hand and industry on the other ranged, during the eighteenth century, from the direct and evident, through the hazy and indistinct to the (apparently) non-existent. We must therefore try to clarify the notion of invention.

We are all, even the dullest and least imaginative, inventors in one way or another, on at least one or two occasions in the courses of our lives. It would, indeed, be difficult to get through life in a complex modern society without a minimum of inventive capacity. On the other hand the familiar, or publicly recognizible inventor foresees a large market for his invention; he will patent it and initiate its manufacture; ideally he will be cost/profit conscious.[25] Essentially, science is not necessarily involved in the inventive process, although it may have been a factor in the invention and production of one or more of the components used by our inventor. One does not have to be a polymer chemist to envisage a new use for plastics.

V. The mechanization of the textile industries

These observations are a necessary preamble to a consideration of one of the most important developments in the eighteenth century,[26] the very spearhead of the industrial revolution, and a development that on the face of it had nothing to do with science. This was the mechanization of the textile industries. Starting from the traditional hand loom, modified by the invention of John Kay's fly shuttle (1733), and the 300-year-old Saxon spinning wheel, the problem that was tackled and solved was to break down each of the complex manual operations in such a way that the entire process could be mechanized in stages, from the opening up of the cotton boll to the weaving of the spun thread.[27] The pioneer was Richard Arkwright (1732–1792), who began his working life as an apprentice to a barber and wigmaker – a vocation that must have given him a good understanding of, and an acquired feel for, thin natural fibres, but little or no incentive to study *Principia mathematica*, or even *Micrographia*. However, Arkwright's training was clearly appropriate. By a combination of the gifts of an inventor, enumerated above, plus remarkable business acumen,

[25] Eighteenth-century innovators were unremitting in their efforts to cut down costs and plough profits back into their businesses. See, for example, T. S. Ashton, *The Industrial Revolution, 1760–1830* (1970 edn), 76f; and *idem*, *Iron and Steel in the Industrial Revolution*, 3rd edn (Manchester, 1973), 209f.

[26] Landes, *Unbound Prometheus* (see note 7), 80f.

[27] R. L. Hills, *Power in the Industrial Revolution* (Manchester, 1970), 73f., *idem*, *Richard Arkwright* (1973).

Arkwright was able to mechanize the cotton industry from his spinning machine (the so-called water-frame patented in 1769) to the scutching machine for opening up the cotton bolls, taking in the carding engine and the roving frame on the way. The final problem, that of the power loom, was surmounted by the Reverend Edmund Cartwright by 1800. In all this scientific research played no part: the basic principles were not deduced from Newtonian, or any other advanced form of mechanics; nor as far as we know, were instruments like the microscope used, even though the structure of fibres might well have been supposed to be a key to the mechanization of the industry.

Nevertheless, it would be unwise to assume that the textile revolution was totally unrelated to the scientific movement that had begun so dramatically and successfully in the previous century. At a more general level than the overtly institutional a common factor can be detected. Franz Reuleaux wrote:[28]

> In earlier times men considered every machine as a separate whole, consisting of parts peculiar to it; they missed entirely, or saw but seldom the separate groups of parts that we call mechanisms. A mill was a mill, a stamp a stamp and nothing else and thus we find the older books describing each machine separately from beginning to end... Thought upon any subject has made considerable progress when general identity is seen through special variety – this is the first point of divergence between popular and scientific modes of thinking.

At a fundamental level, then, it would be reasonable to conclude that the mechanization of the textile industry in the eighteenth century covered an area common with the scientific movement – the recognition and acceptance of general principles. Certainly popular books on mechanics, such as J. T. Desaguliers' *Course of Experimental Philosophy* (1734, 1744) and William Emerson's *Mechanics*, which went through five editions from 1758 onward, lend support to this conclusion. And, more to the practical point, Arkwright himself recruited clockmakers to make textile machinery for, as he recognized, they understood the *principles* of gearing. But more work on the history of practical mechanics is needed before this particular hypothesis can be clarified, much less confirmed or refuted.

Whatever the scientific status of the mechanical inventions that marked the development of textile machinery there can be no ques-

[28] Franz Reuleaux, *The Kinematics of Machinery* (New York, 1963), 9.

tion that science, as commonly understood, came to play an essential part in related technologies that were necessary for the continued growth of the textile industry. The serried ranks of new, high-production machines required buildings that were strong enough to bear their weight, that were reasonably fireproofed and, above all, had sufficient cheap power laid on in compact and easily applicable form.

The limits set to building construction by traditional materials were known through long experience of the properties of timber, brick and masonry. It is quite likely that attempts were occasionally made to construct mills or other industrial buildings of a size and form beyond the limits imposed by those materials, for it seems to be in the nature of man to build up to and sometimes beyond the frontiers of his technological capabilities. But the most likely cause of failure was fire, of which perhaps the most famous example was the destruction of Albion Flour Mill in London in 1786. However, the rapid advance of the iron industry in the eighteenth century, described below, enabled the problems of both strength and fire resistance to be solved at the same time.

The extended use of iron as a structural material made its debut in 1779 when the famous iron bridge was constructed across the river Severn, linking the iron works at Coalbrookdale with that at Broseley. Later in the century the use of cast iron pillars in buildings became increasingly common. In textile mills rows of such pillars, strong in compression and occupying relatively little floor space, could support strong wooden beams that in turn carried the tiled floors on which the heavy new machines stood. By plastering (or nailing thin iron sheets over) the exposed timber, the fire risk could be reduced even further.

Thomas Telford (1757–1834) pioneered an original use for flat iron plates when he designed the iron troughs to carry the Shrewsbury canal over the Longdon aqueduct (1795) and, shortly afterwards, the Ellesmere canal over Pontcyssyltau aqueduct. But the new building material posed problems; traditional knowledge could not be used to predict what size or form of plate or beam would be appropriate for a given load. Calculation was necessary, but calculation by what rules? John Banks, a peripatetic lecturer, brought Galileo's theory of the strength of materials, as set out in *Two New Sciences* (1638), to the attention of British engineers and industrialists.[29] He did not correct Galileo's original error, even though later writers had already done

[29] John Banks, *A Treatise on Mills* (Kendal, 1795), and *On the Power of Machines* (Kendal, 1803).

so. Fortunately this did not matter, for the manner in which the theory was applied by British engineers effectively cancelled out the error.[30]

Galileo had shown also that the optimum shape for a horizontal beam bearing a load at its unsupported end is a parabola. But as the parabola is an unsuitable form for many purposes the engineers, taught by Banks and disciples of Galileo at several removes, compromised by adopting the elliptic or semi-elliptic form. Accordingly the first horizontal iron beams for industrial buildings were semi-elliptic in form. From about 1795 onward elliptic cast iron beams began to replace the wooden beams previously used in steam engines.[31] And the well-known 'fish-bellied' rails of the early railways exemplify the same principle. With the fall in the price of iron and the development of the rolling mill the elliptic form gradually dropped out of favour. The point had, nevertheless, been made: scientific theory from an unimpeachably scientific source had been applied in several major branches of the constructional industry. The new, reformed science of the strength of materials could be applied to structures of ever more complex form (in Britain as well as on the continent, for British engineers adopted the continental science of mechanics in the first decades of the nineteenth century).[32] This culminated in such grand engineering achievements as the Britannia tubular bridge over the Menai straits and, later on, the Eiffel Tower and the Forth Bridge, both of which are made of steel. Accordingly, by the beginning of the nineteenth century the steam engine was ceasing to be the product of collaboration between millwright, blacksmith, mason and carpenter: it was becoming the monopoly of metal workers, a matter of mechanical engineering; a machine made almost entirely of iron with some brass and copper fittings. At the same time iron was being used increasingly in the manufacture of textile machines, and as demand for these machines increased, so rationalized means of production were devised. And this meant a substantially new technology: that of machine tools. But this was not until after 1800.

The problem of the rationale of technical progress occurs in a simi-

[30] A. J. Pacey, *The Maze of Ingenuity* (1974), 229.

[31] *Ibid.*, 228.

[32] Eaton Hodgkinson, who collaborated with Robert Stephenson and William Fairbairn in the design and construction of the Britannia tubular bridge, was one of the first of what we may call the modern generation of British engineers. See B. Warburton, 'Eaton Hodgkinson (1789–1861) and the science of strength of materials' (Unpublished Ph.D. thesis, University of Manchester, 1971).

lar form in the cases of the chemical and metallurgical industries. Chemistry became a predictive and quantitative science as a result of the works of Lavoisier and Dalton and their successors. Prior to the publication of Lavoisier's *Traité élémentaire* in 1789 chemistry was scientific to the extent that there was a mass of empirical knowledge, that some simple rules were known and that an increasing number of new substances and new reactions were being discovered every year. It was on the basis of this collection of factual data and simple rules that the chemical industry began to develop. How can we account for substantial innovation in the absence of predictive theories? Was it all a matter of random experimentation, of trial and error?

VI. Darby and the iron revolution

The problem is not as difficult as it might seem. Let us consider the case of Abraham Darby (1678–1717) and the revolution he effected when, in 1709, he at last succeeded in using coke to smelt iron. For a long time there had been strong pressures to find a substitute for charcoal, which was getting progressively more expensive. Coal was the obvious substitute, but experience had shown that sulphur and other impurities in coal could ruin iron. On the other hand the use of coal for many other purposes had spread rapidly in the first half of the seventeenth century. By the last quarter of the century coal-fired reverberatory furnaces were being used successfully to smelt lead, copper and tin ores. (A reverberatory furnace heats the ores indirectly so that the fumes from the coal do not come in contact with the molten metal; contamination is therefore substantially reduced.) When Abraham Darby went to Bristol in 1699 and set up as a malt-mill maker he[33]

> went to an area where the smelting of copper and lead ores and the refining of the products had been accomplished successfully, and the use of coal and coke in suitable air-furnaces was the essential basis for success.

Thus as coal, and particularly coke, can be used in smelting certain metals it was a reasonable assumption that a way could be found to use them in the smelting of iron. So, although the scientific basis was unknown, by simple association of ideas a process that was known to work in one context could be expected to work in closely similar ones. The appeal is to a deep-seated conviction of the unity, or rationality of nature even though the exact scientific laws of the process are not

[33] R. A. Mott, 'Abraham Darby (I and II) and the coal–iron industry', *Transactions of the Newcomen Society*, XXXI (1957–8), 49–93.

known. We might, for lack of a better term, describe the method as one of associative thinking, and regard it as a first step in the business of establishing a science. It is not necessary to understand the scientific basis of a process if we are concerned with technological innovation; so long as the desired result is obtained that is sufficient – up to a point. Compared with what we may assume to be the random scatter of inventions and the secret mysteries of earlier ages the innovative technology of the eighteenth century may be described as scientific, although perhaps rather more in the spirit of T. H. Huxley's famous dictum that science is organized common-sense than in a way that would satisfy a critical philosopher of science.

In spite of Darby's success the iron industry did not expand rapidly in the following seventy years. Cast iron, with its high carbon content (4 per cent), is strong in compression but brittle under shock and weak in tension; its use, therefore, was somewhat restricted. Wrought iron, which is strong in tension and resists shock, used to be made by hammering out the slag from red-hot 'blooms', but the process was slow and inefficient. However, a second revolution occurred when, in 1784, Henry Cort (1740–1800) patented his 'puddling' process for converting iron with a high carbon content into wrought iron with a low carbon content (about 0.1 per cent). As most accounts of this process are inadequate[34] a short description here is appropriate.[35]

To reduce the carbon content of pig or cast iron Cort melted it in a reverberatory furnace while 'puddlers' with long poles called 'raddles' stirred the melted iron. The puddlers' work was extremely laborious; it was exhausting because of the heat and the glare from the molten iron and it was highly skilled. Using the edge of the furnace as a fulcrum for the raddles the puddlers kept the iron stirred all the time so that the carbon was brought to the surface where it was burned off with the characteristic blue flame of burning carbon monoxide. The melting point of carbon-free iron is about 400 °C above that of high-carbon iron so that as the carbon content was reduced the iron steadily solidified and became harder and harder to work. The skill of the puddler lay in knowing from the appearance of the flames and the feel of the raddles when the time had come to start forming

[34] It would be invidious to mention any particular erroneous or inadequate account here. By far the best account is that given by the Open University film dealing with the history of (iron and steel) bridges.

[35] This description is based on personal observation of the puddling process as carried out at Messrs Thomas Walmsley's Atlas Forge, Bolton, some years ago. The process was stopped as a result of the recent rise in the price of oil. As far as I know the puddling process is no longer practised anywhere in the world.

the bloom. He had to make sure that the solid lumps of red-hot iron were of a convenient size to extract from the furnace and that they had no chance to stick together, or to the side of the furnace. The carbon-free iron could then be taken out, forged and rolled. (The rolling mill was invented by the Swede, Christopher Polhem, in about 1745, and introduced to England by Cort in 1787.) Cort had had his predecessors, those who had worked on similar or related lines, for example Thomas and George Cranage, Richard Reynolds and Peter Onions. But, according to Henry Dickinson:[36]

> What we can say of Cort is that he gathered existing knowledge and technique, absorbed what was useful and necessary for his purpose, rejected what was not needed, and combined it into a system which, on the whole contributed such an advance that it marked an entirely new era in the manufacture of wrought iron.

It was the consequent cheapening of wrought iron that enabled men to convert the steam engine into a machine made entirely of iron, that made steam locomotives and railways possible, that allowed suspension bridges to be built and that enabled a host of special-purpose machines to be made, quickly and conveniently, of iron.

VII. The chemical industry

Turning now to the chemical industry: the invention by John Roebuck (in 1744) of the lead chamber process for the manufacture of sulphuric acid probably belongs to the simplest category of invention discussed above; it was in effect a straightforward scaling up of the historic process, made possible, perhaps, by a reduction in the price of lead. But the invention of the Leblanc process for the manufacture of soda surely illustrates associative method. Traditionally soda had been obtained from vegetable ash, or more particularly from barilla, from Spain, or kelp, from sea coasts. But the process was slow and inefficient and, as with charcoal for iron smelting, demand threatened to outrun supply. Many attempts were made to find a way of making soda rapidly, and from readily available materials. Before the eighteenth century Stahl had known that the base of common salt was soda,[37] and this knowledge was commonly taken as the starting point for experimenters, the usual approach being to heat salt with sulphuric acid and so produce sodium sulphate. Among those who

[36] H. W. Dickinson, 'Henry Cort's bicentenary', *Transactions of the Newcomen Society*, XXI (1940–1), 31–47.

[37] J. R. Partington, *A History of Chemistry* (4 vols., 1962), III, 562.

worked on the problem were Malherbe, Guyton de Morveau, de la Métherie and Lord Dundonald, but final success went to Nicholas Leblanc (1742–1806), who in the eighteenth-century tradition had had a medical training. In the Leblanc process common salt is heated with sulphuric acid and the resulting sulphate heated with limestone and coal to yield sodium carbonate and calcium sulphide; the soda can then be leached and evaporated out. Leblanc seems to have invented his process in 1787 and patented it in 1791. However, owing to his misfortunes in the French Revolution he made no money from it and finally, in poverty and desperation, killed himself in 1806.[38]

To repeat, associative method can be regarded as an early stage of scientific procedure. And this is consistent with the generally acknowledged fact that technology is frequently ahead of scientific knowledge.

Finally, another mode of innovation represented in eighteenth-century chemical technology was the invention of bleaching powder. Chlorine had been discovered in 1773 by K. W. Scheele (1742–1786) and its remarkable bleaching properties noticed by C. L. Berthollet in 1785. 'Eau de Javelle', a convenient bleaching liquid derived from chlorine, was invented by Berthollet but this was soon displaced by the even more effective bleaching powder produced by Charles Tennant in 1788. In this case the invention was developed from an original scientific discovery and it is difficult to see how it could possibly have been made by the associative method. Previous methods of bleaching – exposure to the weather, treatment by sour milk or dilute sulphuric acid – give no clue to the key factor of chlorine and no hint as to its properties. The first synthetic dyestuff (1856) and the first atomic bomb (1945) belong to this same category of scientific invention, although of course at a much more advanced level.

The technological advance during the eighteenth century that had the most general significance was probably the quantification of power and the refinement of the concepts that were later to be subsumed under the nineteenth-century doctrine of energy. According to Galilean mechanics the effort (or energy) put into an ideal machine must equal the effect (or work) obtained. But Antoine Parent, in his seminal paper of 1704, concluded that even a perfect waterwheel, driven by the impact of water on flat blades, could derive only 4/27ths of the theoretical effort of the stream.[39] According to Parent a perfect – i.e. friction-free – waterwheel cannot be an ideal Galilean machine for it cannot work without moving and the very act of moving reduces

[38] *Dictionary of Scientific Biography*, VIII (New York, 1973), art. 'Nicholas Leblanc'.
[39] D. S. L. Cardwell, *Technology, Science and History* (1972), 63.

the impact of the stream on the blades. He did not consider the case of a slow-moving 'overshot' wheel with large buckets in which the water acts by weight rather than impact. However, John Smeaton (1724–1792), a practical engineer, corrected Parent's error, for such it was, in a long series of meticulous experiments.[40] He indicated that the efficiency of a perfect overshot wheel must be 1, in accordance with Galilean principles, while for an undershot wheel it is ½. In the latter case much of the effort of the stream is, he pointed out, wasted in producing turbulence and spray. And, of course, some of the effort will be carried away in the form of the still-moving stream. (Smeaton did not attempt to compare these two sources of inefficiency.) This implied that if all the effort of the stream is to be converted into useful effect, or work, the transformation must take place without abrupt changes, the stream must yield its effort smoothly and completely.[41] Smeaton recognized that with slow rivers the superior overshot mode might be difficult to arrange. He therefore suggested the 'breast wheel' as a compromise. In this arrangement a weir, with a curved slope downstream, raises the river level so that the water spills over, to act largely by weight on the blades of an undershot wheel faired neatly into the curved, downstream side of the weir. If the wheel turns very slowly the water is released with little residual velocity and most of its effort will therefore be given up to the waterwheel.

Surviving material evidence confirms the opinion that few undershot wheels were erected after Smeaton's time. Certainly he was most influential, with many disciples in Britain and the United States.[42] All the great waterwheels that powered the mills of the industrial revolution seem to have been designed on Smeatonian principles.[43] With economic growth the pressure on power resources increased; the best river sites were soon fully exploited and in an intensely cost/profit conscious age economy of power became all-important. By the end of the century there were, in some localities, real power crises[44] and there are well-documented stories of the shifts men were driven to in

[40] John Smeaton, 'Experimental enquiry concerning the natural powers of wind and water to turn mills and other machinery', *PT*, LI (1759), 100–74.

[41] *Ibid.*, *'non-elastic bodies when acting by impulse or collision,* communicate only part of their original power; the other part being spent in changing their figure in consequence of the stroke'. See also his paper: 'An experimental examination of . . . mechanic power . . .', *PT*, LXVI (1776), 450–75.

[42] Layton, 'Millwrights and engineers' (see note 13).

[43] For example, the great waterwheel built by Peter Ewart to power Quarry Bank Mill, Styal, Cheshire, was probably the largest of its kind in Britain; it was a breast wheel, the level of the river Bollin having been raised by a masonry dam.

[44] Hills, *Power in the Industrial Revolution* (see note 27).

their searches for power.[45] It would seem certain, therefore, that Smeaton's work was of critical importance during this phase of industrialization; he may be said to have nearly doubled the effective power available for industry.

VIII. British technology

British technology was substantially geared to the solution of immediate practical problems. It was otherwise in France. French technology was, throughout the eighteenth century, more scientific, more mathematical than British technology, which meant that it was often, perhaps too often, concerned with distant prospects. To illustrate this point we may consider the work of two of Smeaton's French contemporaries: the Chevalier Déparcieux and the Chevalier de Borda.[46] Déparcieux refuted Parent's argument by considering the case of a perfect overshot waterwheel driving an identical wheel backwards so that it worked like a scoop wheel. Such a coupled system, he reasoned, could restore almost all the driving water to the source, just as a weight coupled by a cord over a pulley to a slightly less-heavy weight will slowly lift the latter. The argument is sound and Déparcieux, who was a practical engineer, has the credit for introducing an entirely novel and extremely important concept into technology, and, indeed, science: that of the reversible system as a limiting case.[47] De Borda, on the other hand, demonstrated mathematically that Parent's conclusions were wrong and that the limits of efficiency in the cases of overshot and undershot waterwheels were 1 and ½ respectively. He showed that the conclusions were the same whether the calculations were based on Newtonian or on *vis viva* mechanics. And he went on to argue that an undershot wheel could be made as efficient as an overshot one if the blades were curved to point upstream so that the water could enter tangentially, without shock. He pointed out, too, that if the direction of flow was parallel to the axle of the wheel the peripheral velocity could be as high as one wished while the theoretical efficiency could be kept at the maximum. Borda's paper of 1767 was therefore the

[45] A. E. Musson and E. Robinson, *Science and Technology in the Industrial Revolution* (Manchester, 1969) 369, 446. See also Hills *Power in the Industrial Revolution* (see note 27).

[46] A. Déparcieux, *Histoire et Mémoires de l'Académie Royale des Sciences* (1954), 603–14. J. C. de Borda, 'Mémoires sur les roues hydrauliques', *Histoire et Mémoires de l'Académie Royale des Sciences* (1767), 270–87.

[47] D. S. L. Cardwell, 'Power technologies and the advance of science, 1700–1825', *Technology and Culture*, VI (1965), 188–207.

forerunner of turbine theory,[48] although a practical turbine was not built until Fourneyron made one in the 1830s.

We may, therefore, characterize British technology from about 1750 onward as essentially industrial technology whereas French technology was academic or state technology. A confirmatory instance is provided by the subsequent history of the theory of the beam. Parent had, by 1713, corrected Galileo's error but his work was not taken up, possibly because no one could see a use for it and its academic interest was limited. The problem was finally solved in 1773 by C. A. Coulomb (1736–1806), a military engineer who is well known to physicists as the inventor of the torsion balance and for his work on electricity. But Coulomb's work was ignored for about forty years.[49] Although John Robison brought it to the attention of British engineers in 1793 they still preferred the old, flawed theory of Galileo.[50]

It is pointless to argue which of the two forms of technology is superior. Plainly, industrial technology is more important if you are confronted by immediate problems and want to make money; academic technology is more important if you are concerned about the future. One thing, however, they have in common. It is impossible to find pre-eighteenth-century examples of the kind of scientific engineering so well exemplified by Parent, Borda and Coulomb; equally there are no earlier instances of the distinctive kind of innovation that John Smeaton practised, methodically and deliberately. In Smeaton's method all components, materials or processes save one are kept constant while that one is varied systematically. The effect of each variation on the product or performance is carefully noted. The procedure is repeated for each component in turn until in the end all the data for the optimum result are obtained. This was the method Smeaton used in his experiments on the waterwheel; he used it also to prepare the most efficient hydraulic cement – one that would set under water – for the construction of the Eddystone lighthouse. He used the method again in the long series of experiments he carried out on a small experimental Newcomen engine he erected in 1769. As a result of his systematic work on this machine he was able to draw up a table of specifications for the construction of Newcomen engines ranging from 1 to 75 horse power.[51] With this information he was able to design and build engines, including the huge Chacewater engine in Cornwall, of more than double the 'duty', or efficiency, of previous

[48] Layton, 'Millwrights and engineers' (see note 13).
[49] H. Straub, *A History of Civil Engineering* (1960), 147.
[50] Pacey, *Maze of Ingenuity* (see note 30), 229.
[51] John Farey, *A Treatise on the Steam Engine* (1827), 166f.

machines (12 million foot pounds per bushel of coal burned compared with 5 million).

The origins of Smeaton's method are probably to be found in Newton's works: in the emphasis on measurement and on the inferences that can be made from measurement as described in *Principia;* and also, perhaps, in Newton's famous research on the compound nature of light. We know that Smeaton was interested in astronomy and had read his Newton. On the other hand, his method cannot, in principle, lead to a *new* invention and it is quite independent of the basic laws or scientific principles underlying whatever is being investigated. It is true that a particular variation might produce an anomalous effect and so lead to a new scientific discovery but such an event would be entirely accidental and have no bearing on the main purpose of the investigation. Accordingly, Smeatonian method applied to the atmospheric engine could never have led to the invention of the true steam engine. Smeaton was entirely ignorant of the principles of thermodynamics and did not need even the basic concepts of the science of heat, as then understood, to double the efficiency of the Newcomen engine. Similarly, as we have remarked, he obtained his hydraulic cement without having to know anything about the underlying chemistry. In short, the Smeatonian method tends to the perfection of known inventions, not to the production of new ones. And it is in common use today. As new inventions, or useful scientific discoveries are made so the method is applied to optimize the final product for the market. Occasionally our ignorance of the underlying science, and consequently of some properties of the substances, or processes, involved, lead to disaster.

What, then, is the status of Smeatonian method? It is rational, systematic and comprehensive; subjective and qualitative considerations are excluded. In every case a testable and preferably quantitative answer is given. But the technology it implies is essentially a static, or closed one.

The main achievement of James Watt (1736–1819) is in interesting and revealing contrast to the Smeatonian method. Watt, a Glasgow University instrument-maker, was in close contact with Joseph Black who had introduced the all-important concept of quantity of heat together with the derived concepts of specific heat capacity and latent heat. Equipped with these scientific tools Watt analysed the distribution of heat in the Newcomen engine quantitatively. He found that a wholly disproportionate quantity of heat was being used merely to heat up the cylinder once every cycle. This information enabled him to invent the steam engine with separate condenser that was much more economical than the Newcomen engine. This investigation and its conclusion was, by any standards, applied science on the grand scale. But it was not the sort of investigation that could be systematized in any way; no one could draw up a set of rules for its practice. It was dependent on the individual genius of one man; it was not a team affair. However, the new Watt condensing engine, once produced, was immediately susceptible of Smeatonian improvement as well as of improvement by individual detailed inventions as they were made: better valve mechanisms, improved furnaces, more efficient design of boilers, etc. Watt was always a cautious man and he strenuously resisted the introduction of high-pressure steam engines, right up to the expiry of his controlling patents by 1800. But once his overall control lapsed, small, powerful and efficient high-pressure steam engines were made in large numbers and vast opportunities for further inventions opened up, perhaps the most important being that of the locomotive and the railway system it made possible.

The case of James Watt and the true steam engine exemplifies two basic features of technological progress: revolutionary invention and evolutionary improvement. After Smeaton the atmospheric engine was still recognizably the Newcomen engine; after Watt it was something radically different. The distinction between revolutionary and evolutionary applies in the history of science as well as in the history of technology. It would indeed be surprising if it were otherwise, for unless we want to argue that science and technology are totally different things, having nothing in common, we may well expect the same patterns to appear in the progress of both. We can, depending on our interests, *assume* them to be quite different. The philosopher and the economist, for example, can do so from their different standpoints, but their views are necessarily partial since their loyalties lie elsewhere. But the historian concerned with science or with technology cannot avoid considering the areas of overlap, the similarities, the contrasts and the cross-fertilizations that at times can be of crucial importance.[52]

Watt was a most unusual man: in the individuality of his genius, in the contribution he made to the development of technology and the economy as well as, indirectly, to the growth of science, and in that he was a 'loner', an anomalous individual in the early history of the heat engine. As we have noted, throughout the eighteenth century the science of heat was monopolized by chemists and medical men; the few mathematicians and natural philosophers who showed interest in the subject were effectively ignored. Only towards the end of the century did things begin to change and only with the appearance

[52] D. S. L. Cardwell, *From Watt to Clausius* (1971), 186.

of Fourier's *Analytical Theory of Heat* (1822) could it be said that physicists had taken over the subject. The point is that throughout the eighteenth century the engineers were, with one exception, uninterested in the science of heat even when they were directly concerned with steam engines. Chemists and medical men had little to do with the design and development work of engineers. It is therefore not surprising that even von Rumford, who was uninterested in chemistry or medicine, failed to see the relevance to the steam engine of the science he professed so ardently.

Watt was the exception that proves the rule. And this is an entirely justified use of the cliché, for he was not an engineer, as were Smeaton and his colleagues; Smeaton, indeed, thought that Watt's engine was an interesting machine but far too complicated for practical use. Watt, the instrument-maker, was a man of science, a protégé of the chemist Black. He was therefore uniquely well placed to use the science of heat to improve the atmospheric engine, and as it happened to transform it altogether. But thereafter the technology of the steam engine and the science of heat went their ways independently of each other until they were gradually reconciled by French physicists of the early nineteenth century. In England the divorce was almost made absolute. Up to the end of the century no further attention was paid to heat economy in the design of steam engines; the next radical improvement in this respect did not take place until 1825.[53]

Designers of steam engines had much more in common with water-power engineers than they had with men of science interested in heat. This is hardly surprising. Throughout the eighteenth and well into the nineteenth century the great majority of engineers were accustomed to deal with both steam engines and hydraulic engines; the age of specialism had not yet dawned. At the same time authoritative textbooks and popular accounts dealt with both forms of power. This was the case with the books published early in the eighteenth century, such as those by Stephen Switzer, Jacob Leupold and B. F. de Belidor, and it was no less true of the excellent French textbooks on engineering published early in the nineteenth century.[54] In addition, many of the practical details such as cylinders, pistons, linkages, etc. (used in pumps and water-pressure engines) were common to both

[53] In that year Samuel Grose substantially improved the performance of a Cornish pumping engine (at Wheal Hope mine) by lagging the pipes, cylinder, valves, etc.

[54] See, for example, the textbooks by Hachette, Borgnis, Christian, Riche de Prony, Navier and Poncelet.

technologies and it was perhaps inevitable that concepts should also be shared. If the major design requirement in the case of an hydraulic engine was that water pressure be applied in the most efficient and economical way, then it would be reasonable to assume that the same consideration, with steam substituted for water, should apply to the steam engine. This process of cross-fertilization, or exchange of ideas, no doubt helped engineers steadily to improve the performances of all types of motors. On more than one occasion they did this by using principles upon which science was only subsequently to bestow its approval.[55]

The improvements that we have just discussed tended to be evolutionary rather than revolutionary. Although it is usually fairly easy to decide whether a particular innovation is an evolutionary improvement or a revolutionary invention, and even to locate its origin in time and place, no short and satisfactory definition of the terms can be given. S. C. Gilfillan is one author who has used these terms for a long time and is aware of the difficulties of definition. However his interpretation of evolutionary improvement is essentially biological:[56] the evolution of any particular thing is to be regarded as an evolutionary extension of man, just as the snail's shell has evolved with the creature. The interpretation is teleological and reminiscent of Aristotle's observation that 'if a house had been a thing made by nature, it would have been made in the same way as it is now by art'.[57] Less contentious is Gilfillan's observation that revolutionary inventions tend to be made by outsiders, or comparative strangers to the relevant technology. As we have just seen, the case of James Watt strongly supports this hypothesis.

A major, or revolutionary invention will tend to include components or concepts derived from fields outside the range of that particular technology. The kind of invention that uses only established principles or well-known materials, familiar in the technology, will tend to be an evolutionary improvement. Those who have considerable knowledge or expertise outside the scope of a particular technology will not, as a general rule, be specialists in that technology; they will be outsiders. Watt, Arkwright, Cartwright, Abraham Darby and many others who made revolutionary inventions during the

[55] Layton, 'Millwrights and engineers' (see note 13).

[56] S. C. Gilfillan, *The Sociology of Invention* (Cambridge, Mass., 1963 edn), 14. Gilfillan says: 'Is the biologic analogy anything more than an analogy? Yes. The ship or any invention *is* a biologic organ, in the same sense that a bird's nest is, if not a snail-shell'.

[57] Aristotle, *Physica*, 199a 10. But he would have rejected any suggestion that animals, or the parts of animals, had evolved.

eighteenth century were outsiders to the technologies in which they made their marks. Each one of them brought in new and stimulating ideas to vitalize old techniques or sometimes to found entirely new ones.

Much the same is true of science. It is often the case that a particular branch of science, just before a revolutionary advance, is in a state of complacent self-satisfaction. That is why the outsider, bringing in revolutionary ideas, is often strenuously resisted. Fourier had written, in 1822, that the science of heat had now been completed (by him). And the claim was plausible; certainly it was influential enough to help ensure that three would-be revolutionaries had appalling struggles to get their ideas accepted later on,[58] for none of them was a recognized scholar within the fold. In much the same way the atomic theory in chemistry was not established through the evolutionary progress of that science after the time of Lavoisier; it was introduced through the researches of John Dalton (an outsider in all senses of the word) into meteorology and the physics of gases.[59]

These cases have, I believe, a bearing on the history of both science and technology in the eighteenth century. In complete contrast to those days technology and science are now highly organized: characterized by a large number of specialisms, each one differentiated by an appropriate training scheme, examinations, formal qualifications and specialized professional institutions (the modern equivalents of medieval guilds). Now a formal training tends to limit knowledge and concentrate experience on one particular set of facts, one set of laws or theories, to the exclusion of the rest. No doubt it must be so. But specialism of this sort may, in principle at least, tend to exclude the outsider, and therefore discourage the revolutionary innovator. There is no contempt so withering as that of the recognized expert in a respected specialism, whatever it may be, for the brash young man from outside who has new and revolutionary ideas.

The industrial revolution in Britain was a time of massive innovation in a wide range of industries and the foundation of a series of new industries. Surveying the late-eighteenth-century scene it is difficult to detect any serious institutional obstacles to technological innovation. The state, the Church, the law were either indifferent or concerned to remove any obstacles that might be recognized; internally there seem to have been no obstacles at all. It is true that later on

[58] S. Carnot, J. P. Joule and J. R. Mayer were the three outsiders who had to struggle for recognition. Joule achieved it much more rapidly and easily than the other two; nevertheless it took him six years to win his first influential convert.

[59] See, for example, F. Greenaway, *John Dalton and the Atom* (1966), 105f.

luddites - handloom weavers and the like - might riot and smash machinery. But this was hardly institutionalized opposition backed by authority of any kind. If a technologist or inventor with a sound idea could, like James Watt, find a business man as partner, like Matthew Boulton, a man who knew how to raise money and how to deal with the manufacturing and sales side of the venture, then his problems were substantially solved. In short the field was entirely open to all comers and in theory, at least, the best men would win.

Summarizing these developments, we can recognize several modes of industrial innovation clearly distinguishable in the eighteenth century. Simple invention is a widely distributed faculty and does not, by any means, require scientific knowledge beyond the most elementary and general stage. This form of invention was practised in the eighteenth century as it had been from time immemorial, and was to be up to the present day. However there are grounds for believing that invention of this kind may, under certain circumstances require the recognition of general principles, which may be taken to be the first step towards scientific thinking. Much the same seems to have been true of innovations in the chemical and metallurgical industries before the rapid development of chemistry due to Lavoisier and his disciples. The recognition that if, under specified conditions a certain group of substances behave in a certain way then it was reasonable to suppose that, *ceteris paribus,* similar substances would behave in the same way seems, as a general rule, to have guided the thinking of chemical technologists. In contrast, the systematic method of John Smeaton leads to the optimization of a known product or process but cannot itself lead to a new invention. Finally, we noted that the science of mechanics found effective application in the theory of structures when a new material - iron - was used; and it also played a vital role in the development of power technology. Scientific research was used by Watt to make his radical transformation of the atmospheric into the steam engine; and scientific research in an entirely different field yielded the substance that was to make possible the rapid and efficient bleaching of textiles. But as the outcome of scientific research cannot possibly be predicted these forms of technological innovation cannot be systematized.

IX. Executive engineers

There remains one class of technologists that we have not, so far, considered: the executive engineers; the men who cannot claim credit for particular innovations, whether evolutionary or revolutionary, but who carry out projects using well-known materials and methods on such a scale and in such quantity that industrial growth is significantly hastened. One such man was James Brindley (1716–1772).[60] With the possible exception of the Barton aqueduct, carrying the Bridgewater canal over the river Irwell, it is doubtful if Brindley's name can be associated with any notable invention or technological advance. His most famous work, the Bridgewater canal that linked the Worsley coal mines with both Manchester and the Grand Trunk, or Trent and Mersey canal, was completed some eighty years after the much more impressive Canal du Midi or Languedoc (1681) of J. P. Riquet had linked the Atlantic and the Mediterranean. Even the technique of using 'puddle' to render the canal 'water tight' was almost certainly introduced by the Dutch hydraulic engineers who worked in England in the seventeenth century, and subsequently revived by Brindley.[61] ('Puddle' was carefully selected earth in a semi-liquid state, used to line canals so that they remained 'water-tight' when cut through porous sand or rock.) But the economic significance of the Bridgewater canal was immense. It initiated an era of canal-building that in the next seventy years was to cover England and the lowlands of Scotland with a network of inland waterways. These provided a cheap and efficient means of transport to link the main centres of manufacture with their markets, their sources of supply and their ports for export trade. By the 1790s enthusiasm for canal-building became a mania, foreshadowing the railway mania of the 1840s.

The advantages of a canal over a road for bulk transport hardly need explanation. The complete absence of pot holes and adverse gradients meant that although transport might be slow a minimum amount of effort was required to achieve a given effect; in other words, the system was essentially an energy-saving one. Nevertheless the success of the English canal system did not prevent an extensive development of the road network. Thanks to the turnpike system and in particular to three distinguished engineers – John Metcalf (1717–1810), Thomas Telford (1757–1834) and J. L. Macadam (1756–1836)[62] – hard, smooth and weather-resistant roads were laid down that allowed much faster and heavier traffic. Apart from the immediate economic benefit of these roads they were also a precondition for an incidental but eventually seminal invention made by John Palmer (1742–1818). Palmer, a brewer, a theatre owner and impre-

[60] The most recent biographer is C. T. G. Boucher, *James Brindley, Engineer: 1716–1772* (Norwich, 1968).

[61] Rees, *Cyclopaedia* (see note 22), VI, art. 'Canals'.

[62] Straub, *Civil Engineering* (see note 49), 163–4; Singer *et al.* (eds.), *History of Technology* (see note 6), IV, 530f.

sario was, through his travels and through his business, made aware of the slowness and unreliability of the postal system. The mail was carried by post boy on horseback or conveyed in a slow one-horse cart. Delays were long, robberies and losses frequent. Palmer therefore suggested an organized network of mail coaches to cover the whole country.[63] The coaches were to leave London and various provincial cities and towns at set times; they were to travel at a set speed, making timed stops at intermediate places; a limited number of passengers were to be carried, inside and outside the coach, and their fares would pay for the cost of carrying the mail as well as for the upkeep of the coaches and the wages of the coachman and guard. The officials of the post office were sure that it would fail, but they were over-ruled and the first mail coach ran from London to Bristol on 2 August 1784. Very soon the system had spread over the entire country. A fast, efficient and reliable if not, at first, cheap mail service had been established.

X. Conclusion

Adam Smith, in his *Wealth of Nations* (1776), had surprisingly and disappointingly little to say about technology and innovation. It is surprising because he wrote long after the invention of the Newcomen engine and seven years after the patenting of the water-frame and the condensing steam engine. The case of the steam engine is most surprising of all for he was a member of the small, select group at Glasgow University that included Black and Watt. It is disappointing for Smith's considered views would have been of the greatest interest and value to historians and students of technology generally. Instead he ascribes the great increase in the powers of production to the division of labour, choosing as his illustration of the effectiveness of the principle the manufacture of pins. He shows some awareness of the different modes of innovation when he suggests that the desire of workmen to lighten their own tasks is a major source of new inventions. Unfortunately he puts forward, as an illustration, the absurd story of a lazy boy who, in order to have time to play, made the valves of the Newcomen engine self-acting by joining them to the working beam by lengths of string. He then goes on more reasonably to instance improvements made by specialist makers of machines and to refer to 'philosophers or men of speculation, whose trade is not to do anything, but to observe everything; and who, upon that account,

[63] *Dictionary of National Biography*, art. 'John Palmer'.

are often capable of combining the powers of the most distant and dissimilar object'.[64]

Smith was at least aware of the necessity for skills, and enough was said at the beginning of this chapter to indicate that skills were widely distributed in eighteenth-century Britain and that the level was high. But it may be asked whether skills are enough. The question is intriguing and difficult to answer. From an age of computers, space satellites and nuclear power it may be difficult to recognize any 'science' behind the skills of the eighteenth century. But this may be due to the foreshortening effect of time on our perspectives. Relative to what had gone before science was perhaps a not-uncommon factor in the innovative skills of the eighteenth century. To quote Abraham Rees' *Cyclopaedia* of 1819:[65]

> A skilful engineer should undoubtedly possess a considerable degree of mathematical knowledge. Calculations of which some are of the most abstruse and laborious kind, will frequently occur and he should therefore be well acquainted with the principles upon which all calculations are founded, and by which they are rightly applied in practice. An engineer should also have studied the elements of most or all of the sciences immediately connected with his profession and he should particularly excel in an acquaintance with the various branches of mechanics, both theoretical and practical. His knowledge should comprehend whatever has been written and done by other engineers . . . It is necessary that he should be a ready and correct, if not finished draughtsman.

Seventy years earlier the author of *The London Tradesman* had stated the necessary qualifications for the engineer of his time:

> An engineer requires a very mechanically turned head and should be versed in all the laws and principles of mechanics and what is called the mechanic powers . . . he ought to learn mathematics and designing of which last it is absolutely necessary that he should be a perfect master.[66]

[64] Adam Smith, *The Wealth of Nations* (2 vols., 1776), I, ch. 1. The story about the inventive boy probably originated with J. T. Desaguliers, *Experimental Philosophy* (see note 18), II, Annotation to lecture XII. The suggestion that his motive was to lighten his own labour was probably a gloss added by Smith. No one who has studied the brilliant and beautiful self-acting value gear of the Newcomen engine could possibly believe that it had been devised by a lazy boy with no more equipment than two lengths of string. It is impossible to see how such a modification could have worked.

[65] Rees 'Canals' (see note 61).

[66] Anon., *The London Tradesman* (1747). I am indebted to Mr G. S. Catterall for giving me this quotation.

It would, therefore, be unwise to assume that with a few exceptions the eighteenth-century engineers in Britain had little or no knowledge of mathematics and only a rudimentary acquaintance with the sciences of mechanics and hydraulics. Samuel Smiles made James Brindley out to be a rustic who remained barely literate throughout his life, but Brindley's most recent biographer disagrees and indeed when one considers the many and complex difficulties of building a canal it would seem unlikely that a man of the kind Smiles described could have been the great executive engineer that Brindley undoubtedly was.[67] This is not to deny that many leading engineers and technologists of the eighteenth century came from working-class stock. Arkwright, Brindley, Hargreaves and many others came from the working classes, Watt from the lower middle class, and Boulton, Cartwright and Smeaton from the professional middle classes. Nor was the aristocracy unrepresented: we may reasonably include the Duke of Bridgewater among the innovators of the time.

From all this one might conclude that conditions for industrial innovation were ideal, particularly towards the end of the eighteenth century. But this would be premature. In one respect, at least, Britain, the industrial pioneer, was already outclassed by the end of the century. By that time two distinct types of engineering had become clearly distinguishable: military engineering, which has a long history going back to Roman times, and civil engineering. In France, in contrast to Britain, civil engineers as well as military engineers tended to be state employees, concerned with the building of such public works as roads, bridges, harbours, etc. Inevitably state engineers must have official status; and official status usually means formal training and qualifications. Accordingly the Ecole des Ponts et Chaussées was founded in 1747 and the Ecole Polytechnique in 1794, after the Revolution. These quasi-military establishments provided, by the standards of the day, a first-class engineering education based on mathematics, mechanics and related sciences. As a result the French public service as well as the army could rely on a supply of first-class scientific engineers. But France was not unique in this; indeed she was not even the pioneer. The German states had already established the famous mining schools at Freiberg, Berlin and Schemnitz (northern Hungary). German mines were state monopolies, supplying mainly silver for the coinage. The mine captains and engineers had, as civil servants, to be properly trained for the job. Hence the first recognizable technical colleges in the world appeared in Germany in the form

[67] Boucher, *James Brindley* (see note 60), argues convincingly that Brindley was not a semi-literate rustic.

of the three mining academies.[68] The foundation of these colleges was possibly the first small sign of the German renaissance in science and technology that was such a prominent feature of late-nineteenth-century industrial history. In Britain, on the other hand, the industrial activities of the state were minimal – even the Mines Royal had been wound up in 1684 and handed over to private enterprise – and no such enterprises were conceivable.

Adam Smith had recommended education as one of the (few) things that the state could properly do on behalf of the people.[69] But the sole reason he gave was that education might offset the soul-destroying effects of the division of labour when it was carried as far as in the case of the pin-makers. In the event, therefore, Britain entered the nineteenth century with a superb and successful industrial technology based on do-it-yourself principles. Understandably no one wanted to interfere with a system that was so obviously successful; but as the advance of science and technology accelerated, as the links between them grew stronger and as both became more complex so the system of self-help became steadily less suitable and education, particularly technical and scientific education, became increasingly essential for industrial success. A comparison of the excellent French textbooks on engineering published in the first three decades of the nineteenth century with those available in Britain in the same period will give some idea of the lag that had already developed,[70] and that was to continue throughout the nineteenth century with respect to France and then, increasingly, with respect to Germany and the United States.

It is outside the scope of this paper to attempt to summarize the consequences of eighteenth-century science for later generations. But two points must be made. Firstly, it would be a mistake to assume that technology is a dependent variable of science; that the scientist hands down his discoveries to the technologist who thereupon finds an application for them. This idea was implicit in Robison's accounts of the relationships between Hooke and Newcomen and between Black and Watt. And it is often assumed by the powerful, articulate and usually successful 'pure' science lobbies in the modern world. But it ignores the facts that some technological advances have been

68 D. M. Farrar, 'The Royal Hungarian Mining Academy, Schemnitz: some aspects of technical education in the eighteenth century' (unpublished M.Sc. thesis, University of Manchester, 1971).

69 Smith, *Wealth of Nations* (see note 64), II, bk s.

70 Compare, for example, the works of Banks (see note 29), or Olinthus Gregory's *A Treatise on Mechanics*, 3rd edn (1815), with the books referred to in note 54 above.

independent of science, at least as conventionally understood, and, *per contra,* that technology sometimes stimulated advances in science. The elaboration of the concepts of 'effort' and 'effect' and the formulation of the concepts of power and work were necessary steps towards the foundation of the general doctrine of energy in the mid-nineteenth century, while the detailed studies of the conditions for the transformation of energy by Watt, de Borda and their successors paved the way for Sadi Carnot and the science of thermodynamics.[71] More generally, Coulomb, that exemplary scientist-engineer, can be fairly regarded as the first *physicist* as distinct from natural philosopher, experimental philosopher, or student of 'physics' in the old connotation according to which the scope of physics included the life sciences. And among the founders of geology A. G. Werner, William Smith and James Hutton had definite technological interests. In short it would not be unfair to conclude that science owes more to technology than technology owes to science; and it would be reasonable to claim, against the intellectualist interpretation of the history of science,[72] that science is a distinctive form of philosophy partly based upon and derived from the technical arts.

Secondly, although certain technological developments of the eighteenth century, such as the steam engine and the iron-framed building, had obvious implications for later technology, it sometimes happened that the connections were less obvious but no less important. The rapid extension of the canal system from 1760 onward was necessarily accompanied by a great increase in knowledge of, and expertise in, tunnelling and making embankments and cuttings, as well as in the social skills required to recruit, organize and command large bodies of men (navvies) – all of which was to prove essential when railways were being built in the following century. And John Palmer's stage coaches, running to a time-table, carrying passengers and mail, were clearly forerunners of the subsequent railway passenger services. There was nothing obvious about a public transport system, with time-tables and fares based on mileage covered. The idea could only have been conceived in a technologically advanced society. Similarly it was not an obvious step to combine such a system with the carriage and distribution of mails. That, too, had to be invented. And so every commercial aeroplane, on a scheduled

[71] Cardwell, *From Watt to Clausius* (see note 52), 186f.

[72] For discussions of the intellectualist interpretation of the history of science see D. S. L. Cardwell, 'Science and technology in the eighteenth century', *HS,* I (1962), 30–43; Walter F. Cannon 'History in depth: the early Victorian period', *HS,* III, (1964), 20–38.

flight, that carries mail is, in a fundamental sense, a direct descendant of John Palmer's stage coaches.

The facts that technologies can bring about the circumstances that favour their own replacement and that leading modern technologies can trace their ancestries back to different technologies in the eighteenth century suggest some interesting lines for further research. Of course the historian of science who ventures to identify the outstanding problems in his field gives hostages to fortune. He, of all people, should know that in every generation there have been able men who have pointed out what they considered to be the key problems in science, only to be proved hopelessly wrong a few years later. With this reservation in mind, I believe that the conditions under which the new technologies of the eighteenth century displaced old crafts and the circumstances that led to the replacement of certain technologies by what we may call successor technologies – as road and canal transport were replaced by railways, for example – are questions that are worth detailed technical study. And underlying them is a problem of undeniable importance. It can be put like this: in 1700 the social profile of Britain was that of an 'underdeveloped' country, but by 1800 Britain was set on the path of industrial growth that led to affluence; granted that we are not concerned with the social, political and theological factors that favoured this transformation, what particular features of contemporary technology and science made it possible? What did impoverished Britain have in the eighteenth century that impoverished countries today do not have but could quite possibly acquire? If we can find only a few, partial answers to this question we might have a clue that would help solve the problem of world poverty today.

There can be no simple or single answer to what is, in effect, a whole spectrum of questions. Among these are, I suggest, the following. How important as a practical ideology was the 'mechanical philosophy' of the seventeenth century? How numerous were the peripatetic lecturers of the eighteenth century, and how effective were they in spreading technical knowledge? To what extent did innovation depend on relatively advanced skills such as draughtsmanship, surveying and a knowledge of elementary mechanics and hydraulics? How numerous were the engineers and how did their profession evolve in the eighteenth century?[73] On the other hand, the specifically technical questions are bedevilled by the

[73] It is indicative of the work still to be done that the first full biography of John Smeaton is only now being prepared, under the auspices of the Institution of Civil Engineers and the editorship of Professor A. W. Skempton, F.R.S.

fact that much of the development of technology was never chronicled. For various, fairly obvious reasons full accounts of inventions and their development were not always published in contemporary books and journals. One way over this difficulty is, as Layton has shown,[74] through the study of reports of legal actions concerning patents. Another, as indicated by R. L. Hills[75] and by A. R. Williams,[76] is through the use of laboratory and experimental techniques. Indeed, the growing interest in 'industrial archaeology' reveals an awareness of the importance of the study of material, as opposed to documentary, sources for the history of technology. It seems likely (at least to the writer) that these techniques will be increasingly used by historians of technology.

These are only some of the questions that the student of this particularly creative epoch may wish to ask and some of the problems that he may face; others have been mentioned elsewhere in the chapter. There is, in addition, the general question of the effects of eighteenth-century technology on nineteenth-century science. No sweeping claim is implied by this; all that is asserted is that the two-way relationship between science and technology gave science insights and concepts that would otherwise have been missed. The general problem is to elucidate them. After all, it would be unreasonable to suppose that the unprecedented industrial transformation that began in the middle of the eighteenth century had no effect whatsoever on man's conception of his universe, and therefore on the practice of his science. At the end, when these and other questions have been answered, one generalization in particular will surely be seen to be correct: it is that the eighteenth century did much of the hard thinking, as well as the drudgery, by which later ages, including the present, self-indulgent one, were to profit.

[74] Layton, 'Millwrights and engineers' (see note 13).
[75] Hills, 'Working model of the Newcomen engine' (see note 17).
[76] A. R. Williams, 'Metallography of some 16th century armour', *Bulletin of the Historical Metallurgy Group*, VI, (1972), 15–23; *idem*, 'Problems in the analysis of armour', *Institut suisse d'Armes anciennes; Rapport* (1972–4), 26–35; *idem*, 'The manufacture of gunpowder in the Middle Ages', *Ambix*, XXII, (1975), 125–33.

Index

Note: This index includes materials found in the main text only, not in the footnotes.

Aaron, R.I., 16–17
Académie des Sciences, 296, 403, 412
Académie Royale des Sciences, 301
Academy of Sciences, Leningrad, 423
Academy of Sciences, Paris, 241, 259–60, 402, 404–5
Ackerknecht, E.H., 231
  History of Therapeutics, 226
Acta, 382, 419
Acton, Lord, 2
Adair, James Makittrick, 207
Adanson, Michel, 265, 269
Addison, Joseph, 194, 292
Adet, follower of Lavoisier, 405
Aegydien Academy, 421, 431
Aepinus, F.U.T., 374, 375
affinity, 396–8
air, 384, 399–400
Alexander, Franz, 200
Althusser, Louis, 73–4
American Philosophical Society, 241
Ampère, André, 414
Annales de Chimie, 412–13
Annales School, the, 148, 251
Anne, Queen, 195, 207, 440
Arabella, character in *The Female Quixote*, 182
archaeology of knowledge, 86–90
Archimedes, 337
Arduino, Giovanni, 310
Aristotelian, 125, 166, 259–60, 270, 288, 292, 392, 394, 398, 401, 411, 419, 458
Aristotle, 270, 357, 473
Aristotle's Master-piece: or the Secrets of Generation, 228
Arkwright, Richard, 459–60, 473, 479
Arnold, John, 441
Astronomisches Jahrbuch für das Jahr 1790, 443
astronomy, 424–46
asylums, regulation of, 190–2
atheists, 315
Atlas geographicus maior, 434
Atlas historicus, 434
Atlas Silesiae, 434
Atlas von Deutschland, 434
Austen, Jane, *Pride and Prejudice*, 205
Auwers, Arthur, 443
Ayers, M.R., 17, 20

Baal-Shem Tov, 193
Babbage, Charles, 450
Bachelard, Gaston, 58, 73–81, 87, 153, 183, 257
Bachelardian, 80
Bacon, Frances, 36, 40, 225, 298, 428, 456
Baconianism, 12, 40, 80, 295, 317, 446, 456
Bailey, Nathan, 146
Baillie, Matthew, 224
Baily, Francis, 443
Bakewell, Robert, 298
Balck, 436
Baldwin, James M., *History of Psychology: A Sketch and an Interpretation*, 152
Banks, John, 461–2
Banks, Joseph, 303–4
Barnes, Barry, 72, 87, 139
Barraud, Paul Philip, 442
Barthes, Roland, 183, 276
Bath, Somerset, 205, 228, 237
Battie, William, 160, 164, 170, 172–7, 190–4, 204
Battie, W., *Treatise on Madness*, 172
Baumé, 405, 412
Baxandall, Michael, 133
Baxter, Andrew, 32, 59, 63, 175
  Enquiry into the Nature of the Human Soul, 176
  On the Immateriality of the Soul, 29

Beaglehole, J.C., 314
Beattie, James, 31–3
An Essay on the Nature and Immutability of Truth in Opposition to Sophistry and Scepticism, 23, 33
Beccaria, Giambattista, 373
Beck, L.W., 47–8
Beddoes, Thomas, 222
Bedlam (Bethlehem Hospital), 176, 179, 194, 234
Beheim, Martin, 420
Behmen, Jacob, 69
Behmenists, the, 68
Bennett, Elizabeth, 205
Bennett, J., 47–9
Kant's Dialectic, 48
Bentley, Richard, 55
Bergman, Torbern, 310, 396, 397
Berk, Dr. Norland, 210
Berkeley, Bishop George, 13, 25, 32, 42, 59, 63, 137, 151, 166, 170, 342, 446
Berlin Academy of Sciences, 241, 301, 371, 374, 378, 420, 430
Mémoires, 445
Berlin Calendar, 424, 427
Berlin, Isaiah, 35
The Age of Enlightenment, 36
Bernal, J.D., 451–2
Bernoulli, Daniel, 328, 337–8, 446, 458
Bernoulli, Jakob, 328, 336–9
Hydraulica, 337
Bernoulli, Jean, 328, 337, 446
Berthelot, Marcellin, *La révolution chimique: Lavoisier*, 403
Berthollet, C.L., 404–5, 413–15, 466
Berzelius, 414
Bessel, 338
Bibliothèque Britannique, 241
Bichat, 184–5, 211, 222
biology, 276–8, 283
Bird, John, 442
Black, Joseph, 66, 81, 391, 393–4, 397, 399, 401, 404, 409, 412, 470, 472, 477, 480
Black, William, 230
Blackmore, Sir Richard, 172, 228
King Arthur, The Nature of Man, Creation: A Philosophical Poem in Seven Books, 176
Blaeu, Johann, 420–1
Blaeu, Wilhelm, 420
Blancanus, 288
Blanchard (Blanckaert), Steven, *A Physical Dictionary*, 146
Blane, Sir Gilbert, 232, 236
bleaching powder, 446
Bloch, Marc, 252
blood-letting, 226–7
Bloor, David, 139
Blumenbach, Johann Friedrich, 322
Board of Longitude, 301, 443
Boehme, Jacob, 69
Boerhaave, Hermann, 56, 67, 156, 216, 218–19, 221, 238, 243, 259, 273, 384, 412, 457
Elementa Chemiae, 457
Institutiones medicinae, 217
Boerhaavean principles, 128, 238
Bologna Academy of Sciences, 301
Bonnet, Charles, 255, 260, 265–6, 282
Borsieri, G. B., 225
Boscovich, Roger Joseph, 27–9, 32, 45, 51, 54, 56, 64, 84, 137, 432–3
A Theory of Natural Philosophy, 56
Bose, Georg Matthias, 370–6
Bostock, John, 213
Bouguer, Pierre, 56, 423
Boulanger, Nicolas Antoine, 300, 314, 320
Boulton, Matthew, 475, 479
Bourguet, Louis, 273–4
Bowlby, John, 249
Bowles, Geoffrey, 16, 19, 24, 61–2
Boyer, Carl B., 342–6
The history of the calculus and its conceptual development, 342
History of Mathematics, 331
Boyle, Robert, 17–20, 55, 99–104, 108–12, 116–19, 125, 135, 146, 155, 166, 217, 260, 392–3, 398
Experiments and Considerations Touching Colours, 436
The Sceptical Chymist, 392
Bradley, James, 429, 443
Astronomical Observations, 443
Brahe, Tycho, 421
Brandes, Georg Friedrich, 420
Bridgewater, the Duke of, 479
Bright, Richard, 179
Brindley, James, 476, 479
Brisson, Mathwrin Jacques, 270
British empiricism, 15–34
British technology, 468–73
Broc, Numa, 309, 316
Brooke, John, 321
Brosses, Charles des, 315
Brown, Dr. John, 221–2
Elementa medicinae, 221
Brunonianism, 221
Brunonians, the, 222
Brown, Theodore M., 124–8, 130–2, 219
Browne, James, 425
Browrigg, William, 401
Bruce, James, 308
Bruhns, Christian, 442

Brunet, P., *L'introduction des théories de Newton en France au XVIIIe siècle*, 55
Bruno, Giordano, 112
Brunonianism, *see* Dr. John Brown
Buache, Philippe, 309, 315, 417–18
Essai de geographie physique, 312
Buchan, William, 228
Domestic Medicine, 227
Buchdahl, Gerd, 49, 107
Metaphysics and the Philosophy of Science, 15, 49
Buer, M.C., 245–6
Buffon, Georges Louis Leclerc, 4, 44, 85, 260, 262–6, 268, 273–9, 280–2, 294, 299, 309–17, 320, 322–3
Epoques de la nature, xi, 280–1, 283, 312
Histoire naturelle, 262, 280, 289
Bulkeley, Richard, 194
Burnet, Thomas, 288, 319
Telluris theoria sacra, 293
Burton, Robert, 172
Anatomy of Melancholy, 207
Burtt, E. A., 106
Metaphysical Foundations of Natural Science, 13
Busching, Anton Friedrich, 435
Butterfield, Sir Herbert, 147, 392
Bynum, William F., 158–60, 210
Byrom, John, 61, 69
Journal, 69

Cabanis, P.J.G., 213, 225
Calvinism, 34
Calvinists, 12
Cambridge Group for the History of Population and Social Structure, 248
Cambridge, University of, 12, 27, 29, 98, 175, 239, 243
Camisards, the, 193–5
Campbell, Thomas, the poet, 179
Canguilhem, Georges, 73, 79, 84, 153
La connaissance de la vie, 153
Essai sur quelques problèmes concernant le normal et le pathologique, 153
La formation du concept de réflex aux XVIIe siècle, 153
Cannon, Susan F., 1, 309
Cantlie, Neil, 231
Canton, John, 373–4
Cantor, Geoffrey, 57, 58, 72, 87, 139
Cantor, Georg, 327
Cantor, Moritz, *Vorlesungen*, 330–1
Capaldi, N., 18
Carkesse, James, 176
Carlyle, Thomas, 161
Carnot, Sadi, 481
Carpenter, the geographer, 288
cartography, 302, 308–9, 421–32
Cartwright, Rev. Edmund, 460, 473, 479
Cassini, Giovanni, 424
Cassirer, Ernst, 113
Philosophy of the Enlightenment, 300
The philosophy of symbolic forms, 300
Castiglioni, Arturo, 253
Catcott, Alexander, 309
Cauchy, Baron Augustin Louis, 327, 335, 337–8, 343
Cavendish, Henry, 374–5, 393, 399, 401, 404, 406
Cellarius, Daniel, 420
Celsus, 230
Cervantes, 156
Chambers, J.D., 250
Vestiges of the Natural History of the Creation, 282
Chaptal, Jean Antoine Claude, 414
Charke, Charlotte, *The History of Henry Dumont esq.*, 198
Charles II, 98
Charleton, Walter, 125
chemical industry, the, 465–7
chemical nomenclature, *see* Berthollet; Fourcroy; Guyton; Lavoisier
chemical revolution, the, 389–90, 402–13
chemistry, 362, 389–416, 463
Cheselden, William, 176
Cheyne, George, 61, 62, 69, 164, 172–5, 178, 263
The English Malady, 172
Philosophical Principles of Natural Religion, 62
Chiarugi, 174
Chlysti, the, 193
Christie, J.R.R., 402
Christ's Hospital, 234
chronometers, marine and other, 441–2
Clairaut, Alexis, 337, 418, 423
Clarke, Samuel, 26, 35, 60, 103, 175, 279
A Demonstration of the Being and Attributes of God, 26
classification, theory of, 266–71
Cleland, John, 197
Coffee House Physical Society, 364
Cohen, I. Bernard, 59
Franklin and Newton, 55
Colden, Cadwallader, 57, 60
Cole, William, M.D., 125
Coleman, William, 219
Coleridge, Samuel Taylor, 48
Collège de France, 415
Collège Royal, 259
Columbus, Christopher, 289, 358
Commentarii, 382, 439

competitions, scientific, 385
Comte, Isidore, 286
Condillac, E.B. de, 36–8, 41, 46, 85, 390
An Essay on the Origin of Human Knowledge, 37
Condorcet, A.N., 403
Connaissance des tems, 427
Connell, K. H., 246
contexts, 382–3, 452–3
contextualism, *see* history and historiography
Cook, Captain James, 295–6, 302–6, 308, 441
Cornaro, Luigi, *Sure and Certain Methods of Attaining a Long and Healthful Life*, 228
Coronelli, Vincenzo, 421
corpuscularians, the, 113
Cort, Henry, 464–5
cosmogony, 288–90
Cosmographical Institute, 435, 437
Cosmographical Mechanical Laboratories, 435
cosmographical schemes, 429–34
Cosmographical Society of Nuremberg, 419–20, 427, 429, 430, 435, 437, 444, 447
cosmography, mathematical, 417–48
cosmology, 134–9
Coulomb, C.A., 77, 374–5, 381, 445, 469, 481
Coulter, J.L.S., 231
Countess of Huntingdon, the, 61
Cranage, George, 465
Cranage, Thomas, 465
creation, the, 278–9
Crell, Lorenz, *Annalen*, 412
Cruickshank, William, 406
Cullen, William, 129, 156, 172, 175, 177, 215–16, 218, 219–22, 224, 227, 232, 393, 412
Cuthbertson, John, 380
Cuvier, Georges, 4, 265, 270, 281–2, 293, 302, 318–19, 322

d'Alembert, J. Le Rond, 11–14, 36, 38, 40, 42–6, 330, 337–40, 343, 382–3, 418
Discours Préliminaire, 41, 43
Oeuvres, 43
Traité de dynamique, 41–4
Dalrymple, Alexander, 315
Dalton, John, 393, 394, 398, 416, 463, 474
Dampier, Sir William, 285
Darby, Abraham, 463–4, 473
Darnton, Robert, 122, 135
Darwin, Charles, 4, 269, 280, 282, 304
Origin of Species, 282
Darwin, Erasmus, 222, 304, 323
Darwinianism, 2
Daubenton, Louis Jean Marie, 396
Daubuisson, Jean François, 309
Daudin, Henri, 267
Davies, Alun, 441
Davy, Humphry, 298, 414
Consolations in travel, or the last days of a philosopher, 322
Davys, Mary, *The Accomplish'd Rake*, 198
Dean, Dennis, 300
death, in the Enlightenment, 244–51
de Beer, Sir Gavin R., 303
de Belidor, B.F., 472
de Borda, J.C., 468, 469, 481
de Châtelet, Mme., 74
Dedekind, Richard, 327
de Dolomiev, Deodat, G.S.T.G., 304, 318
Defoe, Daniel, 194, 197, 205, 453
Moll Flanders, 205
Delambre, Joseph, 389, 425
Histoire de l'astronomie au dix-huitième siècle, 426
Deleuze, Gilles, 183, 201, 209
and Guattari, Félix, *L'Anti-Oedipe*, 201
Schizo-analyse, 201
De L'Isle, Guillaume, 417, 421, 424
De L'Isle, Joseph, 424
Delsenbach, Johann Adam, 422
Déparcieux, A., 468
Derham, William, 263
Desaguliers, J.T., 57, 380
Course of Experimental Philosophy, 460
Descartes, René, 32, 36, 38, 40, 44–6, 87, 167, 202, 216, 219, 270–2, 288, 320, 327, 331, 360–1, 383, 446, 458
Cartesian, 12, 14, 38, 39, 40, 43, 44, 45, 46, 71, 89, 104, 167, 216, 220, 264, 277, 290, 361, 372, 373, 446
Cartesianism, 36, 44, 46
de Thury, Cassini, *La méridienne de l'Observatoire de Paris*, 431
Dibner, J., 368
Dickinson, Henry, 465
Dictionary of scientific biography, 357, 368
Diderot, Denis, 11, 12, 35–6, 38–9, 40–4, 46–7, 87, 121, 314, 363,
and d'Alembert, *Encyclopédie*, 38–42, 80
Pensées Philosophiques: Oeuvres complètes, 39
see also Kemp, J.
Diethelm, Oskar, 239
differentiation, theory of, 349–50
Dijksterhuis, E.J., 106
Dirichlet, Gustav Lejeune, 346
dissenters, 243

Ditton, Humphry, 24-5
A Discourse Concerning the Resurrection of Jesus Christ, 23, 24
Dobbs, Betty Jo Teeter, 2
Dolby, R.G.A., 416
Donovan, A.L., 393
Doppelmaier, Johann Gabriel, 418, 421-2, 431
Atlas Coelestis, 417, 421, 425, 434
Douglas, James, 231
Douglas, Mary, 86, 134, 138, 233
dreams, theory of, 176-7
Dryden, John, 199
dualism, 220-1
Duchesneau, François, 219
Dufay, Charles François de Cisternay, 85, 370-1
Dugas, René, 333, 341
Histoire de la Méchanique, 333
Dumas, J.B., 415
Dundonald, Lord, 466
Dupin, F.P.C., 450
D'Urfey, Thomas, 194
D'Vebre, John, 191

Earnshaw, Thomas, 441
East India Company, 441
Ebersperger, Johann Georg, 434
Ecole des Mines, 301
Ecole des Ponts et Chaussées, 301, 479
Ecole Polytechnique, 353, 386, 479
ecology, 287
Edinburgh Medical School, 238-9
education, of people, 480
Edwards, Jonathan, 193
effluvia, theory of, 371-2
Eiffel Tower, the, 462
Eimmart, Georg Christoph, 421-2, 424
Eimmart Observatory, 421-2, 431
Einstein, Albert, 333
elasticity, of the nerves, 172
electricity, 358, 366-82, 387
Eliot, T.S., 186, 244
Ellis, John, 302
Emerson, William, *Mechanics,* 460
Encyclopaedia Britannica, 425
Encyclopédie, 453
Endeavour, Captain Cook's, 296, 302
energy, theory of, 466-7
Engels, Friedrich, 451
engineering, 462-83
engineers, executive, 475-77, 482
Enlightenment matter-theory, 111-23
Entralgo, Pedro Lain, 143, 226
Mind and Body: Psychosomatic Pathology, 143
environment, concept of, 294-6
Epicureanism, 102, 119, 137
episteme, the, 4, 258
Erbery, William, *The Mad Mans Plea,* 194
Essays and Observations, 234
Euclidean geometry, 47, 327
Euler, Leonhard, 70, 87, 328, 330-50, 386, 418, 423-5, 429, 438-47
Differential Calculus, 331
Eulerian principles, 345
Euler's fables in *Opuscula varii argumenti,* 424
Introduction to the Analysis of Infinites, 331, 345
Mechanics, 336
Opera Omnia, 334
evolution, 281-2
Darwinian evolution, 286
Ewing, A.C., *A Short Commentary on Kant's Critique of Pure Reason,* 48
experience, 52-4
exploration, 302-8

Faber, Knud, 216
Fabricius d'Acquapendente, 263, 270
Faraday, Michael, 27, 48
Farrington, Benjamin, 452
Favre, Robert, 196
Febvre, Lucien Paul Victor, 148
Ferguson, Allan, *Natural Philosophy through the Eighteenth Century and Allied Topics,* 3, 285
Fermat, Pierre de, 327
fire, theory of, 72
Fischer, Johann Carl, 365, 383
Fischer, K.P., 46
Fisher, Susanna, 441
Flamsteed, John, 424
Historia Coelistis Britannicae, 421
Flinders, Matthew, 303
fluxions, 342-3
Fontano, Alessandro, 144
Fontenelle, Bernard de, 36, 256, 272
Forman, Paul, 2
Forster, E.M., *Howard's End,* 143
Forster, J.R., 302-3
Förster, M.C., 378
Forth Bridge, the, 462
fossils, 281-2, 290
Fothergill, John, 237
Foucault, Michel, 4, 58, 73, 79, 86-9, 158, 163, 182-9, 200-2, 209-13, 238, 257-67, 290, 321, 323
archaeology of knowledge, Foucault's, 290-1
L'archéologie du savoir, 290
Histoire de la Folie, 159, 163, 201-2

Foucault (cont.)
Histoire de la sexualité I: la volonté de savoir, 158
Les mots et les choses, 212, 257, 290
Maladie mentale et psychologie, 158
Naissance de la Clinique, 143, 183–5, 212
Power, Truth, Strategy, 144
Surveiller et punir, 158, 192
Foucaultian method, 185
Fourcroy, Antoine de, 403–5, 414
Leçons élémentaires d'histoire naturelle et de chimie, 396
Fourier, Jean Baptiste, 474
Analytical Theory of Heat, 472
Fourneyron, Benoit, 469
Fowler, Richard, MD, 229
Frank, Johann Peter, 220
Frankists, the group of, 193
Franklin, Benjamin, 57, 76–7, 82, 368, 372–4, 379, 382
Franklinists, the, 373
Franz, Johann Michael, 419, 423–4, 427–38, 444–7
Fracastorius, Hieronymus, 420
Frederick the Great, 430
Freke, John, 68–9, 77
French, Roger, 218
Freud, Sigmund, 148, 154, 160, 167, 175, 208
Freudian interpretations, 249
Füchsel, George Christian, 310
Fuller, Frances, 172

Galenic concepts, 125, 270
Galenism, 225, 230
Galileo, 3, 327, 333, 358–9, 393, 461–2, 469
Two New Sciences, 461
Galilean concepts, 76, 466–7
Galois, Evariste, 327
Galvani, Luigi, 375
Garrison, Fielding H., 179–82, 186–8, 253
History of Medicine, 178–9
Garrisonian method, 179
gases, 398–401
Gaubius, 220
Gauss, Carl Friedrich, 425–6
Gautier, E., 248
Gay, Peter, 11, 35–6, 39–41, 46–7, 58–9, 198
The Enlightenment: An Interpretation, 35, 36
Gay-Lussac, Joseph Louis, 414–15
Geer, Baron Carl de, 265
Gehler, J.S.T., *Physikalisches Wörterbuch*, 365
Gelfand, Toby, 236
General Land Office, the, 302
generation, theory of, 268–70
Gentleman's Magazine, 241, 372
gentleman physicians, 176–7
Geoffroy, Etienne François, 397
geology, 285–324
Geological Society of London, 301, 321
strata, 310–11
George II, 207, 247
George III, 159, 430–1
German mental asylums, 180
Gersaint, the French merchant, 261
Geymonat, Ludovico, 361
Gibbon, Edward, 36
Gilfillan, S.C., 473
Gillispie, C.C., 72, 86, 94, 121–2, 135, 285
Glacken, C.J., *Traces on the Rhodian shore*, 291
Glaser, Christoph Jacob, *Epistola Eucharistica*, 422
Glasgow University, 477
globalization, 312
Godwin, William, *Caleb Williams*, 170, 176
Goerke, Heinz, 269
Goethe, Johann Wolfgang, 322
Goldsmith, Oliver, *History of the Earth and animated nature*, 310
Vicar of Wakefield, 169
Gombrich, E.H., 133
Goodall, Charles, 125
Goodwin, Philip, *The Mystery of Dreams, Historically Discoursed*, 176
Göttingen Academy of Sciences, 419–20
Göttingen Anzeigen von gelehrten Sachen, 445
Göttingen Observatory, 438, 442
Göttingen Society of Science, 383, 419–20, 439
Goubert, Jean-Pierre, 250
Malades et medecine en Bretagne, 237
Gough, J.B., 401
Gower, B.S., 16
Graham, James, 243
Grange, Kathleen, 199
Grattan-Guinness, I., 346
gravity, 373, 381
Gray, Stephen, 370
Gray, Thomas, 155–6
Great Chain of Being, the, 286, 312
Green, George, 375
Greene, John C., 314
Greene, Robert, 19, 27–32, 69–71
The Principles of the Philosophy of the Expansive and Contractive Forces, 27–8
Gregory, Dr. John, 222, 227, 229
Gren, Friedrich, 406
Grew, Nehemiah, 272

Griffith, G.T., 245–6
Grimaldi, Francesco, 383, 425
Grimaux, E., 403
Grimsley, R., 41–2, 46–7
Grinnell, George, 97–8
Gross, Dr. Gloria, 210
Guerlac, H., 57, 94, 107
Guyton de Morveau, Louis, 396, 405–6, 411–14, 466

Haase, Johann Matthias, 421, 429, 431, 435
Sciagraphia integra tractatus de constructione mapparum omnis generis . . .,421
Habakkuk, H.J., 246
Hadley quadrant, the, 434, 440
Haen, Antoni de, 220
Hales, Stephen, 29–30, 219, 260, 273, 391, 393, 395, 399
Vegetable Staticks, 399
Hall, A.R., 72, 285
Hall, Diana Long, 253
Hall, Sir James, 304
Hall, R., 30
Hall, T.S., 219
Haller, Albrecht von, 213, 220–1, 223, 240, 255, 259, 271, 274–5, 278, 282, 372, 419
Halley, Edmond, 293, 309, 429, 442
World Chart, 446
Hambridge, Dr. Roger, 210
Hargreaves, James, 479
Harré, Rom, 6
Harrington, Robert, 66
Harringtonians, the, 112
Harris, John, 61–2
Lexicon Technicum, 61
Harrison, John, 418, 441
Hartley, David, 80, 151, 173, 175
vibrations, doctrine of, 173
Harvey, Gideon, *Morbus Anglicus . . . To which is added, some Brief Discourses on Melancholy, Madness, and Distraction, occasioned by Love*, 182
Harvey William, 3, 167, 214, 256, 270, 278
Hasidism, 193
Haslam, John, 159–60
Hauber, Eberhard David, 420
Hauksbee, Francis, 370–1, 380
Hautefeuille, Abbé, 454
Haygarth, John, 226
heat, theories of, 391–411, 415
Heberden, William, 225, 230
Hegel, Georg Wilhelm Friedrich, 202
Hegelianism, 286
Heimann, P.M., 16, 27, 58–60, 63, 66–8, 70, 88, 107, 115–18, 135
Heinroth, Johann C.A., 179
Heister, Lorenz, 232
Helmont, Jean Baptiste van, 398
Helmontians, the, 125
Helvétius, Claude Adrien, 300, 315
Henry, G.W., 150
Henry, Louis, 248
Herder, Johann Gottfried von, 281–2, 320
Herschel, William, 64, 66, 83–4, 297, 444
Hessen, Boris, 106, 452
Hevelius, Johann, 421
Selenographia, 425
Hewson, William, 223
Heylyn, Adrianus, 288, 291–2
Heyne, Christian Gottlob, 420
Heywood, Eliza, *Life's Progress through the Passions*, 198
Love in Excess, 198
The secret history of the present intrigues of the Court at Caramania, 198
Higgins, Bryan, 66, 70
Hill, Christopher, 101, 135, 194
Hill, R.G., 152
Lunacy, Its Past and Its Present, 152
Hills, R.L., 483
Hippocrates, 221, 225, 230
Hippocratic concepts, 216, 225–7, 230, 233, 299
Hippocratism, 225, 233
history and historiography
chronological history, 149–50
contextualists, 111, 133, 158–60
coping with history, 161–2
diachronic history, 256
dialectical view of history, 148–9
historiography, 11–15
historiography of science, 105–11, 147–8, 156–8, 341–2, 354–5, 358–65
History and Theory, 149
history of ideas, 1–6
history versus historiography, 148
hypotheses, 3–5
internalist–externalist debates, 209–10
marginality, concept of, 242–3
Marxist concepts and methodologies, 4, 47, 74, 114, 138–9, 157, 183, 189, 451–2
metaphor and historiography, 82
mysticism, 192–8
nationalistic aspects, 153–7, 415–16
nature and history, 278–82
periodization, 148–51, 282–3, 387
Proteus, as an image, 65, 399, 407
reconstructionism, 131, 150
rhetoric and the history of science, 182–7
Romanticism, 3, 153

history and historiography (cont.)
secularism, 146–7
structuralist history, *see* structuralism
teleological history, 153–60
thematic history, 151–3
thinker-oriented history, 150–1
threshold, theory of, 209
tunnel history, 285–6
uniformitarianism, principle of, 318
'Whig' history, 147, 285, 291
Hobbes, Thomas, 174–5
Hobbesists, 100, 119, 137
Hobbs, William, 290
Hodges, Nathaniel, 125
Hoffmann, Friedrich, 216–19, 229, 259, 273
Fundamenta Medicinae, 217
Hofmann, J.E., *Geschichte der Mathematik,* 331
Hogarth, William, 180, 214
Holbach, Baron Paul von, 44, 113, 300, 315, 320
Hollingsworth, T.H., 246, 249
Homann, Johann, 421, 434, 436–7, 447
Cartographic Bureau, 421–5, 427–8, 431–5
Erben, 419
Homannisch-Hasische Gesellschafts Atlas, 427, 434
Home, Henry, Lord Kames, 154–5, 161, 174
Home, Roderick, 57
Hooke, Robert, 263, 293, 393, 480
Micrographia, 459
Horrebow, Peter, *Operum mathematico-physicorum,* 443
Horton, Robin, 86
Hoskin, Michael, *xiii*
hospitals, 234–7
Hughes, Thomas P., 452
humanitarianism, 156–7
Humboldt, Friedrich Alexander von, 293, 306, 309, 312, 317, 322, 448
Hume, David, 11–13, 17–22, 29–42, 48, 54, 59, 151, 166, 317, 446
A Treatise of Human Nature, 27, 34, 37
Humean principles, 17, 30, 35
Humphrey, Charles, 61
Hunter MD, John, 223, 225, 236, 239
Hunter, Richard, 180, 187, 199
and Macalpine, Ida, *Three Hundred Years of Psychiatry 1535–1860,* 151
Hunter MD, William, 231, 239
Hutcheson, Frances, 174
Hutchinson, John, 66–8, 114, 115
Hutchinsonians, the, 27, 87, 114–15, 119, 121, 137
Hutton, James, 57–8, 62–9, 81, 116, 135, 298, 300, 312, 317, 320, 322, 481
Theory of the Earth, 279, 298
Huxham, John, *Essay on Fever,* 230
Huxley, T.H., 286, 464
Huygens, Christiaan, 293, 327, 333, 458
hydraulics, 479
hylozoists, 113

iatromechanism, 124–8, 219–20, 273
iconographical evidence, 223
Imhof, A.E., 251
and Larsen, O., *Sozialgeschichte und Medizin,* 251
Industrial Revolution, the, 449–83
industry, 413–14
infinitesimals, 344–5
Ingenhousz, Jan, 282
Inkster, Ian, 243
inoculation, *see* medicine
insensibility, 172–3
Institutum Cosmographicum, 435
instrumentation, 379–80, 447
instrument makers, 385, 447
internalist-externalist debates, *see* history and historiography
invention, 456–9
irritability, 261, 274
Isis, 357, 368

Jacob, James R. 99, 100–1, 104–5, 111, 123, 132, 135, 139
Jacob, M.C., 98–9, 102–5, 111–14, 118–19, 123, 132, 135, 139, 219
Jacobins, the, 121–2, 135
Jacobinist science, 121–2
Jacobites, 61
Jacyna, Stephen, 139
James MD, Robert, 229
James II, 98
Jameson, Robert, 309
Jansenists, the, 196
Jansson, Johann, 420
Jardin du Roi, the, 259–60, 396
Jenner, Edward, 247
Jessop, T.E., 30
Jevons, William Stanley, 286
Jewson, N.D., 212
Johnson, Samuel, 174–5, 199, 208, 213, 228
Jones, Kathleen, 191
Jones, William of Nayland, 27, 57, 66–7
An Essay on the First Principles of Natural Philosophy, 27
Jordanova, L.J., *xi*
Joule, James Prescott, 74

Journal de Médecine, Chirurgie et pharmacie, 241
Juliusspital, in Wurtzburg, 179
Jung, Carl Gustav, 154
Jupiter, the motions of, 439
Jussieu, A.L. de, 259, 265

Kaestner, A.G., 448
Kant, Immanuel, 12–13, 15, 19, 27–8, 32, 45, 47–54, 202, 300, 313, 317, 382
  Anthropologie, 180
  Critique of Pure Reason, 13, 48–9, 50, 51, 53
  Kantian ideas and the historiography of science, 47–9, 71
  Metaphysical Foundations of Natural Science, 47, 49–50, 51, 53
Kargon, Robert, 107
Karsten, W.C.G., 363, 382
Kaulbach, Wilhelm von, 180
Kay, John, 459
Keevil, J.J., 221
Keill, John, 60
Kelvin, Lord, William Thompson, 286
Kemp, J., *Diderot: Interpreter of Nature,* 38, 39, 45
Kiernan, C., 36, 45–6
King, Lester, 171, 178, 217
  The Philosophy of Medicine: The Early Eighteenth Century, 164–5, 171
Kircher, Athanasius, *Mundus subterraneus,* 292
Kirwan, Richard, 404
Klein, F., *Vorlesungen,* 331
Klein, Jacob Theodor, 270
Kline, M. 332
  Mathematical Thought from Ancient to Modern Times, 332
Klinkerfues, Wilhelm, 438
Knight, David M., 16
Knight, Gowin, 57, 66–7, 69
knowledge
  growth of, 305–12
  politics of, 203–10
  problem of, 22–3
Knox, Ronald, *Enthusiasm,* 198
Koch, J.A., 443
Körner, Stephan, 47
Kosmographische Nachrichten und Sammlungen auf d. J. 1748, 425, 429, 438
Koyré, Alexandre, 106, 257
  Koyréan ideas and the historiography of science, 107–8, 110–11, 115
Kratzenstein, C.G., 374
Krieger, Leonard, 149
Kronecker, Leopold, 327
Kubrin, David, 107–8, 110
Kuhn, T.S., 2, 58, 73, 78–82, 84–5, 168, 257, 351, 402
  Kuhnian methodology, 80
  Structure of Scientific Revolutions, 2, 257

Laboratorii Cosmographico-Mechanici, 435
La Caille, Nicholas Louis de, *Astronomiae Fundamenta,* 444
Lacépède, Bernard, 270
La Condamine, Charles Marie de, 308, 423
Lacy, John, 194
Laennec, René, 211, 225
Lagrange, J.L., 328–30, 332–3, 337–43, 350, 418
Lakatos, Imre, 351, 402
Lalande, J.J.L. de, 329, 429
  Astronomie, 426, 444
Lallemand, Claude, 184–5
Lamarck, Jean Baptiste, 4, 260, 268, 277, 281–3, 300, 311, 320–3
  Hydrogéologie, 279, 289
Lambert, Johann Heinrich, 381
  Photometria, 383
La Métherie, Jean Claude de, 405, 466
La Mettrie, Julien Offrey de, 44, 220
La Peyrouse, Stanislaus de, 296
Laplace, P.S., 44, 279, 359, 398, 404–5, 414, 418
Larsen, O., 251
Larson, James, 269
Lavoisier, Antoine-Laurent, 84, 86, 261, 390–8, 400–16, 463, 474–5
  Lavoisier, Guyton, Berthollet, and Fourcroy, *Méthode de nomenclature chimique,* 404, 411
  Traité élémentaire de chymie, 410, 463
Law, William, 61, 68–9, 87
  Appeal to all in doubt, 68
Lawrence, Christopher, 128–32, 227, 253
Layton, E., 452, 483
Leblanc, Nicholas, 413, 465–6
le Cat, Claude Nicolas, 311
Lecourt, Dominique, 74
Lee, H., 25
  Antiscepticism, 24–5
Le Fanu, W.R., 220
Lehmann, J.G., 310
Leibniz, G.W., 14, 28, 38, 45, 48, 51–2, 54, 87, 97, 105, 116, 273, 279, 288, 320, 327, 331, 342, 344–5, 348, 446, 458
  Leibnizian ideas, 25, 60, 71, 275, 343, 348, 457
Leigh, Denis, *The Development of British Psychiatry: II, The Eighteenth and Nineteenth Century,* 151

Lémery, Nicolas, 76
Cours de Chymie, 412
Lennox, Charlotte, *The Female Quixote*, 182
Lepenies, Wolf, 278
Leslie, Patrick, 66
Lesser, Friedrich Christian, 263
Lettsom, John Coakley, 230, 237
Leupold, Jacob, 472
Levine, David, 249
Lévi-Strauss, Claude, 110
Lewes, George Henry, *The Biographical History of Philosophy from its Origin in Greece down to the Present Day*, 151
Lewis and Clark expedition, 295
Libes, Antoine, 365
Lichtenberg, G.C., 363, 377, 382
Opera Inedita Tobiae Mayeri I, 438
Liebig, Justus von, 414
Lilley, S., 452
Lind, James, 226, 232–3
Lindeboom, G.A., 218
Linebaugh, Peter, 59
Linnaeus, Carl, 4, 157, 216, 259, 265, 268–9, 270, 279, 282, 309, 312, 317, 396
Linnean taxonomy, 216
Amoenitates academicae, 269
Nemesis divina, 269
Lister, Martin, 231
Liverpool Hospital for Lunatics, 192
Lloyd, C., 231
Lobachewsky, Nicolai Ivanovich, 346
Lochiel, *see* Campbell, Thomas
Locke, John, 13–27, 32, 35–44, 46, 50, 59, 63, 151, 166, 170, 173, 175, 225, 264
An Essay Concerning Human Understanding, 17, 20–3, 41, 215
Lockean epistemology, 16, 23–4, 27–9, 36, 38, 41, 60, 70, 225, 233, 317
Reasonableness of Christianity, 17
Lomonosov, M.V., 70, 416
London College of Physicians, 125
longitude, problem of the, 353–4, 440–3
Lorry, Charles, 160
Louis, Pierre, 230
Louis XV, 417
Lovejoy, A.O., 257
Lovett, Richard, 68
Lowitz, Georg Moritz, 419, 423–3, 428–9, 431–8, 444, 447
Luc, Jean André de, 299, 304, 317–20
Ludolff, C.F., 371
Lunar Society of Birmingham, 63, 241
Lyell, Charles, 3, 323
Lyonnet, Pierre, 265–6, 282
Macadam, J.L., 476
Macalpine, Ida, 180, 187, 199
Macbride MD, David, 216
McEvoy, J.G., 63–4, 71, 118–21, 135
McGuire, J.E., 2, 16, 19, 27, 58–60, 63, 66–7, 88, 107–11, 115, 118–21, 135
Mach, Ernst, 332–3
Die Mechanik, 332
machines, 272
see also Gilles Deleuze, *L'Anti-Oedipe*
MacKenzie, Sir Alexander, 295
MacKenzie, Donald, 139
Mackenzie, Henry, *The Man of Feeling*, 156
McKeown, Thomas, 246
McKie, Douglas, 394, 406
Mackie, J.L., 17
McLaren, Dorothy, 248
Maclaurin, Colin, 45, 60, 342
McMullin, Ernan, 107, 110–1
Macquer, Pierre Joseph, 403
Dictionnaire de chymie, 399
magnetic forces, 444–6
Maillet, Benoit de, *Talliamed*, 289
Maire, Christoph, 432
Maitland, Charles, 247
Malebranchists, the, 57
Malherbe, 466
Malpighi, Marcello, 272
Malthus, T.R., *Essay on the Principle of Population*, 280
Malthusian concepts, 312
Manchester Hospital for Lunatics, 191
Manchester Literary and Philosophical Society, 241
Manchester Science Museum, 454
Mandelbaum, Morris, 173
Mandeville, Bernard, 174, 178
Mandrou, Robert, 148
Manley, Mary, *The Adventures of Rivella*, 198
Manuel, Frank, 2, 97
Marat, Jean Paul, 76
Margetts, E.L., 152
marginality, concept of, *see* history and historiography
Mariotte, Edme, 260, 293
Marsigli, Luigi Ferinando, 293
Martin, Benjamin, 385
Martin, G., *Kant's Metaphysics and Theory of Science*, 49
Marx, Karl, 133, 451
Capital, 451
Marxism, 1, 139, 286, 295, 451
Marxist, 4, 47, 74, 114, 138, 139, 157, 183, 189, 451, 452

Marxist concepts and methods, 4, 47, 74, 114, 138-9, 157, 183, 189, 451-2
Maskelyne, Nevil, 442
Masonic order, the, 198
mathematics, 283, 327-54, 363-4
matter-theory, 394-400
Maupertuis, Pierre, 56, 274-5, 281, 317, 423
Maxwell, James Clark, 48
Mayer, Tobias, 378, 419, 424-8, 431-47
  Collectanea geographica et mathematica, 433
  De refractionibus astronomicis, 442
  Mathematischer Atlas, 447
  Nova theoria declinationis et inclinationis acus magneticae, 445
  Theoria Magnetis, 445-6
Mayow, John, 30, 393, 401
mechanical models, 272-4
mechanics, rational, 327-54
Meckel, J.F., 184-6
medicine
  inoculation, 247-8
  interaction of medicine and society, 234-42
  Medical and Chirurgical Review, 241
  medical economics, 253
  medical education, 239-40
  Medical Essays and Observations, 241
  Medical Observations and Inquiries, 241
  medical practice, 226-33
  Medical Society of London, the 230
  medical students, 240-1
  medical theory, 215-25
  medicine and philosophy, 163-9
  midwifery, 232
  nutrition, 246
  quackery, 243-4
Mehrtens, H., 352
melancholic medicines, 205
melancholy, 207-8
Melanchthon Gymnasium, the, 421
Mercator, Gerhardt, 420-1
mercury, 400
Merton, R.K., 106, 452
Mesmer, Franz Anton, 122
Mesmerism, 122
Metcalf, John, 476
Methodism, 198
Metzger, Hélène, 59, 70, 393
  Newton, Stahl, Boerhaave et la doctrine chimique, 55
Michaelis, Johann David, 420
Michell, John, 12, 29, 32, 57, 64, 310
Midriff, John, 172, 182
midwifery, see *medicine*
Miles, John, 190
Mill, John Stuart, 18, 286
Millar, John, 230
Miller, Genevieve, 247
Miller, Perry, *The New England Mind,* 188
mineralogy, 396
Mineral Water Hospital, Bath, 237
Mink, Louis, 149
Mirror for Magistrates, 151
Molières, J.P. de, 57, 70
Moll, Hermann, 421
Momigliano, Arnaldo, 149
Monceau, Duhamel du, 260, 273
Monge, M., 404-5
Monkemoller, Eduard Otto, 180
Monro *secundus,* Alexander, 219
Monro MD, John, 160, 164, 170, 172
Montagu, Lady Mary Wortley, 206
Montesquieu, Charles Louis, 315
Monthly Review, 241
Montucla, J.E., 329-32
  Histoire des mathématiques, 328
Moore, Cecil A., 199, 208
Mora, George, 200
Morgagni, G.B., 224
Mornet, Daniel, 261
Morris, Stephen, 210
mountains, 290
Müller, Maria Clara, 422
Münchhausen, Gerlach, Adolph von, 420, 431, 436
Murhard, Friedrich, *Geschichte der Physik,* 365
Murr, Christopher Theophilus, von, 422
Musschenbroek, Pieter van, 77, 82, 365-6, 376-7, 381
  Essai de physique, 365
  Introductio ad philosophiam naturalem, 365-6, 368
Mykovini, 430
mysticism, *see* history and historiography

Namier, Lewis, 88, 203
Napoleon I, 44
Napoleonic Wars, 213
Narrenthurm, in Vienna, 179
Nassau Senior, 450
National Maritime Museum, 441
natural history, 262-70, 395-6
naturalism, 44-5, 313-16
naturalist ideology, 315-16
natural philosophy, 55-91, 386
nature, 278-82
  aesthetics of, 299-30
  complexity of, 310-11
Naturphilosophie, 448
Nautical Almanac, 441
nescience, 23-50 *passim*

neurology, historiography of, 168–73
Neve, Michael R., 253
Newcomen, Thomas, 454–7, 469–71, 477, 480
  Newcomen engine, the, 454–6
Newton, Isaac, 1–3, 16–18, 20, 35–6, 40–1, 43–5, 55–8, 60, 65, 68–9, 77, 94–8, 103–11, 113–18, 127, 137, 157, 218–21, 272–5, 279, 319, 327, 331–9, 342–3, 345, 359, 361, 363, 373, 380–1, 386–90, 393–4, 396, 398–9, 423, 447, 457–8, 470
  Newtonianism, 16, 18, 57–8, 64, 71, 87, 95–6, 98, 101–3, 105, 111–15, 118, 218–19, 320, 360, 365, 393
  Newtonianism and the history of science, 18, 25–7, 35–40, 45–6, 53, 56–62, 66–71, 74, 86–9, 95–9, 102–8, 112–24, 128, 132, 146, 172, 203, 275–9, 290, 320, 342, 360–1, 371–3, 380–1, 393–9, 418, 446, 457–8, 460, 468
  Newtonian matter theory, 58–71
  Opticks, 55, 103
  Principia, 55, 97, 103, 106, 114, 334, 338, 380–1, 459, 470
Newton, James, 189, 191
Nicolson, Marjorie Hope, *xii*
Niebuhr, Carsten, 308
Nockholds, Susanna, 441
Nollet, Jean Antoine, 362, 371, 382
nomenclature, 407–11
Nopitsch, Christian Conrad, 421
Norris, John, *A Philosophical Discourse concerning the Natural Immortality of the Soul*, 175
North West Passage, the, 302
nutrition, 246

Observations sur la Physique, 412
observatories, 423
O'Connor, D.J., 17
Oersted, Hans Christian, 48
Oldenburg, Henry, 55
Oldroyd, D.R., 323
Onions, Peter, 465
optics, 383
  Georgian, 72
Ordnance Survey Office, 301, 433
organicism, 45–6, 312
Ortelius, Abraham, 420
Ortous de Mairan, J.J. de, 57, 70
Oxford, University of, 12, 239, 243
oxygen, 400-10

palaeontology, 323
Pallas, Peter Simon, 270, 296
Palmer, John, 476–7, 481
pantheists, 315
Pantin, Carl, 6
Papin, Denis, 454
Paracelsus, 253
  Paracelsian concepts, 125, 392
Parent, Antoine, 466–9
Paris Academy of Sciences, 370, 417, 423–4
Paris Company of Surgeons, 236
Parry, Caleb, 225
Partington, J.R., 393–4, 406
Pascal, Blaise, 327, 337
Pasquino, Pasquale, 144
passions, the, *see* psychology
Passmore, John A., 149
Pecheux, M., 73, 76
Pelling, Margaret, 243
Penelhum, T., and MacIntosh, J.J., *The First Critique*, 47
Penther, Johann Friedrich, 431
periodization, in history, 148–51, 282–3, 387
Perrault, Claude, 260, 272
Peter, Jean-Pierre, 250
Peterson, Jeanne, 237
pharmacists, Enlightenment, 412–13
philomaths, the, 352–3
philosophes, the, 314–16
Philosophical Transactions, *see* Royal Society of London
philosophy, *see* natural philosophy
phlogiston, 396–7, 405–11
physicist, the first, 481
'physick of the soul,' 144–7
physics, 360–4, 367–87, 391, 458
physiology, 124–9
  see also medicine and philosophy
Pinel, Philippe, 151, 154, 156–63, 167, 174, 177, 204, 216, 222
  Pinelian ideas, 157–8
Piquér, Andrés, 157
Plato, 163
  Platonic ideas, 43, 280
  Platonism, 110
  Platonists, 175
Playfair, John, 65, 75, 322
Pluche, Noel Antoine, Abbé, 263
  Spectacle de la nature, 305
pneumatics, 364–5
Pocock, J.G.A., 133
Poggendorff, J.C., *Handwörterbuch*, 359
Poisson, Siméon Denis, 375
Polhem, Christopher, 465
politics and science, 112–13, 203–8
Poncelet, Jean Victor, 327
poor, the, 237–8

Poor Laws, 190, 237
Pope, Alexander, 94, 98, 156, 172, 176
An Essay on Man, 156, 172
Popper, Karl, 73, 351
Porter, Roy, 4, 210, 253
Portsmouth Naval Hospital, 236
Power, D'Arcy, 213
Price, H.H., 18
Priestley, Joseph, 11–12, 14, 16, 19, 29, 31–2, 57–8, 62–6, 69, 71, 77, 80, 82, 84–5, 135, 261, 383, 390–3, 399–406, 411
Disquisitions relating to Matter and Spirit, 32, 64
Priestleyianism, 170
Pringle, Sir John, 232, 233
prize-competitions, 385
professional scientists, 260
psychiatry, 158–201
see also psychology
Psychologia anthropologica, the German, 146
psychology
anatomic bases, 169–71
hypochondriasis, 207–8
hysteria, 162, 182, 207
melancholia, 162, 199, 208
melancholic medicines, 205
passions, theory of, 155
philosophical components, 163–8
psychology and neurology, 168–73
psychosomatic medicine, 168–71
ruling passion, the, 172
social history of, 189–92
Ptolemy, Claudius, 291, 424
Ptolemaic concepts, 288–9, 312, 419
Public Dispensary, the, 237
Purcell, John, 172
Puschner, Johann Georg, 421
Putscher, Mariclene, *Pneuma, Spiritus, Geist*, 218

quackery, *see* medicine
Quaker, 152, 195, 243
Quaker Asylum, the York, 179

Rankenian Club, the, 59
Raspe, Rudolf Eric, *Specimen historiae naturalis globi terraquaei*, 310
Rattansi, P.M., 2, 107–8
Ravetz, J.R., 346
Ray, John, 263, 265, 269, 280
Razzell, Peter, 246–8
Réaumur, René Antoine, 76, 223, 260, 265, 266, 272, 283
Récamier, Joseph C.A., 184–5
reconstructionism, *see* history and historiography
Rees, Abraham, *Cyclopaedia*, 478
Regency, the, 159
regulation of asylums, *see* asylums
Reid, Thomas, 18, 59
Reil, Johann Christian, 156, 180
Reuleaux, Franz, 460
Revue d'histoire des sciences, 357
Reynolds, Richard, 465
Ricardo, David, 450
Riccioli, Giovanni Battista, *Almagestum novum*, 425
Richardson, Samuel, 174, 182, 197
Clarissa, 205
Richmond, the Duke of, 260
Riemann, Georg F.B., 327
Riquet, J.P., 476
Risse, Guenther B., 221
Ritter, Johann Wilhelm, 48
Robertson, Abram, 443
Robertson, Robert, 232
Robinson, Abraham, 342
Robinson, Thomas, 288–9
Robinson MD, Nicholas, 57, 172, 175, 178, 207
Robison, John, 123–4, 374, 394, 458, 469, 480
Roebuck, John, 465
Roemer, Olaus, 443
Roger, Jacques, 2, 4, 6
Roller, D.H.D., 85
Romanticism, *see* history and historiography
Ronchi, V., 383
Rosen MD, George, 187–8, 193, 195–6, 198–9, 202–3, 209
Madness in Society: Chapters in the Historical Sociology of Mental Illness, 187, 202
Rosenberger, Ferdinand, 359
Rosicrucians, the, 198
Rothschuh, Karl E., 217
Rouell, G.F., 412
Rousseau, G.S., 6, 253
Rowlandson, Thomas, 214
Rowning, J., 57, 64
Royal College of Physicians, 242–3
Royal Salop Infirmary, Shropshire, 234
Royal Society of Göttingen, 241
Royal Society of London, 17, 97, 99, 125, 260, 321, 370, 385
Philosophical Transactions, 404
Royal Society of Stockholm, 385
Rudé, George, 157
Rumford, Count von, 472
Rush, Benjamin, 156, 221

St. Bartholomew's Hospital, 234
St. George's Hospital, 236
St. John's College, Cambridge, 210
St. Luke's Asylum, 179, 190–1
St. Petersburg Academy of Sciences, 301, 382, 437
Saintsbury, George, 195
St. Thomas's Hospital, 234
Saltykov-Schedrin Public Library, 422
Sandler, Christian, 419
Saturn, the motions of, 439
Saucerotte, Nicolas, 160
Saussure, Horace Benedict de, 260–1, 282, 293, 304, 318
Sauvages, Boissier de, 216, 224
Schaffer, Simon, 4, 6, 210
Scheele, K.W., 412, 466
Scheidt, C.L., 431, 436
Scheuchzer, J.J., 319
Schmettau, Field-Marshall von, 430
Schneider, T., 352
Schofield, R.E., 16, 57, 61–4, 218, 223, 393
  Mechanism and Materialism, 57
Schröter, Johann Hieronymus, *Selenotopographische Fragmente*, 438
science
  negation of, 72–85
  nescience, 23–50
  and politics, 112–13, 203–8
  popularization of, 305
  scientific revolutions, 392
  syntax of science, 82
sciences of man, the, 259
Sechendorf, Veit Ludwig von, *Deutschen Fürstenstaat*, 433
secularism, *see* history and historiography
Séguin, Armand, 405
sensationalism, in epistemology, 36–40
sensibility
  concepts of, 129, 208–9, 221–2
  cults, 204–8
  tone of the nerves, 172–3
  see also insensibility; irritability
sentimentalism, concept of, 156
Serafimer Hospital, Stockholm, 238
Sergeant, John, 25
  The Method of Science, 175
Seutter, Matthias, 425
Seven Years' War, the, 437, 440
'sGravesande, W.J., 363, 365, 369–70, 376, 380
Shaftesbury, Anthony Ashley Cooper, Earl of, 174–5
  Characteristics, 174
Shakers, the, 193
Shakespeare, William, 156, 176
Shapin, Steven, 6, 87, 253, 305
Shaw, Peter, 56, 65, 75
Shryock, R.H., 211
Siegfried, R., 401
Sigerist, Henry E., 252
Simpson, Thomas, 443
  Mathematical Dissertations, 442
Singer, C.J., 160, 285
  Short History of Medicine, 160
Skinner, A.S., 34, 35
Skinner, Quentin, 133
Sloane, Sir Hans, 227
smallpox, 247–9
  see also inoculation
Smeaton, John, 297, 467–72, 472, 479
Smeatonian method, 467, 470–1
Smeaton, W.A., 416
Smiles, Samuel, 450, 479
Smith, Adam, 34–5, 81, 151, 174, 177, 279, 477–8, 480
  An Inquiry into the Nature and Causes of the Wealth of Nations, 34, 477
  Theory of Moral Sentiments, 177
Smith, Kemp, 30
Smith, Peter, *Cosmographie*, 431
Smith, William, 297–8, 481
Smollett, Tobias
  Humphry Clinker, 169
  Life and Adventures of Sir Launcelot Greaves, 181, 191, 208
  his heroine Aurelia Darnel, 181, 191
social history, 178–98, 452–3
Société Royale de Médecine, 250
Society of Apothecaries, 242
Society for the Improvement of Medical Knowledge, 241
Socinianism, 16–17
Socinians, the, 14, 32, 115
Spallanzani, Lazzaro, 223, 255, 261, 282–3
Spencer, Herbert, 282
Stahl, G.E., 160, 213, 216–19, 222, 272, 395, 405, 465
  Stahlians, the, 277
Starobinski, Jean, 185
Stavdt, Carl Georg von, 327
steam engine, the, 470–7
Steele, Richard, 194
Steffens, H.J., 383
Sterne, Lawrence, 197, 199
  Tristram Shandy, 169
Stevin, 337
Stewart, Dugald, 59
Stone, Lawrence, *The Family, Sex and Marriage in England 1500–1800*, 192
Strabo, 291, 292
Strawson, P.F., *The Bounds of Sense*, 48
structuralism, 157, 183

Struik, D.J., *Concise history of mathematics,* 352
Stukeley, William, 172
Sturm, Leonhard Christoph, 377
Swieten, Gerhard van, 220, 232
Sulivan, Richard Joseph, 311
Sutton, Daniel, 247
Sutton, Robert, 247
Suttonians, the, 247
Swift, Jonathan, 156, 174-5, 194, 292
  Mechanical Operation of the Spirit, 169
  A Tale of a Tub, 156
Switzer, Stephen, 472
Sydenham, Thomas, 156, 182, 215-16, 233
  Epistolary Dissertation to Dr. Cole, 162
Symes, Richard, 68-9
sympathy, doctrine of, 129

Taton, R., 285
taxonomy, 266-71
technology, 450-83
  French technology, 468-71
  history of, 453-4, 456-83
teleology, 153-6, 214-15
Telford, Thomas, 461, 476
Temkin, Owsei, 231
Tennant, Charles, 466
textile industries, 459-63
Thackray, Arnold, 18-19, 59, 61, 96-7, 219, 394
theology, and natural history, 279-80
Thenard, Louis Jacques, 414-15
Thompson, E.P., 96
  Albion's Fatal Tree, 192
threshold, theory of, *see* history
Thury, Cassini de, 302, 428, 433
time, theory of, 322
Tissot, Samuel Auguste, *Advice to the People,* 227
Toland, John, 112
Toulmin, George Hoggart, 222, 315, 320
Tournefort, Joseph Pitton de, 259, 265, 269
Trembley, Abraham, 260, 265, 272, 274, 282
Trevelyan, George O., 195
travel and exploration, 295-7
Treviranus, Gottfried Reinhold, 277
Tröhler, Ulrich, 230, 253
Tronchin, Theodore, 225
Truesdell III, Clifford Ambrose, 334-41, 346
Trumbach, Randolph, 249
Tuan, Yi-fu, 287, 289
Tuke, Samuel, 151, 158-9, 163, 174
Tuke, William, 152
Tunbridge Wells, 205
Turgot, Anne Robert Jacques, 400
Turlington, Robert, 191
Turner, Frank, 137
Tyson MD, Edward, 160

Underwood, E.A., 160, 243
uniformitarianism, principle of, *see* history and historiography
Unitarians, 243
universities, in Enlightenment, 375-8
  contemporary, 259
  professors, Enlightenment and natural philosophy, 375-9
University of Berlin, 180, 386
University of Copenhagen, 374
University of Edinburgh, 129, 165, 220, 230, 243, 404
University of Göttingen, 377-9, 419-20
University of Halle, 217-8, 363, 378, 447
University of Leicester, 212
University of Pavia, 377-8
University of Pennsylvania, 139
Uranus, 297
Ure, Andrew, 450

Vallisneri, Antonio, 273
Varenius, Bernhard, 288, 292
variable quantity, the, 347-9
Vartanian, Aram, 36, 43-6, 219
Venus, the transits of, 303
Vesalius, Andreas, 166-7
Vess, David M., 231
vibrations, doctrine of, 173
Vico, Giambattista, 320
Viète, François, 327
vitalism, 126-8, 223, 275-6
Vives, Juan Luis, *De anima et vita,* 155
Volta, Alessandro, 374-8, 386, 414
Voltaire, François Marie, 35, 41, 44, 56, 386
  Voltairians, 46
  Pangloss, Dr., 35

Waddington, Ivan, 242
Walker, Adam, 66, 84
Wallace, A.R., 137
Wallace, W.A., 18
Walpole, Robert, 190
Walsh, W.H., 149
Warburg Institute, scholars of the, 134
Warburton, William, Bishop of Gloucester, 175
Ward, Joshua, 243
water, 406-7
Waterhouse, Benjamin, 226
waterwheel, the, 466-7

Watson, R.I. *(cont.)*
Eminent Contributors to Psychology, 151
The Great Psychologists: from Aristotle to Freud, 151
Watson, Sir William, 57
Watson Jr., William, 83
Watt, James, 458, 470–5, 477–81
Webster, Charles, 243
Webster, John, 194
The Displaying of Witchcraft, 193, 194
Weierstrass, Karl Theodor, 327
Weigel, Erhard, 421
Werner, A.G., 296, 309–10, 481
Wesley, John, 198, 228
Primitive Physic, 227
Westfall, Richard, 107
Westminster Hospital, 234, 235, 237
Whewell, Sir William, 48
Whiston, William, 26, 35, 194, 288, 319
Astronomical Principles of Religion, Natural and Revealed, 26
White, Gilbert, *Natural history of Selborne,* 305
White, Hayden, 149
Whitefield, George, 195, 198
Whitehead, A.N., 163, 452
Whyte, Lancelot, 152, 154–5
The Unconscious before Freud, 154
Whytt MD, Robert, 129, 164–5, 219, 223, 272
Observations on the Nature, Cause, and Cure of those Disorders commonly called Nervous, Hypochondriac, or Hysteric, 164–5
Wilcke, J.C., 374
Wilde, C.B., 58, 66, 114–15, 118, 132, 135
Williams, A.R., 483
Williams, Guy, 212
Willich A.F.M., 228
Willis, Thomas, 125, 153, 167–8
Willughby, Francis, 263
Winslow, Jacob, 239
witchcraft, 192–8
Withering, William, 226
Wittgenstein Ludwig, 133
Philosophical Investigations, 133
Wolf, Abraham, 383, 387
History of Science, Technology, and Philosophy in the 18th Century, 359
Wolff, Caspar Friedrich, 282, 446–7
Anfangs-Gründe aller mathematischen Wissenschaften, 447
Wolffian concepts, 71, 446
Woodward MD, John, 61, 70, 172, 309
Woolhouse, R.S., 20
Wright, Peter, 139
Wrigley, E.A., 248
Wurtz, Charles Adolphe, 415
Wurzelbau, Johann Philipp, 422

Yates, Frances, 2
Yolton, John, 16–17, 20
York Hospital for Lunatics, 192
Young, James Harvey, 229
Young, Robert M., 2, 321
Young, Thomas, 1
Youschkevitch, A.P., 346–7

Zevi, Sabbatai, 193
Zilboorg, Gregory, 149–50, 199–200
History of Medical Psychology, 149